The Great Earth Puzzle

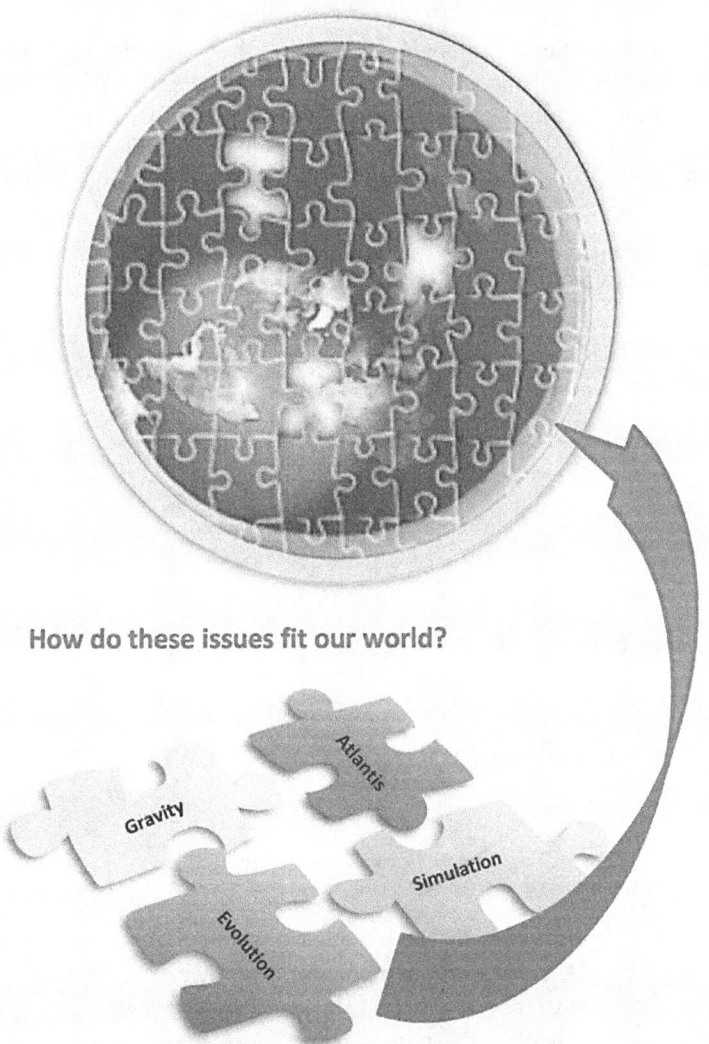

How do these issues fit our world?

This book summarizes what the Earth Realm really is and whether these issues, and more, fit our reality. Are we really living on the planet we think we are?

Categories: Metatags: Anunnaki, Nephilim, Djinn, interdimensionals, ET, UFO, Mankind, origins, Genetic engineering, DNA, Quantum Physics, Maya, Egyptian, extraterrestrial influences, soulless, sociopaths, auras, OPs, NPCs, creation, evolution, Darwin, reptile, Sumeria, Catholic Church, Religion, Fortean phenomena, Jacques Vallee, John Keel, Stuart Wilde, Anatoly Fomenko, Robert Monroe, OBE, Control System, Virtual Reality, Simulation, holograms, Holodeck, Angels, Beings of Light, demons, souls, Scripts, karma, reincarnation, recycling, Déjà vu, Earth Graduate.

Cover design: **The Great Earth Puzzle**
source: Bing Images/GettyImages
and Bing Images/fotosearch.it

Book text in Garamond 12 font.

Author may be reached at TJ_cspub14@yahoo.com

ISBN – 13: 978- 1544030203

Other Books by the Author*

VEG	Virtual Earth Graduate
TOM	The Transformation of Man
TEW	The Earth Warrior (docu-novel)
TSiM	The Science in Metaphysics
QES	Quantum Earth Simulation
AL	Anunnaki Legacy

*See last two pages at book's end.

Table of Contents

(continued...)

(*continued…*)

(*continued...*)

Introduction

Due to a number of reader requests (via the **reader-based email address** on the Copyright page) for a coherent overview and update of the major information contained in the first six books, this book will try to summarize and connect the major dots… and how/why they were given to me. The salient points do form a consistent and intelligent description of **what Earth is**, what Man is, what we are doing here, who is here with us and what they are doing, and what is expected of Man. In effect, this amounts to an answer to what appears to be the **Great Earth Puzzle** – this also includes odd structures and cryptozoids around the planet.

Are we really living on the planet we think we are?

Of course many people don't know the question exists and really takes that form. Yet people for centuries <u>have</u> asked why they are here, what Earth is, and what is their purpose here? And what is amazing is that **when you know what Earth really is** all the other questions are immediately and <u>automatically</u> answered. Thus the first question to be asked (rephrased) is: **What is Earth**, really? Or to paraphrase:

What has to be true about Earth for it to look the way it does?

The question above (end of paragraph 1) is stated in such a way as to give the reader a hint that we **are not on the Earth that we think we are**… In fact, Earth probably isn't really a planet… we'll analyze that. And when you are open to learning what that really means, your purpose for being on Earth becomes much clearer. It is because the truth about Earth has been hidden from us, for at least two centuries, that the Powers That Be (PTB) have treated us like a Mushroom – they have deliberately kept us in the dark and up to our neck in …. fertilizer. No joke.

While this book is **not about conspiracy** nor speculation, the fact that since the formation of the **Illuminati** (about 1776), there has been an agenda for those who are more worthy than the bulk of the public (who are "the ignorant masses", also called "the useless eaters"), to run things and take hold of society because they, the PTB, are smarter… They also happen to be very rich. Actually, Earth and Man have <u>always</u> been controlled (guided, manipulated) by the **Secret Societies** to which the Elite and the PTB belong. (Appendix F.)

Note that the PTB subscribes to the **Golden Rule**: He who has the gold makes the rules.

I am reminded of something said in George Orwell's <u>Animal Farm</u> where the pigs ran the barnyard and emphasized a utopian, egalitarian-communistic barnyard for all. So they put up a banner on the side of the barn that said:

All pigs are equal…
but some pigs are more equal than others!

The Illuminati goal as <u>an on-going organization</u> over the centuries is to set up and run the **New World Order**. To that end, a reminder was erected in 1980 in Elbert County, Georgia called the **Georgia Guidestones.**

Think carefully about the 1776 date above… It is also when the United States of America had its beginning, and such an event was commemorated on the back of the One Dollar bill – the pyramid and the 'all-seeing eye' of the Egyptians and the Freemasons. Most founders of the US and the framers of the Constitution were all Freemasons.

Read the Great Seal on the left side carefully (below) – **Novus Ordo Seclorum** – New World Order. Even the date on the bottom of the pyramid is MDCCLXXVI … 1776. There are 13 levels of stones going up the pyramid (13 original colonies and today's 13 Elite Families) until one comes to the All-seeing Eye of … the Grand Architect. Or the Egyptian Ra (where Freemasonry had its origins)?

The text above the pyramid says, **ANNUIT CŒPTIS**… perhaps referring to the Grand Architect (God? Providence?) who "[he/she/it] has favored our undertakings."

The intent, according to Wikipedia was

> The Eye over it [the pyramid] and the motto *Annuit Cœptis* allude to the many signal interpositions of Providence in favor of the American cause.

The Freemasons knew of the Illuminati and adopted some of their advanced principles (in level 33 and above). That does not smack of a dark conspiracy – it started as more of a proactive union to establish **order based on higher principles** in a world of occasional chaos. And that is not even saying that the Illuminati or the Freemasons had devious intentions. Personally, if truly smart and enlightened people of integrity could establish order, peace and brotherhood on Earth, who would be against it? The Church has tried it for centuries…. How is it working so far?

But the point of this book is that the **Fat Cats, who have the gold** (most of whom are not Illuminati), have manipulated the Media, Church and Education in the USA – and I repeat: the real Elite Families are not Illuminati. The Illuminati dance to the tune played by the 13 Families… who are bloodline descendants of the rulers originally set up by the Anunnaki (think: *Dieu et mon droit*).

The Media has seen to it that the average person does not know what was known just 200 years ago. To govern the Sheeple, they have to be ignorant of a lot of things (so they can't object), and so today's public is entertained with sex and violence in the movies and on TV, and sometimes the public is treated to "fake news" and ponored street demonstrations to creatively guide their opinions… and at best to spin the news so the public does not know what really happened, but will be told what to think by the News Media. Such was said to be Pearl Harbor, JFK's murder and 9-11 for starters.

> Again, this book is not about politics or conspiracies, but this inserted information is included because the Earth really is a Great Puzzle and secret societies are part of it. Book 6 (AL) examined the secret societies and what they teach. Be clear that most Freemasons are not part of the esoteric knowledge. If it can be accurately sourced, and that is hard, it tends to point to the PTB and the Illuminati (backed by the Anunnaki Remnant Dissidents as examined in VEG and AL).

This book recognizes that the public has been lied to for about 200 years, about what Earth is, where people are, and why they are here. **Science seeks to replace God…** Thus was the **Great Earth Puzzle** amplified. That was necessary **to keep the**

public (aka Sheeple) from trying to pursue meaningful spiritual development… and from discovering where they are! And when you start a deception that far back, about late 1700s, by the time you get to 2017 if you tell the public the truth, it sounds like a lie because they all 'know' the truth that they learned in our controlled schools. The Science <u>lies have been compounded</u> for about 200 years and trying to peel them back, like the skin of an onion, is something the average person cannot handle without laughing at you, or getting angry because they think <u>you</u> are lying to them!

Hitler's minister of propaganda Joseph Göbbels, said it:

> A lie repeated often enough will be taken as the truth.

And that has happened to the point where two major <u>and related</u> items in Chapter 10 cause people to laugh and ignore any new info around them:

> Just mention UFOs and/or the Flat Earth.

The public was initially curious about those two items, then the 'experts' (scientists) and the Government said the UFOs were swamp gas, the planet Venus, weather balloons (Roswell, NM) …. And the public was also taught to laugh at the Flat Earth (FE) scenario – we all 'know' the Earth is a globe, don't we? Gee, people were so stupid 400 years ago… So stupid, as a matter of fact, that the FE scientists in 1800 said there was no such thing as Gravity – and today's scientists still can't find it or measure it! (Chapter 4 and Appendix D.)

So the Great Earth Puzzle exists because **there was a deliberate intent to conceal Truth from the public** (i.e., souls) starting in the 1700's with Sir Isaac Newton so that the New World Order (i.e., One World Government) could be brought into reality – probably in our time. Newton was a mystic really into Alchemy and Jewish Kabbalah... not so much Science (see Chapter 4).

This book's theme is not meant to be melodramatic, nor is it BS. The **Earth is what it is**… the problem is that there is a lot of Science **deception** circulating as truth on Earth and it is the responsibility of each soul to wake up, see it all for what it is, and overcome challenges to personal growth – which often means **thinking outside the box** that has been created for us by the major institutions: Media, Church and Schools. Is that a conspiracy? No, it is simply the nature of the Greater Drama engineered by Astral 4D Beings who run this **Earth School** (i.e., "the gods") who <u>permit deception because it is a</u> **catalyst for growth** against which souls are measured.

Fail to see the lies, fail to overcome the world, fail to stand for Truth and you will be recycled (Book 2, TOM). That is not my idea, it is what I was given and is why the gods are so concerned that **deception is spawning permissiveness and hedonism** <u>today</u> and is taking a major hold of people in our time. People mistakenly think that permitting anything and everything is Ok … and that is what got Sodom & Gomorrah, and Mohenjo-Daro and Harappa destroyed. Think about it.

Thus this book is an attempt to connect the dots, bring several "Aha's" into one's awareness, and equip people to handle their lives with <u>better insight</u>. It is not preaching a new religion or philosophy, nor is it teaching any froo-froo from the New Age.

Earth is what it is and we don't get to vote on it.

And the major points shared by the previous six books are not my concoction, opinion, or something I pulled from off the Internet. As I will explain shortly, in Chapter 1, the information in VEG **was merely transcribed.**

Now I have to write this 7th book to <u>add clarification</u> to some things on which I have received the latest insights, such as the UFOs, the **Flat Earth scenario versus the VR Sphere**, and the Anunnaki aka the Watchers vs the Ancient Ones. The hope is that you have assimilated the earlier information (principally in VEG and TOM) and are now ready for the "cornerstone" in this book.

I prefer Earth as a VR Sphere (Chapter 3), but the Flat Earth evidence is too fascinating… and Earth probably used to be flat – the ancient peoples were not as dumb as we like to think. In some ways (societally), we have gone dumber than they ever were… rationalizing anomalies and ignoring odd events around us. By Chapter 10 we should have a better grasp of that situation, and a resolution. And **Appendix C** should clarify Simulation, **Appendix D** should clarify Gravity, and **Appendix E** should clarify the Firmament. (Details of these 3 issues are appendicized so that the larger picture is not lost in excessive explanation.) **Appendix F** will amaze you.

Earth School

Earth is a School for educating and testing Souls and we are expected to learn why compassion, patience, brotherhood, humility and respect are <u>the</u> appropriate behavior (i.e., STO focus – see Glossary). And then we are expected to gain **Knowledge**: to learn who and what we are, what Earth is, and then **Serve** from our individual strengths and giftings. Some souls do not want to do that – and that is why they are still here. You could say that Earth is often a **School for the Rebellious**, an **Asylum for the wayward**, and they are here to learn that rebellion doesn't really work.   The rebellious souls are birthed into a context of their equals and the like souls play off each other thus reinforcing the need for appropriate thinking and behavior. That makes it also **a Prison** because they can't get out ('graduate') until they change their thinking and behavior.

Earth is many things to many people: School,
> **asylum,**
>> **prison,**
>>> **and zoo.**

So much for the soapbox statement on Reality. Ok, Earth <u>is</u> a School – but what you don't know is <u>what form</u> the Earth School takes (and that is important) and why it was designed and built the way it is… like an **IMAX Theatre**!

It is just important that the reader understand that **Earth has a purpose** and the reader has a purpose and what it is all about… to hopefully **maximize the time spent in the Earth Realm. The PTB hope you waste your time. VEG was written to counteract that agenda… so you can graduate from Earth School.**

Author and Book Insight, Part I

Thus the gods of the InterLife (last time I was on the Other Side) wanted me to write VEG to help clarify that **Earth is a School** and learning, loving and service (STO) are the reasons for the School.
Yes, <u>They chose **me**</u> because I hit the InterLife on fire, angry as hell –
> due to my outrageous murder and I wanted vengeance!
> I had fought Corruption head-on and lost.
They also chose me because I would learn to write well in college, studying to be a Foreign Correspondent… with 5 languages.

The Masters converted my anger into a more proactive way to try to lessen the Darkness on Earth. They also said that They may have to **reset** the School because **this version of Man is not working out.** (This is reexamined and clarified at the end of Chapter 11.)

> Earth is currently less than 55% positive…
> when it hits the 49% positive level … danger!
> (See Glossary for **Wipe & Reboot** vs Timeline Shift…
> examined in VEG and TOM)

And then I was told I can't promote/advertise the 6 books, and so I have been puzzled … **Why write the books if I can't promote them?** They said others sent here (inserted) will do that part… see the end of Chapter 1.

And then I chose to write this final book to connect the dots since I was given several new insights that <u>clarify things further</u>: the truth about

UFOs, Gravity, Simulation, Secret Societies, the true shape of the Earth... and what my part was in all this.

...continued in Part II below....

FYI: Description of the Six Preceding Books

Prior to this seventh book, in the order of creation:

1. **Virtual Earth Graduate** (VEG) – This is the seminal book that led to the other 5. It covers a wealth of information on History, Science, Religion, and explains why Earth is not our home and **we are in a School** and need to Graduate. It spends the last two chapters explaining <u>how</u>.

2. **Transformation of Man** (TOM) – overflow from VEG – examines Timelines, the **Interlife**, energy healing and Hybrids. This book was originally part of Book 1 that was essentially "dictated" (no voices, no automatic writing and no trance) but the original book was 868 pp long. Hence the split into TOM and VEG.

3. **The Earth Warrior** (TEW) – this is a novel with real people, real places and real events and is a 'continuation' of Chapter 4 in VEG. The heroine, Maria Orsitch, was a real person who guided the Germans in electro-gravitic propulsion and eventually saves Earth and mankind from a Galactic SuperWave... the space war and SFX would be great on the silver screen. The book also exists as a screenplay (WGA West Registry 1782827). But, Zombies are currently more important.

4. **Quantum Earth Simulation** (QES) – this is the continuation of Chapters 12-13 in VEG. It examines the current Quantum Physicists' and Philosophers' (including Nick Bostrom) teachings where Earth may be a **Simulation**. Also examined is the Flat Earth scenario (detail) with some surprises in Ch. 11.

5. **The Science in Metaphysics** (TSiM) – this book was the extension of Chapter 7 in VEG, and yet was soundly rejected* by local New Age leaders. It examines the **nature of Vision**, Health, and how the subconscious interfaces us with the world (at a layman level). The point was to **show the science behind the metaphysical teachings** – e.g., we attract to us what we dwell on, how violent video games harm our brains, and how habits are made/broken.

6. **The Anunnaki Legacy** (AL) -- This book is an extension of Chapter 3 in VEG and examines **what was left around the world** (teachings & structures) by advanced beings as they visited and taught mankind the same things – mathematics, writing, agriculture, etc. In addition, the contribution of **goddesses** to mankind is examined, as well as an examination of Metaphysical principles that we should know about… a restatement of the last two chapters of VEG.

*Book 5 was rejected not because it was wrong. The New Age leaders do not want anything but their version of truth presented to people. In the same way, talk radio and TV shows on UFOs and Aliens have ignored my books (and they have copies of them) because

they want to keep the mystery going.

The 6 books are presenting the truth. The Media Others feel they would lose their large audiences if my books resolve the mystery of what the UFOs are and where they come from, and what Earth really is, so I am discredited and ignored. And that jibes with what I learned at the end of Chapter 1, from Baldy.

Corrections

I remember three basic pieces of information while writing VEG…and New Jay (explained in Chapter 1) was told to transcribe these three items at that time:

Earth is a **3D Construct**, created with 3D Laws but residing in 4D so that the greater power of 4D can empower the Earth Realm;

There is an **energy shell** around Earth, effectively quarantining it so that humans are not interfered with by denizens of 4D – this shell is what the *Gegenschein* reflects off of; (See also Appendix E, NASA found a Shield at 7200 miles up)

Earth is a **very sophisticated Simulation** so that objects and beings can be **inserted** as needed, and if things go wrong, it is easy to remove the souls to a holding area, suspending their memories, and **reset** the Simulation, then put the souls back as if nothing had happened…
This starts a **new Era**, and has been done several times (see Chapter 3).

I intuited from that that the Earth was a **sphere**, with a shell (Quarantine) and it was simulated for our benefit as an Earth School. A sphere can be inferred but may not be true. Their info was not as complete as it could have been (on purpose?) but it was <u>all that was transcribed</u> – at that point.

No author likes to address the issue of clarifying prior books' statements, but it is necessary to clarify two issues that are <u>not wrong</u>, but they now occupy an expanded context. If a reader studies QES and the Flat Earth scenario, all of a sudden

the **VR Sphere** and the real nature of **the Anunnaki**

spring into view with a question mark. These **two items** have attracted more attention since 2008 when they were first written, and some clarification can now be added.

> **VR Sphere** – this was the result of <u>three</u> pieces of information that were written in 2008: the Earth Simulation, the *Gegenschein* , and Earth as a 3D Construct. Thus QES was originally written (2009) based on the information in Chapters 12-13 of VEG (2008) wherein it <u>appeared</u> that the Earth was a sphere – no one said anything different. Then a mysterious email came in from a reader (whom I could not answer – it was a one-way email) who said I should look at the Flat Earth scenario and discuss that in QES – so I added Chapter 11 to QES -- **intending <u>to prove the Flat Earth wrong</u> – to substantiate the VR Sphere (which was my assumption – in the absence of Their telling me specifically what Earth was)**. I failed.
>
> The clarification point is this: Earth in many ways <u>does</u> appear to be flat, but it is still a Simulation and such are at the whim/design of the gods (see Glossary) who run this place.
>
> The point now suspected is that as Man grows and needs more space to expand, including Outer Space, **the gods may have morphed the original Flat Earth into a VR Sphere** – still in protection, but Man can now travel farther and explore more in the **Cspace – see Chapters 3 & 9**).
>
> Chapters 3 and 10 will explain why it is important for you to know whether Earth is a VR Sphere or a Flat Earth. Chapter 9 picks up the theme again, drawing an unexpected conclusion in Chapters 10-11.
>
> > **I will show you the evidence for the VR Sphere and the Flat Earth, and let you make the call. I think the choice is obvious.**
>
> …and the second item…

Anunnaki -- it was in VEG that the evidence (ancient records and what New Jay wrote) said that the Anunnaki were originally **reptilian**, then over time, with genetic modifications they (e.g., Marduk, Inanna, and Ningishzidda) birthed as human/mammalian bodies instead of in a reptilian form. Still true.

The original Anunnaki (Enlil, Enki, Ninharsag, Anu…) stayed reptilian and that is why they resided up in their ziggurats and pyramids as humans found them repulsive [see Ch. 3 in VEG], and so Enki was charged with genetically modifying the future Anunnaki offspring on Earth. That part has not changed.

What appears to be new information was discovered during the writing of AL (Book 6): some advanced beings were flying around Earth (in fire-breathing, smoking 'dragons,' and smaller rocket ships [AL, Ch. 2]) educating mankind. Zechariah Sitchin had said the Anunnaki were "those who from Heaven to Earth came" and human cultures all over the planet speak of **"the Skygods"** who came and went in their flying craft….
Rocket ships? Smoking, noisy, flaming dragons (i.e., the description of the Skygod craft in AL and by Zechariah Sitchin) are not ETs.

> Flying thru the air and landing among humans could be seen as gods coming from … another world? But, they might have taken off in Egypt and flown to the Indus Valley, and **coming down from the sky does not make them ETs**… Sitchin was wrong. But they did earn the title 'Skygods' in their Skycraft.

Excuse me, but smoke and flame does not describe ETs with advanced electro-gravitic (antigravity) Skycraft. The Skygod craft being described in the Sumerian cuneiform accounts are **Earth-based flying machines** --- rockets don't work in space as there is nothing to push against! That in turn means (1) **the Anunnaki were an Earth-based group of advanced beings** (some reptilian, some human – they were also created here before Man) and (2) the advanced humans (**Ancient Ones**) lived up in **Hyperborea** at **Mt. Meru** (in the Arctic). Then about 600 BC the Greek and Roman gods left Man, and the reptilian component went underground (as the Nagas in AL) and because they were all advanced, both groups earlier took compassion on humans after The Flood (about 8,000 years ago) and helped Man rebuild Civilization.

All that means is that **we are not alone and never have been**.

The Nagas are the **Remnant Anunnaki (VEG, Ch. 3, see Glossary) and they were also called The Watchers (IGI.GI in orbit) who created Nephilim with Earth women**. The same Watchers that Enoch was shown. And The God of the Universe might have put both reptilians and humans on Earth originally, but separated in distant locales, to see which lifeform would work out. Known: the reptiles created the humans as slave workers as Sitchin said.

More likely: **the reptilians were the first on Earth**, and they created the mammalian humans, so that Earth is really theirs and Man is the newcomer. Shock. That all suggests that Sitchin invented some aspects of the 250,000 real Sumerian tablets and what they said…. For example, Anunnaki are not from Nibiru, their Skycraft are not ET (interstellar) craft, and **there is no planet Nibiru**. And that is further exposed in VEG Ch. 3, and in AL, Ch's 1 and 2.

Summary

There is another point to this Puzzle. I was counseled to not try and promote or advertise the 6 books… and that was my personal puzzle: Why? And Chapter 1 finally explains why, much to my continued disappointment.

Ch. 11 in QES shows why the public thinks the Earth is round, and goes into the proofs that it cannot be a 3D round rock circling the Sun – but it might be a **Simulated Globe (which could be the VR Sphere)**. This book goes into the possible fraud that has been perpetrated and WHY it was done. You have been deceived. UFOS do exist and the Earth is not a 3D rock spinning around the Sun. (Chapter 4 and Appendix F.)

> You must read Chapters 10-11 in this book to see how it all fits together. Then make your decision afresh as to what UFOs are and whether the VR Sphere is a reality or not. The evidence is not BS — it will astound you as much as it annoyed me…and **I was trying to prove the Flat Earth wrong.** I think I succeeded….

> I also added some clarification in three Appendices – C, D, and E.

I wanted the UFOs to be flown by ETs and I believed Earth to be the VR Sphere – a real Earth globe in Quarantine (the energy shell off which the *Gegenschein* reflects) as VEG Ch. 12 says. **My understanding in 2008 was incomplete**… I did not know about the Earth-based UFOs and I was not told specifically whether Earth was flat or a globe. **The complete picture was then withheld – forcing me to dig.** It turns out that I was not given some deeper pieces of the Puzzle until January 2017. Hence this book. A major part of the Earth Realm info had been withheld until I

(Old Jay) was able to assimilate what had already been given, and could then absorb more. Chapter 1 makes this clearer.

> **Thus: (1) most UFOs are <u>built on Earth</u> and flown by humans, and Chapter 10 explains why, and (2) the Earth appears to be a VR Sphere for reasons given in Chapter 10 (and also in QES Ch. 11). Most Quantum Physicists agree.**

> **Note: the Anunnaki never had electrogravitic craft – they flew rocketships (Book 6, AL – Ch. 2: Dragons vs Skyboats). The true UFOs in history were not flown by ETs, either – it was the technology of the Ancient Ones … who later gave it to Man (1936) perhaps anticipating their use if the FE morphs into the VR Sphere… No ETS are currently visiting us.**

Author and Book Insight, Part II

A very important aspect of this book is about to be disclosed in Chapter 1. And sometimes I wish I didn't have to explain that part, but if I don't, you will not understand how the 6 books came to be, <u>why they are true</u>, and why this final one, Book 7, is different. I also want to explain the whole thing as much as I can.

I had an agreement from the **InterLife** to do something proactive and did not remember what it was at any stage after being born. But as you will see in Chapter 1, I was **protected and guided** so that the 6 books could be written. It just would not be me (Old Jay as born in the body) that wrote them. So the 3D (third dimensional) Jay who was born here was just the vehicle for <u>transcribing</u> the 6 books.

As Chapter 1 shares, after my Karma from the last lifetime was paid by age 54, I was visited on the night of **October 25**[th]**, 1998** and while I had the amazing descending dream and remember being protected in a swirling green energy field, <u>I was not told</u> what that was all about. They just did it, and I assumed I had been **'upgraded'** or as I thought, **'rewired'** to be more aware, or more clearly: connected with my **Higher Self**. (All of the above as it turns out.) Rewired yes... and most likely, I also had an unseen **Associate**, later called Baldy... (Chapter 1).

What I later <u>conjectured</u> as years went on was that I had been 'merged' with a higher soul aspect from my same Soul Group. I may be wrong… somehow "I" was not to know then what had happened, and that is partly because I was very resistant to the idea of **Walk-ins**. [1] I was also resistant to the idea of a **Soul Merge**, as well. I had read about them and was not comfortable with either idea.

The gods (Appendix F) are always very polite and not only do not tell you what to do, They will not share things with you to which you are resistant. That is called the

Law of Confusion (see Glossary). I can only assume that "not knowing" was done to protect me…? (I still really don't 100% understand it, but the truth is in there somewhere.)

But the agreement had been made somewhere to do this (InterLife), so the Old Jay who was born here, was rewired and upgraded with a higher consciousness – see **Appendix B**) and that New Jay would do the books as **he had the greater knowledge**. And yet "I" today would think it was just the old me with new insights… and I had no clue.

I know it sounds weird, but it really happened. That is why I know the 6 books are **true and trustworthy** (besides vetting VEG for 4.5 years), and when the books were done, in August 2016, my higher consciousness left…or anointing, or higher soul aspect, or whatever it was that accompanied me… It left early Aug 2016 before the Aug. 24th Presentation!

Thus I (Old Jay) readjusted to the body in late August 2016 and felt a need to clarify things, writing this final 7th book, as much as to leave **a brief autobiography** of the historical process. I also received two more **"1-second drops"** of information in January 2017 in response to my wondering what to do with the short, anonymous email of December 2016 which asked me what I knew about the Flat Earth scenario.

This is not to be melodramatic or evasive. I really do not know 100% what was done to me October 1998, and I sure wasn't buying into the Walk-In or Soul Merge scenarios. But I was 'rewired' that night to 'receive' info given to me. Since I did have my Interface Guide show up several times, physically in person, I called him **Baldy**, it might just be that **Baldy** directed the writing of VEG. He was/is a powerful, polite and knowledgeable Associate or Interface between me and the White Brotherhood who accompanied me (for 18 years), and the Walk-in and Soul Merge scenarios are *non sequitur*.

Baldy also was responsible for guiding me, **protecting me** and keeping me perfectly healthy – all to do the books – VEG was his book, not mine. VEG was the seminal book that led to the other 5. BTW:
I sometimes get a headache reading parts of VEG as it is a style and depth that I am not comfortable with and cannot replicate.

Chapter 1: Genesis of the Books

This is the final book, for reasons that are clear by the end of this chapter. It might now be appropriate to briefly explain how these books came to be. It is only in retrospect that there is a perspective on the issue that I was not aware of… I had no idea I would ever <u>write</u> these books…. and didn't… I <u>transcribed</u> the 1ˢᵗ one (VEG) and the other 5 were extensions of VEG…. but I am jumping ahead of the story.

In high school and college I had learned to write such that my term papers always got A's and B's. In college I took English, composition and **journalism**, and it "just happened" (you know how They work) that my high school English teacher drilled us on new vocabulary and grammar. She also taught a Creative Writing class but I really blew that and stuck to documenting what things were, or how they worked, or why a person was famous – standard term paper stuff. And in reality, each chapter in VEG was really a term paper, and then they were all collated together – and that is called a book.

So in 1958 (I told you I am a dinosaur) I went into the Easthampton (Massachusetts) Public Library, a quaint old building – gothic style – and I browsed and "just happened" to find several books on UFOs by **George Adamski**. I read them all and wrote to him, got on his mailing list, and he sent me some pictures he took of UFOs and Orthon, the Venusian, and I really liked the **Cosmic Consciousness** idea he promoted.

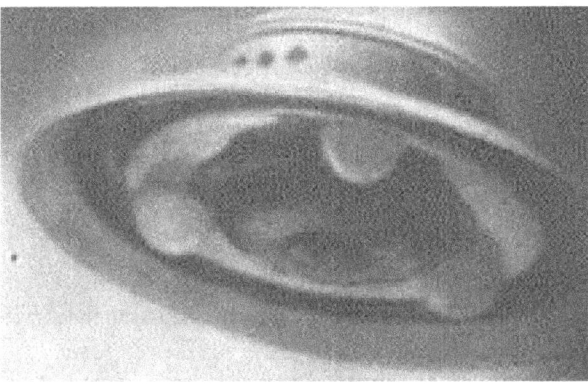

Left is one of the pictures George Adamski took of a UFO as it flew around his home at Mount Palomar, Southern California.

He had a very big telescope and could attach his camera to it.

(Also see pages 353-356.)

I shared him and some of his material with my loving grandmother (Mamo) who had raised me till I was 11 years old, I was now 15, and Mom had remarried husband #3. Mamo wrote back sharing some of her Rosicrucian teachings, and her fascination with Atlantis, Lemuria and the Great Pyramid. She recommended George Hunt Williamson's books.

Later, Adamski would discover he had been 'had' and when my grand-parents went to hear him speak in 1963, he let it all out and blasted those who had deceived him. I got their first-hand report. The UFOS were real but not flown by ETs.

Germans and UFOs

I was fascinated with UFOs and yet there wasn't much on them back in the 1950's and 60's. And then Dad (#3) got transferred to Huntsville, Alabama in 1964, to the Redstone Arsenal, working with **Wernher Von Braun** and the Saturn V rockets. Mom knew how to cook German meals (that had been due to German husband #2) and so Dad invited the German engineers home for dinner – a little PR work there, but I also got to practice my German with them (thinking that with journalism, economics and some languages I could be a Foreign Correspondent in Europe). And one night while chatting at the dinner table, I remembered my UFO interest of past years, and I had read that the Germans had worked on electro-gravitic propulsion systems (allegedly that was what The Bell (*Die Glocke*) was.

So I asked if they knew what Die Fliegende Untertassen were? After they stopped laughing, I was politely told that they were called **Flugscheiben**. And yes, Joseph had worked on some of them in Pilsen and I thought Dad was going to have a heart attack. Dad said he didn't have security clearance for any of that, and Joseph said, "You do now" and gave him a big smile. Joseph was his boss, and Dad went from a Confidential to Top Secret clearance the next day, and Joseph told him that what we heard at the table was to go no further. (Dad had to sign off on a new agreement form, too.) We heard all about Die Glocke, the Vril and the Haunebu (see Chapter 10), and what the US Navy had recovered around the US. I thought it odd that the Navy was managing the UFO recovery efforts instead of the Air Force, but the Navy had begun the process before 1948 when the Air Force was created, and so seniority had precedence.

At one dinner, with German guests, I shared that I had seen a UFO above our house (about ¼ mile straight up) and it had multicolored lights running around it so I thought it was a blimp. I was standing on the back porch, doing some star-gazing, and looked down to grab my binoculars, and when I looked back up it was gone. Blimps don't move that fast so I climbed the ladder to the roof of the house (Dad had one there as he was earlier adjusting the CB antenna up there) and I could not see it anywhere. I mentioned this to Hubert who was another invited guest, and he just smiled and looked at Dieter and they both smiled, and Hubert said that UFOs often check out the **Redstone Arsenal**. We lived about 3 miles from the base. But he refused to elaborate, and I wondered then if he was kidding me. (He wasn't.)

Dad was transferred a year later to another big GE installation in Syracuse, NY, and the rest of my college years were uneventful. I dropped studying UFOs and became interested in Ancient Earth History. In 1987 that would connect me with the **Zechariah Sitchin** material. And yet, my job history was going down the tubes – I'd start with a company and they'd go out of business, or merge or layoff… and I was often out of a job. This was Southern California in the 1970s to 80s. I secured two Data Processing Manager positions in later years only to lose them to the vicissitudes of the economy, too. In 1991 a hypnotic Regression told me why.

Angels Watching Over Us

> All of the following events are absolutely true, with no exaggeration… I don't want to be 'caught' making stuff up when I am age-wise close to making a transition back to the InterLife. That I was guided and protected is putting it mildly.

1956 TWA Flight 2

If you are old enough, you may remember that TWA (Trans World Airlines) was a common carrier airline in the 40s-50s… a major airline along with Pan Am but later bought out by American Airlines (around 2001). **TWA Flight 2** was to be my flight from Los Angeles to NYC in late June 1956… I was rejoining my family in Massachusetts; Mom had just remarried and my new Dad was transferred to Pittsfield, MA. I had stayed behind in Riverside, CA to finish 7th grade.

I had spent several years with my grandparents, which was fortunate as Mamo (my grandmother) was a Rosicrucian adept, up on Edgar Cayce, Atlantis and Egyptian pyramids, as well as a Practitioner with Religious Science and she introduced me at age 11or 12 to Dr. Ernest Holmes, its founder, so I was exposed to "alternate history" and metaphysics early on, and developed a fascination for both. Some things she taught me (from her Rosicrucian Inner Teachings) made it into Book 6, <u>The Anunnaki Legacy</u>.

The day of my flight, June 30, we had to drive to LAX from Riverside, CA, and the traffic was so bad, we missed the morning flight and had to stay in a motel that night. The next morning we made it to the airport on time, and got me aboard a plane – without me seeing the *LA Times* morning papers! My grandfather had bought and read one and was shaken by something he had read on the front page, but kept it to himself.

It was when I landed at La Guardia and Mom was all over me, unusually affectionate and attentive (which she rarely ever did) that I knew something was up. My new Dad said nothing but he knew, too, and left a copy of the *NY Times* on the back seat of the '54 Chevy as we drove to my new home in Pittsfield. Sister was asleep in the back with me so I picked up the paper, glanced at the front page and a cold chill went thru my body… Dad was eying me in the rearview mirror…. and said, "Are you Ok?" I stared at the paper….

TWA Flight 2, June 30, 1956.

Dateline: Grand Canyon, AZ – yesterday late afternoon, a TWA Lockheed Constellation, Flight 2, was hit by a United Airlines DC-7 crashing in the Grand Canyon, killing all 128 on board both planes. The wreckage would take days to reach and analyze… [2]

Holy Cr*p, Batman! That was to have been my flight! That was the first evidence I had of being protected in my life. Mamo was impressed and continued to write me over the next 5 years, sharing deeper metaphysical teachings... she kept telling me I was headed for something. Neither of us knew what it would be.

1965 Hospital Visit

In December 1964 I had had it with trying to get my college degree at the University of Alabama (they were not offering the courses I wanted, nor the specific degree), so I re-enrolled in Syracuse University and relocated to Central New York. I got an apartment and a basic car ('59 Renault CV4) and prepared to finish my college degree. It was not to be…

Just 1 week before I started my classes, I came down with what my former New York doctor called The Grippe (aka Flu). Really bad, sapping my energy and spreading in my lungs… medication was not helping and so I had to leave New York and return to Alabama. My folks came up to help me make the move.

The day we left Central New York, it was starting to blizzard and so we holed up in a local motel for the night, and Mom being a devotee of Unity (profiled in Chapter 5) called **Silent Unity** and made a prayer request for my healing and safe passage back home.

The next morning it was still snowing and the NY State Thruway was beginning to clog with snow, but we had to get out of there! Mom called Silent Unity again. And then she said we all had to get in our two cars and hit the road before they locked down the Thruway. She was following a snowplow headed West (toward Buffalo NY) and I was right behind her, being buffeted by the wind and snow… and the car's heater stopped working. I could see the blizzard closing in right behind us, in my rearview mirror, but the weather ahead of us was strangely clear. We made it to the Pennsylvania State Line and headed southwest, but bad weather socked in again, and we again holed up for the night at 4pm. I was beginning to have trouble breathing (the mucous was starting to fill my lungs). We could not make many more of these stops or I might die on the road back to Alabama… but I didn't know just how serious my condition was!

We were making the transition from I-90 to I-71, cutting south thru Central Ohio… and with better weather we were pushing it at 70 mph (top end for the Renault I was driving). My lungs were beginning to ache, and breathing was getting harder, but we finally made it back to Huntsville, Alabama. The next day I was taken to the doctor who listened to my chest and immediately walked me across the street and into the hospital. He had prepared Mon for the worst… one lung full and the other was ¾ full.

I later learned that Mom that night had again called Silent Unity to heal me with their special Prayer Team. So January 1965 I was in a hospital room, all to myself, with the door locked. I didn't know I was dying at age 22. Sometime during the night I was awakened (and didn't open my eyes) but was aware that there was <u>a very bright light at the foot of the bed</u>. I thought it was the night nurse. (I wish now I had looked!!)

I was asked "How do you feel?" and I said Ok, and then I was asked "Do you want to continue?" I said Yes.

I awoke the next morning about 8:30 am and was hungry, so I got up and went to the door and opened it. I called out to the nursing station across the hall, and their response was to scream. Stunned, I didn't know what to say, and one of them came over to me, and I asked about breakfast. She was a big black nurse, and I asked if she was the one who looked in on me last night. She looked at me strangely, and said "Honey, ain't no one went in your room last nite!"

I still didn't know. One of the other nurses called my doctor, who came in about an hour later, white as a sheet…. He ordered many tests, and listened to my chest… nothing. He even x-rayed me. Nothing. It would be months before it dawned on me that **nurses don't ask you if you want to continue (living!)**… And over the years, I would discover that I had not only been 100% healed, <u>but I would never have respiratory issues again for the rest of my life.</u> I had had pneumonia 7 times before

I was 18. The doctor by the way, started drinking and smoking… he could not say "miracle." (He kept saying 'remission' like a good robot.)

When the Higher Beings or even an angel heals you, you are healed!

I was being watched over. But I didn't know why. I went back to the U of A and majored in something else, and it took months to get my stamina back up – meanwhile I was classified 4F by the local Draft Board. I finished college back at Syracuse when Dad was transferred back to Syracuse, GE Electronics Parkway in 1966. I graduated in 1967, English and Economics, with some Journalism courses.

1979 Est Training

I had moved from Rochester, NY by 1972 and was back in Southern California already a proficient Programmer Analyst, working in a high-paying Lead Programmer position (which would later morph into Data Processing Manager by 1985).

It was during my exposure to the many **Enlightenment Movements** in Southern California that I was invited by one of my coworkers to see what the Est Training was. Having been New Age and New Thought for several years, whatever Est was during the Intro Seminar totally fascinated me. I signed up, paid $250, and got ready to attend the two weekends at the Orange County Fairgrounds, in a totally blacked-out huge building – no clocks, no windows, no water (but our group DID have scheduled bathroom breaks)! Our trainers were Roger Dillon and Charlene Afremow. I wanted Stu Esposito or Werner himself, but no such luck.

So the first weekend with Roger passed, semi-interesting, and we were out at 11:30-midnight both nights – the staff commented that they had never seen that before. So far it was boring. We were all there to "Get It" – whatever It was.

Then came the second weekend with Charlene and all hell broke loose… 300 people in the room and a lot started crying, yelling, peeing their pants…. And I wondered what I had signed up for! One guy tried to punch Charlene and had to be restrained and "spoken to" on the side of the room… he opted to leave. People who had no clue what New Age and Zen and Gestalt Therapy were all about were all semi-traumatized. I sat there observing it all, a little unnerved but rolling with it all…

And then it was my turn. Somewhere in the middle of the 3rd day I was sitting in a seat on the center aisle, when Charlene walked up behind me, and stopped, staring at me. *Oh Shit*, I thought… what now? And she kept on staring… and staring… and she finally said "Why are you here?" I said I was curious and did she want me to playact something with her? (She had been doing that all around the room.) She laughed and said I was 'almost there' – wherever there was, and she moved on…

What the heck did she see? I had the feeling she was looking right thru me – Roger was nothing (sorry) compared to her – She had a presence and an awareness that was shaking up the attendees in the room.

Later that day, semi-bored, I sat back in a new seat (after the break) and all of a sudden, I had **the strangest sense of watching myself** – as if a larger part of Me was watching the Jay that had walked into the room. I was aware of two of me and the larger one was more powerful, more calm and for what seemed to be minutes on end, I was aware that this larger Me knew everything and I picked up on the fact that I had a purpose that would later be revealed to the smaller, daily version of me (later in 1991, as a matter of fact).

Charlene had walked down the aisle again and was staring at me, just 4 seats away. She smiled and said "Welcome. You got it." I had snapped back into my smaller self, stunned, but with an awareness I had never experienced – not even in subsequent Peak Experiences that followed the Training. How did she know and what did she see that gave my new awareness away? (I later learned that **she saw auras** and mine lit up the room around me when I had that larger Me experience.)

Chocolate or Vanilla?

One of the interesting moments in the Training was when Charlene asked a lady in the front row to stand up. She was given a microphone. Charlene then asked her to choose between two ice cream cones, chocolate or vanilla. The lady chose chocolate. Charlene asked why she choose chocolate. The lady said because vanilla was bland and… she never got to finish, Charlene took the mike and told her to sit down. This went on around the room, and everybody had a reason for their choice, and one guy even chose strawberry… Charlene laughed and told him to sit down.

Somehow, no one was doing the exercise correctly. She remembered me and came over and told me to stand up, gave me the mike, and said "This better be right… after your latest experience!" She asked me to choose chocolate or vanilla. I looked at her, and because of all the other 'wrong' answers, blurted out, "Chocolate!" She asked why. I said, "Because that is what I chose!" She smiled, said thank you, and I sat down. The room still didn't get it… you don't need a reason to choose something – the mind just thinks it does and needs to be reasonable and justify your choice(s)! No choice was wrong, she just wanted us to stop being 'reasonable' and realize that another part of us chooses! (In fact, the larger Me chose.) She said, "Get out of your mind… and live who you really are!" In short, come from the Heart.

One of the key issues in the Training was to see that **we are biological robots,** doing things because we are programmed to do them, making choices that are often pre-programmed into us by our parents, teachers, society, etc. They wanted us to see (as I did) that we are more than our mind, more than our body, more than our choices… a key statement was to **Be Here Now** and realize what is really in front of you, not what you have 'expected' to see or hear… to really 'get' that Rocks are Hard, and Water is Wet… to stop pretending that our precious, developed (adult) act is something that we have to protect and sustain. **And see that we do need to be responsible for our choices**… Most humans don't want to be responsible (see end of Chapter 11).

In the meantime, I was dating many women in Orange County and not being very responsible (or discerning), and so I had a 'come-uppance.'

1980 HIV

The Southern California lifestyle was such that I feared getting married – too many women had bumper stickers (in Newport Beach and Irvine) that said:

If you're rich, I'm single!

I dated a lot but it wasn't until I was transferred to Texas (in Spring of 1987) that I contemplated getting married at age 44. And as a lot of the California women were beautiful, hot and 'liberated', I had intimate relations with many of them… God only knows which one gave it to me!

In the Spring of 1980 I went to see the doctor for what he termed Chronic Fatigue, then Epstein-Barr, and then "Gee, we don't know what it is!" I was a mess… I felt like I had a **severe flu**, often had thrush and was treated several times for fungus infections… and this went on so that I had to live on my savings as I could not hold a job for 9 months, until my body began to get the upper hand… whatever it was!

It was at this time that I enrolled in Nutrition classes, and began eating right, stopped drinking, started weight lifting, and even stopped dating…. No energy! And classes and supplements were eating up my money for dating anyway…

Remember, this was 1980 and they had not 'discovered' HIV, and that was what I had. Too many sex partners, and I learned that my partners could have HIV and not have any symptoms since it is a **stealth disease** – it enters the body and lies dormant for some years, before finally blossoming… but can still be passed on thru sex, kissing, even if your partner has HIV and sneezes – some strains of HIV are airborne.

My doctor later said this was logical as someone with HIV, <u>a virus</u>, could also get an airborne <u>flu virus</u> and the two viruses would swap RNA (trading DNA), and *voila* – one HIV virus is now airborne!

It took about 1 year to completely overcome the HIV virus… and it wasn't until 1998 that the doctor in Texas discovered how I did it. I have a genetic 'mutation' called **CCR5delta32**… I acquired it from my Norwegian father… (so was Mom marrying him just a coincidence, or did the gods know something about me and what I might get involved with in my future years?) HIV attacks white blood cells which are **coated with protein** and destroys them. Because of the mutation, a large part (50-60%) of my white blood cells do NOT coat for protein and those 'naked' white blood cells attack HIV and kill it. It just took some time and felt like I had a nasty 6-9 month flu!

Again, the gods seemed to know what I would do and what They had to do to **protect me**…. because that is the way it happened. They knew I had a task to do and were protecting me… And today, all that the doctor in Texas found were antibodies from the fight against HIV and no trace of HIV in my body today.

> In addition, just 5% of the world population has the CCR5delta32 mutation – 32 base pairs of DNA are dropped from the gene so that it cannot code for a protein coat. Most of these fortunate people are native to Northern Europe and Scandinavia.

1987 Alabama Audit

Things went fine health-wise until I began flying around the US doing computer audits. I was sent to **Huntsville Alabama, again** – scene of the 1965 hospital healing. Again, Winter…. but this time was different. I was attacked by something that tore up my sinuses and I had to suspend the audit. My boss in California threatened to fire me… even though officials at the aerospace company where I was auditing had explained how sick I was and how I could not concentrate, and they did not want me sneezing all over the place! It was bad and I was getting desperate. It was so bad I could not get on a plane with the congestion and plugged-up ears!

So, now 22 years later in February 1987, I looked up my old doctor and paid him a visit. He was still in the same building, and as I entered the waiting room, I was overcome with cigarette smoke – not only did he smoke but he permitted his patients to smoke as well. I guessed I might be in the wrong place…. He remembered me, but he was shocked to see me, and he just said I had an allergy to a fungus native to Alabama. Nothing he gave me worked, by the way, and I went back to the motel room. It was a Friday night and I said a very desperate prayer for healing, and turned in about 10:30 p.m. I had planned to visit the old house we used to live in, but the weather report at 10 pm said heavy snow all weekend…

I awoke the next morning to sunshine coming in thru the partly closed drapes. And the real mystery began: I was on top of the covers, sideways, and my pajama bottoms were on backwards! Even more weird was that my sinuses were perfectly normal. I took a shower and came out and turned on the TV, and the weather report that morning said that the weather was indeed strange, and no snow this weekend, just sunshine. So I later visited the old house… much smaller than I remembered!

Even more interesting nowadays is that I took what had happened for granted… I tried to rationalize that maybe the allergy was conquered by my immune system, and yet I could not explain the inside-out pajamas and waking up on top of the covers… I have wondered if I was somehow 'prevented' from analyzing what happened because I had done the same thing after the hospital healing in 1965… I didn't examine any supernatural events until 1991.

I was watched over again. I finished the audit and flew back to California.

1987 Mom Dies

An interesting sidenote, Mom had kept me on Probital and Librium for 17 years (1962-1979) since I had a peptic ulcer at the end of my first college year – very stressful with late nites and poor nutrition. It was a healthnut girlfriend who weaned me off the Rx, and Mom then hated her and said it had been an excellent tax-writeoff. Such was my mother's view of me and the world (no, she did not have a soul… she had no aura but I didn't know what that meant at that time – See VEG, Ch. 5 and below: "2006 Training").

But it didn't stop there. Even though I was a successful MIS Manager in Newport Beach, Ca, Mom took my sister and went to the bank and had the grandparent's large estate converted in 1985 to benefit just her and my sister in a **California Inviolate Trust**. I was not consulted (being in Texas at the time) and Mom lied to the Trust Department and said I had approved that all be left to my sister…! She and Nancy then spent thousands in travelling all over the world.

But there is justice. Mom and Nancy spent 3 ugly days in a part of Scotland where the air is very bad, laden with soot and smoke (city name withheld on purpose), and 10 months later Mom came down with an **inoperable brain tumor**. After the operation, I went to visit her in a convalescent home and learned she had had brain mass removed from the left side of her brain, the speech center. She was awake and listening to me, she just could not speak. (This I got from her roommate.)

So there I am sharing with her the success of my job and how I had met a new woman and it might end in my finally getting married – Would she like to be a grandmother?

She glared at me and said, "That'll be the day!" Her roommate was overjoyed – "Oh, glory be! She is healed! She spoke! Praise God!" I was stunned because it was <u>not her voice</u>… it was deep and guttural. I bid her adieu and left the room, and left the building, asking God to remove her if He was not going to heal her. The prayer was answered. She passed two days later. (I made the mistake of telling Nancy, my sister, that someone spoke thru her and what I had done.)

My sister and I never spoke again. She even changed her name and address but I went thru public records and found her, and called. Was she happy? No. Was I welcome to visit? No. What she said was that she would take out a restraining order if I showed up and tried to get my part of what was left of the inheritance.

> As luck would have it, she married the janitor in the hospital where she worked and thru drugs and travel, he took her thru what was left of the inheritance.

> In 2007 I got a letter from the executrix of the estate informing me that Nancy had died – of the same brain tumor as Mom, and she had left me $2000. The larger share of the estate was left to a lady who sheltered stray cats. (Seriously.) Such was our dysfunctional family and that is why (thank God) I was raised until I was about 12 by my more loving grandparents.

1989 Sedona Food Poisoning

Several years later, I was vacationing in Sedona, AZ and awoke one morning with major diarrhea … something in the Mexican dinner I had had the night before which tasted great but the sour crème (enchiladas) must have collected bacteria (when they leave the containers of sour cream open on the kitchen counter all day) or maybe it was the guacamole…? I had to check out of the motel and move on to Flagstaff but could not stay off the pot. And I had no Kaopectate or Pepto Bismol.

11:35 a.m. I was calculating paying for another day's stay under these circumstances, when I made a desperate prayer… "Heal me, and get me out the door!" or something to that effect. Immediately, I kid you not, there was a "download" into my consciousness (no voice) that said stand in the middle of the room and hold my arms out to my sides, making a "T." Bam…Bam……….Bam – three times I was hit with some sort of energy wave and was completely healed… but I waited another 30 minutes just to be sure. Checkout was 1 p.m.

Later, after 2008, They would occasionally use a similar communication technique to answer questions I had from time to time and I called these "insights" **1-second drops** (in the Glossary). It was like receiving an insight from something that I just now remembered, but this was not memory – most of it was new information and I would spend 15-30 minutes looking at what I had been given. (This was what They would do to me the night of October 25, 1998.)

I say all of that to emphasize that **we are not alone**, we are watched over, and if you have a job to do in the future (even if you don't know it at the time, and I didn't), the Angels or **Beings of Light** as I call them (because they don't have wings), will guide, heal and protect you.

The Great White Brotherhood (Appendix F) was not thru with me.

Transfer to Texas

In May 1987 I was transferred to the Dallas-Fort Worth area and 6 months later was **laid off again**. I found another job in 1988 and did well until about 1991 when it all went awry again. By this time, I was beginning to be suspicious and blaming the gods for doing it to me – whoever they were, and I didn't know if they existed or not, but I knew I was innocent. What irritated me was thinking that they healed me, and protected me, just to keep me alive as a 'whipping boy' – heal me and then remove the job, or cause me financial ruin – I was **forced into Chapter 11.** It seemed a funny way to bless me then curse me (I later discovered it was Karmic). I was considering suicide to stop the nonsense.

And on a chance visit to a Holistic Fair one weekend, I met a psychic who was the real thing – <u>she</u> called me over to her table, <u>she</u> told me about me, and I had not said anything but she strangely knew a lot about me. And all I did was sit down at her booth, and boy, did she 'read' me. She said I needed to spend a different session with her (she was a licensed clinical hypnotherapist and that scared me), but I agreed. She was also a Christian and I was desperate for answers, so I thought I'd take a chance. Right move.

1991 Insight

The **Hypnotic Regression** was supposed to be 1 hour, it was 3. We had already spent 20 minutes praying for **protection and guidance**. I came to her that day with a list of issues, so she know what we were working on. The Christian therapist was told things about me and what I was to remember and know. She cooperated with whoever it was speaking thru me (different voice), we recorded the session, and she made notes. The upshot was that in the early part of this lifetime I was doing 'payback' for what I had done in the <u>past three lifetimes</u>, including getting messed up with my soulmate (whom I have had to drop, see TOM). Most of what I was experiencing (in lost jobs and relationships) was due to not doing what I had been asked to do in the last lifetime – I was holding back, and that was a repeated pattern even this time with the VEG book. In addition, working for the Crown, I had cost others their jobs – trying to help the State out (by removing scoundrels)… while it was for a good purpose, <u>it still earned Karmic payback</u>. Surprise!

Karma is more exacting than we'd like to believe.

Doing my own thing, my way, my last lifetime literally cost me my life in a brutal murder, and the lives of my wife, children and my employers who were also murdered due to my carelessness… Karma.

So from the **1991 Regression**, I also saw what happens on the **Other Side** – which I call the **InterLife**, as it is where we go <u>between lives on Earth</u> to evaluate and make plans for subsequent lifetimes. Our lives are all there, like on **videotape**, and my life was replayed with a very loving Master, and then I would counsel with some of the advanced Teachers. It was one of the Masters speaking to the therapist as well as my Higher Self at times (different voices thru me). I saw only <u>major parts</u> of my three past lifetimes, but it was not a complete 'playback' – it was just enough to understand what the Master was saying. <u>That information largely went into TOM</u> (Book 2).

And the upsets and job losses <u>here</u> continued for another seven years. The only consolation was that I now knew <u>why</u>, and had the assurance that if I cooperated and got the message from the ongoing experiences, it would stop in 1998. Such information kept me from further considering suicide – events were that bad, year after year, that I had considered it until the 1991 Regression.

I was not told anything about what I had **agreed** to do this lifetime… such was the "agreement" that Baldy asked me about (later in 2013) that I mentioned in the Introduction. 1998 was to turn things around in a big way. So the <u>first 54 years of my life were trash</u> but it had given me some insights, and taught me patience. They finally told me of my agreement in January 2008.

Second Inheritance Fiasco

I mention this as it also shows how I was being watched over, how **some themes repeated themselves**, and the event corroborated the information from the 1991 Regression regarding me and money in this lifetime. I knew that Brian had saved much money in his lifetime (20 years with GE) but I didn't know how much, but in his letters he had always said that upon his death that he would "take care of me" unlike what Mom had done. He knew all about her and Nancy.

Although I had had a good relationship with Mom's third husband, Brian, he still lived in Upstate New York and while we exchanged letters every month, he had not shared that he was dying, and living in the house of his girlfriend near Oswego. So one day I got a certified letter from a law firm in the small town where Janet (not her real name as she is still alive) was nursing Brian, and it informed me of his death and that I had been left $40,000. I don't know why, I just let that go, I didn't contest it; I should have travelled up there and demanded to see the will (I would recognize Brian's handwriting) but I could not get away from my job at the County, and I had no idea his estate was very large. And it would not have done me any good as Janet had had the will rewritten upon his death leaving her $1.2 million (after greedy NY State Inheritance tax!) and I got the $40,000 that he had left her. And she wrote me several letters subsequently crowing about her 'windfall.'

You get the point. But there was justice again— Janet had put most of the money into building a new barn (she ran a dairy farm) and she expanded and renovated her farmhouse. She offered to buy me a campershell for my pickup, which I declined. (Guilt offering?)

Months later I received a letter from her saying she was moving to Florida with what money she still had left – a **freak lightning storm** had hit both the house and the barn and it **all burned to the ground**! She sold the cows and was out of business.

> I used the $40,000 to pay several big bills and pay off the truck, and took a brief vacation to Sedona.

October 1998 Visit

I did not know for sure IF the 'payback' would stop (had I paid my dues yet?), and living in Dallas Ft Worth was beginning to wear on me (I missed deserts and mountains), so I decided to move to Tucson. Nine hours from DFW to **El Paso** – Sheesh, driving thru Texas the longest way I could! I holed up for the night, my truck full of boxes and stuff for the move – which I had to unload and stick in the motel room. I hit the sack about 10:00 pm and fell right to sleep. October 25th.

Next thing I knew I was **out of my body**, 'standing' among the boxes by the window, with a **swirling green energy cloud** around me and a voice said "Don't move. Do not leave the cloud." I could see my body over on the bed, and some beings working on me, doing something ... which was puzzling as the room was pitch dark but I could see— as if I had night vision. Then nothing.

The next thing I remembered, after I woke up was a very unique dream – had it 3 times in succession (as if to say, "remember this!") as a matter of fact. When I woke up the clock said 3:10 and I thought, dang it, I overslept – gotta get going – I thought it was 3:10 p.m. I rushed to the window (which had opaque drapes) and it was still night... 3:10 **a.m**. I was really disoriented, but the dream fascinated me and I went over to the desk and pulled out some paper and documented the dream… then made sure my portable alarm clock was working, and went back to sleep.

> I dreamt I was in space, all blackness with brilliant colored stars all around. I saw the Earth below me and began descending a beautiful, wide mahogany staircase. As I descended the red carpeted steps, the bright stars disappeared, the carpet became dull, the air became heavy and 'grayish.' I exited at a level I knew to be 3D Earth and approached a man who appeared to be asleep and I entered him thru the head. I threw out a red book (!??) and some other things, and "cleaned house."

I have never understood what the 'red book' was and I would love to tell you what happened that night, but I still (2017) do not know. I theorized it was a **Walk-in** scenario but I have never liked that and have resisted it – even after being led to Ruth Montgomery's <u>Strangers Among Us</u> [3] book in February 1987:

> It was during the February **1987 Audit in Alabama**, and I was watching TV in the motel room, and at 15 minutes to 9p.m., I <u>was urged</u> to go down the street (in snow, ice and all!) to the Mall and into the small bookstore. I resisted… snow and icy road, but something said I <u>had to go</u>, so I put my jacket on, wondered if I could get there before they closed… and hit the road... front-wheel drive rental car, so the traction was fine. I got there <u>thanks to all green lights</u> and parked the car, went into the Mall, not knowing where the bookstore was, but actually went right to it! I went in, down the main aisle, hung a right, crossed two rows, turned right, and turned around, and <u>the</u> book (I swear to God) fell of the shelf at my feet. Puzzled, I bought it as they were dropping the steel gate over the store opening, and went back to the motel … and devoured the book. Fascinating – all about Walk-ins. But <u>I didn't like the idea</u>, and it would be the next week when that serious fungus allergy would hit me! Of course, I have wondered if the book was 'preparing me' or 'asking me' to permit something…? But I was against it. (See Endnotes 1 and 3.)

Years later, about 2008-2010, I would wonder if the 1998 Visit was a **Soul Merge** (my higher 5D soul aspect from 5D would merge with me, 3D Jay), but that didn't feel right either…(see Appendix B, and VEG, Ch. 7) so I was 'stuck' with a greater awareness and the ability to ask/know things which is what fed the writing of the VEG book. Did a 5D version of me Walk-In <u>anyway</u>…?!?! Did I have a prior (InterLife) agreement to do that? And what/who "zapped" me August 27th 2016?

In any event… my resulting vibration was a bit higher and I had access to more Knowledge and some abilities with energy for the coming years. The higher energy also **kept my body perfectly healthy** – and whatever it was, <u>was all to keep me healthy to do the books.</u> If I was a Walk-In, I didn't know it (and didn't want to), but my aura was no longer blue – it was now white/gold (I kept the aura pix).

2001 Miracles

At the risk of giving too much info, I repeat: **all of this chapter is absolutely true**, it is not to impress anyone as I am no one special. The following two events really set me back on my heels… and I began in earnest to look at what was being done to protect and guide me… all the while wondering *who the heck am I and why are They doing this?* Yet I learned in the 1991 Regression that **I <u>did</u> have a pre-birth agreement** (although I didn't know what it was) and the gods (see Glossary) were just upholding their part of the agreement – seeing to it that I got to the point where the 6 books would be done (I just didn't know that the Old Jay [pre-1998] would not be writing them!).

As you will see, this was an on-going thing.

I-75 Accident

Driving up the I-75 in Dallas, Summer 2001, which is a bad driver area (sorry, but North Texas has the 3rd highest auto insurance rates in the US), I was getting ready to make the transition to the I-635 East and out to Rockwall for a church seminar at 1p.m. I was driving a green **1993 Toyota T-100 pickup**.

The traffic had slowed to about 35-40 mph, I was in the curbside lane, and out of the corner of my eye, to my left, I saw a white Nissan sedan coming closer as if to force its way in front of me, and moving faster than I was, but the driver misjudged, or something, because while I gently braked, he slammed into the mid-left side of my pickup. (I mention that because the T-100 side panels, doors, etc. were all made with Jap beer cans… the metal was so flimsy you could push and flex it with your hand!)

I signaled him to pull over, really angry with the idiot road behavior that passes for "driving" in the Dallas area. I expected him to have no insurance on top of it, as they often don't. I noticed that he and his passenger friend were both about 18-20 years old, and Middle Eastern.

We pulled over and I had decided to punch him out. I got out, didn't look at my truck and I slowly walked closer to the driver, now standing on the shoulder. His friend exited their car and ran down the outside around my truck and up behind me… WTF! I was getting madder all the time. Great, I thought, his friend is attacking from behind! (No fear, I have a 2nd degree black belt.)

All of a sudden the shorter one, who came up behind me, was yelling something in Arabic (?), and pointing to my truck. Neither spoke English, and the driver was gesturing that I not hit him. The attack from behind me did not happen. From behind me, occasionally I heard *Allah…. Allah…..* and so I had to turn to see what he was yelling about.

There wasn't a fricking mark on my truck!

A 40-mph impact and there was no white paint, no scratch, no dent… nothing! **Just as if it hadn't happened.** I looked at his car and the right fender was dented in and his bumper was bent back. My anger switched to puzzlement, and now both Arabs were looking at the truck and touching where the damage should have been…. Even my driver's door was not bent or sprung! How could I be angry when there was no damage?!

Before I could say anything, both ran back into their car and sped off. Unreal.

That same day I attended the church seminar, hosted by healing evangelists from Florida… I didn't know the woman carried the great English healer **Smith Wigglesworth's** anointing for healing. I was to get my second surprise…

Physical Healing 2001

I gave a testimony to the large group I was in (about the non-accident) and then the blonde lady minister, dressed in a white silk Nehru-like outfit, looked at me and gestured to me to come up front. The seminar was about energy healing, and I was going to tell her that my adrenals (and cortisol output) needed healing. I never got the chance.

She said nothing, just took me by the shoulders, faced me forward (toward the audience) and put her hands directly over the small of my back. Damn, her hands were <u>hot</u>! She noted my flinching and said I would not be burned, and to just relax. Twenty seconds later I collapsed on the carpet, sweating like a pig. Before this, I didn't believe in the **Laying on of Hands, and Slain in the Spirit** (aka "Carpet Time") but it happened – I could not stand up.

As I lay there, she kept an eye on me but went on with her presentation… every now and then she'd glance over and cock her head to the side (as if listening to something), and point at me and actually shoot me more energy! I felt the tingle and energizing and started to sweat some more. Because I was embarrassed lying there in front of all those people, I tried to at least sit up and could not manage it. She chuckled and hit me with more energy!

About what must have been 20 minutes later (or more?), I had lost track of time and maybe passed out… most of the audience was gone, and she came over and helped me to get up. I had a hundred questions… How did she do that? <u>What</u> did she do? And she just smiled and said <u>she had been told to heal me</u> and exactly what my problem was as I walked into the room. I thanked her, not knowing if I was healed and said I'd come back next week and give her a report. She smiled.

I went to my doctor the next day and he ran a **cortisol test**. The range for cortisol in the body is 5 – 25, and I had been a 7 for two years: **Addison's Disease**. A few days later in his office, the results said I was a 14. He said that must be a fluke, so we redid the test. Three days later, it was a 16. Similar to the doctor in Alabama, he could not say 'miracle.'

Robin had healed me (or <u>They did</u> thru her as she claimed), and I had more stamina and was less susceptible to stress. I gave her a copy of the doctor's report and she had me document and sign a medical release form – her group was compiling a book of miracles and mine might be used as one of the stories.

2003 Baldy Visits

I was a teacher in the local area, and was subbing during the Summer. I was seated in the local high school cafeteria in June 2003 in an effort to sign up more kids for the second Summer Session – 6 weeks of intensive course credit – in many subjects. I

was manning the Foreign Language table and about 20' across from me were the Science and Math sign-up tables. History and English tables were also there – it was a busy cafeteria. It was about 20 minutes to Noon.

I was looking down making some notes and became aware of someone standing on the other side of the table. I looked up and there was a tall, perfectly built man, about 6' 8" in black slacks and a **loose-fitting but long-sleeved white shirt** (the shirt reminded me of Errol Flynn in some of his movies, see **picture left**) and he was smiling at me.

I instantly forgot Errol Flynn and thought he looked more like Mr. Clean – at which point he laughed – as if reading my mind! (I later learned that he was.) His teeth were perfect, and his irises were very blue and bigger than most people's.

I asked if I could help him, and he asked me if I had an interest in Physics and Quantum Mechanics. I said yes and gestured to him to have a seat, and instead of

using one of the chairs, he perched on the edge of the table, moving some papers out of the way. This was not a standard cafeteria table with built-in benches; it was an 8' fiberglass Steelcase ® table – and it really creaked under his weight!
At this time, there were no more signups for my table – we had the space all to ourselves.

I was aware that we chatted for some minutes, and strangely did not recall much of what was said, and then he stood up and wished me well, and walked across the aisle to the Science table, took the arm of a very pretty brunette (about 6" herself!) and they exited thru the side door. I figured he had to be the school's Physics teacher, and as there were no more kids, I walked over to the Science table and asked what my visitor's name was. They looked at me like I had 3 heads. I asked if the tall man with me was their Physics teacher. They told me that I had been alone at my table for the last 45 minutes. Forty-five minutes! I glanced at the cafeteria clock, and sure enough, 45 minutes had gone by, it was 12:25 pm, and I thought Baldy and I had chatted for about 15 minutes! And the papers he had moved were still moved aside!

What had transpired during that time? – and why didn't they see Baldy?! I almost expected Rod Serling (*Twilight Zone* fame) to show up. That was **Baldy's first manifestation to me**… and ever afterward I have had a fascination for and attraction to Quantum Physics. (Book 4 QES makes use of what I was given.)

2006 Training

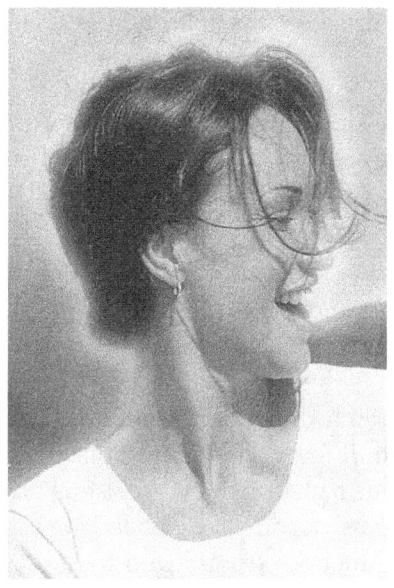

Because this is an historical account of what They were doing to prepare me to transcribe VEG, the following is also relevant – even though it is more fully examined in VEG, Ch.5. I agree that it may seem bizarre, but is absolutely true, and I can still see today what I was shown back in the period 2006-2008. It is part of the Great Earth Puzzle… and is addressed in Chapter 9 on Simulation in this book (as well as in VEG).

This should not freak anyone out, but not all humans that we see around us on any given day have an **aura**… that means they have no soul. The aura is the 1" energetic 'glow' around the body (left) that is the **reflection of the soul**… what the Bible calls the "glorified body."

People who do not have auras do not have a soul, and the absence of a soul means that they are <u>not</u> connected with anything spiritual… they have no spiritual component and thus **do not have a conscience**. While not evil, they <u>can</u> cause problems in a marriage: if the wife has a soul and the husband doesn't… That is called being "unequally yoked." They may rarely agree on anything. The soulless husband can cheat on his ensouled wife and not have any remorse… no conscience.

VEG Ch. 5 goes into <u>much more detail</u> on the ways to spot these people, what it means, why they exist, and explains that they are also **called OPs** and NPCs – as in a video game (see Glossary). It is very real and is one of the sources of dissension in today's world. As Chapter 9 shows, their only purpose is to drive the Father of Light's Greater Script for the Earth School, like bit-part actors. (Think: Richard Ramirez, Charles Manson, Stalin… and criminals with 'steroidal behavior' are often OPs.) But not all soulless humans are bad – many are here <u>earning</u> a soul!

The reason for inserting this into this book is that it relates to the Earth Realm as a possible Simulation, and such needs "bit players" (NPCs) to drive the School Script – giving lessons to the ensouled humans….explained in VEG, Ch. 5. The purpose herein is to show how this knowledge came about.

I was sitting in a bookstore near the café part, sipping on a Border's Freeze, and reading a fascinating book on the Sumerians, when I glanced up across the room and did a double-take. This was the first of many such events for me. I saw several people with that aural glow around their head, and three others who had no such thing – in the case of the latter, they had what appeared to be "heat waves" above their heads. This fascinated me.

Then an <u>actual voice</u> behind me said "Pay attention to the differences." And it was so real, I turned around to see who was there… and of course, like at other times, there was no one. This was the last time this happened – I got rather irritated with voices coming to me at times, while in church, while in the bookstore, while shopping, while driving, and I said **a quick prayer request that the voices stop and another way (for Baldy or Angels only!) be found to speak to me**. It had been real unnerving… as well as startling.

> Sure enough, the voices stopped and They started doing the **"1-second drops"** instead. (See Glossary.)

And then over the weeks that this went on, I started keeping a count, a record of how many people had auras and how many didn't… and I just assumed my ability was part of my October 1998 Visit and 'rewiring'…. Or maybe Baldy was assisting me?… No matter, I checked this phenomenon out for the next 2 years, while shopping, while <u>at church</u>, and of course in all visits to the bookstore… and I discovered that **at least 50% of all people around me at any time had no aura**, and it sometimes went **as high as 60%** -- and this is significant when we get to Drs. Bostrom and Greene in Chapter 9 where they suggest high headcounts IF we are in a Simulation, because there would have to be NPCs – **Non-Playable Characters** –

carrying the Earth Drama as if we were in a play (Yea, Shakespeare!) or a huge video game…. So <u>there is corroboration</u> even though the average person cannot see auras.

The other part that really caused me to start asking questions and do further research was seeing the non-aura (soulless) people <u>in church</u>. In VEG Ch. 5 it is stated that these people who do not have a soul, therefore have no connection with anything higher than themselves – **no connection to God, or a Higher Self**, or even a Soul Group. Thus they would not be interested in spiritual matters. **So why are they in church?** And not just 2 or 3 of them – I usually see 20-30 of them, at rough count.

Answer (and this came by another 1-second drop): The soulless have the unusual opportunity to **'earn' a soul** on Earth **if** they are interested in what their ensouled friends share with them. If they want to develop the lifespark within themselves and are genuinely curious about the spiritual realm, what their church friends say, and even have the slightest curiosity about what they are, where they are, what Earth is… that can lead to them becoming **a Baby Soul** (see **Appendix B**). I was also told that a 1ˢᵗ time Baby Soul has a very weak aura, very hard to see, and some may <u>also</u> have the vestige of the "heat waves" above their head for a while. That surprised me… but is another example of "God's Grace that none should perish."

Of course, if the OP (or NPC) does not go on to develop a soul, when they die, it is dust to dust… there is no soul to move on.

> So all of that to say, **we are not alone, and there is more going on than meets the eye… the gods are real.**

2008 Transcribing Their Book

So the years from 1998 to 2008 appeared, in retrospect, to be learning and documenting. I had made notes from key books I read – some books They would not let me read – and some seminars and special people I met along the way – all got documented in the folders. By January 2008, it amounted to about 12 folders on my PC, and coincidentally, they would be additional material for the first 12 chapters in VEG. But I still had no clue what I was doing – Why was I collecting all this information? (I was not aware that I might be a New ["upgraded"] Jay… after 10 years, it seemed that I was just the same Old Jay that I had been…)

Then one morning at 3 am (one of Their favorite times I would discover) I was awakened with the **Table of Contents for VEG** very vividly in mind, and I got up and went to my PC and typed it in. I could not go back to sleep… I was wide awake and felt like ("coincidence" again?) doing the first chapter. I finished it about 2 pm later that day. **No voices, no automatic writing, no visions, no trance**…. It just flowed – later I had to realize that some "higher" part of me was typing the book.

And occasionally there were special **1-second drops** to make key points in the book – there would be a total of 7 of those for VEG.

> **1-second drops** turned out to be from Baldy, (not my Guardian Angel) whereas the New Jay was just <u>transcribing</u> the book...

> I also did not trust the Guardian Angel concept, much less the Walk-in or Soul Merge aspect... I was afraid I might be tricked by Astral beings... (but I wasn't; it was just a remnant fear from my Christian days).

Ten to twelve hours typing for several days per week (which can result in "<u>internal</u> hemorrhoids"!), and I was never tired. Somehow I was being energized and sustained to do the **transcription**. Starting in January, we were done in July. **Seven months and 868 pages**. I knew that was too big for one book, so I felt Ok with breaking it into two books, and the 'overflow' from VEG became TOM.

However I still had a problem. I was using **MS Word 2004** and it could not handle a huge 80 Mb file as one document, so I could not assemble a printfile for the publisher. Thus I had to wait until **MS Word 2010** came along. And that version also had a Docx-to-Pdf converter which the publisher wanted. And still I delayed.

I sure as heck was not going to put **my real name** on the book's front cover – there was some confrontational material in the book and I didn't want some fanatic readers looking up my real name on Google, discovering my home address (it was there and Google would not remove it), and then some fanatics might drive by my house and add to my collection of rocks!

And yet, **I was concerned with vetting what was in the book**, and so I sat on it for what turned out to be a total of **4.5 years**. What had been transcribed in a sort of Flow initially had very few footnotes, and most of them were later added when credible material or confirmation was found of some key points, and the Endnotes grew. That was to add credibility <u>for the reader</u>, but I later realized it looks like I orchestrated the whole book, whereas **it was merely transcribed**... Perhaps that many Endnotes was a mistake.

So by December 2013 I had basically updated Endnotes and vetted the book, and with MS Word 2010 had formatted the huge document (now 14 chapters), and yet I still had **two serious 'showstoppers'** – it was two issues that I wanted, to connect two more dots to the rest of the book, and complete it, and as the transcriber, I still didn't have all the answers.

If you are wondering about how I could write the book and then question it, 'vet' it, that was because I seemed to be the same Old Jay (analytical, data processing programmer, logical me). The Old Jay was reading what the New Jay wrote and that was the difference. But the New Jay was different and no one noticed any difference – even to those who knew the Old Jay. No one ever questioned my words, thinking or behavior. And yet my taste in clothes had changed ... and I could no longer program a computer, so I lost my job. I was different somehow.

2013 Prodding

I was sitting in my study in December 2013, knowing I had two big 'showstoppers' that I could not resolve dealing with VEG, and I could not release the book until I was satisfied that it was correct – I had spent months making sure the text was accurate, adding pictures, and I didn't want to put my real name on some froo-froo. (I used a penname anyway when I did publish it in 2014.)

While I was sitting there wondering what I could do, into the room came this beautiful scent, like **orange blossoms**, and all of a sudden I became aware that Baldy was visiting me again! (**Baldy** later turned out to be an Interface – and the significance is revealed in Appendix F.)

> He asked, "Did you have an agreement to do a book?"
> (*Don't even think about lying to a Higher Being! Of course I did and he knew it.*)
>
> I said, "Yes, and I have two major showstopper issues…"
>
> He interrupted, "I know. Would you like the answers?"
> (*Sheesh, he already knew what I was thinking and what I needed.*)
>
> "Yes," I said.

Poof, he disappeared but the orange blossom scent lingered. And I had just been given something that I have called a **1-second drop** (see Glossary) of information, better now called a **Download of Information**. I grabbed a pen and paper and began to write what I had been given – 30 minutes worth – and it filled 4 pages of my writing tablet, and I spent hours reflecting on what it said and what it meant. It was obvious this material had to go into the book and it resulted in **Appendix D** and the inserting of **Chapter 13** – into VEG. And it was obvious that I did not know or have access to everything! And Baldy turned out to be quite trustworthy.

That was always Baldy's style – he never told me what to do, <u>he always asked me questions</u> to which I knew the answer.

He gave me the 'download' and I sat there analyzing what he gave me, really evaluating what the information from him was, and I made many notes and worked out how I would insert those into the VEG book. I had wanted to know if **Jesus** was really real, or was an invention of the Church based on Apollonius, and secondly I wanted further clarification of the **VR Sphere**. (Those became Appendix D and Chapter 13 in VEG.)

> Later I would find that they had not told me whether Earth School was a **round globe or a Flat Earth,** and the *Corrections* section of the Introduction outlined 3 things they did give me … apparently it wasn't important what the Earth structure really was, but late in 2016 it would come back to haunt me, the Old Jay … and that is addressed in Chapter 11 of QES as well as in this book. It is a very significant reminder that **we are not alone**, we have a purpose, and Someone is watching over us.

Baldy also told me in 2013 that I would not be advertising nor promoting the books. That irritated me in late 2016 – **Why write them if I can't promote them?** And yet attempts to do so have resulted in wasted money, time, energy, and failed book promotions. I was puzzled, then upset, and felt like I had been used… but somewhere at some time I must have agreed to my part (even though I don't remember it). Maybe in the InterLife, or during the 1998 Visit…?

> **In early 2017 as I began to realize that the 6 books contained a lot of knowledge, I thought might not be able to remember and coherently explain <u>all</u> that had been written, and thus it is not surprising (in the following seminar event) that I was blocked.**

About mid-2016 I had an opportunity where <u>I was invited</u> to promote the TSiM, Book 5, and I took it, thinking if I was asked, it was Ok. No it wasn't.

2016 Seminar

A New Age church in Dallas invited me in February 2016 to do a presentation on TSiM and even stocked a few books in their bookstore. I thought things were maybe changing…. And the evening event finally arrived. On **August 24th** I spent over an hour with an overhead projector (and $32 worth of slides) and a detailed outline to present and explain the book. The event was poorly attended, despite church advertising – and even the church promoters did not show up. Maybe 12 people were there out of the 60 who had signed up during the months leading up to the presentation.

I had prayed about the event, that it would go well, that what needed to be said and presented would be said and presented, and I assumed that I could do this, so that the presentation would be dynamic, <u>and bless people</u>! After all, this was to benefit the attendees and was not about me – the seminar was free.   There were major points I wanted to make about the Earth School, overcoming challenges, our Scripts… and I brought two people with me who already knew the basic teaching.

What happened? Disaster.

I felt 'blocked' and not able to stay on topic. I have done public speaking for years, even having been a teacher for 4 years, so I didn't freeze up – it was just disjointed and ineffective – <u>despite having made</u> **an outline** <u>and following it</u>. At the end, I asked my two friends what they thought, and John said, "What the heck happened? You were all over the place, little coherency, and we knew what you were trying to say… but it did not come across!" **I had been psychically manipulated.** And sabotaged. But why? I was not asked back to the church, I was embarrassed and left it anyway, and in my opinion the seminar was a fiasco. Six months of preparation, and outline, super slides and …. Zip.

> **What I later found out was that the church was home to witchy groups who practiced Black Magic, Wicca and Santeria. I did not know that and even John and Sonia said the energy felt weird in the building. Astral entities in the building did not want my Light, and I found out that 3 staff members in the church had worked to kill the attendance – if not the seminar itself!**

> There is another aspect to this New vs Old "me" as you'll see in the chapters to come: I often forgot that I was New Jay between 1998 – 2016. I had <u>the same memories</u> and it was easy to forget who/what I had become… and after the New Jay **'rewiring' or 'upgrade'** was gone (early August 2016) , I was left with some advanced Knowledge and ability (I could still see auras and manipulate *chi*)… but it is/was often a confusing situation as to what October 25th, 1998 really was.

August 27th, 2016: It Leaves

I was still upset with being 'blocked' from enlightening the seminar's attendees…. And just <u>three days later</u>, after supporting some friends in managing their successful Saturday evening event, I was laying on my bed noting how I had been able to make others' dinner-dance a success (helping with logistics) but I had been abandoned when trying to do <u>my</u> presentation. Why did I not get protection and guidance? And then, lying on the bed, staring at the ceiling, the anger really boiled up: I said

> "Damn it, I am pissed with whatever or whoever interfered with me Wednesday and if it is still with me, in Jesus' name I want it gone!"

Wow. CRACK! There was a slight pop and my body jumped on the bed about an inch and I now felt very alone… a weird feeling compared to 1 minute earlier. What was that? A psychic whack? I wondered if I <u>had</u> been a Walk-In and now some part of me was gone…? It was like an **electric shock** hitting my body, as if I had stuck my finger in a light socket. Damn it, that was nothing like the original, gentle dream (in 1998). Then it dawned on me – something had attached itself to me, manipulated my presentation, and just now left! It wasn't a higher soul aspect.

A week later (September 4) I went to the guy with the <u>real</u> aura camera in Dallas to see just who I was – my aura had been light blue before all this started (1998), and then over the years (1998- 2008) my aura had been consistently white/gold… What was it now (2016)? Did I dare consider that I had been a Walk-In and because I hated the idea, so I had been one and wasn't told? (No.)

Daniel's aura camera had always taken an unusual picture of me –touching nothing– **white and gold**. Now I would have to see if that was still the same, or what color my aura was now. If it wasn't white and gold, then… what to make of it all?… but a $30 aura picture might be interesting.

> Daniel has a real aura camera and appears at Dallas holistic fairs. He has appeared on **Good Morning America** and the camera uses the UV spectrum to photograph the person. Daniel touches nothing, just a remote button, there is no color wheel, and **the subject touches nothing.** The film is a standard Polaroid film pack in a Box Camera-like device. And whomever he photographs gets the same result every time – I did 2 in a row one day (1998) and it was always white and gold. Then another person had her photo taken, getting red, and I redid mine and it was white and gold again. In 2012 it was still white/gold. (Also see Appendix F for significance.)

My aura this time came out very pale yellow and light brown. **Yellow** reflects intellectual, mental, intuition aspects and **light brown** is 'grounding' so that the person doesn't get too carried away into any psychic/froo-froo issues. So from white/gold to pale yellow/brown, <u>a close likeness</u>, I did grow over the 18 years association with higher Knowledge and it moved me from my **original aura of light blue** (1998). White/gold aura is what one would expect from a more evolved being from the higher realms… was this Baldy's effect on me? (Yes.) White is a color of completion; it contains all the other colors, and white is the color (with some violet) of the Crown Chakra – a very high vibration. We tested my aura again 2 days later and it was still pale yellow/brown. (I still have the pix.)

But there was more. And beside any higher aspect being gone, and with it the higher consciousness, **my perfect physical health was also gone!** But I could still see auras… with some effort. November – December 2016 was weird, bad health!

As the New Jay, the body had been kept in perfect health for the last 10 years – my doctor constantly ran tests, CBC's, yearly checkups, and would say "You have the body of a 45-year old man…" (I am currently 73)…. And now since the zapping ("departure?") of August 27th, I am back to some of the normal issues plaguing men who are seventy years old…. Great!

> Now I have to watch blood pressure, aging bones & joints, tinnitus
> and the future potential of eye & ear issues.

That electric shock was a kick in the pants from an entity I picked up in that off-beat church. I never went back (fortunately) and am glad I bitched about the Wednesday night seminar or that entity might still be with me…. After all, I was teaching against the Darkness in that church, and the entity had the right to afflict me as I was on its territory. I had seen similar problems among people we set free while serving in the **Deliverance Ministry** (1999-2000). So all that to say that the guidance and protection is back with me now (August 2017) but any higher soul aspect may have departed when the 6th book of Their set was done (early August 2016)… coincidentally just before the August 24th Presentation – which I was told to not do! Perhaps my disobedience allowed the negative 'attachment' to me to be done – and was also the start of the **9 months of weird health problems** (outlined below in the insert)… which stopped when I asked Baldy to restore my protection.

Summary

And that was the background to the writing of the 6 books. This one (in your hands) is on me, Old Jay, trying to document who I am, and what it was all about.

The issue of not being able to promote the books appears to be related to an insight I had several weeks ago: there would be **another one after me who picks up the books and does something with them.** I understand that I cannot accurately explain all that New Jay wrote…. And I was originally foggy on the VR Sphere, as you will see… so I transcribed, but can't promote. Aargh!

And the worst part of the issue is that Old Jay may have to face a possible higher aspect of myself at some time in the InterLife – I was not to promote the books and tried it anyway, and it appears that the New Jay was somehow working with a higher soul aspect while transcribing the 6 books (a <u>positive</u> version of support and protection – compared to the <u>negative</u> attachment of some Negg in that low-vibration church.) Apparently I had a lesson to learn…

What is important in **this final book** is the addition of some of the major information that was given me over the last 9 months, enlarging on VEG. And that includes the VR Sphere, Jesus info, Ancient Ones, the White Brotherhood and the Flat Earth scenario.

Some issues have cleared up, and I see more how it all fits now. In fact, without my discrediting VEG, and in concert with something the AL (Book 6) said, the **Anunnaki** issue, the **VR/Simulation-Young Earth** issue, and the **machinations of the PTB** are about to be made clearer (Appendix F). I will not be contradicting what VEG said, just clarifying and dealing with a major issue raised by Chapter 11 in QES – the **Flat Earth scenario**. All the pieces <u>that I am permitted to see</u> have come together and this book will hopefully provide more unification of the overall scenario.

Rewind: Overall Scenario

I emphasize this because it is so important. Please don't speed-read the last two pages. **New Jay + Baldy wrote and transcribed what was given**…but it is clear that New Jay either did not know enough about some issues, or certain <u>information was withheld</u>. Old Jay (me) upon reading it later (even late 2016) had questions, and I did not completely understand what New Jay had written, but it all seemed to hang together… suggesting **credibility**. Just some of the information in VEG was confrontational, shocking sometimes, blunt and sure to put some people off… yet it was what I was asked to do.

Of course the reader is free to accept, contemplate, or reject what is in the 6 books, but **the 6 books have proven "connected" and sustain each other as would a coherent, credible teaching**. I know some people will not like it – as Chapter 2 next shows, some people are too 'tightly wrapped' in what they think they know to even consider an alternate reality (also see **Appendix C**, Consciousness Levels)… And over just the last 2 years, after writing Book 6 (AL), I was given a bit more, including the **Flat Earth** scenario to research… which raised the issue of **Gravity** (now 90% resolved in **Appendix D**) and the whole is even more unified now.

The Flat Earth issue is clarified somewhat by the addition of **Appendix E** which shows that the **Firmament is a key to <u>both</u> the VR Sphere and the FE Scenario**. Apparently the Firmament was lower, closer to Earth back thousands of years ago…

many humans around the world tried to reach it! Then it was moved higher, and if the gods can do that, they could also morph a Flat Earth into a VR Sphere.

> **Again, why would They tell me all about a VR Sphere (VEG, Ch.s 12-13) if it were irrelevant? It may soon be obvious that the Earth <u>really</u> was flat… the evidence in ancient writings <u>is</u> there. What remains is: Can we <u>conclusively</u> prove that Earth is today a VR Sphere…? ? (Hint: Antarctica is the key.)**

I will not be contradicting but <u>expanding</u> on the information given, largely in VEG, and I can now see where the gods (via the New Jay/Baldy counterpart) were very careful in what They shared initially. There was more, but the first part had to be slowly assimilated. Had I (Old Jay) known back in 2008 what I now see and accept, I might have rejected it all out of hand as too much, too soon… too hard to swallow the whole thing… so I empathize with my readers' wrestling with this material.

Rewind: PTB Involvement

…And that is because the PTB have deftly hidden from the general public <u>what</u> we are, <u>where</u> we are (which emphasizes <u>what</u> we are!) and <u>why</u> we are here (Appendix F). And I have to give the PTB the benefit of the doubt: **there are serious reasons for maintaining a cover-up for 300+ years** (some addressed in Chapter 11 of QES, others in this book's Chapter 10). Some people (young souls) probably cannot handle the truth, so we play a game – the PTB Game is <u>within</u> the Father of Light's larger Drama. And that has to be Ok, since we are eternal souls learning in an **Earth School**. Any game that the Father of Light didn't want would be stopped…

So let's see what new aspects of the Earth Puzzle can be decoded and resolve the **Flat Earth vs VR Sphere** issue, as well as examine the possibility that the Earth Realm is really a **Simulation**. Along the way, we'll touch on the Anunnaki, the Ancient Ones, Secret Societies, Atlantis and UFOs and issues raised in VEG that now have further insight.

> So… why is new information so hard to even consider…??
> Chapter 2 answers this.

This is **an addendum to version 29.7** of this book… a lot has happened since publishing this book. I am not kidding when I say I was upset with not being able to promote the books, and I discovered that my protection was gone…

Reading thru Chapter 1 it is evident that I <u>was</u> protected, and then as the last Book 6 was done, I was 'attacked' August 24th. So what happened?

In October 2016 I had attacks of **vertigo**, inner ear issues, which the doctor cleared up, only to immediately have an 8-week **sinus attack** – similar to the one in 1987 in Alabama… and we cleared that up by January 2017… only to discover that my **blood pressure** was 150/100 and I had to get that under control (with Apple Cider Vinegar, by the way)….

Then to find that I now have a **cavity** under the crown in my back molar and, last but not least, recently discovered that I have three **internal hemorrhoids** – from what?! And now I have a weird **'ringing' in my left ear** at times…

The "Good Guys"` were <u>getting my attention</u> for having broken my agreement to not promote the books (the **August 24th seminar** violated that agreement) by letting sequential afflictions happen to me. It just took 9 months to get Their 'message.' Doing the books was a unique **sacrifice** and the afflictions finally got my attention. So I asked Baldy for an insight and he responded (1-sec drop) that <u>he will restore 'protection'</u> if **I walk away and shut up** – no promoting, no advertising, no teaching the books. (Others from 6D (6th level) will do that in the near future… It is too dangerous for me to get involved in that operation.)

There is a battle for the Truth (and souls) in this world and it is a sort of Game, albeit serious. There are rules that I was dissing. I broke a big one on August 24th, and then paid for it. The higher vibration and Light is something that the PTB and Astral discarnates do not want.

So what is up for me now? Baldy said that if I walk away and keep quiet, They can use me to **Anchor the Light**. That is all – no teaching, no further writing… They will do work via Light and higher vibration <u>thru me</u>… I will not even be aware of what They are doing… Nor can I control anything. I agreed.

Postscript

Everything in this chapter is true and much is included as I am also doing something of an autobiography, for what it is worth. FYI: VEG and TOM are the two key books.

Chapter 2: Humans & Belief

This chapter, like Chapter 1 was, is a prelude to the rest of the book as it is very important that the reader understand WHY humans are so stubborn, fixed in their beliefs, and unable to accept new information. There is serious new information in the rest of this book and it is hoped that the reader can at least think about it, examine it, and evaluate its potential value to him/her. Psychology actually has a name for our resistance to new ideas:

> **Cognitive Dissonance** – when something you hear refutes what you already believe, and even if the old belief is wrong, and is proven to be wrong, it is very hard to change it.

This is why some societies throughout history began "brainwashing" people at an early age, and even the Bible agrees:

> Raise up a child in the way he should go and
> he will not depart therefrom.   Prov. 22:6

The sooner you learn something the better the brain <u>locks</u> it in, <u>almost in concrete</u>, and the belief, or memory, becomes like a fixed part of you. It can be changed but sometimes at a great effort – especially hard to change if one is over 60 years old. Trying to change long-standing memories or beliefs in an older person may create such trauma in the brain that they suffer a nervous breakdown.

> This is why, when a person hears new, contradicting information, s/he will fight it – the subconscious knows that a fixed belief is going to be very hard to **'rewire' the neurons in the brain**, and reacts as if the person's survival is at stake. Sometimes it is better to just let the person have their mindset.

Let's see why this is so... at the layman level.

Brain 101

The following information is examined in greater detail in <u>The Science in Metaphysics</u> (Book 5) and refers to the research done by Dr. Caroline Leaf, a neuroscientist from South Africa.

When forming a new memory, a new habit, or a new belief, the brain makes use of the glial cells, astrocytes, neurons, synapses and axons, etc.... We are largely concerned with neurons and their structures and how they connect to each other.

Neurons and Synapses

Throughout the Cortex, among the many-folded contours of the spongy, soft brain, are found the brain cells: **Neurons**. About **10 billion** of them.

> Contrary to former scientific belief, brain cells die and new ones are born constantly, thus **it is possible** to renew the brain, learn new things, change habits and restore memories.

The many neurons in the brain form the main elements of the **Neural Network** – which is the difficult part to get to reconnect to a new belief.
The other main part is the **10 million Synapses** which are the connections between the neurons. Neurons expand and connect via **Dendrites**.

> **Think of the Neurons as major cities, the Dendrites as freeways between them**. The freeways often connect with cities via loops around the cities – they don't always charge right into the city. Thus the loop is similar to the Synapse… a point of connection but not actually touching the city's core.

The Interstate Highway System:

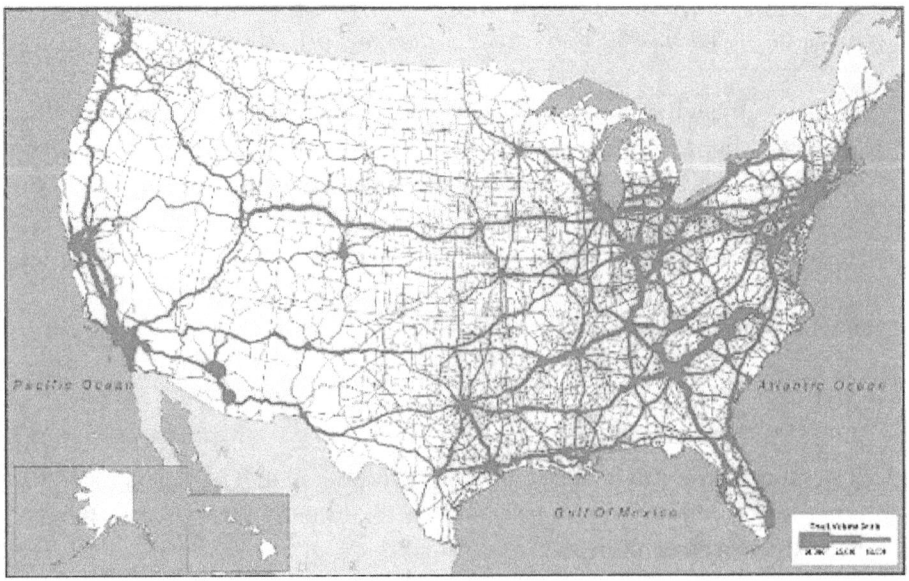

(credit **Bing Images**)

The Brain's Neural Network:

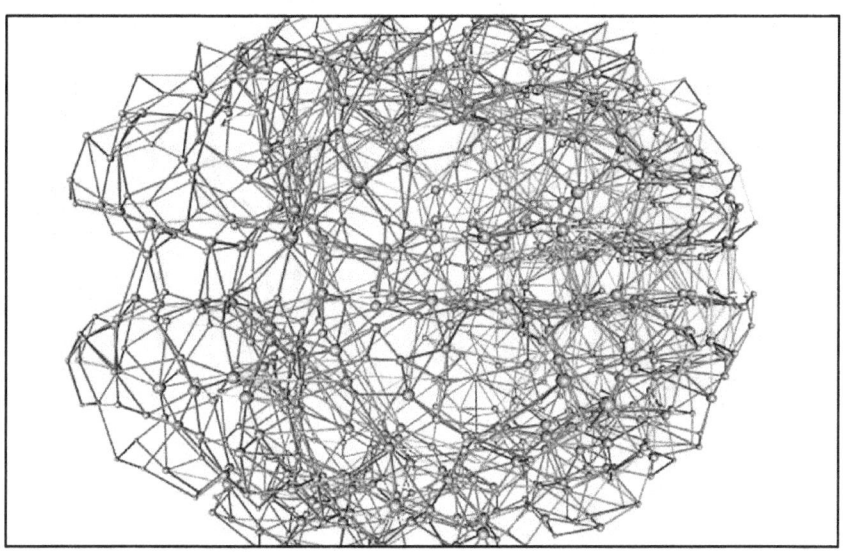

(Credit: Bing Images: scicasts.com)

More specifically, zooming in on a brain connection:

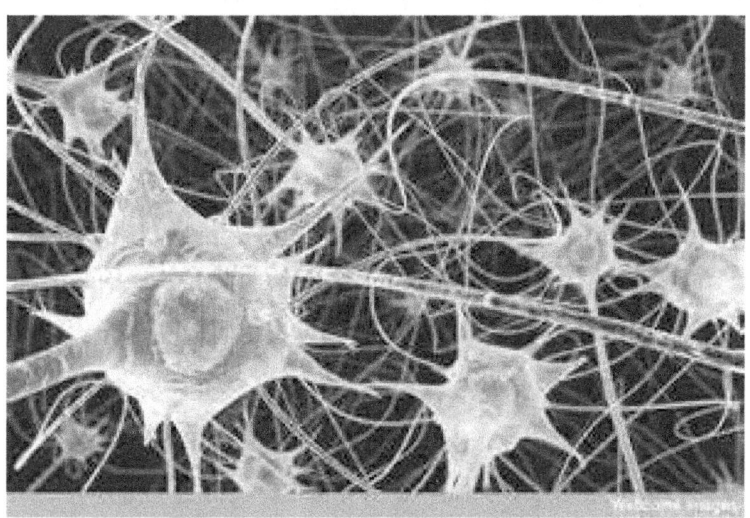

Neurons, Dendrites and Axons
(credit: Bing Images: blog.wellcome.ac.uk)

Neurons and Dendrites a-plenty! The center of the Neuron is the nucleus which has the cytoplasm (cell fluid), the **mitochondria** (power cells) and the Axon's **microtubules** (scaffolding of the cell, also used to transmit information – alleged to be at the quantum level.)

Thanks to Dr. Leaf, we can see the brain cells under a microscope: [4]

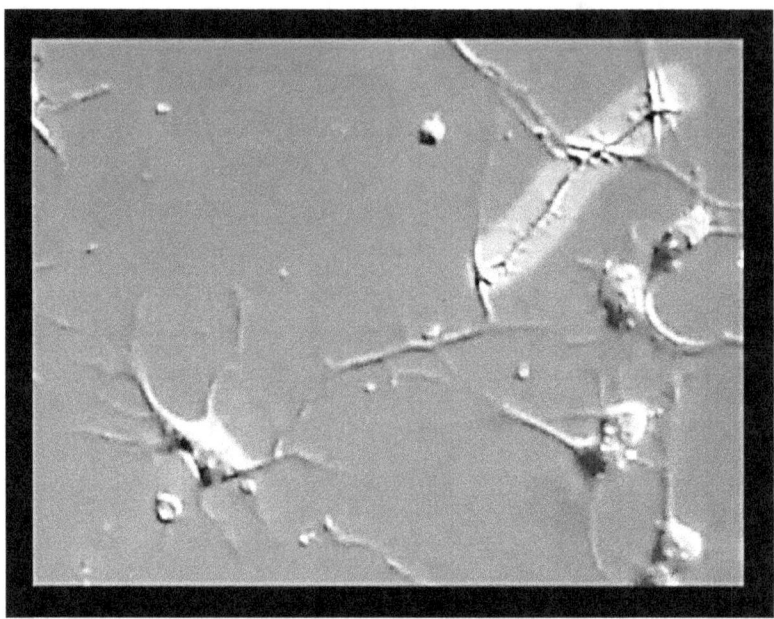

(credit: Dr. Caroline Leaf; Gatewaypeople.com presentation)

The highlighted Dendrite (upper right) is new/current growth.

The **Synapses** are found at the end of Axons aka Dendrites:

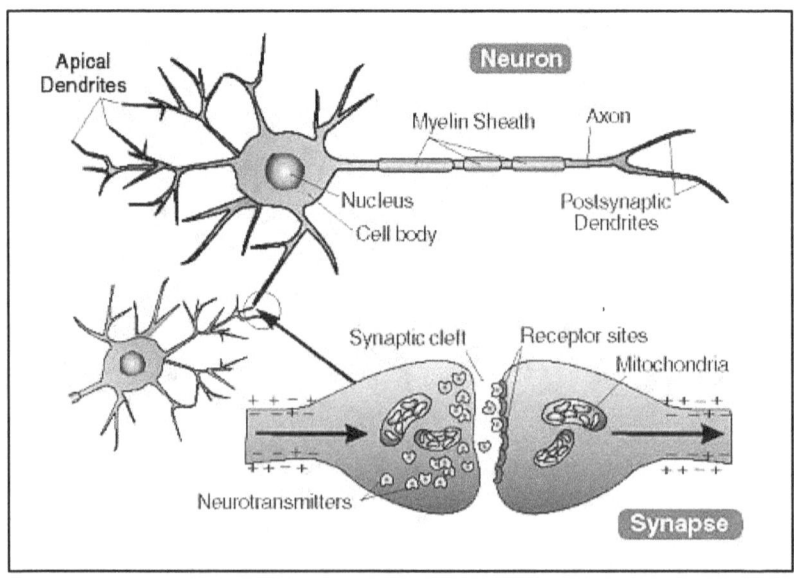

(credit: Bing Images: lookfordiagnosis.com)

If you look carefully at the above diagram, it has both **Dendrites** (which don't connect to another Neuron, they are immature Axons), and an **Axon** (ensheathed) which does lead to another Neuron (not shown). The function of the Axon is to transmit information to different neurons, muscles and glands. Dendrites are restricted to a small region around the Neuron body while Axons can be much longer and link Neurons, and Dendrites usually receive signals while Axons usually transmit them.

Dr. Caroline Leaf

(credit: www.drleaf.com)

Dr. Leaf is a pioneer in the field of **Neuroplasticity** from South Africa. She has worked in the area of **cognitive neuroscience** since 1985 and is one of the latest people to discover that **the way we think actually affects the brain.**

And changing our minds, changing our thinking, will change our lives – especially useful to remove bad habits and program good ones in the brain. If we are strong enough to do it.

If we change our thinking, it will also change our behavior as we tend to act out what we believe, say what we believe, and deactivate old habits and form new ones. One of her most dynamic books is called Switch on Your Brain wherein she first explains the science of **neuroplasticity** (changing your brain) and then shows the reader how easy it is to form a new, positive habit – within **21 days**. Her *modus operandi* is to establish the science behind what she says works… so you will believe it works!

It is done unto you as you believe.

Neuroplasticity

> What you think with your mind changes your brain and body, and **you are designed with the power to switch on your brain.** Your mind is that switch.[5]

She further maintains that if you change the programming of your brain, you will also change the chemistry. **Negative, depressed thoughts suppress the immune system,** <u>and</u> release negative chemicals into the bloodstream which further depress the person, and can actually cause health problems.

Exactly what Dr. Holmes said in his <u>Science of the Mind</u>.

That is good news: it means that **attitude can modify DNA**. And Ch. 9 in VEG presents evidence from foreign research to that exact effect.

> This state of mind is a real, physical, electromagnetic, **quantum** and chemical flow in the brain that switches groups of genes on or off in a positive or negative direction based on your choices and subsequent reactions. Scientifically this is called **Epigenetics**…. The brain responds to your mind by sending these neurological signals throughout the body, which means that **your thoughts and emotions are transformed into physiological and spiritual effects**, and then physiological experiences transform into mental and emotional states.[6] [emphasis added]
>
> And she has the hard evidence to prove it and often demonstrates it during her presentations.

Changing Beliefs

Our interest in this science is what happens when you form a belief/memory, what does it look like and why is it so hard to change?

The following are pictures taken thru a microscope of the way thought causes neurons to begin to structure Axons, or paths, to interconnect with other neurons.

The beginning is in Fig 1 and continues thru Fig. 3 below...

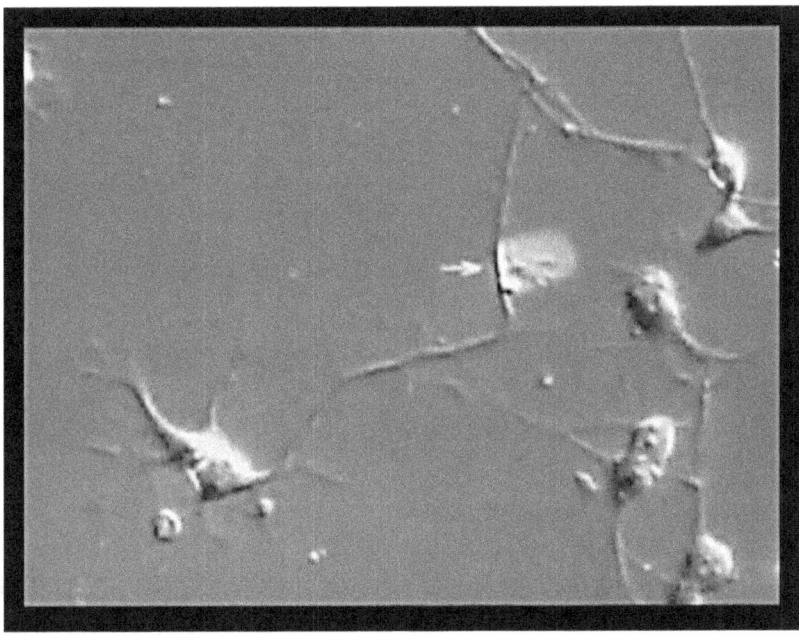

Fig. 1 Note the arrow where an Axon is beginning to grow in response to a thought.

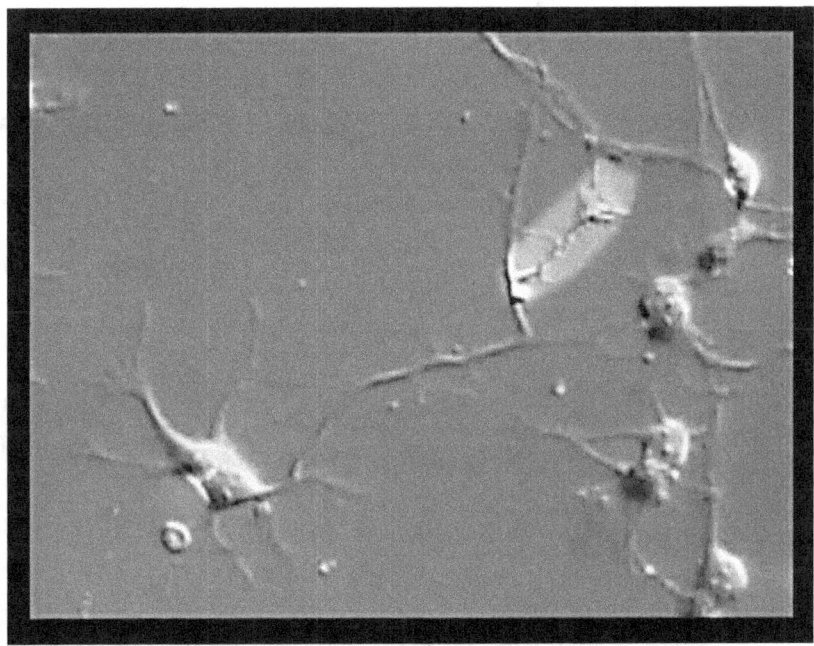

Fig. 2 Note the highlighted area where the Axon is still growing.

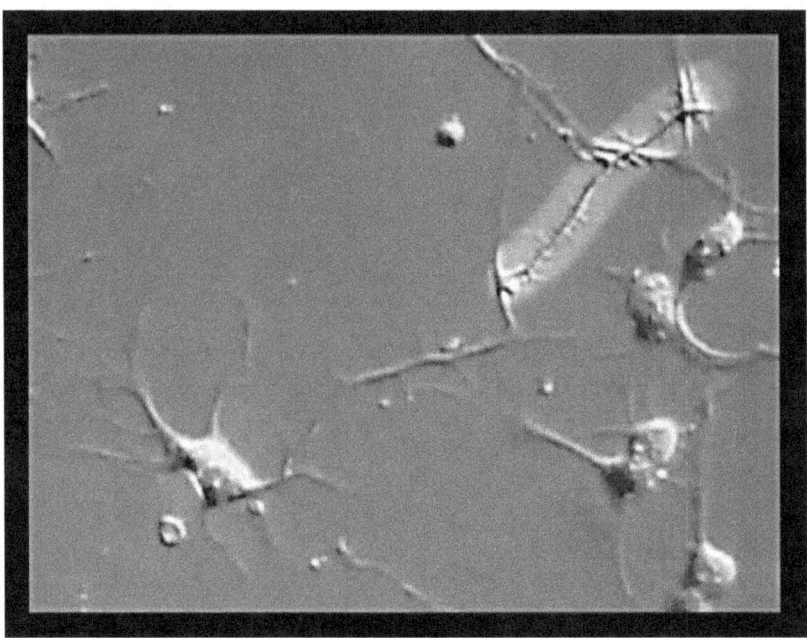

Fig. 3 Note the extended Axon area has enlarged.

The above is also called **Neurogenesis** wherein the brain is able to grow new neurons as well as linkages between them – and that is the **secret of building new habits or memories or beliefs.** In fact, the **brain is changing all the time** and with repeated effort, new structures are formed, while some old ones are withering thru disuse.

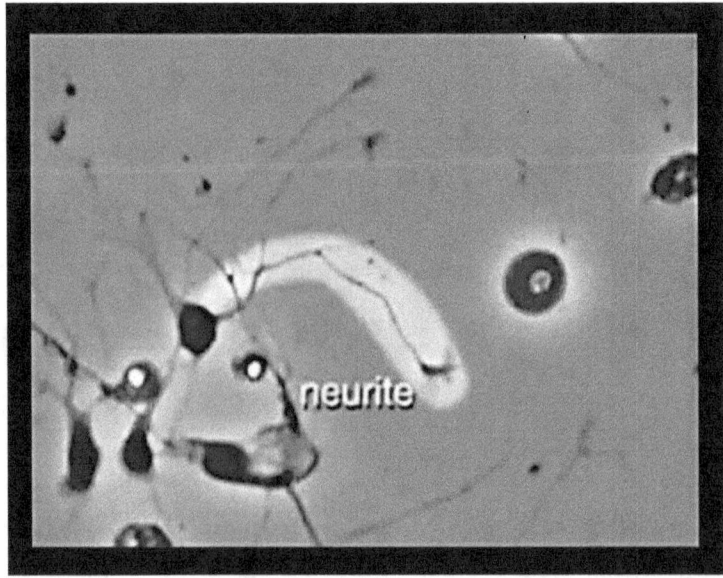

Fig. 4 A picture of a neurite. A growing, moving appendage to a neuron.
(A neurite can be an Axon or a Dendrite.)

Now here is the key part of her research that pertains to this book.

Bumps, Lollipops & Mushrooms

Dr. Leaf has provided graphic displays of what structures can grow in response to our thoughts, and she very aptly calls them Bumps, Lollipops and Mushrooms – depending on the size – which depends on how much energy and thought has been given to promote the growth. They are shown below in the circle.

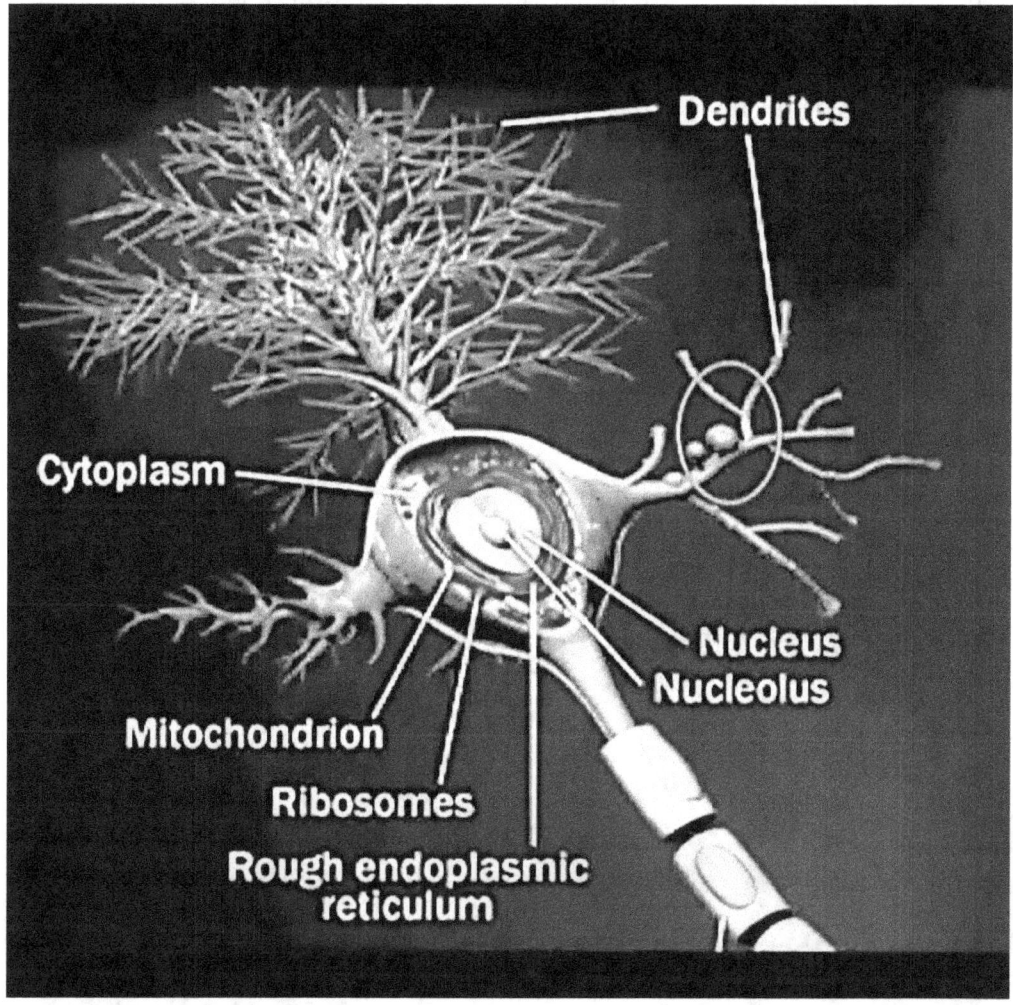

(credit: preceding pictures and above diagram from the Dr. Leaf presentation at www.GatewayPeople.com, 7/12/15.)

Note the 3 structures by the circle: the first is a Bump, the second is shaped like a

Lollipop, and the third is like a Mushroom (or some say, an Umbrella). These are three stages in the beginning growth of little things called ***spines.***

> These spines change shape, from a bump at around 7 days, to a lollipop shape at around 14 days, to a mushroom shape at around 21 days as the thought becomes stronger…. The proteins change progressively by day 21 …. To become self-sustaining proteins, which are like a long-term memory. [7]

Thus, if a person has been diligent and every day for 21 days practices a new thought, belief, or even a new skill for a sport, the action produces a new structure in the brain which then can be part of one's lifestyle. If one really continues practicing, over the next 21 days, the structure is <u>reinforced,</u> and if the person practices for a third 21 days, the habit is really **concreted** – Thus it is recommended that 3x21 (=63 days) is a required minimum to really <u>fix</u> a new habit or belief into one's lifestyle.

Rewind: Changing Beliefs

The opposite also applies – it requires some energy and persistence to **"rewire"** the brain and undo old beliefs and set up mew ones. The problem is, old beliefs have quite a bit of strength to their more solid structure, being reinforced over the years, and the older a belief is, the more effort it will take to change it.

(credit: TheWorldWeLiveIn: https://youtu.be/s0AOj01yzII)

Faulty Rewiring

While we are talking about rewiring the brain, there are **two aspects** that must be covered in the interest of supporting our kid's physical and mental health. What some of our kids are doing is very dangerous to their health, and if they get 'programmed' the wrong way, both they and society will pay for it.

What an I talking about?

In simple terms:

Garbage In, Garbage Out

So here is the first aspect to be aware of…

Thinking Activates Genes

In a further exploration of Dr. Leaf's findings, we will learn that **our consciousness activates our genes**. Science has elsewhere proven that thoughts affect the body and cause specific biochemical messengers to be released into the bloodstream and into the brain…. Dopamine, serotonin and adrenaline are common examples.

What has not been commonly known is what the effect is of **negative thinking** on the body and brain. Negative thoughts have been found (scientifically) to not be the norm. [8]

> Science has shown that our thoughts, with their embedded feelings turn sets of genes on and off in complex relationships….We may have a fixed set of genes in our chromosomes, but which of those genes are active and *how* they are active has a great deal to do with how we think… [9]

Taking all this to a deeper level, **our DNA actually changes shape according to our thoughts. Toxic thinking will change your brain wiring in a negative direction** and put your mind and body in a state of stress. [10]

The **Heartmath Institute** did experiments on DNA with people who projected fear, anger, negativity and changed the shape of the DNA (in Petri dishes) – the DNA responded to the negative input by tightening up and becoming shorter, switching off several DNA codes. The negative shutdown was reversed by projecting positive thoughts of Love, joy and gratitude. [11]

Positive thoughts result in a positive neurochemical rush in the body/brain complex. The body is constantly creating proteins to feed cells in the body and rebuild tissues, and our thinking affects how they are shaped. Regarding the **Epigenetic effect** of thinking, it has been discovered that **proteins formed in the body during toxic thinking are misshapen** (hint: *prions*) and act differently than normal, proactive proteins. [12] (See Chapter 9: 'Violent Video Games.')

> When somebody says 'prions' think Mad Cow Disease... misfolded proteins that corrupt and deform other proteins when they come in contact with them.

The pictures below show a healthy brain and one (right) ravaged by prions...

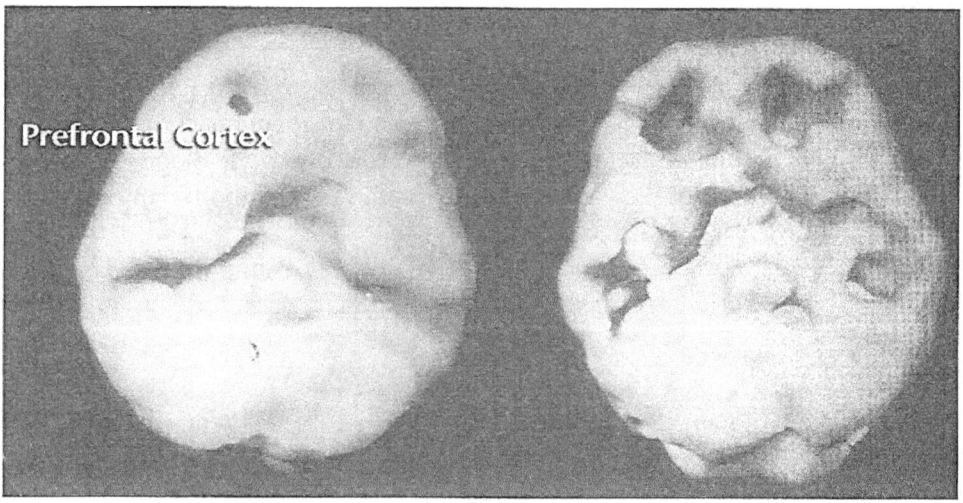

Toxic Thinking

The physical result of thinking can thus take its toll on the body.
Doesn't this give a new meaning to the ancient teaching:
"The sins of the fathers are visited on the children"
Whatever the parents have created in their bodies can genetically
be passed on to the next 4 generations. (Ex. 20:5)

And yet, Dr. Leaf would quickly point out that although you may have inherited a physical deficit, genetically, you are not condemned to live with it... and Drs. Holmes and Maltz would wholly agree. You can change your thinking and change your life. **Your thinking affects your DNA**. You can rise above any tendency to alcoholism or diabetes, for example.

The day will come when Man can reverse color-blindness and sickle-cell anemia, which are also inherited genetic defects. If we knew more about the Russian genetic research (last chapter), perhaps we could also fix blindness, color-blindness and sickle-cell anemia – just by the power of the word! After all, ascended masters have done it, and Jesus healed people, so it can be done… (Or by beaming perfect, healthy DNA to a body's sick area… as Was done by Russian scientists in Ch. 9 of VEG.)

The sins of the parents are thus a predisposition, not a destiny cast in concrete.

The Prion Problem

There is a certain type of protein created in the body, called a *prion*, which can operate like a Dr. Jekyll and Mr. Hyde. Shaped like a grain of rice, when a prion folds over itself it can lead to awful degenerative diseases, such as **Mad Cow Disease** – and that can happen in humans, as well. Scientists have now found that the healthy protein abounds in Synapses and they help form long-term self-sustaining memories.

They are also involved in neuroplasticity (brain flexibility).

The key point in Dr. Leaf's findings is that this protein does amazing things in response to good signals, and **goes crazy when it receives negative signals**.

> **Toxic (negative) thinking sends negative signals thru the Synapses and negative prions can result.**

Hence, negative thinking does affect the body/mind complex. [13] And there is another thing that we think we can do, that aggravates the brain…

Multitasking

This is a persistent myth. **The human brain was not designed to multitask** and persisting in **it can cause brain damage** (due to unnecessary stress and its biochemical effects). Says Dr. Leaf:

> Our brain responds with healthy patterns, circuits and neurochemicals when we think deeply, but not when we skim only the surface of multiple pieces of information.

This is **"milkshake" multitasking** (her term) in that it creates a jumbled mixture of biochemical neurotransmitters and **promotes a "hurry up" mentality that is**

similar to toxic thinking in its stressful effects on the body/mind. [14] Even worse is the potential for multitasking while on social media that can be addictive and thus more harmful in the long range than occasional multitasking.

And that brings us to the second aspect that endangers our kids…

Mental Health

Since this is a chapter on rewiring the brain and making proactive changes in one's thinking and beliefs, our **brain health** is affected not only by what we think, but by things that occupy our mental focus. Since it is very important to see how what we think and mentally rehearse does affect us, so a word should be said about **violent video games**.

To the degree that they overstimulate the brain and promote stress, that is a cause of **Excitotoxicity** (inflammation) and another source of damage to the brain. Overstimulation of the neural network, via overuse of the glutamate neurotransmitter, can result in an excess of the CA^{2+} ions which damage cells and kill mitochondria (the 'batteries' in our cells).[15] The CA ions cause the formation of superoxide production and death of the cells/neurons. That is the technical explanation.

Violent video games are not good for your child's mental health.

> **Let it be emphasized right up front that the author is not categorically against them. However, there are children with anti-social tendencies, even with autism, and so-called Crack Babies, and children who are taking doctor-prescribed medicine to stay with a balanced mood, who should NOT be playing the Stalk-It-And-Kill-It stressful types of video games (referred to herein as the 'violent video games').**

While there is a lot of literature that says it is really OK and kids know the difference between real killing and fantasy killing, the words of **Dr. Caroline Leaf** should be heard above it all … because Adults also play the games and are also at risk:

> …when an individual pays attention to a stimulus, the neurons in the cerebral cortex that represent this object show increased attention.

And

> …the way you focus your attention [has] a direct effect on how your proteins are synthesized, how your enzymes act, and how your neurochemicals act together.

And

>...the things going on in the environment get into the mind, changing the brain and **having a [biochemical] impact on the body.**[16]

Intense focus on the screen, **multitasking**, playing the game, trying to stay alive and **kill the other player(s)** (which is toxic thinking – remember what was said about that earlier!), puts the body under **stress.** It also produces a focus to KILL, or at least cause harm to another person – even if you don't know that person. It doesn't take much imagination to see that playing such a game repeatedly is going to **desensitize** the player to shooting and causing harm with a weapon (usually a gun).

Behavior: Reinforced and Automatic

>It is a **response action** to being threatened (in the game), and the more **instinctual, automatic** the self-defense response becomes, the more successful the player becomes, AND **the more ingrained becomes the behavior**... What we do and think, DOES create neural pathways.

According to what Dr. Leaf has shared, this amounts to **rehearsing an aggressive part of one's psyche**, and **the neural network will respond and build neural patterns consistent with that negative behavior** practiced (over days) while playing the violent video game.

To repeat:

>...negative thinking [including killing and causing harm to others] creates **atypical responses in the brain**, which will result in atypical manifestations [disease and/or asocial behavior]. Such as mal-formed prions...

And

>...lack of quality in our thought lives is the **complete opposite of how the brain is designed to function** and causes a level of **brain damage**. [17]

> Dr. Leaf also emphasizes that negative thinking is not how the brain was designed to function – especially as it unbalances the biochemistry and can form prions. She found that **the human brain was designed to think positively** as it releases beneficial chemicals into the brain and bloodstream.

Playing violent video games is not a positive experience… unless you are the last one left standing, having killed everyone else, and you are the (questionable) Winner! And what have you won?

According to Dr. Leaf, you now have a reinforced neural pathway of focused aggression and an anti-social way of solving problems (especially after 21 days of game-playing).

Think not?

Adam Lanza, the **Sandy Hook** shooter, whipped himself into a frenzy playing a video game in the months before his murderous rampage.[18]

And

It was the 1999 **Columbine High School** shooting that got many Americans thinking about violent video games. After the attacks, victims' families sued more than two dozen game makers, saying titles such as Doom, a first-person shooter **that the two teen gunmen played, desensitized them to violence**.

[Are you ready for this?]

A judge dismissed the lawsuits…[19]

Would you like to know what the writer of the quoted magazine article concluded, based on his use of <u>special</u> statistical data?

…as violent video games proliferated in recent years, the number of violent youthful offenders fell – by more than half between 1994 and 2010. [20]

Gee, that obviously makes it Ok! – the kids (allegedly) used the games to get violence out of their system. Wonder how they will respond in the future when bullied…. or frustrated because they don't get their way? Let's see what they do when cut-off on the highway…

And then again, another <u>funded</u> study found that

… higher rates of violent video game sales actually coincided with a drop in crimes, especially violent crimes.[21]

I wonder who funded the study and what was the sample size; Was it across the US, or just in Squaw Butte, Idaho?

The point is summed up by an astute comment, known to a lot of us:

As a man thinketh in his heart, so is he.

(Proverbs 23:7)

Equal time that supports Dr. Leaf comes from psychologist **Douglas A. Gentile** of Iowa State University:

> …whatever we **practice repeatedly** affects the brain. If we practice aggressive ways of thinking, feeling and reacting… then we will get better at those….Gentile and his father, psychologist J. Ronald Gentile, found that children and adolescents who played more violent games were likelier to report "aggressive cognitions and behaviors." They concluded **that violent video games "appear to be exemplary teachers of aggression."**
>
> They also found that eighth and ninth graders who played violent video games more frequently displayed **greater "hostile attribution bias"** (being vigilant for enemies) and got into more arguments with teachers. [22] [emphasis added]

Has anyone done a study to correlate discipline problems in school with the rise in violent video game playing? Or would that not be "politically correct?"

Negative Thinking Creates Prions

Remember Dr. Leaf's warning about **prions** that cause Mad Cow Disease (and its human version **Creutzfeld-Jacobs Disease**):

> …this protein [prion] does amazing things in the brain in response to good signals and **goes crazy in response to negative signals**. A chaotic mind filled with … rogue thoughts of anxiety, worry, and any and all manner of fear-related emotions [including fear of being 'killed' in the video game?] sends out the wrong signal. [23]

There is no way that playing a violent video game is relaxing, peaceful and just good fun. Exciting, yes, but so is skydiving with a parachute that may or may not open. And the video game player is unwittingly 'programming' himself for future problems

– if not psychological or behavioral, then physical health issues (due to a change in biochemistry). There may be no 'chute to keep him from falling into trouble.

Parents are just happy to let their kids play any game that keeps them corralled to the PC/TV so that they know where the kids are, and they don't have to worry about whether the kids are getting into trouble outside, on the street. PC games are great baby-sitters…. Or so it would seem.

In the final analysis, is it just the video game manufacturers (and sellers) who benefit from pandering to the youthful search for excitement? Perhaps the writers of articles supporting the violent video games also ca$h in??

Back to the magazine article.
This is the article's final shot, and it should rouse you to disgust:

> In a way, we are pointing fingers at the wrong people [at the video game makers]. When we worry that a violent video game is going to turn our kids into **serial killers** [extreme, unfounded conclusion], aren't we the ones who can't tell fantasy from reality? Kids already know the difference. [24] [emphasis added]

Remember:

Knowledge Protects
Ignorance Endangers

Those who want to know more about the Dr. Leaf findings can either get her book, <u>Switch on Your Brain</u>, or read <u>The Science in Metaphysics</u> (my Book 5) which explores Dr. Leaf's information with a tour of related, non-technical brain info, Epigenetics, Male and Female brain differences (yes, there are some!), the Brain and Optical Illusions, how Vision works, Nutrition for the Brain, the new field Quantum Biology, Consciousness and Torsion Waves, and useful Affirmations.

Ok, back to the Cognitive Dissonance issue….

Accepting New Information

The really sharp reader will read this book's new information and NOT dismiss it out of hand, but look at it, wonder about it, and even research it – such as when we get to the Flat Earth (FE) scenario… We all 'know' that FE is something to laugh at and dismiss. We all think that people centuries ago were all fools and didn't know any better, but it may intrigue you to realize that **there is no way science today can prove that the Earth is a sphere revolving about the Sun**. Seriously – that was just an hypothesis by Galileo, Copernicus and even Newton bought into it… and no one knew if it was really true or not, <u>and said so</u>. **We still can't <u>prove</u> it**…we all just accept it, and there is an outrageous reason for sustaining the round-Earth-spinning-around-the-Sun belief among the public. (We'll get to that and NASA and the Moon in the next chapter, and again in Chapter 10.)

Therefore, I am asking the reader to reserve judgment until all the evidence is submitted in the next chapter, as there are some serious things to consider. I would not stick my neck out and risk looking like a fool if I didn't know that the FE scenario was <u>at one time true</u>. Remember, I tried for three months to prove it wrong, and just became more and more convinced… so I had to combat my own **cognitive dissonance. Of course I can't tell a lot of people about it, and have lost two friends because of it (humans can be very resistant)….** So if you see the truth or value in knowing what Earth really is, be very careful in sharing it when you do see it.

The natural thing would be to seek out like people with whom you can discuss the issue, but beware – I have been told by several FE Truthers out there to NOT join the Flat Earth Society as that group is not what they appear to be… they are invested in making the subject look silly and are behind the bozo FE proofs. As I was cautioned, so I advise you.

Just FYI, there was a recent (Feb. 19, 2017) sports figure who endorsed the Flat Earth Theory (the following is a cut-and-paste from the online MSN sports page:)

LeBron: If Kyrie decides Earth is flat, 'that's OK'

© Ben Golliver/Twitter

Kyrie Irving can seemingly say or do no wrong in the eyes of his Cleveland Cavaliers and Eastern Conference All-Star teammate LeBron James.

The news article read:

> Expressing his opinion that the planet is actually flat and not round
> during a recent podcast appearance has turned Irving and his
> controversial take into one of the major talking points this weekend
> in New Orleans.

Source:
http://www.msn.com/en-us/sports/nba/lebron-if-kyrie-decides-earth-is-flat-thats-ok/ar-AAn5uPS?li=BBnb7Kz&ocid=AARDHP

And when Kyrie was pressed for clarification (USMagazine.com 2/20/17 reported):

> "This is not even a conspiracy theory," Irving, 24, said during an
> interview on the NBA's *Road Tripping with RJ & Channing*, podcast
> on Thursday February 16. "**The Earth is flat**. … It's right in front
> of our faces. I'm telling you, it's right in front of our faces.
> They lie to us."

> A day later, the athlete repeated the comments during an interview
> with ESPN. "I think people should do their own research, man,"
> he said. "Hopefully they'll either back my belief or they'll throw
> it in the water. But I think it is interesting for people to find out on
> their own…. **I've seen a lot of things that my education system
> has said that was real that turned out to be completely fake**. I
> don't mind going against the grain in terms of my thoughts."

And he is right…**the evidence is there**, but he is not aware of the VR Sphere
alternative, nor is he aware of a possible or potential FE to VR Sphere Conversion.
So check out the proofs in this book, and if that isn't enough, there are 96 pages in
Ch. 11 in QES to chew over.   And yet -- some of the FE evidence also substantiates
the VR Sphere. That is why it is a real puzzle.

> You won't have to wait long to meet and trust others as it appears
> that the FE scenario is catching on, gaining notice (hits on FE videos
> on YouTube are really up in the last year) , and with any luck, the
> truth will out – for all our benefit. And then it is but a small step to
> see that the Flat Earth could be morphed into a VR Sphere –
> Has that been done, or is it just a potential?

In the final analysis, being open-minded and proactive is in the reader's best interest as far as brain health and spiritual growth is concerned. As was said

> You shall know the Truth
> and the Truth shall set you free...
>
> ...but first it may p*ss you off!

Free from what? Being deceived, living in fear or anger, and spending time, energy and money in the wrong direction.

Chapter 3: What Earth Is

The preceding chapter was necessary to prepare the reader for what this chapter shares. Earth is <u>not</u> a round rock, spinning at 1000 mph, and rotating around the Sun at 67,000 mph. That is what Science says (see page 97 statistics).

Based on what I was given, Earth should be a VR Sphere, but before we get into that, there are some aspects of Earth that are true regardless of the VR Sphere or FE Scenario. Let's look at those briefly first – they were covered in more detail in VEG.

Geologic Timetable

A lot of what we call Earth History is based on traditional suppositions fed to us by the often atheistic scientists who believe in and promote the erroneous idea of Evolution. Geologists, archeologists, and historians to name a few, want to sustain the notion that the planet is 4.5 billion years old, and that Man went thru stages of evolution from a basic (Ape) hominid to what he is today. To prove this, the geologists who believe in Evolution also devised a geologic timetable to support the **Theory of Evolution** -- but the timetable (below) is based on the <u>assumption</u> that the planet is 4.5 billion years old.

<u>Period</u>	<u>Time (mil. of years)</u>	<u>Life Forms</u>
Quaternary	0 - 1	**Rise of Man**
Tertiary	62	**Rise of Mammals**
Cretaceous	72	**Seed-bearing plants**
		Dinosaurs
Jurassic	46	**First Birds**
Triassic	49	**First dinosaurs**
Permian	50	**First reptiles**
Pennsylvanian	30	**Shells, insects**
Missippian	35	**Crinoids**
Devonian	60	**Plants, fish**
Silurian	20	**Earliest land animals**
Ordovician	75	**Early bony fish**
Cambrian	100	**Invertebrates**
Pre-Cambrian	???	**Bacteria-algae-pollen?**

Simplified Geologic Time Scale
(source: *The Young Earth*, J.D. Morris, p.8)

The timetable does not reflect any hard, scientific fact.

Anybody notice that **the years don't add up to 4.5 billion**? They total 600 million, and no one knows for sure how long each period was, so there are some real 'guesstimates' in the list.

Also, the fossil record shows that **Dinosaurs and Man were here together**, [25] and there is a human footprint form in Permian rock! [26] The timescale was arbitrarily decided on, and subsequent students in Biology and Geology have come to accept the Chart as gospel.

The following considerations are all based on scientific facts, and do not spell out an old Earth… Earth is younger than we think.

Young Earth Evidence

To avoid making this a long chapter, it is sufficient to list only 9 issues that today's scientists have discovered about the Earth that make them think that it is as young as 10,000 – 20,000 years old, and possibly as old as 2 million years old. The difference in age arises in the aging/dating methods (Carbon-14 versus tree-rings) which are still subject to criticism.

The main arguments in favor of Earth being somewhere between 10,000 years to 2 million years old contradict the Evolutionary model. The old Evolutionary model needed 4.5 billion years to allow for evolution from a primordial sea of goop to evolve into Man; the Evolutionist thinking being that that is enough time for *something* evolutionary to have happened. Young Earth scientists, and their number is growing, believe that the evidence presented below suggests other than an old Earth.

Fossils

First, **there are no intermediate stages of dinosaurs** which would show a dinosaur species evolving into a new species. In addition, there are eggs and adults but **no juvenile forms of the same species**. The absence of 'transitionary fossils' kills Evolutionary theory [27] because over a very long period of time, there should be transition fossils. The fact that there aren't any means the dinosaurs weren't here for that long, and thus there are some real 'holes' in the Evolutionary Theory and the Geologic Time Scale. (Chapter 7 goes more into the dinosaur issue.)

> We will also see in Chapter 7 that some of the dinosaur findings have been faked.

The fossil record shows no evidence that any basic category of animal has ever evolved from or into any other basic category. [28] In short, there are so "transition" fossils.

and…

…the curious thing is that there is a consistency about the fossil gaps: **the Fossils go missing in all the important places**. When you look for links between major groups of animals, they simply aren't there…[29]

And lastly,

All paleontologists know that the fossil record contains precious little in the way of intermediate forms; transitions between major groups are characteristically **abrupt**. [30] [emphasis added]

"Abrupt" suggests **intervention**, as if the new group had external help to produce an 'upgrade' or variegated replica of the original group – like hummingbird to condor.

If the dinosaurs were here for millions of years (Triassic – Jurassic – Cretaceous), then why are **there no intermediate forms or transition skeletons** found indicating that Evolution did take place? And why are there <u>so few</u> dinosaur bones found?

Findings in the fossil record contradict **the Darwinian view that species arise gradually over long periods of time**. In fact, there is no evidence for the gradual appearance of one species out of another. There is no evidence of intermediate species, linking one form to another… [31] [emphasis added]

In addition, there is no fossil evidence to 'document' the transition from reptile to bird (Triassic to Jurassic), nor is there any evidence to show how the finned fish evolved. [32]

If Apes evolved into humans, why is it not still happening?

Population

Second, calculations regarding **Earth's population statistics supports a young Earth**. It is noted that at a population growth rate of 2% per year, which has been observed for almost a century, and given a current 6+ billion population, it has been calculated that it would only take 1100 years to reach the present population from an original pair of humans. And that isn't counting wars and death. [33] Even if wars and death are included, and plagues, wars, mass weaponry, crowded cities, abortion rates and famines are included, the population growth rate hasn't changed much. [34]

Even more interesting is to do the math from the other end: assume Man has been here for a million years, and use the 2% growth rate per year, and then estimate how many people would be on the planet today:

> The number is so large, it is meaningless, and it's approximately the number which could fit **inside** the volume of the entire Earth…. there should be about 10 to the 8600th power, or 10 with 8600 zeroes following it…. [and if that's true,] **where are their bones?** [35]

Anthropologists tell us that **Man has <u>not</u> been here for a million years**:

> ….civilization dates to only five thousand or so years ago, at the beginning of human history…. Archeologists have shown that in a variety of places around the world, **very advanced, modern cultures sprang up suddenly, almost simultaneously…. Human culture from its very start was advanced, and humans have always been intelligent**. [36] [emphasis added]

Zechariah Sitchin would agree. And according to the Sumerian accounts, Man <u>as we know him</u>, civilization and all, was begun right after the Flood – about 8,000 years ago.

No Bones

Third, if Man has been evolving for millions of years on the Earth, **why are there so few bones found?** Starting just 1 million years ago and using a 2% growth rate, yet finding today a population of 6+ billion, where are all the millions of bones from those who must have died? [37] The same question can be asked of the dinosaurs, also supposedly here for millions of years.

Magnetic Field

Fourth, the **Earth's magnetic field has been decaying at a constant rate** since it was first measured in 1835, and using that rate (with 1400 years half-life) and extrapolating backwards, it can be determined that the magnetic field must have been much stronger in the past. Such a field is necessary for life on earth as it stops harmful cosmic radiation from killing lifeforms. Yet if the field were too strong, using the doubling factor every 1400 years, just 100,000 years ago life would have been life living on a dense neutron star here – impossible. [38]

So, either the decay rate is not constant, or if it is, the planet is not more than 10-12,000 years old.

Helium

Fifth, **the amount of helium found in the atmosphere is a clincher for a young Earth**.

Helium is produced below the Earth's surface by a process of radioactive decay. According to the latest scientific measurements, 13 million helium atoms escape into the atmosphere <u>every second</u>. And the amount of helium atoms escaping the Earth's atmosphere into space is 0.3 million per second. Thus helium is accumulating at a very rapid rate... doing the math, based on the amount of helium in today's atmosphere, delivers a figure of 2 million years old maximum as an age of the Earth. That is, <u>assuming that the rates of escape have been uniform</u>... but what is weird is **that helium is in abundance in the rocks, and not in the atmosphere**, so the Earth is much younger than 2 million years old. [39]

Sediment

Sixth, sediment on the ocean floor is another good measure of age of the Earth. Since the Earth has been covered with water from Day 1, and water is constantly eroding the continents, **there should be a <u>lot</u> of sediment on the ocean floor. There isn't**. The rate of sedimentation has been found to be a fairly consistent 27.5 billion tons <u>per year</u>. Yet the amount of measured sediment on the ocean floor today stands at 410 million billion tons. Simple math arrives at a maximum possible age of 15 million years (assuming a uniform rate, and no biblical Flood).[40]

Salty Ocean

Seventh, **salt in the ocean should be getting saltier** with the years and if the original ocean was salty, 3-4 billion years ago, shouldn't it be <u>too</u> salty now? The scientists studying the ocean asked themselves the same question and set about determining the 11 types and amounts of salt input to the ocean and the 7 types and amounts of output – i.e., the ways that the ocean can gain and lose salt. These were quantified and, using the minimum and maximum values that they developed, the maximum age of the ocean can only be 62 million years old. That is not saying that the Earth is 62 million years old, just that it couldn't be any older than that. [41]

And that is not considering whatever The Flood of Noah's day may have done to alter the rate of salt gain/loss... (see Chapter 7 for items #6 and #7 in Young Earth Summary section). And consider the underground rivers and Fount of the Deep, also in Ch. 7.

Meteor Dust

Eighth, meteoric dust from space has been accumulating on the Moon and Earth since the beginning, and it was feared in the early 1960's that if we tried to land on the Moon, there **should be a foot or more of dust** and that could adversely affect landing and takeoff. Measurements back in the '50s and '60s indicated that meteorite dust was coming onto the Earth and Moon at the rate of 14 million tons per year. That dust includes a lot of iron and nickel. It was inferred

> ...that if the Earth has been here for 5 billion years, then there should be enough such material here on Earth to form a layer over 150 feet thick. No one expected to find such a layer, of course, since the Earth's surface is continually mixed by rain, wind, erosion, etc, but it *did* bother scientists that **nickel is so rare on Earth.** If the Earth is old, and the rate of accumulation [per the Uniformitarian assumption used by Evolutionists themselves] has been the same throughout Earth history, there ought to be more! The **Earth's nickel content is much more compatible with a young Earth** than with an old Earth. [42] [emphasis added]

So what about the Moon? There is no rain, wind or erosion comparable to Earth – what falls on the surface stays there. Later when Man went to the Moon, the dust layer was allegedly found to be **only an inch or so....!**

> Given the measured rate of influx, this small amount of dust could easily accumulate in a few thousand years, but if the Moon is old, something is wrong. [43]

Of course this argument has been revisited with better and more modern scientific measurements of dust influx, and it has been found that the rate of fall onto Earth and the Moon varies quite a bit, and is <u>not</u> consistent. It falls alternately in both heavy and light cycles. Nevertheless, if Earth is supposed to be 4.5 billion years old, and even if the influx varies, there would have been many cycles of influx resulting in <u>more dust and nickel</u> than we have today on Earth.

Continental Erosion

Ninth, and lastly, there is **the erosion of the continents** (via streams and rivers) which has been consistently measured and calculated to be 27.5 billion tons per year – see 'Sediment' in #6 above. Now it is known that the total land mass above sea level for the last 70 million years (since the last geological upthrust) is 383 million billion tons. Simple math tells us that "[at] present erosion rates, all the continents would be below sea level in 14 million years!" [44]

This last point may not be a valid way to evaluate the age of the Earth as much as it levels a devastating critique against the story that **Uniformitarians** tell. The assumption that decay and erosion processes have proceeded at a uniform rate throughout the Earth's history is just as spurious as the Evolutionary assumption that given an incredible amount of time, all life on Earth could have evolved from a single cell organism in an on-going, uniform process. What really happens over a long period of time is called **entropy** – natural decay – and it has been institutionalized as the **Second Law of Thermodynamics**. Thus,

Evolution's worst enemy is time and is not its ally.

And there are other tests, but the true significance of <u>what appears to be</u> a young Earth was discussed in VEG Ch. 12. It is also interesting to note that the Jewish culture has been counting years since their inception, and according to them, this is the year 5775 (2014) which tends to suggest that the Flood might have occurred around 6,000 years ago. Atlantis was 8,000 years ago. (See Chapters 7-8.)

One may also be struck by the fact that all 9 evidences above do not yield a consensus where all 9 point to the same, or roughly the same, age for the Earth. It is suggested that such would be the case if Earth went through **different eras** and between each there was a **"Wipe and Reboot"** or terraforming done before starting the next Era. As weird as that sounds, resets have been done, and VEG Ch. 12 offers the rationale for that.

Earth Eras

At this point, it would be relevant to clarify what has been meant by 'Era' inasmuch as the periods that Man has been living in are generally consecutive, time-wise, but the Eras are separated by the oft-suggested **"Wipe and Reboot"** thus inserting gaps in the chronology.

This is easy to do if the Earth Realm is indeed a Simulation – as a later Chapter 9 all but proves.

The following Chart 5 is a basic diagram showing a number of past Eras, principally characterized by a dominant race/civilization. Eras have no fixed length. What happens in one Era is often left for the archeologists of the succeeding Era to discover – hence the Anunnaki <u>were</u> here, but that was another Era. Eras are terminated when it gets out of hand: either too much violence or too much pollution, or the Greater Script (God's Plan for Man) is tending to go off-track.

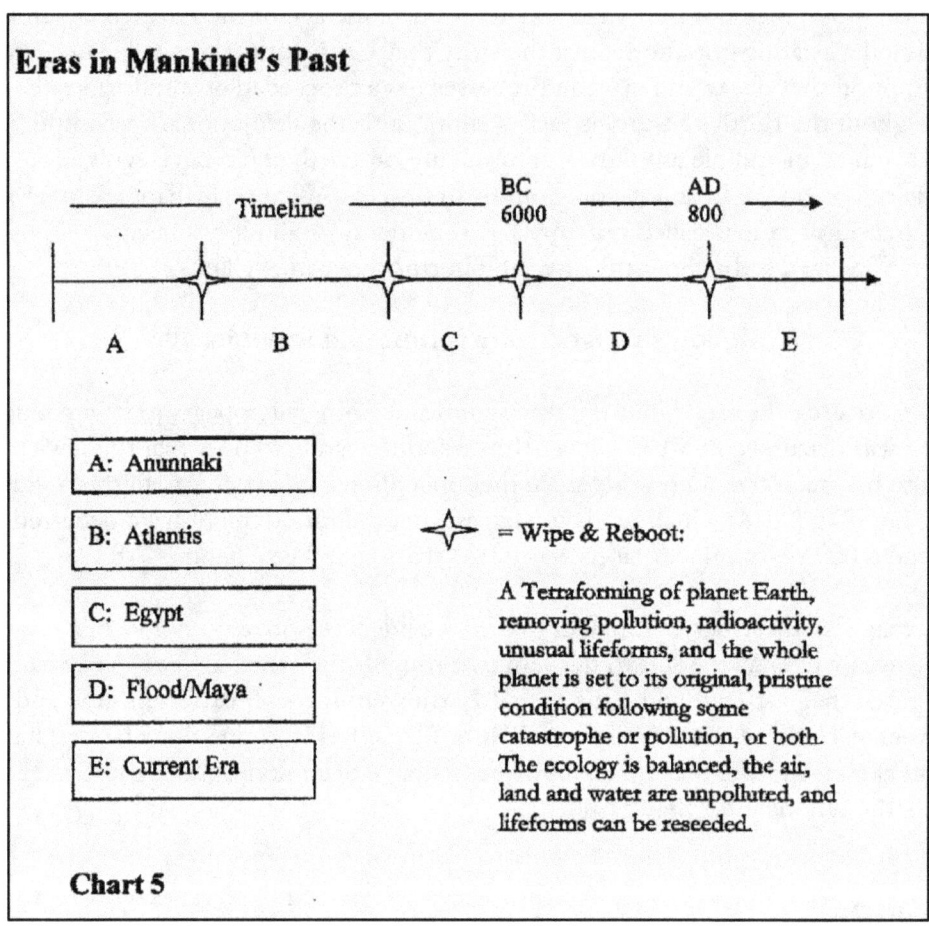

Eras in Mankind's Past

A: Anunnaki

B: Atlantis

C: Egypt

D: Flood/Maya

E: Current Era

= Wipe & Reboot:

A Terraforming of planet Earth, removing pollution, radioactivity, unusual lifeforms, and the whole planet is set to its original, pristine condition following some catastrophe or pollution, or both. The ecology is balanced, the air, land and water are unpolluted, and lifeforms can be reseeded.

Chart 5

In Chart 5, note that when the current Era E was begun, around AD 800-900, by AD 900 Man had lost several hundred continuous years of history, and the history that pertained to the end of Era D, which began with the Flood and ended with the Maya disappearing, left a curious gap – which **Dr. Fomenko** explains in VEG Ch.10.

> A history was provided for Man by back-dating and <u>copying</u> salient history from the beginning of Era E (AD 900-AD 1500) to make it look as though we had a more continuous history with the past. This resulted in duplicate historical elements, called **"phantom parallels"** which is what Dr. Fomenko discovered, and proved mathematically..

The **dinosaurs** (if real – see Fraud in Chapter 7) were removed at the end of Era B. The **Flood** was the "Wipe and Reboot" that closed Era C. The start of Era D is approximate. December 2012 was not the end of the world. The Aztecs, Maya and Hopi have all spoken of preceding 'Worlds' or Eras that ended by fire, water, calamity, etc. and they like to say that we are in the 4th or 5th World – depending on source. So the concept is not new.

What modern Man needs to realize is that the Earth is watched over and if things get out of hand, a Wipe & Reboot (or reset) is <u>very</u> likely.

What is Earth?

So if Evolution is not how all life came to be on Earth, because **Entropy** kills everything that is not constantly sustained energy-wise (and that suggests <u>Intelligence and purpose</u> – hence Creation again!), and dating methods have some issues, rendering them semi-credible, and The Flood happened about 6-8,000 years ago (proven by geologic strata and historical records [*Atra Hasis, Enuma Elish and the Epic of Gilgamesh*] giving the ages of people we know to have existed, like Noah and Sargon), just how old is Earth and when was it first able to support flora and fauna?

The Bible in Genesis gives such a high-level account of the Earth's creation that people don't take it seriously... <u>but</u> it says that Earth was **created**. The Koran also says the Earth was **created** ... and flat. The **Book of Enoch** also says the Earth is flat and he was taken up and shown the **"ends of the Earth"** (round, spherical planets do not have "ends"), and he saw the Firmament... as Genesis says.

If, as QES has examined in some detail, the **Earth is a Simulation**, a powerful God (or gods) could have created that in 6 days, as the Bible says... yet, do we know how long a "day" was? So the significant data given in the nine points in the Young Earth section above suggest that (1) either Earth <u>is</u> very young (and has never been thru a Wipe & Reboot), or (2) Someone was doing several Wipe & Reboot events and a "cleanup" (terraforming) of the planet... so maybe the concept of days really doesn't apply... Perhaps Earth was created in **6 phases**...?

All we have is what we have now. And what do we know about it, in addition to the nine aspects listed earlier?

Anomalies

While most of these were covered in 3 sections in QES, their significance means that Earth is not a round rock circling the Sun.... (see page 97).

Weather Anomaly

This is a doozy and cannot be explained but was a real event in Australia. It suggests that weather is caused by inserting energy 'wheels' into the Simulation... it is heavy evidence that the Simulation DOES exist and **interferes with our satellites, and**

our weather. The picture below was taken over Australia in 2010 by a weather satellite, and many times a strange disk (or 'wheel') appeared over the regions being photographed... and each was followed by extreme weather.

Image: credit: http://www.news.com.au/lifestyle/real-life/bueau-of-mereology-cant-explain-mysterious-patters-on-radar-system/story-e6frflri-1225848774377

And in every case, the appearance of the 'wheels' was followed by unusually severe weather – Each wheel type seemed to correspond to a particular type of weather... but if it was a warning, the weathermen didn't get the message.

Black spokes with clockwise motion.

Over S/E Australia

> There is a theory that the 'wheels' were inserted into the Simulation to **create the weather**, and each type of weather corresponded to a specific wheel design.

All types of wheels were shown in QES, Ch. 3.

What does that mean?

Charles Fort and the *Gegenschein*

Fortean Sphere

The famous **Charles Fort** often questioned the scientific dogma of his day, 1912. He wondered where all the strange objects falling from the sky came from, and he wrote 4 books documenting the oddities and speculating on their source. During his writings, he also kept coming back again and again to the idea of **a 'shell' surrounding this Earth**.

> ...whether there be a shell-like, evolving composition, holding the stars in position, and in which **the stars are openings**, admitting light from an existence external to the shell, or not, all stars are at about the same distance from this Earth as they would be if this Earth were stationary and central to such a shell, revolving around it. [45]   [emphasis added]

This does give a new meaning to Copernicus' idea that the Sun did not revolve around the Earth – What if he was wrong? Intriguing, but nonsense you say. And Charles Fort would agree with you – with a wink! It was an idea which really fascinated him, and he was close to the truth, as we now close in on what even the Quantum Physicists of today are considering: **a 3D Construct run in a very sophisticated Simulation.**

Fort was no uneducated man, totally mystified by science and the world. It was his unique ability to go toe-to-toe with the astronomers and scientists of the day – demanding answers to the oddities he had recorded. He was considered an *enfant terrible* of science – questioning everything. In math and astronomy, he could hold his own. And yet, he repeatedly questions the presence of <u>something</u> surrounding the Earth:

> The *Gegenschein* -- Now we have indication that there is such a shell around our existence. The *Gegenschein* is a round patch of light in the sky. It seems to be reflected sunlight, at night, because it keeps position about opposite the Sun's position. The crux: Reflected sunlight – but reflecting from what? That the sky is a **matrix** in which the stars are openings, and that, upon the inner, concave surface of this celestial [transparent energy] shell, the sun casts its light, **even if the earth is between**... [46] [emphasis added]

Note his use of the word, 'Matrix' – 100 years ago, and the Wachowski Brothers had not even conceived of the 1999 movie, *The Matrix*. (In fact, no one had conceived of <u>them</u>, yet!)

The *Gegenschein*

The *Gegenschein*
(credit: NASA: <u>*http://apod.nasa.gov/apod/archivepix.html and below*</u>)

It is recommended that the reader check out the above NASA link to three samples:
2008 May 07: The Gegenschein over Chile. (sunlight)
2006 December 26: The Gegenschein . (sunlight)
June 25 1999: The Gegenschein . (sunlight + Sun)

Interesting that in his book <u>New Lands,</u> written in the 1920's, Fort used the word **matrix**. But he is making a very interesting point that science still today cannot answer:

> Suppose the *Gegenschein* could be a reflection of sunlight from anything at a distance less than the distance of the stars. It would have **parallax** against its background of stars.

> *Observatory*, 17-47: "**The *Gegenschein* has no parallax.**" [47]

> Parallax is the apparent change in shape or size of an object if the observer changes his position.

So Fort is saying that since its perceived shape and size does **not** change with any change in our position from which it is viewed (i.e., a parallax), that it must be **reflected off the surface of something consistent in shape** (the Firmament). If it were reflected off dust in the atmosphere, it would have size and shape distortion depending on the viewing angle – <u>but it doesn't</u>.

Fort was no neophyte to astronomy and certainly would have known the difference between sunlight reflecting off dust in the upper atmosphere, and swamp gas, Moon dust, or whatever the standard argument of the day is for what he personally observed.

Note that **NASA** has pictures of the *Gegenschein* (above) – Fort wasn't imagining things. As will be seen shortly, Fort was on to something, and while he knew the Shell wasn't a metal thing, and that there were no holes in it to simulate stars, it <u>is</u> real, and is also what bounces back our radio and TV transmissions:

> …not enormously far away, there is **a shell** around this earth… According to data collected by the Naval Research Laboratory [1925], there is <u>something</u>, somewhere in the sky, that is deflecting electro-magnetic waves of wireless communications, in a way that is similar to the way in which sound waves are sent back by the dome of the Capitol, at Washington. The published explanation is that there is an "ionized zone" around this earth… the [term] "ionized zone" is not satisfactory… From Norway [there were] short-wave transmissions… reflected back to earth… as if from **a shell-like formation, around this earth, not unthinkably far away.** [48] [emphasis added]

And yet, we keep up the ruse that we are looking for extraterrestrial life with our SETI radio telescopes… [mostly shut down in 2011 – Ed.]   Wouldn't any ET radio transmissions <u>to us</u> be reflected back, too, from their side of the same 'barrier?'

> *As Robert Monroe shortly discloses from his OBE ventures, there are no* **3D** *lifeforms out there… most sentient life in the universe is in the 4D and above realms. And that makes sense if you understand that the soul has almost unlimited potential 'out there', yet is constrained here on Earth.*

There <u>is</u> a very advanced (energy) Dome around this Earth as we will see (Appendix E), functioning in 4D+ but reflecting light in 3D. It is interesting that an open-minded skeptic of the 1912-1932 Era first voiced the concern that Science was ignoring the *Gegenschein* , and not giving us all the answers.

In fact, knowing Fort from his writings, he was 'baiting' the scientific establishment with the idea of **a Shell** whose holes were the stars, but the bait was not taken. And as will be seen, there **is** something around the Earth, and another researcher, Robert Monroe, ran into the Barrier around the Earth – literally (see VEG, Ch. 12).

Fortean Clues

Here are some of the relevant things that Charles Fort discovered and documented.

1. **Stone throwing**.
Since 1841 people in the English countryside have reported stones being thrown at their houses and windows by unseen beings. Sometimes the stones just fall from the sky onto the roof, and at other times they enter the house, breaking a window.

   Despite being vigilant and watching each others' houses for a while in a small, rural town where the stones were commonly 'thrown', no one ever saw anybody doing it. To argue that it must have been the "little people," or fairies and elves, raises issues beyond the scope of this book.

2. **Falling Fish and Toads**
Sometimes fish, alive and sometimes dead, were found falling on rural houses in parts of England… too many to have been thrown there by vandals. In addition, some hailstones were found to contain frogs, some alive, some dead, when they were picked up and thawed out. Sometimes only frogs fell and hopped away.

   The usual explanation is that some sort of cyclone scooped up the fish or frogs and blew them inland. Fort did not accept that since some locales were hundreds of miles from the nearest water – <u>and</u> some of the fish were fresh-water types and there had been no tornadoes or such reported in connection with these anomalies.
   The two above are recorded in Mr. Fort's <u>Book of the Damned</u>.

3. Disappearing People

This one is a real oddity.

A man is sitting on his front porch, watching the Sun set. His wife goes in to get her sweater, and when she gets back 20 seconds later, her husband is gone. The rocking chair is still rocking, the pickup truck was not taken, and nothing is missing except her husband. He was never seen again.

On another account, a farmer out in his field calls to his wife who is on the back porch of the farm house, she waves, he waves, and he starts across the field, as it is supper time. As she watches, right before her eyes, he disappears. She runs out to where she last saw him, and there is no hole, no evidence that he was ever there.

In a third account, a waitress leaves a restaurant, steps out to the street to put another quarter in the parking meter for a customer, and she is never seen again. And as was reported in Ch 14 of VEG, sometimes lakes and whole rural villages have disappeared – a lake in Peru and a village in Siberia.

These above accounts are principally from Mr. Fort's book called <u>Lo</u>! They are perfectly explainable if Earth is a Simulation (see QES programmer glitches).

As if that weren't enough, consider the following:

The Observer Effect: Anomalons

Scientists admit to the issue of **anomalons** – subatomic particles that the physicists around the world seem to 'create' because they are looking for them, and the subatomic particles appear with the parameters that the scientists are expecting to see. Are the subatomic physicists <u>creating</u> the subatomic world instead of discovering it? [49]

There does seem to be a certain aspect to the **"observer affecting what is observed"** as Physics maintains, and yet, as a 1960 radio station discovered, with the help of its listener audience, 10,000 people could not blink out a light on top of a building. Perhaps **we only affect the subatomic world** because our thoughts carry more energy than do quarks, electrons, etc. (See Appendix C.)

> Indeed if the universe is a holodeck [as Dr. Tiller suggests], all things that appear stable and eternal, from the laws of physics to the substance of galaxies, would have to be viewed as **reality fields**, will-o-the-wisps no more or less real than the props in **a giant, mutually shared dream**. All permanence would have to be looked at as illusory, and only consciousness would be eternal, the **consciousness of the universe** [or may we suggest, the consciousness of the Simulated universe?]. [50]

What do we really know about our ability to see? Do we really see the world around us as it actually is? We don't. (See Book 5, TSiM.) And because of that, it is plausible that this whole reality we live in is an **Orchestrated** Simulation.

Is Someone watching and creating what we are looking for?
Is that same Someone manipulating the Physics Double-slit Experiment? (See Appendix C for Double-slit Experiment.)

And as was said in VEG, memory appears to be holographic and DNA processes and communicates via waves of light or biophotons – interacting with the cells of the body, including those in the brain. So **vision is holographic because the real world 'out there' is holographic,** <u>and</u> we store and retrieve memories holographically. So, we are part of the hologram.

And now some feedback from Science:

I.   Antimatter Mystery

The Big Bang should have created matter and antimatter in equal amounts, or so our best theories have it. If that were truly the case, though, then the universe would have disappeared in a big puff of self-annihilation almost as soon as it began. The fact that we are here to ponder it tells us something is wrong with the Big Bang picture *(New Scientist, 12 April 2008, p 26)*. The question is: what? What if there was no Big Bang because the Earth was created in a Simulation (as FE or VR Sphere) and what we think is the Universe around us is just the sophisticated "planetarium" display off the Firmament connected with the *Gegenschein* ? (Don't laugh, Quantum Physicists are really serious about that.)

II.   Noise From the Edge of the Universe

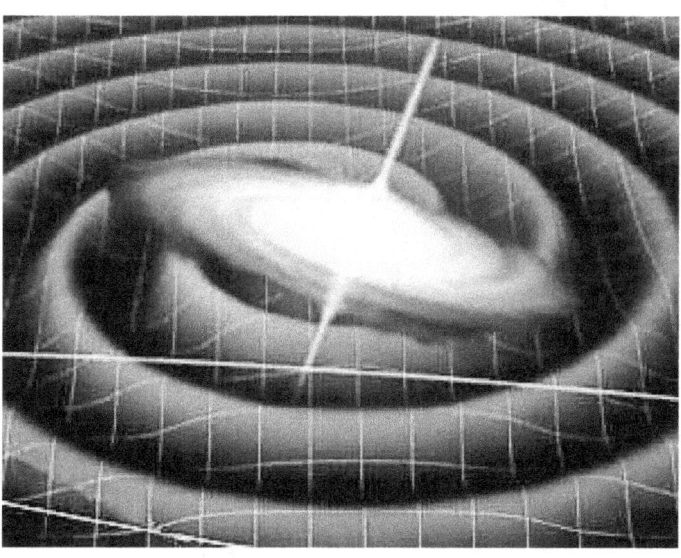

Are dud signals from a gravitational wave detector **evidence that the universe is a holographic projection? (Image: ESA)**

> **Gravitational waves** are ripples in space-time that are emitted by cataclysmic cosmic events such as exploding stars, merging black holes and/or neutron stars, and rapidly rotating compact stellar remnants.

The **GEO600** gravitational wave detector in Hanover, Germany, has not yet detected any gravitational waves. As a consolation prize, **it may instead have uncovered the ultimate nature of reality.** In 2008, physicist Craig Hogan at the Fermi National Accelerator Laboratory in Batavia, Illinois, was trying to work out how we might test the idea that **everything we see as physical reality is the result of a kind of projection** from the boundary of the universe. This is known as the **holographic principle**.

The information held at the boundary is not smooth, but composed of "bits", each one occupying an area that corresponds to the most fundamental quanta of distance in the universe. This is **the Planck length**, around 10^{-35} metres – far too small for us to see the individual bits. When this information is projected into the volume of the universe, however, each bit gets magnified. That means **we might just be able to see pixellation in space-time** (see QES, Ch. 7). Hogan worked out how the **pixellation** might manifest itself for GEO600 and sent his result to the researchers there.

By strange coincidence, the GEO600 team had been having problems with "noise" in their detectors. But **here's the kicker: the noise had uncannily similar characteristics to Hogan's anticipated signal. Is it indeed the result of information that resides at the edge of the universe?** "The issue is still unresolved," says Karsten Danzmann, principal investigator for GEO600. **"The noise is still there and we have no explanation."**

QES Ch. 7 introduced Dr. Hogan's report that **sound was coming from the edge of the universe,** and wouldn't that be strong evidence of the energy being used to generate the **Konstruct** and the **Cspace**? (Next section). Even if you say that the sound is just radio waves bouncing back at us – what are they bouncing off of?

> As of 2016 there has been no further clarification from GEO600.

And this one:

III.   The Kuiper Cliff

If you travel out to the far edge of the solar system, into the frigid wastes beyond Pluto, you'll see something strange. Suddenly, after passing through the Kuiper belt, a region of space teeming with icy rocks, **there's nothing.**

Astronomers call this boundary the Kuiper Cliff, because the density of space rocks drops off so steeply. What caused it? The only answer astronomers can come up with is there seems to be a 10th planet sweeping the area clean... but they can't find it. What if the empty space is the Cspace or the end of the "stars and planets" display? (Next Chapter.)

And an engineer from today's world of Science, who can think outside the box, tells us this:

Jim Elvidge, Electrical Engineer

This scientist agrees that we live in a programmed reality which reflects **intelligent design**. [51] As support for that thesis, he offers:

> **The parameters of our world are fine-tuned for our existence –**
>> Just the right distance from the Sun, water, air, and a Moon to control the tides and a precise eclipse of the Sun,

> **There is a non-random, or pre-planned aspect to the events in our reality –**
>> This is due largely to the Control System – inserts designed to teach and inspire, (Dr. Jacques Vallée would love it – See VEG)

> **The Programmers (gods) make frequent modifications to fine-tune the program and its data structures –**
>> Note the anomalistic changes in the Laws of the Universe... Ch. 8-9 in VEG, where decay rates are not constant, the Sun is now heating the Earth, and *anomalons* which appear as the observer expects,

> **They have included "easter eggs" for our enjoyment –**
>> Hammers made by Man found in old coal strata, the Antikythera mechanism, ancient models of planes from Peru, and the Nazca lines... ,

> **There are those pesky anomalies –**
>> In metaphysics, physics, philosophy, geology, anthropology, and psychology "can all be explained only by the **programmed reality model**." [52] (See Anomalies I, II, and II in QES.)

> **There is a special anomaly in the Observer Effect –**
>> Why does observing sub-atomic quanta result in them changing or

behaving differently? Classic case in point: The **Double-Slit Experiment** where a photon thru one slit produces a dot on the wall, and a photon thru two slits produces a wave pattern on the wall – and then, Step 3, install a measuring device to see which of the two slits the photon went thru – and it goes back to behaving like a single particle! Spooky.

These things are easy to do if we are in a Simulation, created for Man as a learning, experiential realm – suggested for soul growth. And there are more examples (in QES), but the foregoing were enough to get one to think that the Earth Realm is special and unique things happen here… by design. And if there is a design, there must be a Designer. (Elvidge also elaborates his **Digital Consciousness Theory** in **Appendix C**.)

So if Earth is a Simulation, what kind of simulation is it?

As was said in the Introduction, I was given 3 basic facts about Earth and was not told whether it was spherical or flat. So being educated (viz., 'brainwashed') like everyone else, I <u>assumed</u> that the 3 aspects had to be applied to a spherical Earth, but one that was a Simulation (due to the foregoing anomalies), or a **Virtual Reality** – thus a **VR Sphere**. My assumption might have been wrong (Ch's 3 and 7).

If you are still resisting the Simulation idea, here is one more anomaly:

Professor James Gates

… a professor at Cornell University, while analyzing the Superstring mathematics dealing with quarks, branes and subatomic activity that have accurately defined their operation, he noticed something very unusual in June 2012 which suggests a design to our universe. It was corroborated by **Neil deGrasse Tyson:**

> …theoretical physicist S. James Gates has discovered something extraordinary in his String Theory research. Essentially, deep inside the equations we use **to describe our universe** Gates has found **computer code**. And not just any code but extremely peculiar **self-dual linear binary <u>error-correcting</u> block code.** That's right, error correcting 1s and 0s wound up tightly in the quantum core of our universe. [53] [emphasis added]

This is almost the smoking gun but it does not prove conclusively the existence of an Holographic VR Simulation – yet. Physicists have yet to be able to demonstrate an experiential model of Strings in the laboratory– all they can do is postulate based on observations and known facts. Yet, again, the discovery is shocking as <u>there would be</u> **self-correcting code within a Simulation** just as there were error-detection and self-correcting routines within all the computer programs that this book's author wrote in 35 years in data processing. Such routines show **intelligent design** and would not have evolved naturally in the fabric of space. (See also Dr. Gates in QES, Ch. 7.)

> With this discovery, we are 90% home in establishing that our world/Realm is a Simulation. Because this data is very significant, these last two scientists are visited again in Chapter 9.

But, first, what is meant by the VR Sphere?

VR Sphere

What we see in the night sky is probably the <u>replicated</u> 4D universe that surrounds our **3D Construct** (VR Sphere or Flat Earth). It is reminiscent of a VR Game where you can see the background but you can't go there. The Stars, Solar System, our Galaxy and other Nebulae are a replicated part of the real 4D Realm… simulated <u>around</u> the 3D Construct – **which means it is contained in a larger Konstruct.** (That was a simple, logical way the gods could do it and QES, Ch. 5 goes into the why and what of the setup more in detail…) <u>assuming</u> that Earth is a VR Sphere.

The VR Sphere concept is this: at the core is the 3D Earth Construct itself with about 7200 miles of altitude to the **first 'Shell'** – which often reflects the *Gegenschein* . The Sun and Moon operate close within the **Cspace** which extends several million miles to the **outer 'Shell'** also called the **Konstruct** – in which are the simulated stars and constellations.

Sphere Within a Sphere at the Vatican: Two Shells

The Earth or 3D Construct is inside the first Shell (aka Firmament).

The outer Shell is the **Konstruct**.

The **Cspace** is the distance (shown not to scale) between the two Shells.

(credit: Bing Images: Panoramio.com)

The <u>superposition</u> of the **3D Earth Construct within a Konstruct** can be thought of as similar to shifting the Earth into another Timeline. The effect of phase-shifting the whole Konstruct has the same effect as shifting the Konstruct into an adjacent Timeline – It is not available in normal 4D where it originated.

Rewind: The 3D Earth Construct is contained within the inner Shell (aka FE Firmament) which is contained within the larger **Konstruct** Shell and between the two Shells is the **Cspace**. The simulated solar system and stars of the universe are pictured on the inside of the Konstruct (outer Shell). **The Sphere within a Sphere in the picture above is not to scale** – there is quite a bit of space (3-10 million miles) between the inner and outer Shells. And that space will be referred to as **Cspace** (for ease of reference).

When we send **space probes** off into the "solar system," the (benevolent) gods that run this Simulation might permit the probes to travel into and transmit from the Cspace area where time/space are **fractally** manipulated. Quite a large Simulation.

Scenes outside the Earth Sphere in the Cspace can be simulated much as in a Video Game where the player moves thru many scenes, but they are constructed and deconstructed (as needed) as the main Video character moves thru the settings. Sun and inner Solar System, for example, may be 4D *simulacra* surrounding us (in the Cspace) and some objects may be real, but we on Earth are contained by an (inner) energy 'Shell' (Firmament). Thus Sun and stars appear to be part of the energy 'Shell' containing the 3D Earth and on occasion, the <u>transparent</u> <u>inner energy</u> Shell reflects the Sun's light and is called the ***Gegenschein***. Charles Fort figured that out and his discoveries and conclusions were summarized above.

And yet there are problems with just wrapping a spherical Earth in an energy cocoon: How does the **heavy ocean water** stay on a globe? Why has today's Science not found nor been able to measure Gravity? Where is the Sun – near or far? Why did the US Govt carry out **Operation Fishbowl** and launch nuclear-tipped missles at the Firmament? And why are Southern Hemisphere distances between continents much farther apart than shown on a globe? (Chapter 11 addresses some of these with partial resolutions.)

FE Scenario

So, for equal time: Could the original Earth really have been flat <u>and</u> a Simulation?

Flat Earth -- Serious Evidence <u>Approaching Proof</u>....

> Again, I know the risk for even suggesting that there might be something to the evidence for a Flat Earth (FE)… and yet, when I started out <u>to prove it wrong</u> in QES what a shock I got. Some arguments <u>for</u> the FE are substantial, and **I still cannot disprove them** because <u>they can also apply to the VR Sphere</u>. Keep an open mind for the time being… we'll resolve this issue.

While QES Ch. 11 goes into this in much greater detail, basically finding more proof that the Earth is (and was?) flat rather than a globe, the major proofs are offered here…. And yet, we need some on-going explanation which requires familiarization with the standard mathematics regarding the Earth. Thus, the following figures are what Science says about the Earth:

Preliminary Scientific Data

As we will be coming back to the numbers again and again, it is useful to present the alleged facts about Earth, as everyone knows them and mainstream Science gives us:

Distance from the Sun	93,000,000 miles
Earth's circumference	24,901 miles
Earth's diameter	7,926 miles
Earth day	24 hours (23 hr 56 m)
Earth speed around Sun	67,108 mph
Earth axial tilt	23°
Earth rotational speed	1038 mph (24,901÷24hr)
Oakland, CA – Norfolk, VA	3,000 mi (2989 mi)
US timezones	3

The math: if the circumference of the Earth is about 25,000 miles and it takes the Sun 24 hours to return to its same position, then the Sun effectively 'traverses' the Earth at about 1000 miles per hour. D = R * T (Distance = Rate * Time) or in this case, R = D/T.
Said in terms of <u>today's</u> Science: the Earth is said to be rotating on its axis at (about) **1000 mph.**

Another interesting way to look at the ocean's often flat horizon…

If the Earth's circumference is 25,000 miles, and there are 360° in a circle, then each 1° = 69 miles. Thus, if we see a horizon that covers 500-600 miles, we should see an 8° curve to the horizon… Do we?.

This does not prove the Earth is flat… the VR Sphere also shows 'flat' ocean horizons **depending on distance and angle of view**… In fact, a 300 mile horizon that appears to be flat is just 300/25000 = 01.2 % of the circumference… We need to review this aspect a bit more…

In fact, an easier analysis is that of the "hotspots" on the clouds and oceans from what appears to be a Sun that is closer than 93,000,000 miles…

Hotspots

(Credit: TheWorldWeLiveIn: https://youtu.be/vknnZoBrAP8)

(Note that the above picture, from space, also shows the edge of the Earth to be hundreds of miles long and flat…)

The photo above shows a "hotspot" on the Earth – if the Sun were really 93 million miles away, there should be no hotspot, but a general diffused light, shown below…:

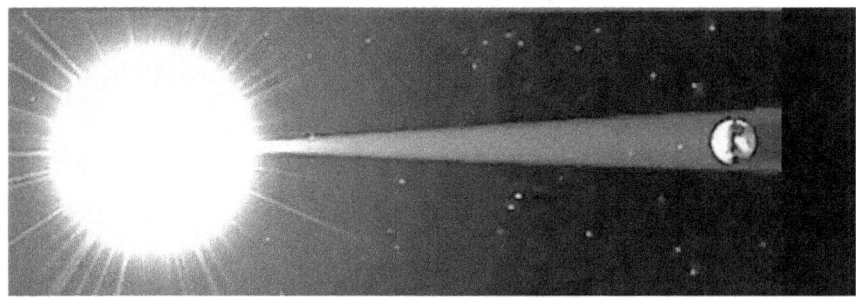

The hotspots happen because the Sun is much closer to the Earth than we are told. And the Sun is <u>inside</u> the Firmament of the Flat Earth, and just outside the Firmament (in the Cspace) in the case of the VR Sphere.

A very similar thing happens with the following pictures…

I wonder how is it the hot spot stretches from the horizon all the way to my feet

The FE Theory suggests that the **ocean is flat** since a curve in the Earth would curve the ocean and <u>not</u> transmit a **'hotspot'** as shown above. Any curve in the ocean would absorb or block the straight flow of the light. <u>Unless</u> the Sun is actually closer than 93 million miles! So where is the Sun?

This is a perfectly normal glare – the Sun is <u>above</u> the horizon. There does seem to be a problem with a "hotspot" when the Sun is <u>on</u> the horizon (previous picture, last page)…. NOTE:

> The **glare line cannot curve with the water**…. So is it flat?

Water Level

Be assured that water always seeks its own <u>level</u>. In other words,

> **water on the ocean is not curved or convex – This is the 1st Biggie!**

Below is <u>not</u> what we see out on the ocean: **the resting water doesn't curve**.

This picture was obviously made with a convex lens….. and they are doing the same thing in orbit around Earth to show the 'curvature' of the Earth??....

Many pictures of Earth from a satellite or ISS, show a flat horizon, and an equal number show a curve... Which is correct? (See end of Chapter 10.)

When we are out on the ocean, what we see is this:

(credit: TheWorldWeLiveIn)

And this one:

(credit: TheWorldWeLiveIn)

The above picture is looking at 300-400 miles of ocean/horizon… why is there no curve? Remember the earlier calculation that said 300-400 miles of horizon should have an 8° curve to the horizon…

I am deliberately creating doubt and questions in your mind, in an attempt to free you from the 'brainwashing' we all received when we sat in 1st Grade and saw the globe of the Earth on the teacher's desk. No one said the Earth was round; it just sat there on her desk… no one had to say anything. Thus we 'knew' the Earth was a sphere.

NASA Exposed

NASA Flight Path Insight

Whenever people watched the Space Shuttles (<u>any</u> orbital flights) make **Earth orbits** on TV, on the large NASA Houston wall screen, there was a curious "sine wave" (or 'S') layout to the path followed, and that never made sense – until the Flat Earth layout showed it to be **really a circle**:

This is the **2nd Biggie!**

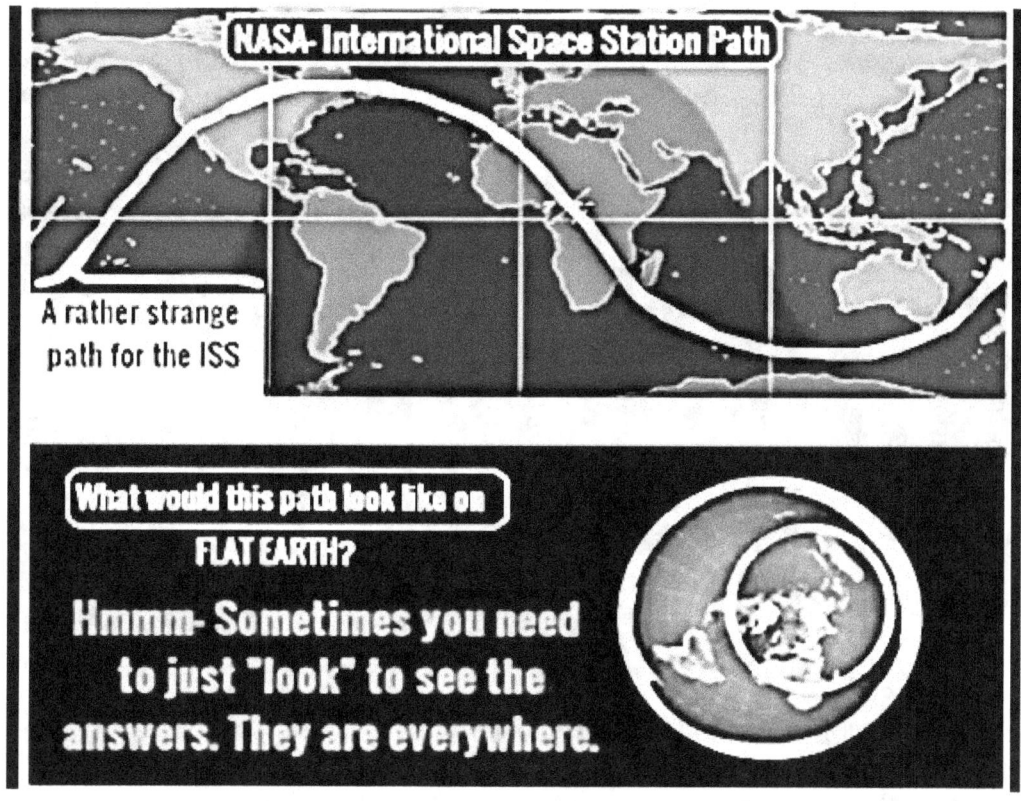

(credit: Odiupicku: Flat Earth – Guess Why We Can't See Antarctica...)

This discounts the VR Sphere if Earth is not a globe. The NASA wall screen would have made more sense if the ISS or any Shuttle path across the world map (upper half of the above diagram) had been pretty much a straight line, even at the equator...

Plausible answers are given in Chapter 10.

However, NASA Mission Control knows something...

This is further examined below....... NASA Mission Control – see center wall tracking screen (next page)...

(credit: NASA http://spaceflight.nasa.gov/gallery/images/shuttle/sts-115/html/jsc2006e40472.html **Public Domain**)

To replay that, because it is very significant, here is the Mission Control Screen by itself –

Note that the lower S-curve crosses America, Africa and goes under Australia.

Now compare that with the FE map below:

The "S" orbits become a circle....

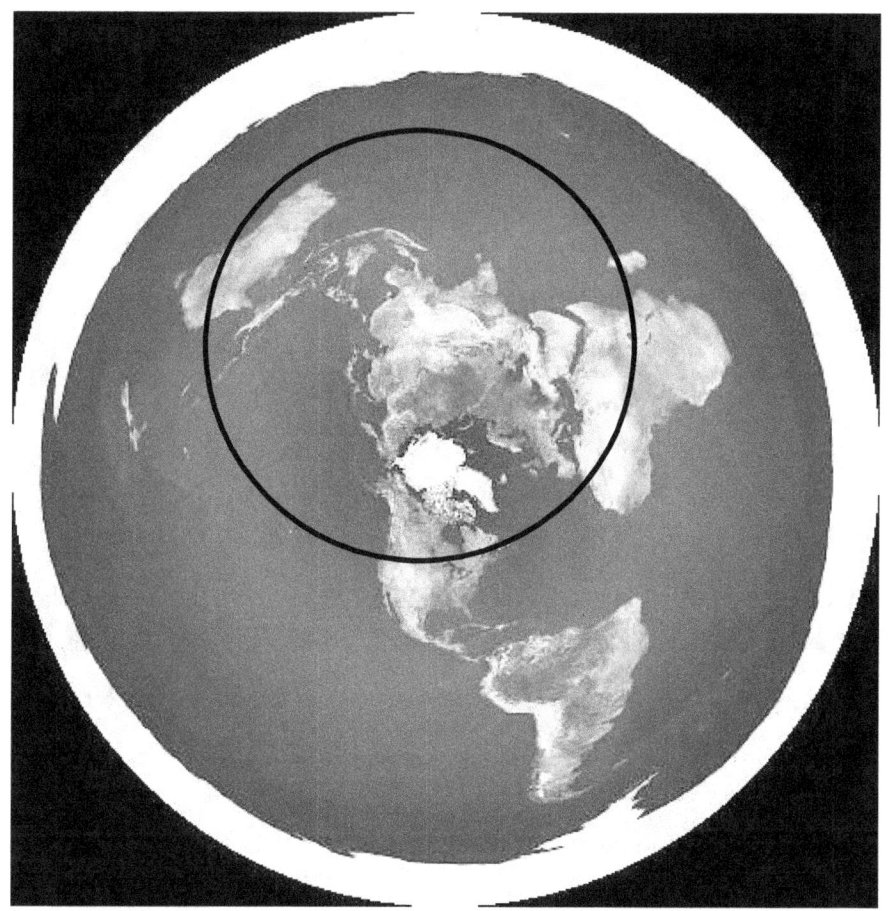

The orbital "S" paths describe a circle on Flat Earth.... and it makes more sense than an 'S' path across the world.

The orbital path is the same for the ISS, a satellite, or a Space Shuttle. And **Chapter 10 (page 396) offers another plausible explanation besides the FE Theory** – yet this Wall Screen issue is a doozie!

What an interesting 'coincidence.'

There is another issue, a serious one, where Gravity and the Centrifugal Force are found to be at odds with retaining the oceans' water, and the result is analyzed in QES Ch. 11 (96 pages of FE analysis).

Gravity: this is the 3rd Biggie!

Explain that. Gravity is further examined in Chapter 4: Newton was wrong.

Corollary Evidence 1

There is another 'proof' when taken with the sunset photo earlier: Because the Sun is actually closer to Earth than we have been told, it often leaves a **"hotspot"** on the ocean – which could not be seen if the Earth were curved (even a VR Sphere) because the water on curved Earth would be curved (convex) and the sunlight would not be a **straight line from the horizon** to a person standing on the beach.

Again, the Earth would have to be flat, because standing water (i.e., the ocean) is <u>not convex</u>. It is flat. And the Sun has to be close to create a hotspot, because a hotspot is concentrated, focused light. How does the diagram below concentrate light and create a hotspot?

And again:

The Sun is shooting a straight line from the horizon to the beach. No curve. It cannot be 93 million miles away to do this… the light would be either more diffused or it would not be able to 'ride' on curved (convex) water at all!

…as shown on the next page…

And again, consider where the Sun has to be to make this real picture:

At 93,000,000 miles away the light would look more like this:

Corollary Evidence 2

Has anyone seriously considered: If **space is a vacuum** and Earth has an atmosphere, what keeps the atmosphere from being sucked off the Earth and into the vacuum of space? (Answer: the Firmament.)

(credit: aplanetruth.info)

Science says "Nature abhors a vacuum" and we can't use the Gravity argument here since Gravity is not strong enough to pull the clouds (heavy water vapor) to the ground, nor does it pull smoke to the ground, let alone pull on the much lighter air molecules! A **Firmament** would contain the atmosphere, however.

Earth Shape

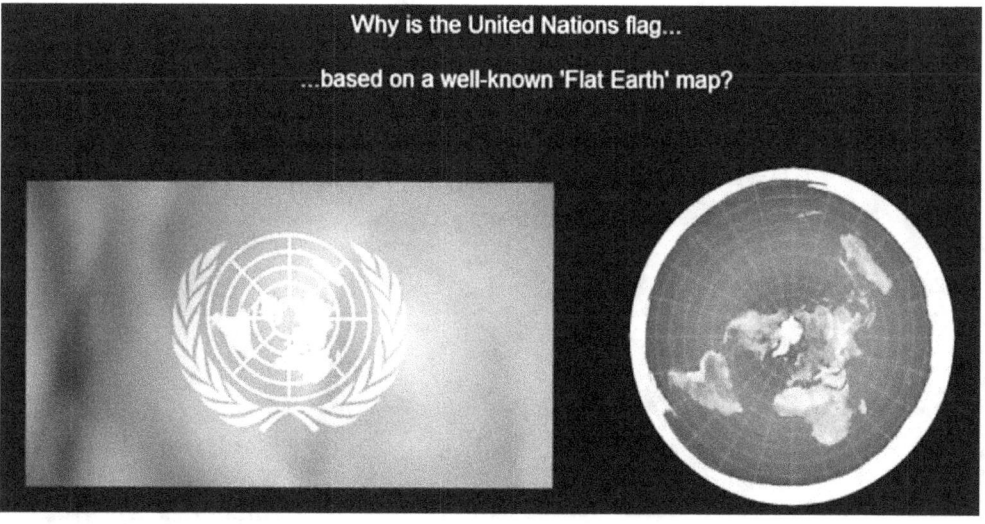

Another in-the-know world agency, besides NASA, also knows what the layout is of the Earth (above). Notice **Antarctica is not a continent on either diagram**… and looking at the UN flag, above, you have to wonder why they didn't picture Earth as follows:

But they didn't because they <u>know</u>. And as you will see, the above map is wrong. It looks like a flight from Capetown, South Africa is a short hop to Australia. The actual distance is much different (see FE map above), and an airplane does not fly as you would think… because it is in the **Southern Hemisphere.**

Five Major Evidences

Let's look at a shorter route in the Northern Hemisphere….

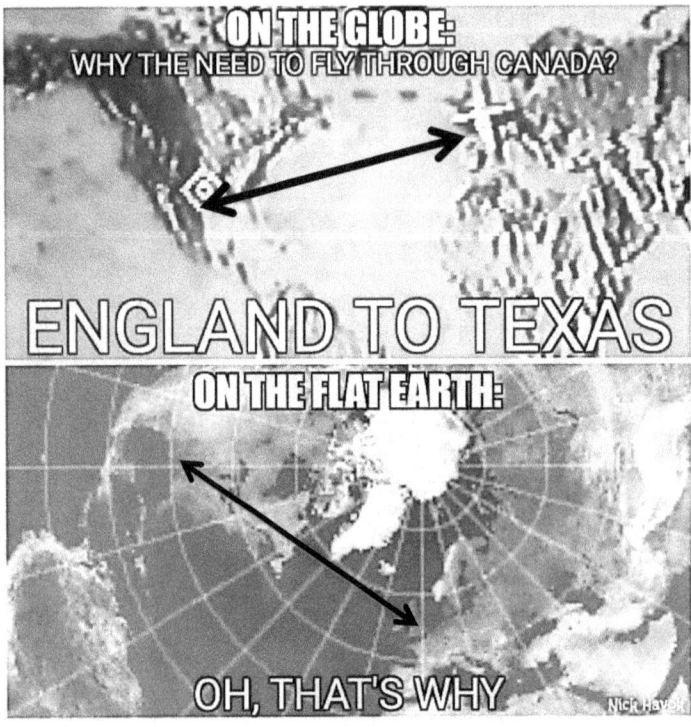

1. **England to Texas Flight Path** The picture left says it all -- **Why do planes fly thru part of Canada to get to Texas?** On a globe, one can draw a straight line/path from England to Texas (arrow at left) which does not touch Canada, however the Flat Earth map tells us why:

(credit: Odiupicku: **Flat Earth – Guess Why We Can't See Antarctica…**)

110

Airlines all want **the shortest, most direct flight path to somewhere**, and curving up or down (as one would on a globe) spends more aviation fuel. But if you know that the Earth is flat, and can plot the most direct route, it will take you thru Northeastern Canada.

Southern Hemisphere Routes

And shown below are flight routes in the Earth's Southern Hemisphere that look Ok on a globe, but in fact **do not exist**…

This is the **4th Biggie!**

(credit: Odiupicku: Flat Earth – Guess Why We Can't See Antarctica…)

<u>Why</u> do they not exist? They are the shortest routes from Australia to South America, and New Zealand to South Africa – they should be used – even though it means flying over the South Pole…. Unless the South Pole doesn't exist. Look again.

This issue is visited again in Chapter 10 with a plausible answer.

There are <u>no</u> non-stop flights from Australia to South Africa or to Buenos Aires. Unlike the non-stops in the Northern Hemisphere, **none** exist in the Southern Hemisphere.

Let's say we want to fly from Auckland NZ to Capetown SA, so using a standard world map, that should be as follows (left: double arrow).

Simple eh?

It doesn't work that way—

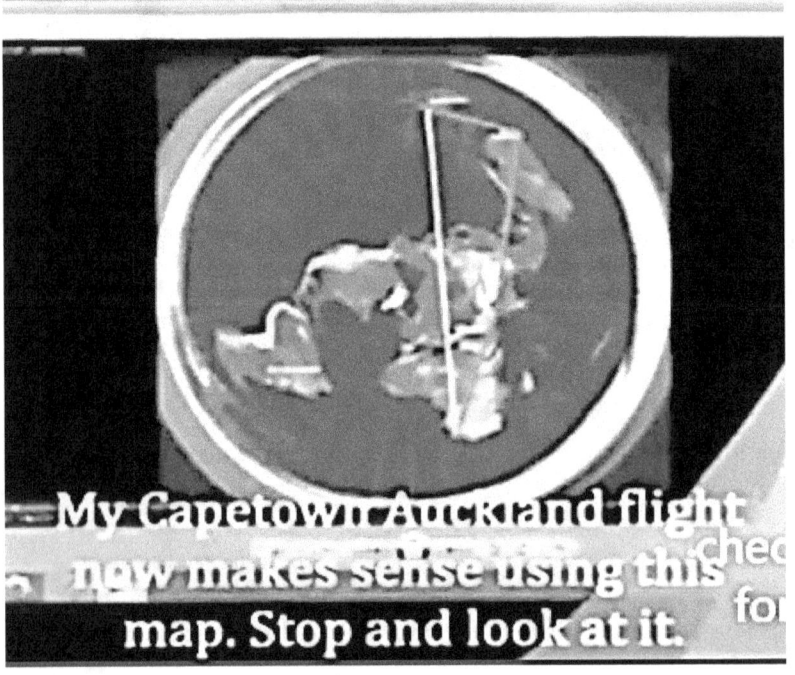

where they really send us is shown below (because they use an FE map). Non-stop route NOT taken – use fight path on left:

(credit above: two pix: YouTube:TheHippie Moderne: Don't Believe in the Flat Earth Theory?)

Also this:

2. Polaris, the Pole Star, is Always Straight Up, Overhead

As was said earlier, this one is a killer to the rotating, weaving and bobbing Earth as it allegedly moves around the Sun, and moves with the solar system around the Galaxy. How would a star **billions of Light Years away** (<u>according to astronomers</u>) always stay in the exact same position while the Earth has a 23° tilt on the axis and circles the Sun?

(credit: Bing images: universetoday.com)

The heavens rotate around Polaris… year after year it is always in the same position, directly above the North Pole, and all other stars revolve around it. (In the FE and VR Sphere scenarios, the Firmament is rotating, not the Earth.)

Look closely at the diagram below – the Earth and its tilt are always in the same position BUT the 4 polar axes do not point to the same location in space as the Earth circles the Sun. **How does Polaris move back and forth (as it would have to) so that it stays above the North Pole?**

Polaris does not move, has never moved, and that means that Polaris is directly (permanently) overhead and <u>not far away</u>. If Earth "wobbles" on its axis (i.e., 'precession') how is this possible?   Curious, eh?

This is further examined in the QES book…

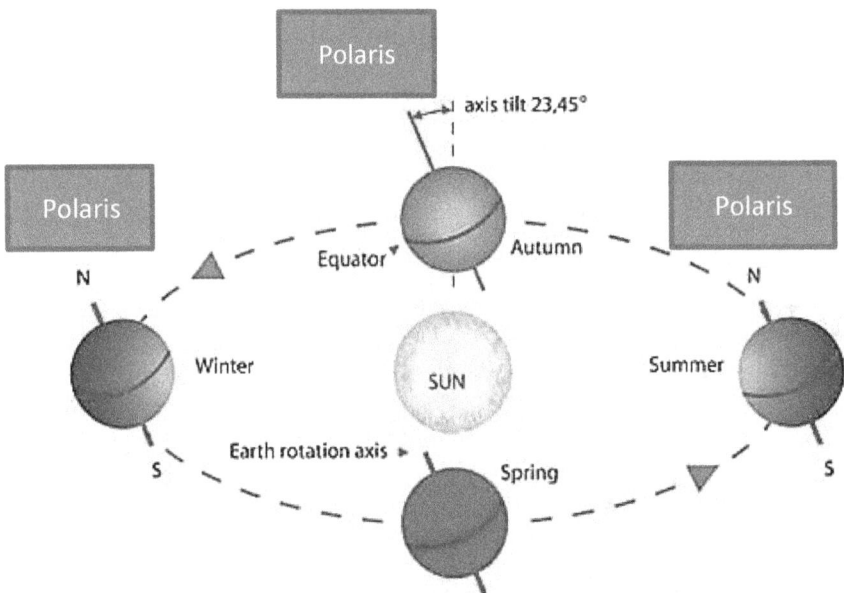

Due to the alleged precession of the Earth (26,000 year cycle), the axis tilt (23°), the Chandler Wobble (433 day cycle), and the polar motion (7 year cycle), the North Pole would not point consistently in the same direction, so how does Polaris stay directly overhead? Does anybody not see this? And how would Polaris move to stay always above the North Pole? (See night sky picture on previous page.)

Not only the 3 above angles are parallel, and do not converge, but the solar system is allegedly moving about the Galaxy as well and Polaris is in the Galaxy… Is Science saying Polaris just happens to keep perfect synchronization with Earth's North Pole in all that gyrating?

If Polaris is 433 LY away (as they claim – so that "the Earth axes would visually converge at that distance" [would they really?]), you would not be able to see the star with your unaided eye.

Who is kidding whom?

3. Sun Never Goes Below the Horizon

This is another "killer" proof: This is the 5th **Biggie!**

So if you go toward the Arctic Circle, at 1400 miles south (in Norway) and set your camera up to track the **Midnight Sun,** you get the following time-lapse picture where the photographer moved his camera over 360° once <u>every hour</u> to capture the Sun in the middle of the picture:

> BTW, there is **no Midnight Sun in the Antarctic** – why not? If Earth is a globe, both poles should reflect the Sun equally, 6 months apart: as the Sun does Winter in the Northern Hemisphere, we get one Midnight Sun effect, then when it's Winter in the Southern Hemisphere, it <u>should</u> have a corresponding Midnight Sun there, too… It is a globe with (allegedly) similar poles, right? (Chapter 10.) Wrong. It is due to the path of the Sun which centers more over the northern latitudes, and does not get below the Tropic of Capricorn (see picture, page 356).
>
> There is an **Aurora Borealis** – in the Arctic and <u>one over Australia</u> – …which changes simultaneously with changes in the northern auroral zone. [54]

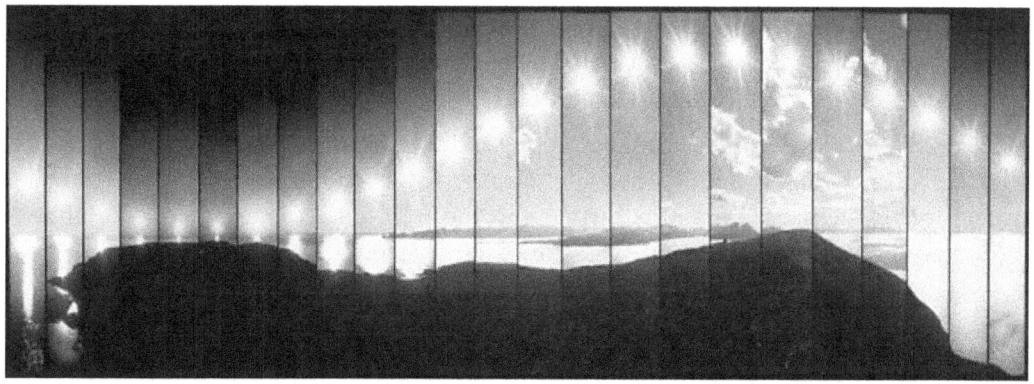

Midnight Sun
(credit: Big Images: gettyimages.co.uk)

Very Important: this essentially <u>proves</u> that the Sun is rotating <u>above</u> the Earth, in a circular path, over the Northern latitudes… not 93 million miles away.

If the Earth rotated, alternately facing the Sun, and then away, you'd get a zig-zag picture. Here the photographer moved his camera <u>once each hour</u> to follow the Sun as it moved over Loppa at **70° north latitude**. … it never went below the horizon, meaning <u>the Sun could not have been on the same horizontal plane as Earth</u> (i.e., not at right angles to the equator) – the **Sun was clearly above** and shining while moving thru the Northern hemisphere! Wow!

4. Rewind: Hotspots

You have to agree with the Flat Earth people… Look at the following pictures.

How is this possible unless the Sun is close to the Earth? It is a picture of Earth and its horizon (flat) from a high-altitude balloon… and an anomalous **"hotspot"**…

(credit: geoshifter.com)

The 'spotlight' effect is not possible if the Sun's intensity is 93,000,000 miles away!

This again is evidence for the Sun being much closer to Earth and supports both the FE Theory <u>and</u> the VR Sphere idea – and don't forget the similar *Gegenschein* which reflects off something above the Earth. They both have a 'concentrated' glare that would not be possible if the Sun's light diffused over millions of miles.

And again, a **hotspot** (concentrated light) where there should be none:

(Credit: TheWorldWeLiveIn: https://youtu.be/vknnZoBrAP8)

5. Dr. Auguste Piccard in 1931

As was shown earlier at the beginning of Ch. 11 in QES – the ancients knew the Earth to be flat. And concurrent with that is the 1931 statement of **Dr. Auguste Piccard…** This was saved for last because it is an **eyewitness account**:

This is fascinating and serious… (see Wikipedia article on Dr. Piccard):

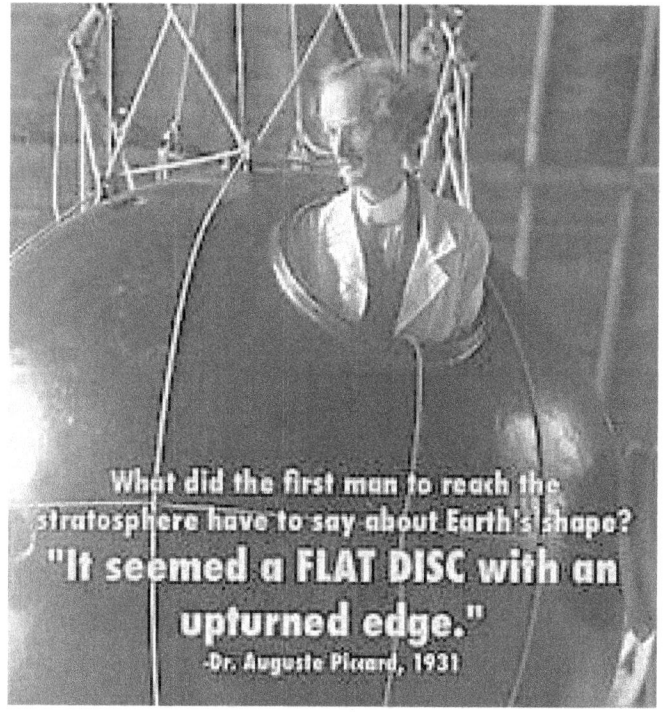

Now, concurrent with VEG's original proposition that the Earth may now be a VR Sphere (as we have orbiting satellites, the International Space Station, and some pictures of Earth from NASA and stratospheric rockets), and considering that the gods who run this place might have had to modify the Simulation to permit Man a larger "field of play" (and raised the Firmament and 'wrapped' [morphed] the Earth into a Sphere) it begins to look as if the **Earth could have been modified from the original flat state** to a current spherical one. But why the "upturned edge"? What did he really see? (He took no pictures.)

Somewhere in 1931, Earth was still flat? That is puzzling – yet Dr. Piccard supports the Flat Earth **because he says he saw it first-hand**! Unfortunately there are no pictures from Dr. Piccard's balloon trip. What the heck did he see? What did his accompanying scientist, Paul Kipfer, see? They were **up only 9-10 miles** – maybe he saw the Alps at a distance ("upturned edge")… or the Urals? No, he took off from

Augsburg (southern) Germany and came down on a glacier near Ober-Gurgl, Austria. So, is the edge visible from 10 miles up (at 52,800 feet)? Satellites are 200 miles up, buy comparison.

And yet, as with so many things pertaining to the VR Sphere and the Flat Earth, it is hard to make a 100% sure call (so far) – and maybe the PTB don't want the real nature of Earth exposed! After all, they are running Earth for their own purposes… …. telling us we went to the Moon… promoting space travel.

> **Those wanting a deeper examination of the FE vs the VR Sphere issue should spend some time with Book 4 <u>Quantum Earth Simulation</u>… in Ch. 11 (96 pp).**
>
> **This issue is further explored herein in Chapters 7, 9 and 10.**

Summary

There is much more in QES, Ch. 11, but the above should suffice to raise some logical issues – **the major set of FE evidence is in QES Ch. 11**. Whether it is a VR Sphere (meaning Earth is still in Quarantine (Firmament), the Sun and Moon are close to the Firmament (as with the FE scenario), and Antarctica is a continent), or whether Earth is a Flat Earth, the two major differences are:

> around the FE layout where the Firmament attaches to the **Ice Wall** (no Antarctica continent),

and

> the oceans are flat, water is always level, the Sun and Moon are under the Firmament (just as Enoch said he saw them).

In fact the last point would be <u>the</u> major objection to the VR Sphere… how do the heavy oceans stay on a sphere? (Possible answer: Simulation holographic programming : Chapter 11.)

Regardless, if there is a *Gegenschein* Quarantine or a Firmament, we did not go to the Moon, and no ETs are going to visit us. No one gets thru the barrier.

Real food for thought.

Why Hide What Earth Is?

If Earth is not a rock rotating at 1000 mph and spinning around the Sun at 67,000+ mph, then it is a FE or a VR Sphere. In any event **Earth was created** and is proof of a God who created it for Man, what has been called the Earth School. It means we are watched over.

Atheists don't like that, and many people prefer to do whatever they want with no one watching, and thus no accountability. **Scientists** are hung up on Evolution – God wasn't necessary in their version – again, they do not want to be **responsible** for their actions. (This issue is further reviewed at the end of Chapter 11.)

And many scientists are so narrow-minded: "We can't see God, therefore He doesn't exist!" They say the same thing about the **soul**, but VEG tells of the doctor who weighed dying patients and found they weighed an average of **21 grams less** at death (and a movie was made about it, *21 Grams.*)

> Should we let the godless, ignorant and 'politically correct' tell us what to believe?

However, there is a downside to knowing what Earth really is: **all mystery is gone** – no more mystery about UFOs, Hollow Earth, Visiting ETs, Anunnaki from Nibiru/Planet X, Moon landings, a trip to Mars, etc. but once you know, you cannot be lied to – you will know whether what you hear is *plausble*.

> *And once you know, you can get out of here.*

> *That means: Graduate from Earth School.*

Chapter 10 goes into another salient reason for hiding the truth. One you need to be prepared for. Chapter 10 goes into the **6th Biggie**: What is Antarctica?

Chapter 4: Science Errata

Recall that the Introduction asked:

> Are you really living on the planet you think you are?

While the previous chapter questions the Spherical Earth myth (and Chapter 11 in QES explodes it), this one continues the exposé – showing that **we are not on a rock spinning thru space.** Current-day Science <u>cannot prove</u> the Earth revolves around the Sun… despite the supportive CGI pictures of Earth in space done by NASA.

Now in addition, we examine four major names in Science and show that they also made errors. Whereas it is a normal thing for a scientist to make a hypothesis about what he sees, and then polish it to a Theory, either upgrading it or discarding it later, there are three mainstays of current-day Science that are still Theories and they still miss the mark.

> **Note**: I am not being a smart-ass here in this chapter. If YOU have the patience and objectivity to consider what is said in the following pages you may actually see what is going on, and it IS rather weird compared to what we think we know (see Chapter 2).
>   It is always wise to question – even what I am saying.
> What is really going on in our world is so odd, that I had to write QES about it, and I am still stunned, and I only 98% lean toward the Flat Earth, much more preferring the VR Sphere – although it has a problem with the oceans, too!

This issue of what Earth really is, which was begun in the last chapter, will be wrapped up in this one. What you might find interesting <u>and relevant,</u> is who started the whole Spherical Earth business to start with. Whereas people knew for millennia that Earth was flat, the Bible said so, the Koran said so (Ch. 11 in QES), the Book of Enoch written by an eye-witness said so, and the Church also supported that scenario, so **who** overturned the FE concept, originally credited to Copernicus, and why did he have so much power to make the Spherical Earth scenario take hold and persevere?

Any time the issue is brought up, five names come to light: Ptolemy, Galileo, Copernicus, Giordano Bruno, and Tycho Brahe. Galileo was not the first, but because the Church liked him, and he recanted after the Inquisition found him guilty, he was put under permanent Church (House) Arrest, but Bruno was burned at the stake (AD 1600) for his heresy, which involved more than what the Earth's position was in the universe. That leaves **Copernicus** (AD 1543) who was influenced by **Ptolemy** (AD 150). So Copernicus was not the first – but somehow he made the revised Heliocentric Model stick.

And again, Ptolemy was not the first. **Pythagoras** (500 BC) and **Aristarchus of Samos** (250 BC) were among the first to suggest something like the **Heliocentric Model.** And Ptolemy did not get it right, either – the flaws with his model are what provoked further analysis and discourse. Yet **Anaxamander** (540 BC) had a flat view of the world...

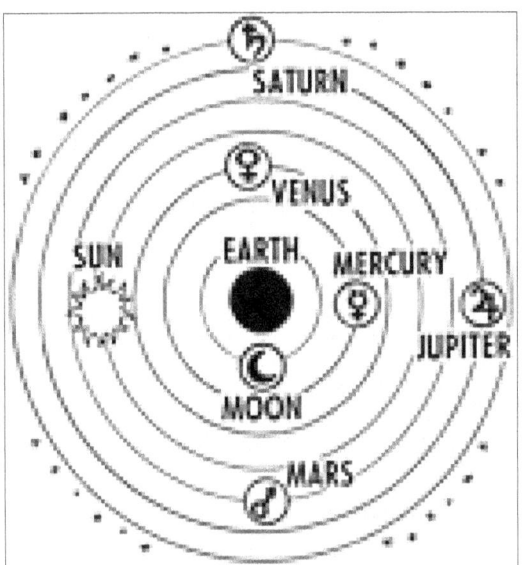

Ptolemaic View of the World

Note: the planets revolve around Earth, as does the Sun and Moon.

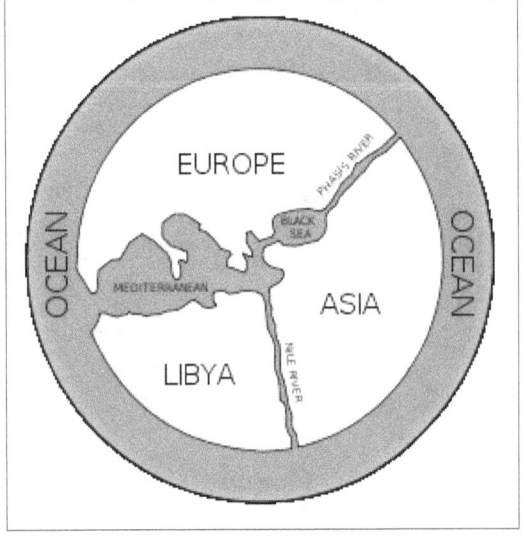

Anaxamander's view of the world.

This is closer to the Flat Earth model being re-promoted in today's world.

So what does **Copernicus' system** look like?

While it is in Latin, the Copernican Model describes what we today believe to be our solar system, **Sun at the center** and the planets in orbits around it, in the order we recognize today.

The forgoing is just to emphasize that Man was really beginning to analyze the world around him and try to figure out whether he was the "center of the universe" as the Church had always said, or whether Earth was something else.

As a side note, recall that <u>Anunnaki Legacy</u> showed the advanced beings travelling around the Earth, dispensing knowledge and skills to the growing human population. In addition to writing, medicine and agriculture, they also shared some astronomical concepts with them, as was shown with the Dogon, the Zulu and the Maya. Or is it possible that the Ancient Ones did NOT want Man to know where he really was, and let him come up with his own ideas (as seen in the accompanying diagrams)?

Remember: if Earth was always a Flat Earth with a Firmament, even the Ancient Ones could not leave it, but they would have known what Earth was as it was their job to 'manage' the flora and fauna. And would they want Man to think he was 'trapped' under a dome (Firmament), being watched, and could not escape? Would it be better to present Earth as a free rotating rock, spinning around the Sun, to make Man think he was not being watched, and could someday reach the stars that he saw in the night sky?

Hint: you need to juxtapose that with the same gods taking Enoch up and showing him what Earth really was.

Before proceeding with Copernicus, we need to take a quick look at Tycho Brahe, even though <u>Copernicus preceded Brache chronologically</u> ...

Tycho Brache

Brache worked to combine what he saw as the geometrical benefits of the Copernican system with the philosophical benefits of the **Ptolemaic system** into his own model of the universe, the **Tychonic System**, about AD 1600.

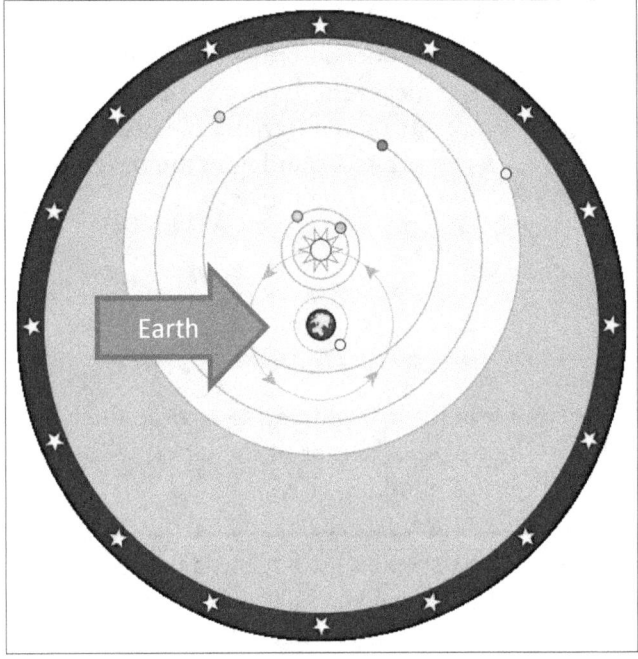

Basically, the objects on blue orbits (the Moon **and the Sun**) **revolve around the Earth.** The objects on orange orbits (Mercury, Venus, Mars, Jupiter, and Saturn) revolve around the Sun. **All is surrounded by a sphere of fixed stars** (though they are fixed only with respect to each other, **for the sphere revolves around the earth**). The system is essentially geocentric, though everything except for the moon and the fixed stars and the earth centre itself revolves around the Sun. Distances are of course just generalized, though it is important that the minor planets are always "tied" to the Sun while the major planets can be on either side of the Earth. Note: the path of the Sun's orbit intersects with the path of Mars' orbit, causing a problem for any astronomer thinking of the mechanism as incorporating nested physical "spheres." [55]

Note that Brahe also has an **outer shell** (like the VR Sphere's Konstruct – Ch. 2 in QES) which displays the stars. Charles Fort would love it.

Brahe and his system are very similar to that of Copernicus…

> The Tychonian system is mathematically equivalent to the Copernican system, except that the Copernican system predicts a **stellar parallax**, while the Tychonian system predicts <u>no stellar parallax</u>. Stellar parallax was not measurable until the 19th century, and therefore there was at the time no valid disproof of the Tychonic system on empirical grounds, nor any decisive observational evidence for the Copernican system.[56]
> [emphasis added]
>   And parallax is a very important aspect of our Earth realm.

Remember that **Charles Fort** (in Chapter 3) discovered the *Gegenschein* which reflects off something surrounding the Earth. It also has **no parallax** – meaning that what you are looking at does not change when you change your physical perspective.

The significance of that says that Brahe got one aspect of the stars correct: **the stars show no parallax.** Copernicus, on the other hand said the stars would have a parallax (e.g., change). This is coincidentally related to the Flat Earth model which also says that the stars have no parallax... shown below:

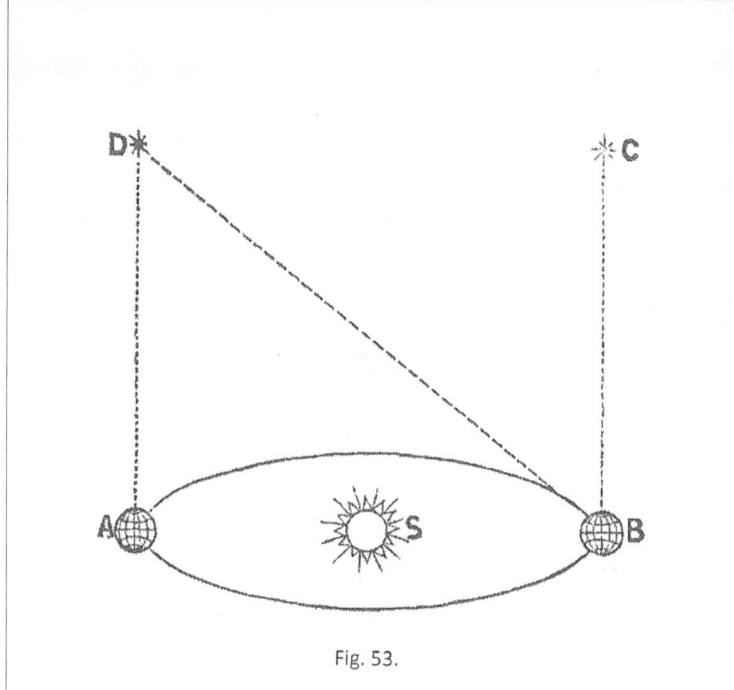

A is Earth in June
B is Earth in December

AD is where the star was seen in June.

A parallax was expected in December -- **BD**

BC is how the star was seen in December – no parallax.

Fig. 53.

(credit: Rowbotham: <u>Zetetic Astronomy</u>, pp 64-65.)

The No Parallax situation is proof that the **Earth is not rotating** around the Sun, and the Sun and stars are fixed above, <u>IN the Firmament</u> which <u>does rotate</u>.

Be that as it may, we need to examine Copernicus very closely so see why he was not burned at the stake, nor subjected to Inquisitional penalty... although he was **imprisoned by Pope Urban VIII for a short time**... but he recanted and influential friends got him released.

Copernicus & The Earth

His greatest work was called ***De revolutionibus orbium coelestium*** (*On the Revolutions of the Heavenly Spheres*) and was initially a set of notes, a partial manuscript which he distributed to friends, less his name, **keeping the manuscript anonymous**; a greater part of it was done by AD 1530, and it was later published in Nuremberg in **AD 1543**. The Church seems to not have noticed as this was all before Galileo (AD 1632) and Bruno were punished (AD 1600) for similar actions. Perhaps it helped that **Copernicus hesitated to publish (and such was done after his death),** and his sister was a Benedictine nun, and his father was a wealthy merchant.

In addition, **Copernicus was well-educated** at the University of Krakow, so his parents had the wealth to put him in college. He was a polyglot [spoke multiple languages], a polymath, he obtained a doctorate in canon law, and was also a mathematician, astronomer, physician, classics scholar, translator, governor, diplomat, and economist. His name carried some weight. That accounts for him being well-respected.

> In 1539 Georg Joachim Rheticus, a young mathematician from
> Wittenberg read Copernicus' manuscript and immediately wrote
> a non-technical summary of its main theories in the form of an
> open letter addressed to Schöner, his astrology teacher in Nürnberg....
> Under strong pressure from Rheticus, and having seen that the first
> general reception of his work had not been unfavorable, Copernicus
> finally agreed to give the book to his close friend, **Bishop Tiedemann
> Giese**, to be delivered to Rheticus in Wittenberg for printing ....
> It was [officially] published just before Copernicus' death, in 1543.[57]

So Copernicus knew the danger of heresy with the Church, and kept the manuscript anonymous, and when the treatise was finally (officially) published, he was within months of dying anyway. But all did not go well...

Martin Luther came by a copy of the book and soundly denounced it:

> People gave ear to an upstart astrologer who strove to show that
> the earth revolves, not the heavens or the firmament, the sun and
> the moon ... This fool wishes to reverse the entire science of
> astronomy; but sacred Scripture tells us [Joshua 10:13] that Joshua
> commanded the sun to stand still, and not the earth. [58]

When the book was finally published, demand was low, with an initial print run of 400 failing to sell out. Copernicus had made the book **extremely technical, unreadable to all** but the most advanced astronomers of the day, allowing it to disseminate into their ranks <u>before</u> stirring great controversy. And, like Osiander, contemporary mathematicians and astronomers encouraged its audience to view it as **a useful mathematical fiction with no physical reality**, thereby somewhat shielding it from accusations of blasphemy. [59]

> Among some astronomers, the book "at once took its place as a worthy successor to the *Almagest* of **Ptolemy,** which had hitherto been the Alpha and Omega of astronomers." In England, Robert Recorde, **John Dee** [the alchemist].... were among those who adopted his position; in Germany....Michael Mästlin, the teacher of **Johannes Kepler**; in Italy... **Giordano Bruno** whilst Franciscus Patricius accepted the rotation of the earth. In Spain, rules published in 1561 for the curriculum of the University of Salamanca gave students the choice between studying **Ptolemy or Copernicus**. One of those students, **Diego de Zuñiga**, published an acceptance of Copernican theory in 1584. [60] [emphasis added]

We can see that it was promoted because it was **very technical** and that much mathematics and science had to mean something worthwhile, or so the story went, and Copernicus' ideas gained in popularity among the astronomers and mathematicians. Not so among the clergy.

Very soon, nevertheless, Copernicus' theory was attacked with Scripture and with the common Aristotelian [flat earth] proofs. In 1549, Luther's principal lieutenant wrote against Copernicus, pointing to the **theory's apparent conflict with Scripture** and advocating that "severe measures" be taken to restrain the impiety of Copernicans. The works of Copernicus and Zuñiga—the latter for asserting that *De revolutionibus* was compatible with Catholic faith—were placed on the **Index of Forbidden Books** by a decree of the Sacred Congregation of March 5, 1616 (more than <u>70 years after Copernicus' publication</u>). [61] Said the Church:

> This Holy Congregation has also learned about the spreading and acceptance by many of the **false Pythagorean doctrine**, altogether contrary to the Holy Scripture, that the earth moves and the sun is motionless, which is also taught by Nicholaus Copernicus' *De revolutionibus orbium coelestium* and by Diego de Zúñiga's *In Job* ... Therefore, in order that this opinion may not creep any further to the prejudice of Catholic truth, the Congregation has decided that the books by Nicolaus Copernicus [*De revolutionibus*] and Diego de Zúñiga [*In Job*] be **suspended until corrected**. [62]

De revolutionibus was not formally banned but merely withdrawn from circulation, pending "corrections" that would clarify the theory's status as an hypothesis. **Nine sentences that represented the heliocentric system as certain were to be omitted or changed**. After these corrections were prepared and formally approved in 1620 the reading of the book was permitted. But **the book was never reprinted with the changes** and was available in Catholic jurisdictions only to suitably qualified scholars, by special request. It remained on the Index until 1758, when Pope Benedict XIV (1740–58) removed the uncorrected book from his revised Index. [63]

All of that to point out that (1) Copernicus' theory was not that well accepted, except by astronomers, and (2) someone was working hard to promote it behind the scenes – **it survived until 1758** (above) – past the time when Brache, Kepler and Newton (all in the 1600s) would encounter it. The deed was done.

What is not obvious is who was doing the promoting, and why.

Germany Influences Heliocentrism

We have to look at Copernicus' land of birth: Old German Empire, called Prussia. It was then the Prussian Kingdom, connected with Poland. It is understandable that Martin Luther (also in Germany) would get wind of Copernicus' manuscript – it was being quietly published in Nüremberg. This is also where the **Illuminati** began about the year 1776 (18 years after Copernicus' book's removal from the Index).

Prior to that, Spain saw the rise of the **Alumbrados** (*the Illuminated Ones*) which flourished during the 15th-16th centuries. The Alumbrados were responsible for the acceptance of Heliocentrism as said above by the **University of Salamanca** and offered the students a choice of studying Ptolemy or Copernicus. Naturally they were the victims of the Spanish Inquisition as their beliefs were heretical.

And Spain also saw the rise of the Jesuits. **Ignatius of Loyola** studied in Salamanca and would have encountered the doctrines of Ptolemy and Copernicus. Ignatius later founded the **Jesuit Order** (AD 1539) and they served the Catholic Church in the capacity of missionaries. As the Spanish Inquisition severely questioned adherents of the Alumbrados sect, they discovered that they also were following the Heliocentrism doctrine. So there was a connection between Alumbrados, Heliocentrism, Jesuits and Salamanca.

What is less known is that the **Alumbrados** had large congregations in Salamanca and Toledo, and communicated with European counterparts, eventually influencing those in France called *Illuminés* and from there moving into Germany as those to be called the **Illuminati**.

The movement (under the name of *Illuminés*) seems to have reached France from Seville in 1623, and attained some following in Picardy when joined (1634) by Pierce Guerin, curé of Saint-Georges de Roye, whose followers, known as **Guerinets,** were suppressed in 1635. A century later, another, more obscure body of *Illuminés* came to light in the south of **France in 1722**, and appears to have lingered **till 1794,** having affinities with those known contemporaneously in the United Kingdom as 'French Prophets', an offshoot of the Camisards... [64] [emphasis added]

And if Spain, France and England were entertaining the Alumbrado ideas, it was but a simple step into Germany and **Adam Weishaupt** who formed the secret Illuminati society. Think not?

Adam Weishaupt … was a professor of Canon Law and practical philosophy at the **University of Ingolstadt...** He was the only non-clerical professor at **an institution run by Jesuits**, whose order had been [officially] dissolved in 1773. The Jesuits of Ingolstadt, however, still retained the purse strings and some power at the University, which they continued to regard as their own.[65] [emphasis added]

The Illuminati were not well-liked, and the government issued an edict banning all secret societies – of course, they just went 'underground.'

The Illuminati were blamed for several anti-religious publications then appearing in Bavaria. Much of this criticism sprang from vindictiveness and jealousy, but it is clear that many Illuminati court officials gave prefer-ential treatment to their brethren. In Bavaria, the energy of their two members of the Ecclesiastical Council had one of them elected treasurer. Their **opposition to Jesuits** resulted in the banned order losing key academic and church positions. **In Ingolstat, the Jesuit heads of department were replaced by Illuminati**.[66] [emphasis added]

There is the connection. And the name Illuminati is so close to Alumbrados as to pay homage to the Spanish sect. Of course there is also an alleged link to the Rosicrucians and German Freemasonry….

Note that the Illuminati agenda was to "oppose superstition, obscurantism, [and] religious influence over public life and abuses of state power..." [67] **Superstition and Religion were opposed** – including the Flat Earth scenario as it is a clear example of God's Grace and proof of His existence. Later, in Europe, when the Industrial Revolution got going, **Religion would be minimized in favor of Science**, while at the same time, quackery and mysticism were smiled upon.

> An interesting social experiment called the Soviet Union tried removing God from every facet of life and prevailed for 83 years... How did that work out?

This is significant as it is alleged that the Illuminati, which has persisted until this very day (under different names) was responsible for promoting the Heliocentrism and **God-less doctrines like Darwinism, Evolution, and the Spherical Earth**. The point is: **anything to NOT see Earth for what it really is** – A world designed by the Creator, home to the human beings also created and watched over by God, and serving as a School for **souls** – which latter is also denied as part of the godless, atheistic doctrine rampant in the 'modern' world today.

Anti-God Sentiment

At this point, we need to do a little segue to make a very important point, and then return with a look at Isaac Newton and Gravity.

The men we are considering as having led the public into a Science-is-King-and-There-is-No-God *meme* are the following:

Copernicus   (1473 – 1543)

Tycho Brache   (1546 - 1601)
Galileo   (1564 – 1642)   }   As can be seen by
Giordano Bruno (1548 – 1600)   the dates, these men
Kepler   (1571 – 1630)   were contemporaries.

Newton   (1642 – 1726)

What I expected to find was that they were atheists, and yet Galileo was a "pious Catholic", Giordano Bruno was an ordained priest, and Kepler was a religious man who used Scripture to substantiate his cosmogeny. And yet the four as shown above, <u>did know each other</u>, Kepler having worked for Brache, and Galileo dismissed Kepler's cause of the tides. Newton was said to be a Christian but as we'll see, he was seriously into Alchemy and the Kabbalah, the Jewish mysticism of the day.

And yet the point remains: these men were thinkers (which is generally healthy – God gave you a brain, use it!) and while some believed in God, design and the Creation, there was a *meme* afoot that said *Could it all be the product of some normal, natural process (set in motion by God or not), that can explain everything on Earth <u>scientifically</u>?*

They hoped to resolve the Earth Puzzle through logic and application of relevant philosophical discipline.

And not everyone agreed… (sorry this is long but it reflects the major reaction to Copernicus at the time…and is still echoed by the Flat Earth proponents of today) [68]

The Dominican **Giovanni Tolosani** of the Convent of St. Mark in Florence… Tolosani sought to refute Copernicanism by philosophical argument. Copernicanism was absurd, according to Tolosani, because it was scientifically unproven and unfounded. **First**, Copernicus had <u>assumed</u> the motion of the Earth but offered no physical theory whereby one would deduce this motion. (No one realized that the investigation into Copernicanism would result in a rethinking of the entire field of physics.)
Second, Tolosani charged that Copernicus's thought process was backwards. He held that Copernicus had come up with his idea and then sought phenomena that would support it, rather than observing phenomena and deducing from them the idea of what caused them….

[In addition] astronomy and mathematics could not be taken as serious means to determine physical causes. Tolosani invoked this view in his final critique of Copernicus, saying that **his biggest error was that he had started with "inferior" fields of science to make pronouncements about "superior" fields.** Copernicus had used mathematics and astronomy to postulate about physics and cosmology, rather than beginning with the accepted principles of physics and cosmology to determine things about astronomy and mathematics….

Thus Copernicus seemed to be undermining the whole system of the philosophy of science at the time. **Tolosani held that Copernicus had fallen into philosophical error because he had <u>not</u> been versed in physics and logic; anyone without such knowledge would make a poor astronomer and be unable to distinguish truth from falsehood.** [which by the way was a valid point made by more than one critic of Copernicus] …
Because Copernicanism had not met the criteria for scientific truth set out by Thomas Aquinas, Tolosani held that it could only be viewed as a wild unproven theory [which was a major point that was also made about Newton's theory of gravitation].

…and it continues:

Tolosani wrote: "By means of these words [of the *Ad Lectorem*], the foolishness of this book's author is rebuked. For by a foolish effort he [Copernicus] tried to revive the weak Pythagorean opinion [that the element of fire was at the center of the Universe], long ago deservedly destroyed, since it is expressly contrary to human reason and also opposes holy writ. From this situation, there could easily arise disagreements between Catholic expositors of holy scripture and those who might wish to adhere obstinately to this false opinion."

[Copernicus was imprisoned by Pope Urban VIII but released when he recanted.] [69]

Tolosani declared: "**Nicolaus Copernicus neither read nor understood the arguments of Aristotle the philosopher and Ptolemy the astronomer**." Tolosani wrote that Copernicus "is expert indeed in the sciences of mathematics and astronomy, but **he is very deficient in the sciences of physics and logic.** Moreover, it appears that he is unskilled with regard to [the interpretation of] holy scripture, since he contradicts several of its principles, not without danger of infidelity to himself and the readers of his book. ...**his arguments have no force and can very easily be taken apart.** For it is stupid to contradict an opinion accepted by everyone over a very long time for the strongest reasons, unless the impugner uses more powerful and insoluble demonstrations and completely dissolves the opposed reasons. But he does not do this in the least.

And the whole quote is very important because Tolosani was not a simple cleric with only an education in Holy Writ. He was learned and trained in the Aristotelian methods of argument and logic, and he shreds Copernicus who, like Newton, gained fame based on a hypothesis that was based on conjecture and supposition (and despite **Newton admitting that**, Science took on his hypothesis as fact… Why were the scientists so desperate?) And yet it is exactly indicative of the arguments brought against the whole **Heliocentric Schema** for centuries by many learned men – not just the Church trying to prove the Bible was right.

And that is **the crux of the issue**: leaned men seeking to define and understand the world wanted to NOT use Scripture and sought to use reason and logic. Hence, **God was suspended in favor of Science which became the New God** – especially as the **Age of Reason** (1620s – 1789)[70] captured men's fancy. Later, Charles Darwin would further champion this idea (because he hated God – God would not heal him… see later section on Darwin)…. And atheists threw the baby out with the bathwater, and we had such notable societal movements as Communism, Marxism – and the wonderful, godless Soviet Union.

The better path would have been to seek to prove Creation via the Scientific Method and thus create a Unity of Science and Religion... but that was not the intent. **The irony in the whole thing is that the Flat Earth or VR Sphere scenarios are <u>both</u> the proof of God and Creation.**

Ok, that is why Science overruled a centuries-old Flat Earth scenario substantiated by the Bible, the Book of Enoch and <u>deeper observation of Earth phenomena</u> – as outlined in Ch. 11 of QES and Scriptural writings – such as Joshua commanding the Sun to stand still (Josh 10:13) – suggesting that the Earth did not move but the Sun did.

On the other hand, playing Devil's Advocate, you cannot use the Bible to prove the Bible. Science <u>was</u> needed ---but it was biased as **atheists sought to prove God didn't exist**.

Hence, Science was seen as an alternative – but its successful use depended in the integrity and true scientific knowledge of the one doing the analysis... which in some cases (Copernicus, Newton and Darwin) was biased if not flawed.

Science Supports Creation

And just for a minute, let's look at 7 science elements that DO support Creation. If we can demonstrate design in various aspects of flora and fauna, then **Design means a Designer** as it reflects intelligence.

1. As will be seen in the section on Darwin, the **human eye** shows design as there are four parts that do not touch each other but if any one does not exist, then vision does not work... the greater point of Evolution being that tissues and organs develop as a result of adjacent/touching structures needing something and thus an appropriate structure is promoted and added to that already existing and that is called "evolution."

2. As will also be seen in Darwin's section, with the **Bombardier Beetle**, if the beetle's two flammable sacs were already mixed inside the beetle, the beetle would be fried... something said 'keep the two incendiary sacs separated so that the beetle survives'.

   The Eye and the Bombardier pictures are later, in this Chapter 4, in the section on Darwin.

3. Another design issue is the long **neck of the Giraffe**… when it bends down, to get a drink at a waterhole, there is a mechanism in the giraffe's neck that transports the water UP the neck (against gravity) and into the giraffe's body.

http://en.wikipedia.org/wiki/File:Flickr - Rainbirder - **Reticulated** Giraffe drinking.jpg

4. Another design issue is the **human ear**. Why not just evolve a tympanum (eardrum) that mimics the eye/optic nerve structure – just have the tympanum send acoustical nerve pulses into the brain – without the stapes, anvil (Incus) and hammer (Malleus) or the cochlea of the 2nd part called the Inner Ear? The ear is a very complex **design**. (see following page)

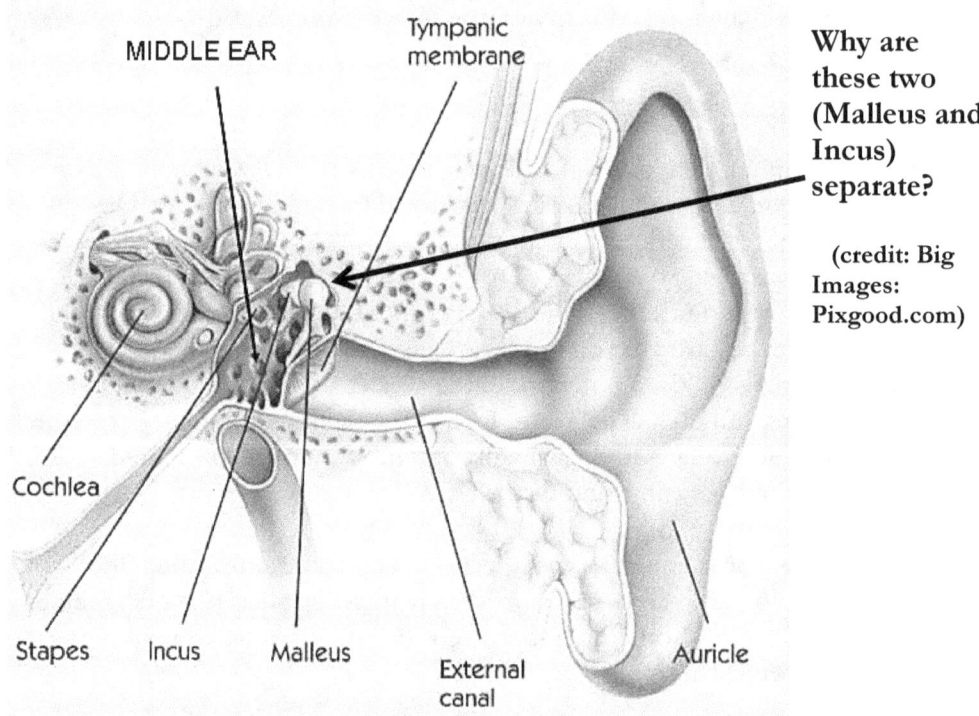

The Malleus/Incus structure is unique to mammals... so why didn't Nature repeat what had been done for the fish... Hey! Man supposedly evolved from aquatic creatures that crawled onto land and morphed into animals! Who intervened?

An even more dynamic example of design is the following…

5. The **neural structures in the Brain**, like **synapses** show design because again the two structures do not touch but are nanometers apart – intended as **a gap for the neurotransmitters to jump across** – based on electrical potential mitigated by such things as Calcium ions (Ca+) and Acetylcholine – and the result is the transmission of signals in the brain.

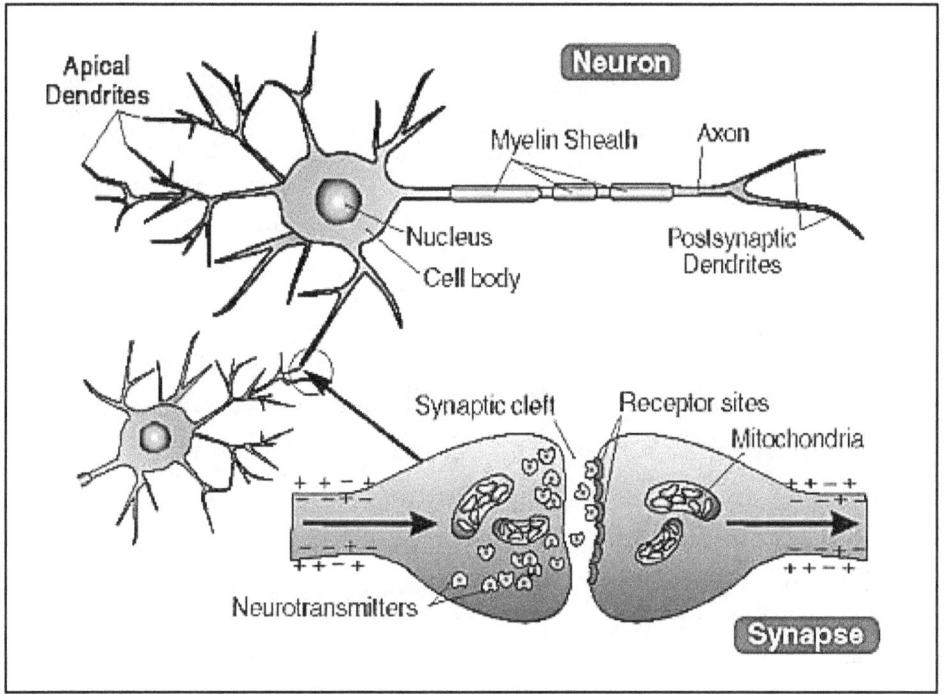

(credit: Bing Images: lookfordiagnosis.com)

How was it 'decided' that there should be a gap between neurons (brain cells) when it was discovered by somebody or something that axons/dendrites can **detach from one neural synapse and <u>reattach</u> to a different one?**

6. Lastly, there is the miracle of **human procreation** – where two organisms come together with what is needed to make a baby – and neither of the two (the male nor the female) has the complete 'package' of genetic material to create the baby. More than that, how is it that the male organ just happens to fit the female in about the same location on the body – so that the two can see each other and communicate – why not mate in her armpit, or why weren't the anus and the vaginal openings switched?

The genetics of the sperm and ova come together as if by design and the sperm is designed to penetrate the egg…and after that penetration, the egg closes off so that no other sperm can enter – and there are thousands of them vying for the ova. And then replication begins – from the one fertilized cell, a billion-celled, multi-organed human is formed … and all the parts are integrated <u>and work</u> (barring some genetic misfire of the DNA).

It has always amazed me that the **human face** is formed in two separate halves, as it were, and they align and form a harmonious symmetry. How does it do that (and what goes wrong when some people's eye are on a markedly different level… as if their face is crooked?)

7. And in a related genetic theme, there are **223 genes** in the human that are not found in any other animals on Earth (see Chapter 8) – so despite a "missing link" how does Evolution explain that? The 223 genes would have come into being <u>after</u> the so-called missing link, and no animal <u>before</u> the missing link has them, so obviously, they were added to the current version of Man…

And yet Science will say things like **Walter Gilbert** [71]   said:

> **We haven't been able yet to determine in terms of genes what makes a human being a human and not another animal.**

And while the original finding of **223 genes** (*Science* issue No. 291) was said to be found without implying that they were due to bacteria (i.e., said to be **panspermia:** meteors bringing bacteria with DNA to Earth), and the journal *Nature* (issue No. 409) found that **these genes involve important physiological and cerebral functions**, so the Science community immediately worked to discredit the finding (in 2001) by attacking what was implied by Sitchin [19] **and others:**

> Rather, they suggest that some proteins may appear to exist only in humans and bacteria <u>due to the loss of genes</u> among non-vertebrate species and the failure to detect human and bacterial genes that are present but mutated in nonvertebrate species….
>
> [The principle:] The **transfer of genes horizontally, or laterally,** between species is a well-documented phenomenon in nature. Bacteria can transfer genes to other bacteria. And mitochondria, [energy] organelles in human cells that once were free-living bacteria, have transferred genes to humans.
>
> But there is **no strong evidence that bacteria genes have been transferred to vertebrates,** say many evolutionary biologists. It

was therefore big news that more than one hundred human proteins are likely to have "entered the vertebrate (or pre-vertebrate) lineage by horizontal transfer from bacteria," as the International Human Genome Sequencing Consortium proposed in *Nature*. [72]   [emphasis added]

The implication was that these 223 genes were inserted in a form of **"assisted evolution"** – and readers of my books know that means that the Anunnaki did it. [73] If bacteria inserted these 223 genes to humans, why was it so localized that no other vertebrates got them? What if the extra genes (whatever the actual number, 223 has been disputed) were not due to "lateral gene [bacteria] transfer"?

But there is a problem with "lateral gene transfer" and the scientists admit it:

The process of lateral gene transfer involves a series of steps. For starters, a gene has to migrate somehow to the nucleus of the 'germ' cells that give rise to sperm and eggs. Otherwise it will not be passed on. How a gene successfully infiltrates the human genome is not clear, say researchers. Unlike bacteria, **humans have relatively sheltered DNA**. Further, **a transferred gene must be in or get in a format that allows long-term maintenance and replication.** [74]   [emphasis added]

And if the Earth is flat, or a VR Sphere with an energy field quarantine, there is a Firmament of some sort over the Earth (the *Gegenschein* proved it) it – so how would a meteor get through the barrier to spread its cargo of bacteria?

This is one reason the scientists do not like the FE theory with its Firmament. They have to have Earth as an open sphere, a rock spinning through space, so that anything can happen. Evolution uses a similar syllogism -- it needs a lot of time for something to happen – therefore Earth is 4.3 billion years old, Man evolved from the Ape, and there is no God.

I realize that this is a long segue from the discussion of the many historical scientists (several pages back) who promoted the **Heliocentric Model**, but it is very important that the reader have this understanding of Design and Creation such that when reading and evaluating the idea that "Science can explain everything" and they say that the Earth has been found to be a round sphere, the reader will question and look at Chapters 3 and 9 with more of an open mind... Science does make errors, and Copernicus, Newton and Darwin did make some big ones.

The foregoing 7 are just a few of the items that appear to be "designed" and that suggests that Somebody had an overall vision of what was the Plan and then built (created) the eye, the ear, and the brain cells for the greatest flexibility and function.

But let's move forward... back to the scientists

The next Scientist deserving a close review is connected with the last chapter – **Gravity**. It was questioned what holds the <u>heavy</u> ocean water to our alleged round sphere, and the standard answer given by Science is that a combination of **Gravity and Centripetal Force** hold it to the Earth (if the Earth rotates!). Chapter 11 in QES went into more detail on how Gravity does <u>not</u> work, but it will be briefly re-examined here and expanded.

Sir Isaac Newton & Gravity

Newton speculated both Gravity and Centripetal Force (CF) into existence, and **admitted it was all a conjecture**, and today's scientists still blindly use it. What was said about him in QES Ch.11:

> **Isaac Newton** went a step further, [regarding Gravity] and said, "allow us, **without proof, which is impossible**, the existence of two universal forces – centrifugal and centripetal, or **attraction and repulsion**, and we will construct a theory which shall explain **all** the leading phenomena and mysteries of Nature." [75] [emphasis added]
>
> Sir Isaac Newton himself does not even attempt to give one proof of the truth of Gravitation; with him it is only *supposition* **from beginning to end**.
> --- Sir Robert Ball, Astronomer Royal for Ireland. [76]

Knowing that **Newton got ½ of his inspiration from Copernicus**, it is appropriate to add what noted, learned men of the time said: [77]

> **Lord Bacon** was completely opposed to the Copernican System as there was no proof for rotation of the Earth in Nature... "let it be asked whether any such motion be found in Nature, or whether it be not rather **a theory fabricated and assumed** for the convenience and abbreviation of calculation..." [many learned men distrusted the burgeoning field of advanced mathematics suspecting that old axiom: 'figures don't lie, but liars figure!] [emphases added]
>
> **Goethe** was a bit stronger: "....where can the man be found, possessing

the extraordinary gifts of Newton, who could suffer himself to be **deluded by such a *hocus-pocus* [of gravitation]**, if he had not in the first instance willfully deceived himself? .... **Self-deception**.... sometimes trenches on **dishonesty**.... To support his unnatural theory Newton heaps fiction upon fiction, **seeking to dazzle where he cannot convince**...."

Goethe also added: "In whatever way or manner may have occurred this business, I must still say that **I curse this modern theory** of Cosmogony, and hope that perchance there may appear... some young scientist of genius, who will pick up courage enough to upset **this universally disseminated delirium of lunatics**."

Dr. Woodhouse, Professor of Astronomy at Cambridge, doubted what he was supposed to be teaching: "However perfect our *theory* may appear in our own estimation, and however simply and satisfactorily the Newtonian *hypothesis* may seem to us to account for all the celestial phenomena, yet we are here compelled to admit <u>the astounding truth</u>, if our premises be disputed and our facts challenged, **the whole range of Astronomy does not contain one proof of its own accuracy**."

When Copernicus was challenged to defend his theory, he confessed that the system of **a revolving Earth was only a *possibility*** and **could not be proved by facts**. And then **Copernicus wrote** (and this is a doozy):

> **It is not necessary that hypotheses be true or even probable**; it Is sufficient that they lead to results of calculation which agree with calculation [what?]... Neither let anyone, as far as hypotheses are concerned, expect anything certain from Astronomy, since **that science can afford nothing of the kind**.... The hypothesis of the terrestrial motion *was nothing but an* **hypothesis**, [and get this] valuable only so far as it explained phenomena not considered with reference to absolute truth or falsehood. [78] [emphasis added]

So why do it? Who is he kidding? Hypotheses that do not bear on truth <u>or</u> falsehood are... nonsense. Yet such was promoted and then Newton stated it as a fact... which he could not prove (and admitted it). This is the work of lunatics, as Goethe said.

[Newton's] work brought him into great repute as an astronomer, and afterwards led to his being made Master of the Mint and **Knighted** [as **Sir** Isaac Newton].... He spent a long life in teaching a false system of Astronomy, **unsupported by any fact in Nature**... [as well as contradicting the Bible – but then, that was the point wasn't it?] [79]

Sheesh – How scientific was that?! Newton was a thinker and a great mathematician, but he was also an **Alchemist and a Kabbalist** -- which is where he got his other ½ of the idea of Gravity idea from.   And so did **Pythagoras**.

> **Gravity is not supported by any scientific proof.** What then was the source of the theory? S. Pancoast reveals that **"the law of attraction and repulsion" in the Kabbalah** was popularized under the name "gravity" by Isaac Newton....
>
> He [Pythagoras] was never permitted to declare publicly what he knew and believed, but taught his immediate pupils all the wonders of his philosophy, under the most binding obligation of <u>secrecy</u>. Pythagoras was forbidden to divulge this knowledge because it would reveal the law of **attraction and repulsion**, which constituted one of the great secrets of the sanctuary. Over a millennium later, Newton was led to the discovery of these forces by his studies in Kabbalah. [80] [emphasis added]

We are not talking about magnetism and its attraction and repulsion. Newton's theory is founded on the premise that all objects are attracted to other objects <u>based on their mass.</u> [81] And **this is nowhere observed in the natural world**.

> There is no example in Nature of a massive sphere or any other shaped object which by virtue of its mass alone causes smaller objects to stick to it… There is nothing on Earth massive enough that it can be shown to cause even a dust-bunny to stick to it… Try spinning a wet tennis ball or any other spherical object with smaller things placed on its surface and you will find that everything falls or flies off, and nothing sticks to it…
> To claim the existence of a physical "law" without a single practical evidential example is **hearsay, not science**. [82] [emphasis added]

Apples fall to the ground because they are heavier /denser than air. And that mass is also what keeps people 'stuck' to the ground. When Science 500 years ago said that the Earth is round and spins (at 1000 mph no less), the public wondered what kept the oceans and people from flying off into space. Isaac Newton saved the day with his <u>Theory</u> of Gravity.   (See **Appendix D** for more complete review.)

Again, the public asked, how that could be so when they all knew that a spinning object would fling things off into space? Again Newton came to the rescue and said that while the Law of Centrifugal Force does throw things off, it is counter-acted by the Law of Centripetal Force. The public gave up. Just invent Laws to explain theories…

Gravitation is only **a subterfuge** by Newton to prove that the Earth revolves around the Sun.[83] The biggest hint we have of that is that **today's Quantum Physicists still cannot find Gravity, neither as a force or a particle**….maybe because it doesn't exist…?

But, didn't **Einstein** (examined in the next section of this chapter) say there were 4 basic forces in the Universe: the Strong force, the Weak force, Electromagnetism, and Gravity? So, doesn't Einstein reinforce Newton's theory?

> What most people don't know is that Isaac Newton was a religious mystic, who was very well studied in Judaism, and Jewish texts….
> Indeed, one writer has revealed that "7500 pages of his [Newton's] theological speculations, written in his own hand, are digitized at Israel's national library at Hebrew University" ….
> It was **Abraham Yahuda** who obtained and maintained Newton's [7500] theological writings….
> **Yahuda was a contemporary of Albert Einstein and conferred with Einstein about Newton's mystical religious writings.**
> Religiously Einstein had much in common with Newton, as both Einstein and Newton believed in the reality of mysticism. [84]

So what does that mean? Simply that Einstein was impressed by Newton's writings on the Talmud and Kabbalah, and both assumed there was a mystical force called Gravity – and again today's scientists are chasing a 'red herring.' **Newton had hidden his theological papers in a box, so they were not to be found, because he did not want them published in his lifetime.** Yahuda found them and shared the information with Einstein – who did not want them published, either.

And then there is this revealing statement (in the same document quoted above) that reveals why Einstein did not want Newton's papers published:

> Why was Einstein so passionately hopeful that Newton's theological writings **not** be published? Because **those writings would give up the game.** Isaac Newton's writings on theology would expose for the world the theological source for his theory of gravity, the Jewish Kabbalah. [85] [emphasis added]

So the history of **Gravity is not scientific at all**, and was just an assumption by Newton, based on his Kabbalistic studies, that the "attractive force" had to exist. Einstein picked that up and inserted it into the Four Fundamental Interactions. [86] How scientific was all that?

> Sir Isaac Newton's study of the Kabbalah is the real source for his theory of gravity….[along with the] Christian Kabbalists such as

Newton, Dee, Kepler, Shakespeare, Cardinal Nicolas of Cusa, and a long list of Theosophists, Rosicrucians, Masons, Crowleyites, etc.... . Newton, in short, insured the acceptance of the Copernican Model, <u>for two centuries</u> with his arcane 'mathematical' concepts; concepts upon which others – **Einstein** through Sagan, *et al* – could erect today's Pharisee Cosmology. [87]

So the point is: **Gravity was not based on Science but on Kabbalistic Theology,** and while theology does have some merit, it would be wise to question Gravity as a concrete Law -- especially if Earth is a Simulation (regardless of shape).

Lastly, regarding Newton, **Sir John Herschel** said:

Newton's theory of **the globularity of the Earth is only *supposition* and *assumption*** and yet by Modern Astronomers it is paraded about as if it had been a true deduction from exact experiment. [88]

While some may argue that these men of renown (quoted above) were resistant to new discoveries, and just *de facto* rejected Gravity and the Globularity of the Earth, that is not the case. Newton's ideas were given due scrutiny but he could offer no proof... so they were rejected as guesswork.
The real mystery is **why the fraud continued for so long,** and was accepted by supposedly other rational men of Science... and the most reasonable answer was presented in the earlier section entitled, Anti-God Sentiment.

The other question is: Why did the Church give up trying to refute and stamp out what they considered **a major, serious heresy**? The Church was never convinced of the correctness of the Globularity, Rotation and Solar Orbit of the Earth... so why did the Church not hold something like a Vatican I to present arguments for and against, and stamp out the assumption by the early 1800s? It would appear that that answer was given earlier in the section entitled, Germany Influences Heliocentrism (viz., the Illuminati were behind it.) Was the Church influenced by the Illuminati?

And while it is not a conspiracy *per se*, it is an **agenda** that has been promoted for almost 200 years... which requires an <u>organization</u> dedicated to a higher purpose.... Such as the New World Order in which the public is kept entertained and unaware of what is really going on. (See Chapter 10, Project Bluebeam.)

Further Gravity Insights

So if Earth is a round sphere, and Gravity is holding the water to the planet, and Centripetal Force (CF) "tends to hold it to the Earth" because it is allegedly **spinning at 1000 mph** (according to Science – the rotation is what creates the passage of time and seasons) …. How does all that work? Wouldn't Centrifugal Force tend to throw the water off the planet? Oh, but Science now says it is a balance of the three forces that holds it all together.

> As a point of reference, the Flat Earth Theory (FE) does not need Gravity – things fall to Earth because they are heavier than air.

> BTW: The Earth is not rotating at 1000 mph – proven by cannonball experiment in QES, CH. 11. **No Coriolis Effect**.

Come on guys! How many different Laws are we going to invent before we realize that as they proliferate, they start to contradict each other – like Centripetal Force and Gravity. Science is inventing (oops, "discovering") new Laws all the time because they have to make the Spherical Earth concept work.

Ok, let's look at this issue one more time. What you think you know about Gravity is going to be further shaken.

If **Gravity** is strong enough to hold the oceans in place, why is it a bird can fly above the ocean and not be sucked into the water? You say it is because the effort of its wings keeps it aloft… well, then what about when the bird is gliding? And what about a fog given off by the ocean… how can that be if Gravity pulls things down?

Oh, I see, you say it is due to the **Mass** of an object. (More invented answers from Science, although Mass of an object is real, blaming it on Mass is not completely correct.) The ocean's Mass is greater thus Gravity has more to grab onto and keep it pulled to the Earth, and the bird has not much Mass… But it does have enough Mass so when it stops flapping its wings, it falls.

Then Science says, Oh, well, it has to be that the forward momentum of the bird, coupled with **Lift** keeps it afloat while gliding in the air…. And it goes on and on, inventing new reasons for each objection…. (and while Lift does exist, it is not totally why the bird flies over the ocean – with ease).

Ok, what about Science telling us that Gravity is **uniform all over the planet**? And if it is strong enough to hold the **heavy** curved ocean water (water is never at rest when it is curved [convex]), then why can a 200 lb. man walk on the ground <u>next to the ocean?</u> …or anywhere else for that matter.

The ocean is <u>several miles deep</u> at many points and that much water weighs <u>millions of tons</u>…. Gravity has to be so strong to hold millions of tons, but a 200 lb. man can walk on the beach with ease?

Clouds are composed of millions of gallons of water – they are also heavy…but they float?

And a biggie that we are coming to: **Gravity** is a different issue for both scenarios. The VR Sphere needs it, and the FE doesn't.

By the way, the VR Sphere as a Simulation simplifies it: the oceans are **programmed** as part of the very advanced holographic Simulation to stay where they are! I like that one; Simulation is what was given to me (in VEG in 2008).
Back in this book's Chapter 3, did they not find "self-correcting code" among the Quarks and Strings? What else does the "code" do?

Playing Devil's Advocate, if the Earth is a Flat Earth <u>and</u> a Simulation, the oceans could also be similarly programmed. So when my Source told me it was a Simulation (VEG, Ch. 12) I still could not tell.... FE or VR Sphere... The answer is coming...

Gravity Analysis

Let's expand the original discussion in QES Ch. 11 and see why **Gravity is very speculative** as the answer.

Let's consider that the Earth is a globe, spinning, and circling the Sun, as Science says. There is a problem with this scenario and it involves a not-so-obvious problem with the interplay of Centrifugal Force and Gravity. (Most laymen do not know or think about the following.)

The **Flat Earth Theory says there is no Gravity**. So, if the Earth is a round ball, then what keeps things like the Oceans from falling off (due to Centrifugal Force) a round Earth? So I have to admit, that part says that Gravity must be joined by something (even Centripetal Force) ... but the real issue is: in places <u>the ocean is several miles deep and</u> that amount of water (i.e., **mass**) weighs millions of tons... How can the same Gravity Force 'regulate itself' to alternately hold millions of tons one moment, and just 200 lbs. the next? Try easily lifting a 5 gal. jug of water... now multiply that by a mile deep water's weight.... what is keeping that water 'stuck' to the Earth?? **Repeat**: And if Gravity is that strong, **and the same all over the Earth**, how can a 200 lb. man walk on the ground?

Oh, you say Gravity is offset by the Centrifugal Force, or maybe Centripetal Force gets involved? Wrong. You really need to see this.

Repeat: If the force of Gravity (with Centripetal Force) is strong enough to hold a two-mile deep ocean to the Earth, how can <u>we</u> even walk on the Earth? This has never made any sense. **Gravity is said by Science to be uniform all over the Earth – the same strength everywhere.**

So again you say, it has to do with **Mass**…. Ok, go to the top of the Leaning Tower of Pisa and drop a cannonball and a book at the same time. They both hit the ground at the same time…. Yet the cannonball is more dense (metal mass) than the book (paper). So tell me again it is due just to Mass.

Let's look at this Gravity and Centripetal Force issue another way…

Rewind: Gravity & Centrifugal Forces

So I initially bought the FE idea that Newton was wrong and there is no Gravity….and that there is the Centrifugal Force (if the **Earth is spinning 1000 mph**) which tries to fling us off the planet. Repeat: Supposedly (according to Science) the Force of Gravity and the **Centrifugal Force (CF)** equally balance each other -- <u>at the equator</u> where CF would be the strongest…. and **CF is lighter near the poles**… so is the Force of Gravity variable all over the Earth, or is it a constant -- as Science says?? Yet **CF varies** -- so a Man walking in Northern Canada has no more difficulty walking than a Man in Egypt, near the Equator. So **CF and Gravity are not balanced.**

We accept that Gravity doesn't vary as CF does, because less Gravity at the poles would mean that ships and men traversing the Arctic, or even the alleged Antarctic Continent, would be flying off into space… Gravity would not be as strong and not hold them as it does at the Equator – so putative Gravity is uniform. CF is not.

Rewind: 1000 Mph Spin

But Science has also assumed a rotating planet for Earth, allegedly rotating at 1000 mph. As Ch. 11 in QES showed:

> If the Earth's circumference is 25,000 mi and it takes 24 hours to do a full planetary rotation, then simple math says a Spherical Earth is rotating about **1000 mph** – also verifiable by the US timezones -- the Sun moves 1000 miles in an hour : the US is 3000 miles wide and we have 3 timezones, with a 3 hour difference between coasts, so the Sun covers 1000 miles in an hour. And since we "know" the Sun is not moving (ahem!), it must be the Earth itself that is rotating at 1000 mph…. Can you feel it?

> Since the clouds are not attached to the Earth, how do they move 'along with' the Earth as it rotates at 1000 mph?! Why does a leaf falling from a tree fall straight down and not a mile off to the side due to the high speed rotation of the Earth?

Secondary hint: if the speed of sound is 768 mph, would we not have trouble hearing an explosion 1 mile west of us – because the rotation of the Earth (from west to east) 'outruns' the sound? You are a mile away – moving at 1000 mph, sound moves at 768 mph.

See, this makes no sense... unless the Earth is Flat and things stay put because **they are heavier than air**, and there is no ocean on the bottom half of the Earth globe (to fall off).... See, if CF is less at the 30th latitude than at the Equator, <u>AND if Gravity is constant</u>, people should find it harder to walk in Canada than they do at the Equator because while CF is lessening in Canada, Gravity stays the same.

In addition, the FE Theory says the Sun moves <u>above </u>the Earth – and that was shown by the earlier 360° time-lapse picture (Chapter 3):

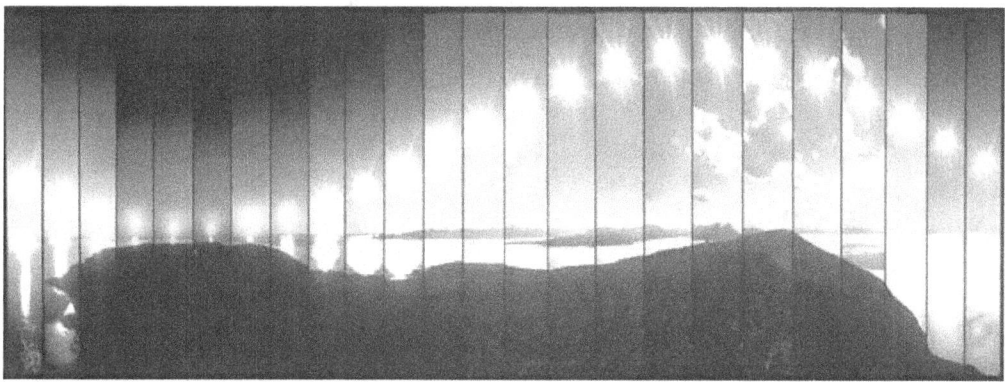

The Earth wasn't rotating, the photographer had to move his camera every hour to follow the Sun.

To me, **this Gravity - CF issue is more evidence <u>for</u> the Flat Earth (FE)**...

Gravity (if it exists) would be constant on a VR Sphere, and there is no danger of water falling off the Sphere, but since the VR Sphere <u>is not spinning</u>, there is no CF to consider... K.I.S.S. Use Occam's Razor.

And don't forget, Newton just a few pages back, said we are <u>assuming</u> the existence of Gravity and "it is impossible to prove." Why are today's scientists vainly trying to prove the existence of Newton's assumption? (The real nature of Gravity is beginning to emerge: **Appendix D**.) Scientists who spend years getting their doctorate in Physics do not want to hear this... "OMG! I spent all that time and $$ for nothing?"

However, something resembling CF does exist -- get on a child's **Roundabout** at the playground, get on and start it spinning, then throw a beachball --not only do you get the **Coriolis Effect** (the ball <u>curves</u> away --not in a straight line) -- and unless you hang on, there is a tendency for the CF to throw you off the Roundabout.... there is <u>no Centripetal Force</u> trying to keep you on, any more than there is on the allegedly spinning Earth.

Further, **Gravity**, as postulated by Newton, may not exist. Newton used the example of the apple falling from the tree to demonstrate a principle that **even today's physicists cannot find, nor prove**. **Gravity** remains elusive because it is not a force. FE says it probably does not exist. (I know that sounds weird, bear with me… I am about to show you what is behind their thinking, and it is a tough nut to crack.)

Rewind: Gravity Illusion

> The problem with the gravitational theory is that ….the gravitational attraction to the earth of all persons and objects **remains the same at all places on earth.** That means the gravitational force at the North Pole is the same as the gravitational force at the equator. That poses a very real problem if the earth is rotating at 1000 mph as alleged….
>
> **Centrifugal force** is allegedly "perfectly <u>balanced at the equator</u> by the force of gravity"… and yet decreases every mile toward the North Pole [where it would be zero] …. Thus as one approaches the North Pole, the force of gravity [less CF] would crush a person, which of course does not happen, so the spinning earth and the mystical force of gravity are thus proven to be **preposterous fictions**. [89]   [emphasis added]

If the apple, when ripe, naturally falls to the Earth because it is <u>heavier than air</u> (not due to Gravity), why doesn't smoke from a chimney also get dragged to the ground? <u>Because</u> the smoke is lighter than air and rises so that "gravitational force" is not even a factor in whether something falls or not.

Thus, the <u>Force</u> of Gravity does not exist (Appendix D) and **both Newton and Einstein were wrong**. And Einstein for his part, was a real character… very good in mathematics according to his university professors (who also called him a "lazy dog"), but not so hot in theoretical Physics.

Albert Einstein & Light

Einstein has been thought a genius and it now appears that he was of average intelligence. So calling someone an "Einstein" when they do something clever or solve something requiring serious thinking, is a misnomer... but I guess we have to have our heroes.

$E = MC^2$

While Einstein is reputed to have been a genius, and this formula is attributed to him, does anybody see the **inherent nonsense** in it? The formula says that mass at the speed of light (c), **squared**, becomes energy. It doesn't take a PhD in Physics to see that squaring the speed of light is an incredible speed – so fast it is hard to grasp: 34,596,000,000 miles per second.

The usually stated speed of light is **186,000 miles per second**. Fast. Earth is allegedly 93 million miles from the Sun. So, light from the Sun reaches Earth in 8.3 minutes. Now it sounds slow, but that's an aspect of relativity for you.

Einstein also theorized that no object could go faster than the speed of light as it would become energy; mass would increase with the speed of light, and convert back into energy.[90] Other scientists do not agree. "Gravitational waves" are said to be faster than light (Ch. 8 in VEG) but have not been found either.

Patent Clerk

Einstein worked as a patent clerk, reviewing and approving patents submitted for new scientific inventions or concepts. One such came from an **Italian physicist** who wanted to patent the idea that Mass at the speed of Light becomes Energy, or $E = MC$ was the original formula. Einstein liked it and rewrote it several years later as $E = MC^2$ – squaring the speed of light which is absurd, it is such a large figure that it has no meaning.

> Now the interesting part. The mass to energy conversion formula was **not** developed by Einstein, but came across his desk as a patent clerk examining the work of an Italian physicist, **DePretto** in 1904.[91] In fact, according to professor C.L. Kervan, "...**it is a mistake that matter can be transformed into energy**.... The nucleons do not disappear, but are found in the fission products. If some neutrons are expelled, they are not destroyed. For matter to disappear, it must be opposed by **anti-matter**."[92] [emphasis added]

So Einstein took DePretto's formula and added a superscript '2' to it... but according to Dr. Kervan, **Einstein is still wrong.**

Let's keep in mind that Einstein did have a background in physics and mathematics, and was pretty good at math. He worked as a scientific **patent clerk** and it was his job to examine in-coming patents for electrical and electromagnetic patents, so he had exposure <u>for years</u> to many new ideas in these fields. His own work in the field of **photoelectric phenomena** did earn him a **Nobel Prize in 1921**, so he wasn't a total dummy, but it has been suggested that he 'borrowed' a lot of his ideas from others' patent applications **and that is why when challenged to defend his ideas, he rarely did** – and he let others speak for him. [93]

It is strongly suggested that he did not debate or answer because **the ideas were not fully his**, he just liked them, and failed to answer Walter Ritz, Georges Sagnac, and Ernst Gherkin, to name a few. [94]   In addition, Paul Gerber called him an **outright fraud** and his long-time teacher in math and physics called him "**a lazy dog.**" [95]

We may have been idolizing a man who was just clever at postulating physics concepts, but was **not the genius** he was reputed to be. He was probably the **"poster boy"** to give credibility to the fledgling field of Quantum Physics around 1905, at that time 5-10 years old, and we all need heroes....

Relativity & Equations

According to **Ben Rich**, who headed **Lockheed Martin's Skunk Works** (where many exotic aircraft were developed), and where they successfully created **electrogravitic craft** (TR-3Bs) that fly, he explained the issue with Einstein, saying

> There is an **error in the [traditional] equations** [dealing with space travel and the speed of light] and we know what it is and we now have the capability to travel to the stars.... We now have the technology to take ET home.... [and] it won't take someone's lifetime to do it. ....they had, for example, determined that **Einstein's equations dealing with relativity theory were incorrect**.... [and he] went on to say that they had ***proved*** [sic] **that Einstein was wrong.** [96] [emphasis added]

Einstein's **General Theory of Relativity** (GTR) has not been very useful in the laboratory because it has not led to the construction of any technological devices that use it. It has not even been updated in the last 100 years "...not because it explains everything perfectly, but because <u>it cannot be developed any further</u> to encompass newly emerging fields." [97]   But **Subquantum Kinetics** (SQK) will replace Quantum Physics... see **Appendix D**.

Recent discoveries [Tesla and SQK] show that there is some new form or "incarnation" of **gravity** – **dark energy**, which manifests itself in the form of a repulsive gravitational force, repelling galaxy clusters …. In fact it represents some three quarters of all the energy equivalent of the Universe. Although Einstein introduced a constant, known as the **cosmological constant**, which is supposed to explain this, there is nothing which could indicate that he himself understood its nature [he didn't even know it existed]….

Einstein did not accept that **gravity may be negative**…

The comparison between "normal gravity" and "dark energy" in the Universe is such that Einstein's theory really describes only around 10% of known gravity….

It appears that this negative gravity is not directly associated with mass because it is literally everywhere…. The only credible… explanation of this fact is that the only possible source …. are the quantum fluctuations of space-time itself (spin of these quanta?). **This is something completely different than what Einstein's equation describes**. [98]

And further analysis of Einstein's work has been done regarding **Gravity** and they now are saying:

…. in my modest opinion, astronomical discoveries also tell us that antigravity [above: 'negative gravity'] is not some exotic interaction…. as physicists imagine. It seems that it [gravity] is a manifestation of some fundamental aspect of nature, and should not be as difficult to generate as we generally think. **We do not really know what it is – and this is the problem.** [99] [emphasis added]

One of the major points of this chapter – **Newton made an assumption and today's scientists are still looking for it.** As was said earlier, when you are examining the **IMAX Theatre** and assume it to be Reality, and build CERN to prove a supposition, problems can arise – and money will be wasted.

According to the theory's [GRT's] predictions, **gravitational waves** should exist. Numerous research projects have been initiated lasting for over a decade, employing various sophisticated technologies, but **not even the slightest trace of such waves has been detected. Nothing at all, which suggests they do not exist.** Therefore the general picture [of gravity] is not as clear as textbooks say.[100]

That is the whole point – and the above quote is from 2013. How many more decades are we going to continue this futile attempt to find the **Force of Gravity** when it does not exist? (See Appendix D) And it is true that a scientist's job is to theorize and many times those theories are overturned, but at least the scientist tried. Yes that is part of the scientific method… my objection to Einstein <u>nowadays</u> is that we still laud him, and his theories have either been disproven or are seriously questioned. As was said, he would never defend his theories – because they weren't his! He was the "poster boy" for the fledgling field of Quantum Physics and as such he was the spokesman for theories given to him by Rosen, *et al.* That way if the theories were wrong, Einstein took the heat and not the tenured academics who gave him their ideas.

Nonetheless, Einstein was clever, and he did earn a Nobel Prize in 1921, and in his defense the following is considered very perceptive, much to his credit:

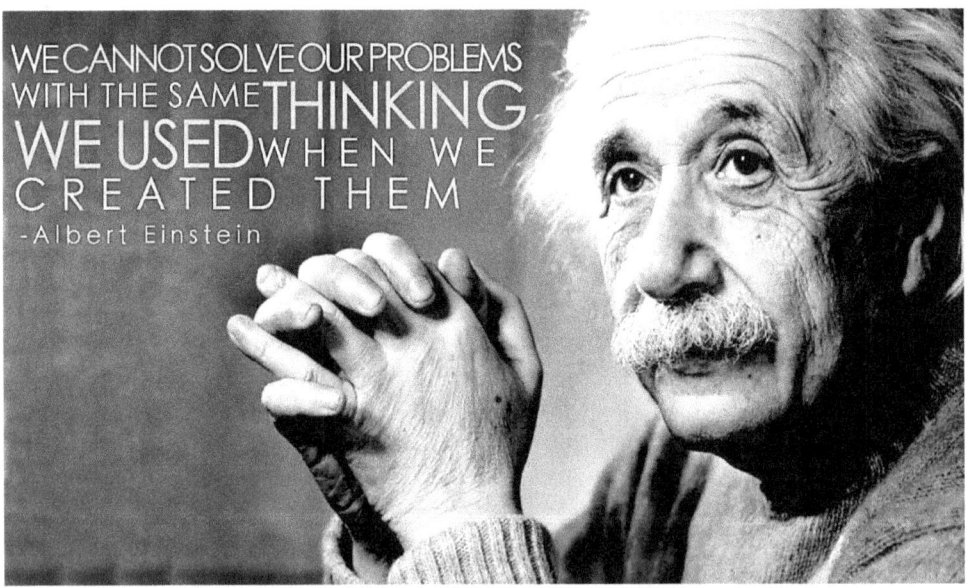

Einstein and his contributions are examined in greater depth in VEG, Ch. 8.

Charles Darwin & Evolution

Charles Darwin and Einstein had one thing in common – they both exaggerated their calculations so that their theories could work. Einstein more than doubled the speed of light such that Mass might become Energy somewhere within that gynormous speed, and Charles, went along with the 4.3 billion years for pond slime to evolve into humans because he felt that given enough time it <u>could</u> happen. Both were still wrong.

Entropy is what happens to biological creations when they are created, then left on their own, <u>and</u> there is no further energy input into the creation.

Entropy is the enemy of Evolution

Logic is also the enemy of Evolution:

    If Apes evolved into humans, why is it not still happening?

and

    If lightning still hits ponds/bogs/oceans around the planet, why are new life forms not still appearing?

Darwin was the victim of a simple mentality:

    Since humans resemble Apes, they must have evolved from them.

So dogs evolved from cats, and because the quadruped form of a dog resembles a horse, horses and dogs must be related... and cows, and maybe even dinosaurs?

You get the point... he didn't.

But he was correct about **Natural Selection** and **Survival of the Fittest**. Today that branch of science is called **Epigenetics.**

Today's younger scientists are beginning to question the Darwinian Theory of Evolution especially as they are finding evidence of a young Earth, and DNA intricacies that suggest an **intelligent design**. Even if a bolt of lightning hit a stagnant pond and created amino acids (the building blocks of life and DNA), unless something is done with them, to keep them living, growing and stabilize them, a few days later, they die. And lifeforms on Earth have a much more intricate design than random evolution would generate – for example, **why does vision see color**? To make an organism so that it can see its surroundings, an eye would be necessary, but (1) why were eyes needed? – bats and salamanders in caves do not have eyes, and (2) **why color** and not just black and white? Some animals do not see color.

Charles would not consider these things and he fought his colleagues vehemently when they suggested he take a look at **the human eye – it shows design** … if 4 parts are not there, there is no vision.

The Human Eye

Three of the four key elements are: the **iris**/pupil (shutter), the **lens** (focuses light on retina), and the **muscles** that allow the eye to adjust quickly.

> In the nineteenth century, the anatomy of the eye was known in detail.… different colors of light, with different wavelengths, would cause a blurred image, except that the lens of the eye changes density over its surface to correct for chromatic aberration. These sophisticated methods astounded everyone who was familiar with them. Scientists.… knew that **if a person lacked any of the eye's many integrated features, the result would be a severe loss of vision or outright blindness**. They concluded that the eye could only function if nearly intact. [101]   [emphasis added]

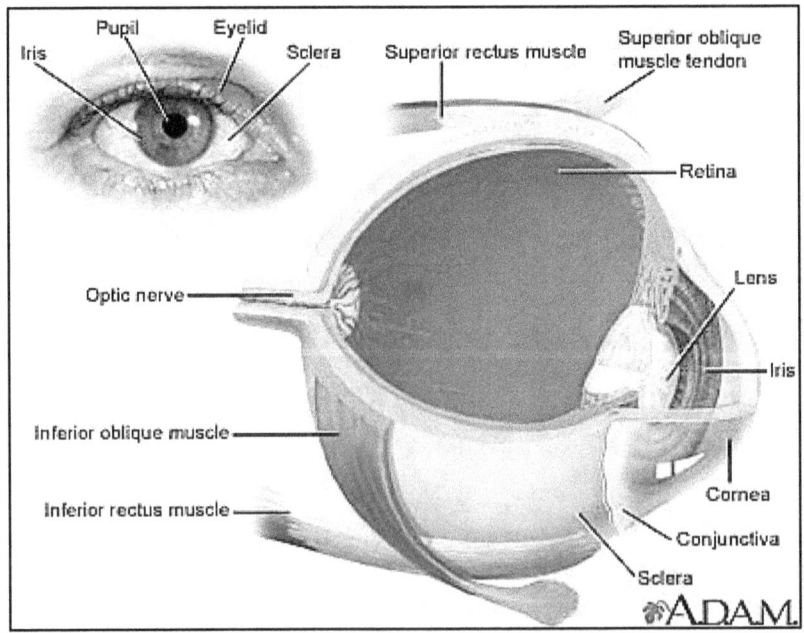

(source: http://apps.uwhealth.org/health/adam/graphics/images/en/1094.jpg)

More than that, the **4th element**, called the **retina**, consists of a patterned dispersion of rods and cones which appear to be designed to facilitate "spatial acuity." [102]   The **rods** facilitate vision at low light levels and the **cones** (3 types, no less) determine color and facilitate high spatial acuity. **Why would two types 'evolve'?** For what purpose? How would the differentiation 'know' to happen?

There are more rods in the eye than cones, and both types are clustered more densely at the back of the eye with the highest cone-clustering (150,000 cones/sq mm) right at **the fovea**, where the eye's main focal point is. There are less rods and cones as one moves from the back of the eye toward the front. There are 150,000 rods per square millimeter – more densely packed than their digital camera counterpart, pixels. In addition, the structure of the individual rods and cones shows **design**, wherein the cell membranes fold in and create multi-layered photoreceptor cells, and each has photopigment molecules. [103]

Interesting that other species also have rods and cones in their eyes, but their physiology is different – wouldn't nature have evolved a simple photoreceptor that worked and then kept that same design? And how would nature **evolve** a photoreceptor cell – how would it know what "vision" was and how to "evolve" to it from a simple cell? Darwin would not initially look at those facts.

And through it all is a **5th element**: a link to the **optic nerve** transmitting the 2-dimensional image which is upside down on the retina to somewhere in the brain, allegedly 'righting' the image by the way, so that we can see things right side up. And according to today's quantum physicists (Pribram and Bohm), if the world around us really consists of **holographic frequency codes**, that makes the brain a "frequency decoder" and we may be projecting our perceived reality "out there," instead of seeing it somewhere "in" the brain. [104] And that is not Evolution but **Design**. (VEG Ch. 12 and Apx. B.) Also see further analysis of Vision in Chapter 9.

Darwin Semi-Confesses

After quite some years, Darwin still sought support for his Evolution of Man, despite his scientist friends advising him that he was in error. And Darwin knew something wasn't 100% solid with his Theory of Evolution, and he said as much in The Origin of Species:

To suppose that the eye with all its inimitable contrivances for adjusting the focus to different distances, for admitting different amounts of light, and for the correction of spherical and chromatic aberration, could have been formed by natural selection, seems, I freely confess, **absurd** in the highest degree. [105]

Nonetheless, he then proceeded to try and figure out how the eye might have evolved <u>anyway</u>. Man always hates new, original breakthroughs that challenge his established ideas (Chapter 2). It is called **cognitive dissonance**. So he wrote to Asa Grey, a famous Harvard professor, for encouragement. Instead Asa wrote back that he seriously doubted that natural processes could explain the formation of the eye, and that no part of the eye is of any use without all the other parts. [106]

Bombardier Beetle

Another killer evidence for Creation was the **Bombardier Beetle**: inside its abdomen are two sacs of fluid which mix just before being expelled from the anal part of the beetle – a hot fluid created by chemical reaction once the two fluids come together. It is very effective for deterring the beetle's enemies.

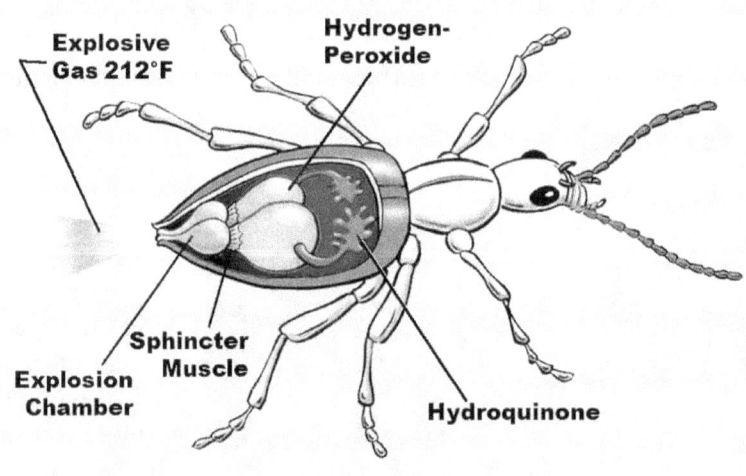

(credit: Bing Images: discovercreation.org)

Again, that indicates <u>design</u>: the <u>fluids are kept separate</u> and cause no problem by themselves – but if stored in the mixed condition in the beetle's body, it would fry the beetle. [107]

Darwin Rejected God

Why was Charles Darwin so antagonistic toward God? His underlying reason for looking for proof of Evolution, and like the scientists before him, for **replacing God with Science**, was due to **his incredible health issues**. While his wife was a good Christian woman who weekly went to church and prayed for Charles' healing, the prayers were not answered… and rather amazingly, when one illness disappeared, another of a different type appeared... leaving him depressed and totally confused… as well as considering suicide.

So was his health really that bad?

To better understand how and why Darwin developed the Theory of Evolution, it is helpful to understand a little bit about Charles Darwin himself and why he so intensely sought to disprove God's existence.

Charles Darwin was a physically sick man whose wife was a good Christian and Charles wanted God to heal him. Charles had a lot of health problems that truly defied diagnosis and cure. **No one could accurately diagnose or heal him**. So Charles turned to God, and his wife prayed with him.

> When God didn't heal him, Charles 'scientifically' concluded that since the prayers were not answered, they must not have been heard because there probably wasn't Anyone up there listening. He reasoned that there must be no God, but he wasn't 100% sure so he remained an **agnostic** until his death. Despite rumors to the contrary, he never converted to Christianity, but he did still occasionally attend church.

Due to his severe health problems, he often could not sit through a church service, and for therapy, he'd walk to church with his wife, and continue walking until after the service when she rejoined him for the walk home again. Sometimes he would not meet her after church and **a search would find him passed out in a nearby park**. For more therapy, and to get away from the sometimes damp English weather, he sailed away to southern seas off Ecuador to regain his health. [108]

While in the Galapagos Islands, he observed nature so keenly that he developed 3 striking theories: The **Theory of Evolution**, The Theory of **Survival of the Fittest**, and the Theory of **Natural Selection**. Nobody argues the last two – they are a given. In fact, it has been said that **Epigenetics** is a modern term in genetics that explains the Selection and Survival theories.

Darwin's Health

> To better understand that Darwin was not unique in his afflictions, and pursuant to a Dr. Shakuntala Modi who was profiled in Ch. 6 of VEG, Dr. Modi discovered that sometimes the negative beings we have called Neggs (see Glossary), can seriously afflict a person – If they have a good reason… and that means it is **Karmic**. So the cause does not have to be due to Exodus 34:7 (or Ex. 20:5) where the children inherit a genetic legacy of the parents.
>
> Dr. Modi's unusual patient was named Ann and Ann was healed by therapeutic release of past life issues stored in the aura/body combination… the Neggs were just aggravating those issues.

Dr. Modi gave a rather long, involved case with the patient she called Ann in her book, <u>Remarkable Healings</u>. Like Darwin, Ann came to Dr. Modi complaining of just about everything under the Sun:

… depression, chronic fatigue, poor memory, sleeping problems, vision problems, skin problems, muscle aches/pains, panic attacks, conversations in her head, headaches, irritable bowels, sinus problems, allergies, and hyperacidity. [109]

Darwin had similar issues. In spades.

Now, let's postulate that **Charles Darwin's contribution to science was colored by his severe health problem(s)** as was his concomitant religious position of an agnostic. It is worth repeating here that this great thinker, who **still deserves a lot of credit and respect**, nonetheless had such severe and mysterious health problems that the physicians of the day could not diagnose nor alleviate them:

For **over 40 years** Darwin suffered intermittently from various combinations of symptoms such as malaise, vertigo, dizziness, muscle spasms and tremors, vomiting, cramps and colics, bloating and nocturnal flatulence, headaches, alterations of vision, severe tiredness/nervous exhaustion, dyspnea, skin problems such as blisters all over the scalp and eczema, crying, anxiety, sensation of impending death and loss of consciousness, fainting, tachycardia, insomnia, tinnitus, and depression. [110]

> Today's Big Pharma would love him – they would have him on at least 45 pills!

With a few exceptions, the description of Darwin's health problems could also be describing Ann's in the earlier paragraph. At least **the extent and mystery of their afflictions is very similar,** even if the exact symptoms are different. As Drs. Modi and M. Scott Peck both point out in VEG. Ch. 6, **if someone has a special proactive task to perform in this world, the Neggs will do their best to stop it,** and it is thus possible that Darwin came here to <u>proactively</u> advance the scientific information on Man, and might have produced a different Creation-based theory, had he not been so afflicted health-wise. He was disappointed in God for not healing him, and thus became a **determined agnostic** whose doubts about God produced an Evolution-based theory instead.

> Or, is it possible that, since the sins of the fathers are visited on the offspring, as was earlier mentioned, that Darwin's problem was genetic and a prior ancestor(s) gave some ground through genetics (a weakness) for him to be afflicted? Such things are called **'generational curses'** and do exist. [111], [112]   [emphasis added]

> *In any event, Darwin is due a lot of credit for persevering in the face of incredible health problems and for being as productive as he was.*

               Yet the bottom line is: design means there is a Designer.

And that same Designer may be responsible for the Earth: as a VR Sphere or even a Flat Earth. Chapters 7, 9 and 10 will attempt to resolve this issue. It is looking more and more like the FE scenario is the easier one to support.

Why?

Again it is important to know what Earth really is – either the Bible is lying (and thus the God who inspired it), **or** it is the Truth – the Bible supports a Flat Earth, as well as Enoch, and recent FE researchers. Science has worked hard to supplant God with their theories and Laws…. What if they are wrong? Would they ever admit it?

VR Sphere vs Flat Earth

And remember, as we go thru the following pages, the Earth may have been created flat and as Man is "spreading his wings" into new activities and territories, he needs more room… and it is **a postulate of this book** that the gods still can 'morph' the Flat Earth Realm into a VR Sphere, if needed… or have they already?

Chapter 5: Religion vs New Age

The last chapter mentioned how the modern tendency is to rely on Science for meaning in the world, and denigrate or remove Religion, i.e., God, from our world view.

In addition, remember it was also said (in VEG):

> <u>Almost</u> everything that you think you know about Earth History, **Religion**, and Physical Science is wrong… by design.

And so far, we have seen how the very geography of Earth and its shape are in credible question, as well as some errors in Physical Science. VEG goes into much more detail with all three of these issues. VEG was concerned with the errors, omissions and inconsistencies in Christian Religion, as well as who created Religion *per se* and why.

Our focus in this chapter is whether Christianity will survive in today's world and how that relates to Man being a watched-over species on Earth (FE or VR Sphere is not important at this point). We do know that **Earth is a created world** and the Creator would not just walk off, therefore He is still around, even though He might have delegated some routine Earth functions to a lesser set of **gods** (still Higher Beings) who now run this place. (See Glossary.)

Since Earth was created, and Man was created, there must be a purpose – it is highly doubtful that Man on Earth is just a "Reality TV" of sorts for some ETs, to amuse them. As has been said earlier, the proposed purpose is **Earth as a School for souls**, and VEG spends two chapters at the end dealing with the scenario – what is expected, how to comply, <u>why</u> one should comply, and what rewards and penalties await the players in the Father of Light's Greater Drama.

One of the main issues examined herein is whether we are free to do whatever we want, ignore the School 'curriculum,' or can we follow the exciting doctrines of the New Age and try to develop our godhood. In short, if Earth is a Flat Earth (FE) or VR Sphere it is nonetheless an isolated realm protected by a Firmament, so what is expected of Man? Or can Man make up his own rules? Or: Are there any rules?

VEG attempted to answer the question: When man and woman birth into this Earth Realm, what is it all about?

Zoo, Circus, Insane Asylum, or Prison

And it was merely suggested: **if Earth is not a School** for souls, what else could it be? Various alternatives were suggested: Zoo, Circus, Insane Asylum, and Prison.

If Earth is a **Zoo,** then we are entertainment for the gods, and Earth would be a sort of "Reality TV" for them… "Wow, Zarg, let's see how they react to this!" – kind of like poking a stick into an anthill and watching to see what happens.

This is not credible given that the human body shows intricate design, the planet is also very carefully designed (just the right temperature, the right breathing balance of oxygen vs nitrogen, the ozone layer, water which has amazing properties [examined in VEG], and very notable "intercessions when souls need help or protection [as in my case – see Chapter 1]). The Quantum Physicists refer to this aspect of Earth as our being in the **Goldilocks Zone**… just the right distance from the Sun to support life. It is also referred to as **The Anthropic Principle** – see QES.

If Earth is a **Circus**, then nothing here has any meaning, no one is watching over us – unless it is for amusement, or scientific curiosity. And maybe this place was once cared for but they left and may come back and terminate the 'experiment.'

As with the Zoo option above, this is an ugly thought and it does not hold water the more you study the unique balance in the environment (that Man is busy upsetting), and if you examine the not-so-obvious internal design of the human body including DNA, the Krebs Cycle, the function of hormones and enzymes, and even the operation of the human eye – as Charles Darwin was reluctant to do.

So Earth is not a Zoo or a Circus… it has purpose and design. So could it be designed as a Prison or an Insane Asylum to hold wayward or dysfunctional souls?

If Earth is a **Prison** or **Insane Asylum**, the basic function of either could be met by <u>containing</u> souls who are wayward, rebellious, violent, petty – and we sure have our share of those! It would be necessary in either case to keep the souls 'collected' all in one place, occasionally send in **("insert") a Master or Teacher** to emphasize correct behavior (Jesus, Buddha, Krishna, Moses, etc…), discipline them if needed to jolt them out of their lethargy and complacency -- make them aware that a higher power is watching what they do (via The Flood, asteroids and earthquakes), and occasionally **insert objects and cryptozoids** (e.g., Bigfoot, Mothman, Loch Ness Monster …) to amaze and entertain them. You might even **insert UFOs** that they could back-engineer to give them hope of reaching the stars… or move faster around the Earth realm.

You would make sure they could not leave the Earth (**The Firmament** or some sort of energy barrier) and yet the Earth Realm would still encourage wonder and exploration with the bottom-line discovery of a design behind everything – to wake those up who may be paying attention.

You would also assign Angels (or **Beings of Light**) to protect, guide and assist those who have waked up and are trying to become better souls – more compassionate, patient, humble, and respectful of self, others and the world. You would also insert special humans who are **observers and healers** to apply positive input where needed – they would come from the higher realms.

Lastly, you would make sure that souls could not use their innate, natural abilities that they normally use back in the 4D Realm from which they came. **Souls would have to function under 3D Laws.** You would create Earth as a 3D realm in which 4D abilities (telepathy, clairvoyance, levitation, etc) are <u>not</u> available to them so that they cannot modify their experiences and escape them.

As can be seen, some of those elements above, related to Prison or Insane Asylum, are in fact part of our scenario here. So what would be the difference between Earth as a School and Earth as a Prison/Asylum?

The difference depends on what the soul does about having been put here – in all 3 cases, **School, Prison or Asylum**, the goal is to learn, become more compassionate and **graduate**! The soul must show itself better than when it was 'inserted/birthed' into the Earth Realm. Along with that you would set up a **Review** at death (or NDE experiences <u>during a lifetime</u> as an 'emergency' evaluation) and evaluate the soul who returns to the **InterLife** (a holding/training area discussed in TOM).
Failing to improve oneself would result in being **Recycled** – back to Earth for more lessons.

If training and evaluation are the norm for most souls, then the overall function of Earth is as a **School**… resistance to peace, love, charity and learning automatically make the Earth Realm a Prison/Asylum for the souls who are intractable and <u>refuse to learn</u>. If they stay rebellious, believe that they can do whatever they want, there is no God and for them Science is all there is. If they are thus too far gone, they will find themselves more than dysfunctional – they may be 'Disassembled' (energy and being reset to zero) and find themselves to be no more.

> This latter option is used when a soul is so damaged that it cannot be infused with new energy and proactive vibrations. Some Earth experiences are so traumatic that the wounded soul has to be 'rebuilt' in the InterLife, and failing that, it is returned to a zero state (i.e., as a new soul with no experiences) and given basic training and preparation for the Earth experience, and then re-inserted to the Earth Realm.

The point is: a soul cannot waste its incarnations and ignore its lessons forever. This is further addressed in the chapter on the InterLife in TOM.

Creation vs Evolution

By now it should be fairly clear that Man and life on Earth was <u>designed</u> and there is a <u>purpose</u> to being here. And a lot of souls intuitively know that, or at least suspect that… so how did the godless Evolution get going?

Charles Darwin

While it was always a subcurrent among humans not wanting to be responsible for what they do, it was **Charles Darwin who gave them an official excuse to be an agnostic or atheist.** And it is sad how that came about. (See Chapter 4.)

To better understand how and why Darwin developed the Theory of Evolution, it is helpful to understand a little bit about Charles Darwin himself and why he so intensely sought to disprove God's existence.

Charles Darwin was a physically sick man whose wife was a good Christian and **Charles wanted God to heal him**. Charles had <u>a lot</u> of health problems (detailed in Chapter 4) that truly defied diagnosis and cure. No one could accurately diagnose or heal him. So Charles turned to God, and his wife prayed with him. When God didn't heal him, Charles scientifically concluded that since the prayers were not answered, they must not have been heard because there probably wasn't Anyone up there listening. Ok you know that from Chapter 4, but you don't know the rest of the story…

Do you see what Charles was really saying? Do you see the ego involvement? Charles was basically mad at God. Then he decided that there wasn't one – but wasn't sure, so he stayed an agnostic. The ego comes in because (1) he wasn't a Christian, and (2) he assumed he was worthy of being healed (because he was married to a Christian woman?). He had health issues but was not a humble man.

He had seen what appeared to be the benefits of his wife's prayers for other things in their life together, so when prayer did not heal him, he reasoned that there must be no God. It never dawned on him that **he might have to meet God at least half-way**... Yes God loves Man, and His Grace does extend to those who humble themselves and ask God directly... in return the best way to get God's attention is to make a decision to walk in the Light, and begin to serve the Creation.

As we know, Charles served Evolution – to spite God.

Charles was a little arrogant and just assumed God would heal him because of his devout wife. But God is not fooled... no matter why Charles had the bizarre health issues, **he never turned to God as his Source**... or accepted Jesus which would also have shown a commitment. **Charles was 'stiff-necked.'** And if you have read how God dealt with the Israelites (in Exodus), and their often mis-behavior then you get the idea...

> The Israelites were also 'stiff-necked' and bitched about the Manna, made golden idols anyway – despite Moses telling them not to... So it is a shame that God let them be **taken into captivity three times** (Assyrians, Babylonians and Egyptians) – and as we saw during the time of Jesus, centuries later, the Pharisees had not really changed, and they militated against Him trying to teach Man a better way!

This is what a lot of souls are here on Earth for – to learn appropriate behavior... Patience, Humility, Compassion, Respect and a sense of Who is Boss. Really. The wayward and dysfunctional souls at some point insisted on doing <u>what</u> they wanted, <u>when</u> they wanted, <u>how</u> they wanted and <u>where</u> they wanted... that behavior is reserved for gods, not humans on 3D Earth. And that is why Earth is a School... and for some it is a Prison, but in any event, souls that wise up and stop demanding what they want, or doing things their way (àla Moses below), will qualify for graduation from Earth School.

The Bible says that **God often disciplines those He loves** and who love/serve Him and walk with him – that is the parable of the **Vine and the Branches** that either bear fruit or are cut off…(see John 15:1) but even the healthy branches are 'pruned' (disciplined). And sometimes the discipline takes the form of not letting his servant come all the way into his reward for doing basically what God asked.

> Case in point was where Moses was leading the Israelites in the desert and they were thirsty. So God tells Moses to go <u>speak</u> to the large rock 'over there' and Moses goes over and <u>strikes it with his staff!</u> Yes the rock gushed forth water, but Moses didn't 100% do what he was asked to do… Nit picking? Well, you either are careful to do what you are asked, or… in Moses' case, he led the Israelites to the Promised Land but was not allowed to enter himself.

And it is possible that Charles' **health issues were to get his attention**, turn to the Light, and do what he came to Earth to do. It did not have to be a case of Exodus 20:5 – involving generational sins. Neither did it have to be Karma. Neither would it have been a genetic predisposition… to 27 different forms of illness? No.

Despite rumors to the contrary, he never converted to Christianity, but he did still occasionally attend church, at Christmas and Easter with his wife. But he never got the real message…

> *As was said in Chapter 4, Darwin is due a lot of credit for persevering in the face of incredible health problems and for being as productive as he was. That said, he could have done more and better if he had not been so stubborn. Persistence is a virtue but not when it is just stubbornness.*

Thus Darwin was reacting <u>against God</u> and we might never have had the <u>Evolution of Species</u> had he had a healthier body. We might have had <u>The Creation of Species</u> instead.

And it is also suggested that today's agnostic/atheistic scientists, who deny God's existence and promote Evolution, are like 'naughty little kids' – they resent parental oversight and refuse to be accountable for what they do – They prefer to believe they are free to do whatever they want, and no one is going to tell them what to do.

They conveniently don't see a God in their world… and as VEG suggested in Ch.5, many of them may well be OPs – **soulless humans** who have no connection to anything higher and <u>thus have no conscience</u>. To them, spirituality and religion make no sense.

Many people resist the idea of soulless humans, and we're not talking about Zombies. **The human body does not need a soul to activate it** any more than dogs and cats, horses, cows, birds, lions, tigers and monkeys (who do not have souls) need a soul to eat, sleep, and procreate.

Besides Evolution and the Earth Globe, this is one of the lies that has been promoted on Earth – i.e., that everyone has a soul and "we're all alike!" NO we are not. I repeat: the Greeks, Mayans and other civilizations knew about this difference and somehow it was lost, perhaps in a misguided effort to "level the playing field for everybody"… just as we are doing today with the LGBT scenario. (VEG Ch. 5 explores this difference, and does not pass judgment.)

So if Religion failed Charles Darwin (or his wife was too passive to convert him), would the New Age or New Thought teachings have been better for him?

Point of clarification: as used in this book,
the **New Age** is a "name it and claim it" get what you want , believe it and see it movement, develop or activate your godhood. Some forms may be full of froo-froo, aliens , UFOs, even Wicca and mysticism.
(Examples: Tony Robbins, Rhonda Byrne, Ashtar Command)
New Thought is different – it is the original, time-honored (often oriental in source) metaphysical teachings --- some of which find their way into the New Age and are used to justify getting what you want.
(Examples: Wayne Dyer, Emmett Fox, Paramahansa Yogananda)

The New Age

So how does the New Age and its teachings fit into the Earth Realm? Is it beneficial to Man and helps him graduate from the Earth School? Or does it retard his progress? Is it the next step in consciousness as its proponents claim?

These questions are very relevant as not everyone on Earth in Political and Religious power wants Man to succeed. Is that saying there is a conspiracy? No, just that **the Powers That Be (PTB) do not support your graduating from Earth School**. What does that look like?

Here is the allegorical scenario, and it reflects a real issue:

Let's say you are a bigwig in the Vatican or a Senator or Congressman or that you are the leader of a Mega Christian church with a congregation in the thousands.

Next, let's say that there is a well-defined Path to Exit Earth School and that it really works, and has been proven. Let's say that it is an energy medicine treatment which when applied, lying on a special metal bed, so enhances normal body energy and empowers the body's Bionet so that the Kundalini rises up the spine and truly, gently enlightens the person and they reach the illusive state of **Nirvana**. And whatever illness they have is also healed. All they long for is Heaven and they have the option now, with higher consciousness, and greater abilities, to voluntarily leave the planet.

> There is no such device, but it makes the present example and conclusions easier to follow. And yet, it is theorized that this is exactly what went on at **Göbekli Tepe**, as explained in <u>Anunnaki Legacy</u> (AL). It was a temple where *Kundalini* was activated.

Suppose thousands of people start using the treatment and they are so enlightened that they willingly choose to ascend spiritually and leave their bodies, and they actually "graduate" from Earth School. And more people catch on and partake of the opportunity to leave a planet of disease, deception and war… and in a few months a few million people have left the planet.

You as the leader of Mega Christian Church, Inc. notice that several hundred of your congregation are gone… and so are their tithes. Your neighbor who is a well-to-do Congressman with three houses, several Ferraris, and a yacht tells you that several thousand of his constituents are gone … and he may not be re-elected … especially as the lack of people in his district means the district will be absorbed into the bigger neighboring one. In Rome, the Pope discovers that Catholics are using the system and leaving the Church… having found a bigger Truth. (And local doctors are not happy because they have lost patients and both the doctors and Big Pharma are losing income, but that is incidental to this allegory.)

You, your neighbor and the Pope (like the PTB) are not happy – they **love being in charge**, they love leading the flock… in short, **they love power.** They must stop the exodus from their little worlds… They have two choices: they can either seize the device and hide it, or (2) they can bad–mouth it and see that it people do not use it. And they could also kill the inventor and destroy the machine.

The point is this: just as in the Medieval Days when Lords ruled from the Castle, and controlled the lives of their serfs (slaves), so too does today's PTB <u>not</u> want souls to graduate from the Earth School – it removes sheep from their little world. So they distract you with electronic geegaws, and sex and violence on TV.

But there is one difference between the PTB and the Pope, Congressman, and rich Church leader in the story above. The PTB are not all 3D flesh-and-blood and the **Astral component** does know who seeks to spiritually grow <u>and</u> they work with their 3D Fat Cats (human PTB) to keep the Sheeple so distracted by electronic geegaws, as well as movies, demonstrations, and just earning a living, that the average person doesn't have the time to seek and pursue spiritual growth. Thus the Sheeple are pandered to with **idiot TV programs** (aimed at the 8th grade level), sex and violence in movies, GMO food, and prescription drugs whose side-effects are as bad as the illness they are supposed to cure… but don't worry, Big Pharma has drugs for the side-effects too!

Notice that during the 60's-70's when young people were very much into Flower Power, Make Love Not War slogans, meditation and living in communes, Joe Average had a job, the family had one car, the wife minded the kids at home, and America was productive and optimistic.

The PTB couldn't stand that…people had too much free time and TV programs were too wholesome and inspirational. That just would not do. Something had to be done. **Control was the goal**.

First the Hippie Communes were chased out of town and disbanded. To do this, fraud Moonie groups were established and then 'exposed' and people avoided such things in the future.

Second, the **bankers** worked with the PTB to change the structure of the average family. Instead of Joe as the sole breadwinner, things were made so expensive that wifey had to go to work, too, and then the family had to have two cars, and that meant more insurance, and the kids were put in Daycare Centers…. And keeping the Middle Class busy earning a living meant they had less time for spiritual growth and wondering who they were, and why they were on Earth. Credit card debt soared, and so did bankruptcies as **Capitalism is an inflationary economy** and working hard could not keep up the rise in prices in goods and services.

The PTB (which now includes the Big **Banks**) had succeeded in bringing the Sheeple (i.e., today's serfs) under control. All that was left was to manage the **Media** and promote consumption (via advertising) of the things the PTB wanted people to buy– which just happened to be the industries in which they had huge investments.

Along the way, it was very necessary to encourage **straight-vanilla churches**… where nothing is really taught, the Sheeple come in, nod to God, drop a few dollars, hum a few tunes, and leave. The Sheeple must not question the Bible or their Faith… So things are kept in a straight and narrow orthodoxy.

Even the "advanced" **New Thought churches** nowadays are Feelgood Centers where not much s taught, again so that no one wakes up and can graduate… What happens is the Sheeple again come in, jump around and dance to the upbeat music, a positive word is spoken, but Truth is <u>not taught.</u> Yes, classes in Truth are offered but only the straight-and-narrow, okayed-for-public-consumption version… Don't go too far and if someone (like myself, with these 7 books) comes in and would like to share, just discuss some new, open ideas… Aargh! Invader! Messenger of Darkness!

Along the way, it becomes obvious that the **Sheeple also have to be entertained**, if not distracted, so that they didn't rebel and riot from their everyday lives. This meant more movies with more sex and violence, Moon landings (How did we get past the shell or Firmament?), small wars here and there (so that family members could participate in Glorious War and the family had another thing to be concerned about), Lotteries and Cash contests, and every now and then it wouldn't hurt to have a Bigfoot, Loch Ness Monster, Mothman, stories of Ghosts and UFOs on TV…

Key Idea: The PTB do not want to lose people to spiritual pursuits and have them become Earth Graduates. Why? Once you graduate, you don't come back and the PTB want a steady supply of sheep (oops, Souls) to manage.

Believe it or not, there is a constant increase in the number of soulless humans on Earth every decade… VEG examined this issue and the current count of souls vs soulless (i.e., OPs) is almost 1:3. That is: 3 OPs for every soul.

The issue is the control and subjugation of **souls**, <u>not just people</u>.

Souls are wising up in the InterLife and opting to not incarnate on Earth, and more and more OPs are found among us.

How do I know? I see auras and it is alarming the scant number of souls with auras on any day in public.

Why is that a problem for the PTB? (Remember the PTB also includes the entities in the Astral who do not want souls graduating and ruling <u>over them</u> in higher realms … that is the <u>promise</u> to all souls, by the way.)

The problem for the PTB was laid out in TOM, but is here repeated:
When a realm or timeline becomes saturated with more OPs than souls, the gods create a new timeline for just the souls, and the OPs and their rulers (the PTB) are dissolved -- that old timeline became dysfunctional and is therefore dissolved. (See Ch. 1 in VEG.)

So the New Age is not lying when it tries to tell you that you have a super potential and should develop it their way, i.e., develop your godhood. And yet that is not for this time, and it's not possible in the Earth School.

So as a smart PTB, you would also want to make sure that any headstrong Sheeple who just have to engage in Truth-seeking, and spiritual growth, are not rewarded.

This is the connection with The New Age.

New Age Centers

What you want is for those seeking to grow spiritually to waste their time – but they must believe they have found something unique that will (eventually) enlighten them. So you would **create false Truth Centers** where the public goes who just has to have a meaningful spiritual pursuit. Some of those that this author visited were not teaching Truth, they were **Jumping Freddy Feelgood Centers** for spiritual living… people singing and dancing, doing a conga-line around the sanctuary. And then their *pièce de resistance* was to play the idiot (but great sounding!) video called Happy by a man who was paid to promote the idea that you should be happy no matter what happens in your life. He says don't handle your problems (tests) – ignore them and be happy! (Also one of the major New Age teachings.) **Many centers are just entertainment... so you feel good.**

And here some people may be shocked and unhappy with what follows…

You have to **set up false teachers who appear to have credibility**, and diminish the public's ability to continually pursue enlightenment -- by diminishing their re$ource$. You charge plenty for books, CDs, DVDs, and seminars where the faithful can hobnob with the 'enlightened.' But you don't teach Truth or what really works… you have to keep them coming back for more – or you go out of business.

You also have to remove Truth Centers that actually have Truth and have worked for people in the past – thus you get rid of **Religious Science** whose founder actually healed people (we can't have that – the public would beat a path to his door!), and you limit the effectiveness of those larger Truth Centers who have too many followers to outright dissolve the organization, like **Unity**.

What you want is modalities that don't work, like Yoga, Light Therapy, Hot Rock Therapy, Aromatherapy, Tarot, Gregorian Chanting, Burning Bowl exercises, Reiki, Wicca, Snake Oil, NLP , and Transcendental Meditation (which harmed people) that will turn people off and they hopefully give up. The really clever artists promote these modalities while making them appear to work.

The 'killer' technique is to promote belief in something that can't be easily quantified, such as teachings that appear to develop one's godhood:

> You Can Create Your Day (promoted in the movie *What the Bleep?*)
> Develop your godhood..... <u>Awaken the Giant Within</u> (book)
> Attract your good (Kabbalistic) – Name it and claim it
> Affirm it, visualize it and receive it
> You'll see it when you believe it
>   (as opposed to "I'll Believe it When I See it!")
> Create green traffic lights and personal parking spaces…on demand
> Ignore problems and they should disappear …
> ACT as if you are happy…ignore your lessons…

And then in those established larger organizations, you promote **mediocrity**. Today this is largely done by **including major New Age authors' books** as part of the organization's/church's teaching… i.e., mixing up the teaching so there is a very subtle level of confusion (it can't be too obvious) – that keeps them coming back for more, but the Sheeple don't ever quite get it together.

All of these and more await the naïve seeker in today's New Age movements.

This is not to be confused with **New Thought**, which is not taught any more. It was genuine spiritual Truth as was taught in Unity and Religious Science <u>decades ago</u>. Its source was the Orient and esoteric teachings (like The Tao, and Zen), not usually for the masses. Some of it today can still be found in what remains of Religious Science, called <u>The Science of Mind Handbook,</u> by Ernest Holmes. Esoteric teachings are also available in <u>The Science in Metaphysics</u>, (TSiM) and the last few chapters of <u>Anunnaki Legacy</u> (AL).

So I leave it to the reader to answer the question: Do today's New Age teachings really help the aspiring spiritual seeker to spiritually grow…. much less <u>become</u> an Earth Graduate? (VEG Chs. 15-16 review what is required and what a Graduate looks like) … and by the way:

You don't have to be perfect to graduate!

If not the New Age teachings, then where can one turn to begin to walk a Path to true spiritual growth and fulfillment?

Religion as Truth

It may seem like we are taking a step backwards by considering Religion out there, or more specifically, Christianity. Ever notice that there is a constant attack against Christians in the Middle East, and even in America among the more progressive New Agers, militant Muslims, atheists… Christianity is discounted, ignored and pooh-poohed. It is said to be old-fashioned and non-functional. This section of the book is going to show you why that is said, and reveal the truer, more powerful foundation of Christianity. Remember, if something works, it is to be discounted so the Sheeple do not go there, find Truth and become empowered Earth Graduates.

> This author has personal experience with the **advanced** aspects of Christianity, which sadly most Christian churches either do not know about, or they are afraid to venture beyond the Nod-to-God-on-Sunday practice. The advanced aspects covered in this section include the **Baptism of the Holy Spirit**, the ability to heal, and having real power to set people free in Deliverance.
> There really is power in Jesus' name and that is why the PTB spend so much time dissing Christianity… making people think they need something more modern… but it takes a special step to get to the advanced level.

Hang on to your hat, everything presented is absolutely true and works.

Traditional Churches

First let's look at what usually passes for church in the world.

A young man or woman is inspired to teach the Word of God, 'save' people, and enrolls in Seminary where the Bible and standard, basic things are taught: how to run a church, tithing, baptism and the basic sacraments, counseling, and having a good grasp of the major parts of the Bible – where key passages are, etc. This person will go far if s/he is 'called' to the profession… those people are often anointed.

The rest of the hopefuls are focused on leading people with sermons and Bible studies so that hopefully the Word itself will do its work in them. And there is another group who were not called, not anointed, and they do church because it is useful in society but these pastors run it as a business – and some are quite successful. (No names.)

The goal of most pastors, called and anointed, is to 'save' people – to bring them into the church, baptize them, and set them on a Path to the **Light**.

That is not New Age: Jesus said He was the Light of the world.

Being Saved

'Saving' people (and that word has quotes around it) can also mean "saving them from going to Hell because of their sins"   and in its simplest form, it just means bringing them to Christ. The more empowered form allows Christians to stand against the tricky Astral entities and get them to stop disease and affliction. Do not forget that there <u>are</u> **beings in the Astral** who do not want souls to succeed at anything. NO – they are <u>not</u> demons but have been called **Discarnates** (ghosts) who when they died, did not go to the Light (see TOM). Along with them are the **Neggs**, "dark angels" who work with the normal Angels to effect oppression in a person's life if that is what it takes to turn their life around.

Regular Angels, or Beings of Light, are only protecting and guiding souls; when a soul needs a lesson, the Negg gets clearance to afflict the human in some way to get their attention, and hope that they will turn to the Light. (Remember Charles Darwin?) **The Angels and the Neggs work together**. This takes a bit of explanation (See VEG and TOM), as the original teaching in Christianity was lost when the Church designed (compiled) the Bible (AD 325) and decided what people would hear and know. Esoteric knowledge was kept from the masses but has resurfaced in some progressive Christian churches .

The Bible says

You shall know the truth and it will set you free.

One of the meanings of 'saved' is to be **set aside** with the knowledge of the Truth so that a person is saved from the ways of the world and its deception. The more standard meaning is that the 'saved' person is now part of Jesus' flock. It also means that the Angels will be watching over their new convert – to protect and guide.

Being saved also means the person has made **a choice** of whom to serve – Mammon or God. Material things or spiritual things. Money and power or God. Lies or Truth.

> The issue on Earth is **Freewill** – souls have free will to choose what they will do and whom they will follow. Thus they must be brought to a <u>point of choice</u> where someone asks them whether they choose Light or Darkness.

Astral Game-players

A soul who does not know that there are **Astral beings who are real game-players** may start playing with a **Ouija Board**, and <u>the Discarnates will oblige</u> (the Neggs serve a higher purpose and are prevented from answering you). The lies will be shared with the game-player human who thinks they are hearing real truth. The average human, having little or no **discernment** (one of the gifts when one becomes a Christian) will tend to believe what the Board says as it comes from the Other Side where (if they are a New Ager) they 'know' that all Truth resides. If the Astral being can add to your stock of false knowledge, you can be led really astray.

> Often it is not what you know that gets you into trouble, it is what you don't know.

And

> The Bible warns against playing games with ("attending to") false spirits and their false doctrines.   (Deut. 18:10)

Possession Danger -- Example

One of the saddest incidents was when a **Discarnate** responded to a group of naïve humans playing with a Ouija Board in England. It quickly told them that it would be able to teach them more truth, and faster, if it was allowed to inhabit the leader's body and speak thru her. They all agreed (again, permission is needed due to the Law of Freewill that even Astral entities cannot break), and **Seth** began speaking thru Jane Roberts. The information <u>was</u> very interesting, captivating and led the group to further sessions.

As time went on, the entity (calling itself Seth) asked if it could once again, as it had done while on Earth, smoke. Jane said yes. Then a few weeks later, Seth asked if he could have wine/whiskey and in exchange would reveal even more interesting secrets. Jane agreed, and Seth began drinking and smoking while in her body. For about a year.

> Higher minded entities do not behave that way – such as RA in
> the <u>Law of One</u> books (channeled from a being in the 6th level.)

The end of the story is predictable: **Seth caused lung cancer in Jane and she died**. A tree is known by its fruit and his behavior was not that of a higher being.

Progressive/Charismatic Churches

There is another type of Christian church where the gifts of the Holy Spirit are recognized (1 Cor. 12-14) and the most common occurrence is for some members of the church to speak in tongues. This gifting has been misinterpreted and abused in many Pentacostal churches.

What does the Christian church a lot of harm <u>from its own members</u> is for someone to start babbling, making some speech that is not English and not recognizable… and no one translates. Some of the sounds are a real joke. Supposedly gibble is **speaking in tongues** and is an ability that is worn as a badge by some naïve members of these "charismatic" churches. Many of these wannabe 'tongue speakers' do not realize that the Astral entities will make a church and/or its member(s) look foolish – if they can.

Case in point: was a visit by a truly anointed pastor from Florida and his wife co-pastor – both gifted for healing. They did a one week seminar in a church in an outlying suburb of Northeast Dallas, in 2001, and they of course met and dined with the local pastor and his wife. The local pastor's wife was so eager to demonstrate her gifting that she started babbling during a workshop led by the visiting pastor's wife.

The visiting co-pastor RH stopped dead in her presentation and stared at the interruption. Even more, she asked the local pastor's wife (BC) to stand up (she was in the first row as was I) to share the experience from the front of the room. No sooner had BC stood up than she started her 'tongues': "B-dip B-dip B-dip B-dip…" at a rapid rate. RH took about 30 seconds of that and shut her down, "In Jesus' name, shut up!"

BC looked at RH with a look that I have never been able to imitate even in the mirror! It was pure hate. RH then said, "Loose her. Come out of her!" and BC collapsed on the carpeted floor … right where I had been a week before (see Chapter 1 account where I was healed). Of course people were standing now, trying to get a better look, and I could not believe what I was seeing.

Just two years earlier I had participated in a Deliverance ministry and learned some very powerful truths – these entities have no real power… the Christian who walks the Walk can command whatever s/he wants and the entities have to do it! I also learned that 99.9% of people who claim **possession** are <u>not</u> possessed… it is illegal for two entities to occupy the same body – that is why Seth (earlier example) had to <u>ask</u> permission. Most often the person is **oppressed** from without because they opened a door by doing something inappropriate… drugs, sex with someone who practices witchcraft, and even getting drunk. Abuse the body and the entities are there!

I leaned forward and BC was now beginning to drool and she was writhing on the floor, yelling "No! No! Leave me alone!" Needless to say, BC's husband, the main pastor of this church was shocked speechless and had no clue what to do.

RH stood right over the writhing woman (she never touched her) and kept commanding BC to be left alone. Suddenly, BC gave a jerk and relaxed. It was over… but not for RH! BC's husband was all over RH attacking her verbally for making a spectacle of his wife. RH's husband, also a pastor, took BC's husband aside and had a serious talk with him…

Meanwhile RH bent down and was checking to see that BC was Ok, that she hadn't hurt herself when she fell to the floor. BC did not remember anything – not the outburst in gobbledegook, nor what had happened on the floor… she was spacey and puzzled.

Fortunately RH addressed the crowd, seeing her husband was trying to calm BC's husband (the local pastor), and she told us what had happened and how she knew that BC was being manipulated – not only to make a Christian look silly, but to discredit the real speaking in tongues!

> RH carried, and still does, an anointing from **Smith Wigglesworth**, the great British healer. Wigglesworth not only healed people, he could simply approach someone who was oppressed and they would collapse on the floor, having been set free.

The local pastor was so embarrassed (and un-Christian) that RH and her pastor-husband were never asked back to the church. I never went back, either.

Real Tongues

Speaking in tongues goes back to the disciples of Jesus who were given the gift to travel to other countries and **minister and proselytize in foreign countries** – they could speak the foreign language without knowing any of it! **That was what it was for**... not to entertain members of one's local church (that is the abuse spoken of earlier). And many of the alleged tongues are so silly as to be obviously fake.

I had an experience with a close friend of my landlady who had come in from the Philippines and was out in the back yard, enjoying some Sun and fresh air. The landlady and I were seated in the breakfast nook where the sliding patio door was open to the backyard.

All of a sudden I heard a wailing, and howling as if the neighbor's dog had gotten into our yard and was making the sound… "Owww…Owww….Owwww…." I got up and slid the patio sliding door open and poked my head out. It was the friend from the Philippines (call her DK) – whose husband also is a pastor. (He was not present.)

I went out and she was smiling and between howls said she was communing with the Holy Spirit! More "Owww…Owww….Owwww…." and I said quietly (under my breath) *Shut up and leave her! You will not get up.*

DK stopped and glared at me… she could not have heard what I whispered, but the one making the noise thru her did! So I engaged it, "What are you doing?" And DK looked like she was ready to kill me but I had given the order to stay seated (having had it happen a few times when I was in the Deliverance ministry, I knew even a 90-pound girl like DK could muster incredible strength and harm me or damage the patio sliding glass door!)

"I'm praying and communing with God!" she said, very indignant… to make me sorry that I had interrupted her worship.

"How does howling like a dog glorify God or speak to Him?" I asked.

"It is what I was given…" she said, smirking and she started again.

"Shut up, and loose her …" and this time I added … "in Jesus' name!"

She glared at me, and this petite 90-pound woman who has always appeared to be so demure, cute, and holy gave me the finger and slumped back in the chair. And 10 minutes later when she gathered her wits about her, she remembered nothing of the exchange… and she was mystified that she could no longer speak in tongues. (I had some teaching to do with her later!)

Just so that you know this stuff is real but has no power over someone who carries an anointing (my 'anointing' came from New Jay back in October 1998). I later was 'upgraded' by RH in 2001.

Deliverance Ministry

On another occasion in 2000 when in the Deliverance ministry, I was called over to work on a young Latino who was seated but writhing in pain – and I saw claw marks coming from <u>inside</u> his throat! It initially rattled me – how could that happen?! And for 10 minutes nothing we (my team and I) did worked, and all we learned was that he was a visiting ministry student from Puerto Rico. He had visited brothels there trying to convert the sinners… and now he was paying for it, <u>and turning blue</u>!

Then a little 80-year old Latina lady came up to us, no one had ever seen her in the workshops in the past few months, and she approached him, and I saw fear in the kid's wide eyes. She pointed at him and said, "Dejalo! Ahorita en el nombre del Señor!"

The claw marks disappeared and the kid was breathing calmly. The team ministered to him, while I followed the little old Latina to the side of the room. I wanted to know what that was all about! She explained that the oppression started in Puerto Rico, a Spanish-speaking country, and <u>the entity was being legalistic</u>: it attached itself to him (in payment for messing with the sinners in the brothels – he had defiled himself!). So it got in in a Spanish-speaking country and was holding out for commands in Spanish! I was stunned. But I noted that was what worked.

She smiled and went out the exit door. I remembered something I wanted to ask her, and 5 seconds later I dashed out the door… it was 100' to the parking lot… and she was nowhere to be seen! I smiled and said, "You sons of a gun!" I knew we had been visited. What a lesson!

The Astral entities are real and are **nothing to fear**. If you know Whom you serve, and you know He can do it, just speak the word. In later weeks, until I got bored of that ministry, I used to sit on the chairs around the edge of the large room, watch the proceedings, and when Big John or Lety could not get anything to happen, they would look at me, and I would speak a release to that person, sometimes pointing at them, and they would always collapse, free.

> Bored? Setting people free?   Yes, it was always the same thing – the entities are not very original, nor clever, and it got tiring dealing with them. Remember that it was not my power; the anointing merely provides a channel for the Angels to work thru the one doing the Delivering.

So one of the last few times I was serving, after about a year, I got the idea to ask the next entity some heavy spiritual questions … I wanted answers to a few things. My teammate was nervous about doing this, and it took three more people before we got one who was oppressed by an entity – people who had problems were <u>never</u> possessed, and only a few were actually oppressed.

I finally got my chance, and the team leader was on the other side of the room, so I bound the entity and demanded answers. I got three answers and then all Hell broke loose!

The young oppressed girl started screaming, we stopped that, she then tried to hit us, and we stopped that. By this time the Head Pastor, the leader, was standing right behind us, and the young girl's entity was trying to tell him what we did – to get us in trouble! Talking with the entities was forbidden.

I gave a command and freed the girl, but the leader had heard a few things he didn't like and I and my teammate were both suspended for a week, during which time I thought it over: the more I relished setting people free, the more it seemed there were people with problems to work on! I got the message and quit. But I had my answers!

> And by the way, after I left the Deliverance ministry, I was harassed and attacked for several months for having 'messed' with the Astral entities' domain. You can't just walk away and everything is Ok. I had to do personal warfare with them until they finally quit.

I share those things to say how banal oppression can be and that there is nothing to fear from the Astral entities. They are just a nuisance. Use **Ephesians 6** and stand your ground. They are permitted to do what they do as part of the Father's Greater Plan for Man. **Intimidation and doubt** are their main weapons. Of course, we began the Freedom & Fullness workshops with prayer and prayer coverage, we weren't foolish.

In addition, **Christianity still has plenty of punch for today's world**. It is still a viable foundation for a productive and happy life. In fact, the principles for becoming an Earth Graduate are those promoted by Jesus in the **Sermon on the Mount (Matt. 5-6-7)**… compassion, patience, humility, faith, and respect.

Modern Christianity

You don't have to do battle with Astral entities, or speak in tongues to be Christian. Those are two of the more **charismatic aspects** to which people are usually <u>called</u>. Those who just show up and try to do battle in Deliverance often fail – and wind up oppressed themselves! – even though it was a noble idea. Being called (uniquely suited) to something means you will succeed where others who have no calling will probably fail.

The obvious inference is that **Christians are good candidates for becoming Earth Graduates.** They may think they are going to Heaven, where they just sit around, sing and play a harp while lounging on a cloud. In fact, from what I saw of the Other Side in 1991, which I call the InterLife (see TOM), it is **a working world** and each person has a unique set of interests and giftings that are of use in the Father of Light's Kingdom. There is a unique place for each person to use their skill… and it is something that you <u>want</u> to do and are very interested in…. the trick is to discover it while on Earth. That way, you can get a jump on learning relevant skills while here.

But we can't serve if we are lazy, argumentative, petty, violent, egotistical…. And that is something we all have to get rid of, and fine-tune our attitude and behavior while in the Earth School.   In fact, **that is the purpose for being here**. **Overcoming is what makes a person an Earth Graduate** – which is highly respected on the Other Side, by the way. They all know how hard the Earth School is, and its graduates are not only respected, they are sought for positions that rule over such places as Earth.

Divine Hierarchy

The Father of Light and Souls exist in an organizational hierarchy – much like any well-run business enterprise. There are the Directors at the top, including CEO, then the Division Directors with their subordinate Section Managers, and below them in each Section is a Supervisor, and sometimes a lead person or Foreman. For example Accounting will have a Director of Accounting, then a Comptroller, then a Treasurer and Chief Accountant, then the staff within each major function. As was said, **Souls cannot interface directly with the One** any more than bookkeepers deal directly with the company's CEO.

The Divine Hierarchy looks like this:

The Father of Light/The One
The Higher Beings
Subordinate Beings including:
Masters, Avatars
Teachers
Oversouls (Higher Self)
Soul Groups

----------Earth Realm ---------- Angels (Beings of Light) and Neggs ------------
Astral entities
Souls

Note: Neggs are Beings of Light (Angels) working the negative side of human lessons. They work <u>with</u> the regular Angels, and are not demonic. (Dr. Lerma in the Houston TMC Hospice discovered them in his work and documented them in <u>Into the Light</u>.)

That concept is loosely represented in the following chart:

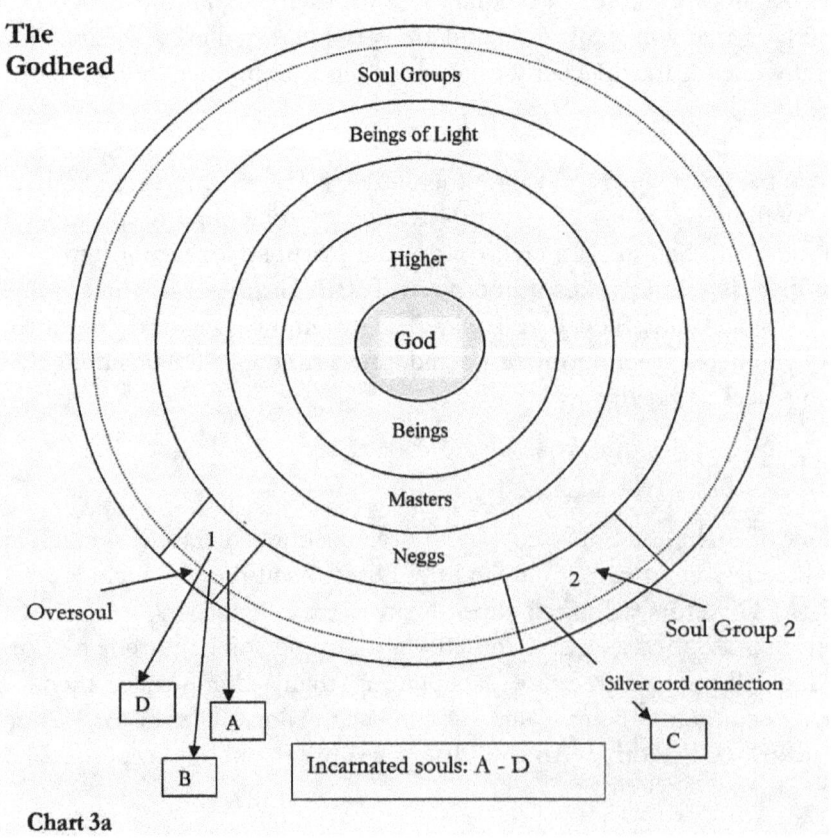

Chart 3a

The above chart is explained in detail in VEG.

The significance of the hierarchy is that there is a lot of space (Realms) between God and the Earth Realm… meaning that **today's Christian and Earth Graduate** can apply for, and find a niche for service somewhere in that range.

> As was said earlier, the Astral Realm will be ruled over (perhaps "administered" is a better word) by Earth Graduates who have the qualifications -- and that is what the Astral entities seek to **prevent** (by their manipulation of the 3D flesh-and-blood Fat Cats who think they run this world and **want to suppress you**. This is where the oppression, thoughts of doubt, fear and even suicide come from – them – to derail a soul's progress. Once you are aware of this, you can block them (VEG and TOM gave this area some attention.)

Potential Areas for Service

Having said that, the Soul who graduates from Earth can fulfill certain basic, initial, functions in the Multiverse, it can be shared that these are some of the areas open to Graduates from the 3D Earth Construct:

> **Bio-plasmic Quantum Computer Techies** – responsible for basic Heavenly Computer support, Scripts and maintenance.

> **Bio-plasmic Computer Programmer** – performs fractal sub-programming (for Recycling) under supervision.

> These last two were demonstrated in Chapter 6 of QES. The Heavenly computers develop and manage Scripts, including the Father's overall Greater Script (TOM). There are also Science quantum computers that simulate new worlds and species to be designed and built.

> **Akashic Records Librarian** – maintains life records' storage/retrieval. (Used during the Life Review reported by NDE survivors.)

> **Gods-in-Training I** – responsible to oversee the Drama/Greater Script Simulation: Man and feedback of the Control System. Many sub-areas here.

Gods-in-Training II – responsible for the Holographic stabilization and interface with the Replicator technology. Sub-areas here.

Soul Counselors – responsible for evaluation, guidance and training of in-coming souls to the InterLife for further development which may include imprinting or vibrational adjustment.
Many levels here, including Teachers.

This 3D Construct science mirrors that of the 4D realm, <u>partially</u>. So as a Simulation, there are basic things to be learned here that apply back in the 4D realm… (Since the Simulation only partially reflects the real 4D world, **what is learned in Science within the Simulation is like learning the IMAX Theatre** and thinking that ALL the laws and processes are the real world. Simulation Science does not count for the following 3 types of positions).

Bio Scientists -- these beings experiment with new lifeforms, ways to engineer them, transplant them, and ways to improve them.

Astro Scientists -- these beings experiment with Galaxies, Suns, planets, comets, etc. to engineer new planets, manipulate orbits, manipulate Dark Matter/Energy, and all the while ensure balance/order in the material world.

MLD Scientists -- responsible for MultiLevel universe and Dimensional interface, including handling Timeline Shifts when necessary.

There are many others, but it is a busy 4D realm over there; no one is sitting around on a cloud playing a harp – unless they're on a coffee break! The reason that the PTB (Astral entities) wants to block Souls from progressing is mainly because of this position:

Gods-in-Training III – responsible for overseeing, managing and controlling the Beings of Light, the Astral PTB, and humans –to make sure that lessons are properly administered (according to Scripts) and it amounts to controlling what the PTB can 'get away with.'
This is as close as the Earth Realm comes to having a "police force."

So who would you work for/with in these positions? You need to realize that due to Jesus' willing sacrifice of self, energy, time and blood, sweat & tears, He is the

majordomo of the Earth Realm. Yes, He really existed and was not a fable. Souls at different levels of personal growth serve below Him. The Angels serve Him, and Earth is a very special <u>created</u> Realm and is the proof of God's Grace toward Man... and even to those who harass and obstruct Man from the Astral Realm – it is all orchestrated as **catalyst for growth**. (VEG went into more detail in Chs. 15-16.)

Must Christianity Change?

This is a key issue in today's world. **Bishop John Shelby Spong** of the Episcopal Church wrote several books, notably <u>Why Christianity Must Change or Die</u>, in which he feared that Christianity was too backward, old-fashioned and could not meet the more inquisitive, critical and demanding public of today. He cited the Bible as being 2000 years old and not relevant to today's world because it dealt with a more primitive type of Man of 2000 years ago. Somehow Man being better educated now and learning more about our world, it is feared that Christianity must be updated to be more relevant. And for a while I supported his idea.

One of the major questions I began to have as I tried to analyze the issue and come up with what Christianity could change <u>to</u>.... sent me back researching ways to **revamp existing Christianity**, instead of changing it into something else. There have been progressive Christianity movements and New Thought **Unity Church** was certainly a new way to present Christianity... And yet some of the earlier progressive churches "devolved" back into standard Christian churches....

The problem was going too far out on a limb to be different but still pay homage to the fundamental Christian foundation. It didn't work because it was necessary, if one wanted to push Christian thought forward a bit, one had to include teachings from Lobsang Rampa, Paramahansa Yogananda, and even George Lamsa and Rocco Errico – the latter are **Aramaic scholars**, going back to the root of Christianity. **(Jesus spoke mostly in Aramaic.)** I was in one of those churches (where else?) in California, and it soon became less Christian and more intellectual, just another form of scholarship.

And that is a problem because the strong foundation of Christianity is compassion, brotherhood, faith and good works – not so much having scholarly knowledge of how everything works and what the original Aramaic meant. And when we did see the original Aramaic idioms for what they were in the Bible, somehow the "magic" or charm of traditional Christianity was lost... and we began to question everything!

Two Aramaic cases-in-point:

The Bible says that it is easier for a camel to go thru the eye of a needle than for a rich man to enter the kingdom of Heaven.

In the Aramaic, the word for camel is gamlá   and the word for rope is gámla…. The same word but with an <u>accent shift</u>! The accent mark is not visible (faded) on the original papyrus.

But the Aramaic is obviously referring to a <u>rope</u> going thru the eye of a needle…. Why it was "mistranslated" is curious.

The second example is based on an Aramaic idiom again. In the Middle East when someone is in a quandary, has a big decision to make and doesn't know which way to go, it is said *"they are in a fish."*

We say something similar even today when we have a problem…
"Gee, I am in a pickle!"

So the story of **Jonah in the Whale** was not literal and people who know the Aramaic idiom just laugh at Christians who think Jonah was really in a whale. He had a big problem, he was in a <u>big fish</u> – a whale!

(Not to say that God couldn't have a whale swallow a man if He wanted to, but it is doubtful (1) that the man's body could get past the constrictive strainer at the back of the whale's throat (for straining out anything but krill), and (2) the stomach acid would have killed Jonah and started dissolving him had he spent more than 15 minutes in the whale's stomach!)

Imagine people 500 years from now finding a text where someone said they were in a pickle – Will our future archeologists think that Texas supplied people with pickles big enough to crawl into?

What happened when our church started hearing from **Dr. Arnold Fruchtenbaum** (a very learned Rabbi) what the <u>Jewish and Aramaic idioms</u> meant, it unfortunately led to questioning other aspects of the Bible and that promoted a slightly negative atmosphere. No one's faith was destroyed, and the information was initially exciting to know what was really being said, but… it just somehow dampened our spirits because we knew that the translators didn't know the whole story and we felt cheated… what else did they mistranslate?

Advanced Christianity, Part I

In deference to Bishop Spong, for whom I have great respect, all that is needed is a return to the dynamic, powerful version where the Presence and miracles were very real. Once one has experienced the power and presence of God doing things in our midst, nothing else can compare… It is <u>we</u> who have moved off of the exciting, dynamic reality of God moving, answering, healing and guiding people. <u>We</u> have lost the connection with the supernatural wonder of the Father of Light operating in our midst – recall Chapter 1 where I list the times and ways in which I was guided and protected!

We have lost touch with Him! He didn't go away, hide or refuse to operate in our midst – <u>we</u> don't call on Him like mankind used to!

> **God is the same yesterday, today and forever. If you can't feel His presence, who moved?** (Heb. 13:8)

So we don't need a new Christianity or an insert of new Truth teaching from The Tao. Yet reading the same old passages in the Bible can get stale… unless one has a **personal experience of Him**. Then reading the same old passages has a vibrant life to it, and many times we see a new something in what it has always said… because we are open to being taught. Because we are receptive. Because we are now expecting Him to show up…. at least in spirit.

> The Christian church must be revitalized… it is called **Revival.**

So how do we 'revive' the church? Do we need to meditate, do Kundalini, or sing more hymns? The question is really asking: **How can we experience Him?** Be careful what you ask for – the Presence can knock your socks off! You will never be the same. It is the Fire of God.

Personal Revival: Peak Experience

I went into and out of everything looking for Truth, and in some seminars I often had what they call a **Peak Experience**…. And once you have had one, you'll find you can generate it again, any time you want, on your own! Connecting with the Presence is the same… God will meet you at least halfway.

> I was walking thru the park on a warm Spring day and was pretty much by myself in my section of the park. I was walking down the

sidewalk past the hundreds of flowers that had just bloomed and my mind was clear (and I think that is a precursor to what happened next)…

All of a sudden, I do not know how it happened, I stopped walking and just stared at the flowers … Damn, they were alive!, and their colors just shouted RED, YELLOW…. And I saw the grass… I had never seen a GREEN like that before! The tree was alive – a massive oak tree and I could have sworn it said "Hi!" while I was in that state. Everything was so brilliant, **vibrant**… and the colors of everything were so rich, deep… I was awestruck.

It was overwhelming. I had never seen the world that way. I knew I was a part of it and it was a part of me. It was a state of being like none I had ever experienced, in any of the Est, Silva, NLP etc. seminars. I had no problems, no questions… it was as if I was standing in Heaven…. why I felt that way I have no idea, who knows what Heaven looks like? Then I realized I not only felt the energy of my surroundings, it was a sense of being **one with it all**, and my energy was flowing out to the trees and flowers there.

Then I had an ugly thought: Oh God, I don't want to drop out of this state! What if I lose it? And immediately I <u>did</u> lose it… one negative 'what if' thought was all it took. But I remembered what it felt like for about 10 minutes and as I walked over to the pond, I got into the memory of what it felt like, and I popped back into that state.

Before leaving the park that day, I played with being able to snap in and out of that state, and to this day, **because I know what it feels like**, I can still recreate it any time I want… but after a while, you have to function in the less exciting real world and I have not done it for a while. I am sure that it is **a state generated by the heart** because I was totally <u>out of my head</u>, and when I wondered if I would lose that state, it put me in my head and I lost it!

All of that to say that **we are a lot more than we think we are**, and can do things that show us a better side of Life. In fact, I felt that the Peak state was the real world and the 'head state' was not the real us.

Fire of God, Kundalini, Peak Experience, anointings… all part of God's world. What would happen if more of us had one-on-one experiences with the **God Presence**? It would dynamically change the Christian church… as it did for the disciples who were in the Upper Room in ACTS and 'tongues of fire' lit on their heads. Until we can handle that, there is no need to look outside of Christianity for God or miracles or the benign supernatural… our problem is **learning to accept**

the depth of the existing Christianity as normal and live with it adding a new dimension to our lives!

There used to be men and women of God walking this country (and world)… Kathryn Kuhlman, **Smith Wigglesworth**, Aimee Semple McPherson, John Lake and today: **Benny Hinn** (verified by personal experience), Robin Harfouche (verified by personal experience), and **Kenneth Hagin**, Agnes Sanford, and Bob Shattles (local Texas healer). These were healers who **walked the Walk** and carried an anointing. There have been none following them. (In addition, in today's world, are found a few Chinese Qigong healers and that is profiled in TOM.)

Could it be that the basics of standard Christian Religion still apply and are a foundation on which to build a working life and behavior that will empower the soul to graduate from Earth School? Again, it appears to be a matter of faith and persistence … if you can see the value in becoming an Earth Graduate, then drinking deeper of the Christian well can achieve that goal.

Baptism of the Holy Spirit (BHS)

What appears to be needed is an immersion in the deeper aspects of Christianity that will empower one to live a life that qualifies for graduation. That is what I call Advanced Christianity. **You ask for it and surrender into it.**

It was said earlier that there is a significant step that will create blessings for anyone brave enough to "go for it!" Having done it myself, I can vouch for it as a real **transformational experience**. It is called the **Baptism of the Holy Spirit** or BHS. This is what is missing in so many drab Christian lives, where the person is just holding on to basic promises and not experiencing them. Many pastors have lost their initial fire and just dole out average messages of faith – and you can see their small churches on the street in many American towns… they are not empowered. They are just barely hanging on and their congregations don't suspect that there is more… no one in that church has ever seen more… People attend on Sundays, nod to God, and drop a prayer request in the box by the door…. Then they go to Sunday brunch. Too many Christians are unempowered to witness, heal, prophesy, and encourage/exhort others. The **BHS confers the giftings**, and while I went thru the BHS, I did not get the giftings I wanted. So it is not like a vending machine… you get what is appropriate <u>for you</u>.

Before doing the BHS, one must have been sincere about wanting to be a Christian. Then there is **water baptism**, just an outward sign to others that you are committed. Then you have to clean up 'loose ends' in one's life… set about **forgiving others** (it detaches you from them [see TOM], and sets <u>you</u> free). Get rid of any witchy stuff, books, pictures, videos… **clean it up**. Spend some time in meditation or reflection on yourself, your life, your job, your church, your goals…. Get a sense of who/what you are. Then you are ready to ask for the BHS… it is best if you can find a (conservative) charismatic church that knows what it is, and someone there has it and can demonstrate some basic giftings for you (not fake tongues), and lastly **you expect it to happen**… and it may not happen on the floor with elders standing and chanting over you – You can just **ask God for the BHS** (in privacy) and if the time is right, you will receive.

When I received the BHS, I started sweating and was fortunately kneeling at the altar of a church I knew had real, gifted people. They laid hands on me, and their hands were <u>hot</u>. What I received was not tongues (that is not always an outward sign) – I discovered I had power to deliver people and occasionally heal others – and yet if the others were skeptical (or really didn't want to be healed – some don't as their current situation gets them a lot of attention!) I could do nothing. That is to say, it wasn't my power, but His power would not flow thru me – but when it does you'll know it!

By the way, even if you have a gift of healing, it cannot be done for just anybody needing a healing. I embarrassed myself real good in 1998 back when I could do it, by driving to Abilene and trying to heal a man who really needed it – and he was a member of our extended family. I worked on him for 30-40 minutes and absolutely nothing happened. We prayed and asked for intervention… nothing. What was going on?

It was his time to go and he was not to be healed.

BTW, the other lesson I got before I quit trying to heal others: if their illness is Karmic, you cannot remove it – it is their "lesson."

Advanced Christianity, Part II

It has to be pointed out that an advanced form of Christianity already DOES exist in the USA, and around the world, and it is called **Unity,** or sometimes Unity Church of Christ. This is <u>not Unitarianism</u>. The founders Charles and Myrtle Fillmore experienced divine healing and amazing answer to prayer by following a simple formula, and in response and gratitude in 1889, they started a study group and later the basis for a church that is now world-wide and there is at least one in most every state in the USA.

The home church and school is in Lee's Summit, Missouri:

The teachings are Christ-based and the Bible is used, but so are the works of Rocco Errico and George Lamsa (Aramaic scholars), the Fillmores having taken pains to include the scholars' works that explain what the Bible really said (Jesus spoke in Aramaic) … as was shown earlier (in the 'Must Christianity Change?' Section) in this chapter. The Church seems to prefer to teach Man the principles of going within (as did the Gnostics) and **discovering the connection that all souls have with God,** and that empowers prayer , health, and living. The Church has 5 basic metaphysical (the larger laws of Life) principles which the Fillmores discovered (Lessons in Truth, a book and yearly classes) that anyone can apply to get proactive results in his/her life.

> **Unity is not traditional Christianity** – but having spent years in it myself – it sure looks like what Jesus would teach if He were a Unity Minister.

Unity Overview

> Unity describes itself as a worldwide Christian organization which teaches a positive approach to life, seeking to accept the good in all people and events. Unity began as a healing ministry and healing

has continued to be its main emphasis. It teaches that all people can improve the quality of their lives through thought.

Unity describes itself as having no particular creed, no set dogma, and no required ritual. It maintains that there is good in every approach to God and in every religion that is fulfilling someone's needs. It holds that one should not focus on past sins but on the potential good in all.

Unity emphasizes spiritual healing, prosperity and practical Christianity in its teachings. Illness is considered to be curable by spiritual means, but Unity <u>does not reject or resist medical treatments</u>.
Unity is accepting of the beliefs of others. [113]

Some of the salient ministers and writers in Unity have been Ernest C. Wilson, Eric Butterworth, Grover Thornsberry, and Catherine Ponder. Today's set includes current graduates from the Unity School of Ministry such as Linda Pendergast (whose NDE turned her into a Unity Minister), Carol Record, Joel & Diana Hughes, Ellen Debenport, Anne Tabor, Steve Baherman [comic turned minister], and such notables as the following have been affiliated with Unity:

Well known persons affiliated with Unity include Maya Angelou, Betty White, Eleanor Powell, Lucie Arnaz, and Wally Amos [founder of Famous Amos Cookies], Licensed Unity Teacher Ruth Warrick, Barbara Billingsley, Theodore Schneider, Erykah Badu, Matt Hoverman, author Victoria Moran, Patricia Neal, and Holmes Osborne. [114]

In addition, Unity also recognizes Truth in the writings of Emmet Fox, Ralph Waldo Emerson, Emilie Cady, and Mary Baker Eddy.

Unity's Five Principles

God is the source and creator of all. There is no other enduring power. God is good and present everywhere.

We are spiritual beings, created in God's image. The spirit of God lives within each person; therefore, all people are inherently good.

Our **thinking tends to reflect itself in our life experiences**.

There is **power in affirmative prayer**, which can increase a person's connection to God.

Knowledge of these spiritual principles is not enough.
People must live them.

Unity Offerings

The Church offers special prayer thru its **Silent Unity** department which is what my mother called in January 1965 (see Chapter 1) when I was admitted to the hospital and supposed to die of pneumonia that night – the emergency prayers of Silent Unity were answered and a Healing Angel was sent to my room. I emphasize that these are very spiritual people and they spend hours a day in prayer 365/24/7. Unity's stand on prayer is thus:

> It has been generally accepted that Jesus' great works were miracles and that the power to do miracles was delegated to His immediate followers only. In recent years many of Jesus' followers have inquired into His healing methods, and they have found that **healing is based on universal mental and spiritual laws which anyone can utilize** who will comply with the conditions involved in these laws. [115] [emphasis added]

> (TSiM examines the metaphysical principles involved in healing, and cites Drs. **Ernest Holmes** [founder of Religious Science…now defunct but his key work *The Science of Mind* still exists], Maxwell Maltz [who wrote *Psycho-Cybernetics*] and even Jose Silva [who founded The Silva Method] discovered the same technique. TSiM also cites **Dr. Caroline Myss** who helps people get their mental focus correct and stresses that ones' biography often becomes their biology and how to stop that.)[116]

Unity Church also publishes the **Daily Word**, found throughout the world in most of the major languages. A daily inspirational guide that one can meditate on, and it gives a Bible reference for each day. And there is the insightful magazine, ***Unity.***

Unity Church is worth checking out – it is a heart trip (versus an intellectual head trip) and very supportive of personal growth.

Thus, if you prefer to remain Christian, and want the same power as the disciples had, and you have no Unity Church to check out, you need to find a real Charismatic church that will help revive your soul and open new possibilities for you.

It is not without **a warning** here that I say there are some Charismatic churches that I cannot handle and do not feel comfortable with – some are really "way out there"… and some do not have the connection with the Presence that they think they do. I was fortunate to have found a pastor who was down-to-earth and yet had the Fire and transformed me and few others so that we know what it is, and on some Sundays, we could feel the Presence… and occasionally gold dust would fall. Seriously.

I scooped some up and stuck it in Psalms 91.

Our church was on fire, and not weird... a rare balance of Presence and conventional behavior from the congregation. I think we were all too awestruck when the Presence fell to do/say anything weird. And only once or twice did I hear speaking in tongues (with translation by a 2nd person).... Most of the time people started getting healed. Most of the congregation hit the floor on their knees... some passed out. The Presence was that awesome!

That was the Christianity of Jesus' day and the one that the disciples knew.

It is hard to go back to a quiet, reserved, straight-laced Christian church after that... and that is why some people think Christianity is dull and doesn't work. That is why some people think Christianity must change or die – because they have never seen what happens when the Presence falls on a congregation. Some people and pastors can't handle it, so they don't tempt the Spirit.

Summary

Between the two options, New Age and Christianity, the Christian life is more likely to get a person living, thinking and acting in ways that will make him/her an Earth Graduate. The New Age has some truth in it, but unless you know what it is, and can spot the deceptions, AND if you can remain humble, patient, compassionate and respect self and others, AND not be egotistical, arrogant or petty – something an Old Soul could do – you are better off being a Christian, even a Zen Buddhist or Hindu and NOT a New Ager.

> And don't go assuming you can do it because there are very few
> Old Souls in our Western (USA) Civilization at this time. Sorry.

Whoa, wait... Did you say Buddhist or Hindu?

Yes, and the reason why is very well demonstrated in the way those two countries behave. Their people are usually very humble, giving, patient and they respect each other and other countries. Do Hindu and Buddhist countries often go to war with other countries? (Japan [Shintoism] was an exception in WW II.)

They also understand **Reincarnation and Karma**... which were removed from the Christian teachings about AD 325. **Father Origen** taught it about AD 100, and was chastised for it, **Apollonius taught it**, and even the Bible still has two spots that the Church missed when they expunged the 'dangerous' teaching (see below).

> By the way, the Church felt that reincarnation was not conducive
> to souls getting their act together in THIS lifetime. The reasoning

was that if a person knew that s/he would come back again, next lifetime, they would argue that they could goof off now and have fun this time, and tell the priest that they'd get serious <u>next time</u>!

The wise Church, knowing human nature, knew that it is important to do well each lifetime, be as good as you could be, face your problems, handle them and deal with others with respect and compassion, and try to die with as much dignity and success as you could. So it wasn't removed from Christianity because it was false, but because it was seen as counter-productive.

Biblical Translations Caution

The two remaining places in the Bible that deal with **Reincarnation** do not label it as such, and Jesus never says the concept is nonsense, he just sidesteps the issue (or His comments were removed)...

> **Matthew 11:14-15**
> **And if you will receive it, this is Elijah, who was to come.**
> **He that has ears to hear, let him hear.**

> Jesus is explaining to the disciples that John the Baptist is actually Elijah and God brought him back for this time. (Interesting that in my KJV Study Bible, <u>this passage is not dealt with</u>. It is not explained.) The second verse indicates that some people will not or cannot hear what is <u>really</u> being said, as they don't have wisdom, or intuition.

> **and the other place:**

> **John 9:1-3**

> **And as Jesus passed by, he saw a man which was blind from his birth.**
> **And his disciples asked him, Master, who did sin, this man or his parents that he was born blind?**

> Note: his parents might be at fault due to the passage in the Old Testament that says: The sins of the fathers are visited on the children.... to the fourth generation.   Ex. 34:7 and Ex. 20:5.

Jesus answered: neither has this man sinned nor his parents, but that the works of God should be made manifest in him.

And he proceeds to heal the man and restore his sight. He was demonstrating **God's Grace to Man**. The point is that the disciples were aware of the **Karma** issue, and if the man had sinned before birth, then it was Karma…. and Jesus doesn't look at them and say "What the heck are you talking about!?" as if they were crazy. He knew about Karma and Reincarnation… but in this case, the blind man was 'planted' by God so that Jesus could demonstrate Power & Grace to mankind.

(And my Study Bible totally avoids the issue, **again**, citing a Jewish belief that one could somehow sin before birth… with no further explanation. My study Bible does not even reference Exodus 34:7.)

As we were noting, the Buddhist and Hindu people know better – they know if they do something stupid or violent to someone else, they will have to make amends and pay it back and they are more peaceful, compassionate and respectful as a result… so did the Western Church err in keeping it from the Christian flock? Western Civilization certainly has a history of war and violence… exceeding that of the Buddhist countries. And just maybe the few souls who would party and err this lifetime, despite Karma, promising to straighten out next time around, might have been so few in number as to make the Church's decision to exclude Karma and Reincarnation an unwise one after all. So maybe the issue was really control….

And by the way – even if Karma and Reincarnation were false, would it not be a great teaching to get people to think twice before doing something negative that they would have to answer for?

If you know you are going to "pay" for your bad actions, would you not stop and think twice… As Buddhists and Hindus do?

Would you not think of what you could do to bless someone else so that you have that coming back to you? (Hint: think about the movie *Pay It Forward*.)

And by the way, since Earth really is either a VR Sphere or a Flat Earth (Chapters 3 and 10), which is proof of God's existence AND His Grace to Man, the emphasis is on knowing that this is an Earth School and graduating is what is important. Christianity can be a better Path than atheism – as you have already committed to the One who runs this place!

Chapter 6: Religion & Antiquity

Chapter 5 dealt with Religion in general and the New Age, and how they contrast. It also dealt with the issue of: Does Christianity need to change to survive (as asked by Bishop John Shelby Spong). It was seen that a major change is not necessary as long as Christianity does not become a mindless nod-to-God experience. There is power in Christianity when one plumbs its depths and original context. Modern Man has lost sight of what his ancestors knew 2000 years ago. Perhaps a revival is in order…?

This chapter takes a step back into olden times and looks at three interesting discoveries. Things that are part of the Christian world but have been ignored.

Those things are two literary works: The Gospel of Barnabas and the Book of Enoch, and a person: Apollonius of Týana.

The Gospel of Barnabas

Don't go looking for this one in your Bible. It was just recently found in Turkey, about the year AD 2000. It was kept secret as the Vatican wanted to authenticate it first and not arouse the faithful because of what it said that puts it on a par with the Nag Hammadí manuscripts.

The Gospel of Barnabas was seized from a gang of smugglers in the Mediterranean area. It is valued at $28 million. According to reports and expert analysis, the book is

genuine and an original. It is written with gold lettering in Aramaic on loosely bound leather pages.

Writing in gold was not for the average author and it means the document was very important.

1500-2000 year old bible found in Ankara, Turkey.

It has turned out to be a bombshell which Christian scholars hastened to discredit.

Yes, Barnabas was a disciple of Christ (Acts 13:1) and the document is dated to the First Century. So that part is not disputed… it is rather what the book says.

The Gospel says that Jesus was not crucified, nor was He the Son of God, but a prophet, healer and teacher.

It also calls the **Apostle Paul "an imposter"** – and that is addressed in the following section on Apollonius. Remember that James and Peter would not accept Paul teaching in Jerusalem and told him to take his message on the road… because he was known to persecute the new church. They didn't trust him. There was something about Saul/Paul that set people off (Acts 8:3) and Barnabas also had issues with Paul (Acts 15:2) and they parted company while doing their missionary work.

The Gospel of Barnabas also says that **Jesus ascended to Heaven alive** and that Judas Iscariot was crucified in His place. The text maintains a vision similar to that of Islam in the Koran where Jesus was a prophet and **did not die on the cross.**

By the way, the Koran shares the information that Jesus did not die on the Cross, but was healed. (**Surah 4:157**). It is rumored that the man called Jesus went back to India and spent the rest of His days there until **dying of natural causes at the age of 117**. [117] {emphasis added]

So if He didn't die, He certainly wasn't a "sin offering" for Mankind. But He did <u>shed his blood</u> as He was whipped, beaten and stabbed, and He did go to the cross because of the sins of mankind.

According to my Source, Jesus <u>did</u> go to the cross about Noon, and was hung on it, but it was a Friday, just the eve of Passover Weekend – a holy weekend and the Jews did not want dead bodies hanging around, so about 3 pm Jesus was taken down and put in Joseph of Arimathea's tomb, and nursed back to health… The Shroud of Turin is His and shows **blood flow into the Shroud**… dead men don't bleed.
If He had been left longer on the cross, He would have died.

At the risk of irritating some Christians, it needs to be said **that Jesus never said He was here to pay for Man's sins**… that was all Paul's idea. Check it out. VEG Chs. 1 and 11 cover this in more detail.

The Gospel of Barnabas seems to worry the Vatican because the 4 canonical gospels of Matthew, Mark, Luke and John were hand-picked by the Catholic Church in AD 325, and the Gospel of Barnabas was rejected along with the Book of Enoch. [118] The Church was formulating its theology and things, <u>even if true</u>, that didn't agree with what the Church wanted to promote among the faithful, were thrown out, seen as heresy – now you know another reason for the 400+ years of Inquisition. People knew about these older teachings…. And many of them knew about **Apollonius of Týana** and saw the juggling of Biblical characters (next section) and questioned it. For their trouble, they were often paid a terminal visit by the Inquisition.

Of course, the Church can label it a "Muslim lie" but that is not fair to the truth. And personally such information (above) does not please me, but truth is truth and is also why I stopped my studies for the ministry… I wondered just <u>how much</u> of what I had been told was a lie?…. albeit done with a benign intent (i.e., don't confuse the sheep, just keep them on the straight and narrow, the simpler the better. Thus was born the K.I.S.S. paradigm.)

Elsewhere I have told the story about coming up with 22 questions during my ministerial studies, and I made a list of them… things I had found that appeared to be contradictions, omissions, and plain errors (mistranslations). I put some of these in VEG (Chs. 1 and 11).

So I went to the head pastor and showed him the list and humbly asked for clarification. Did he suggest I read up on Apologetics? No. Did he have answers for any of the issues? No. He felt threatened.

I was politely told I could take it all on faith or leave the church. Shocked, that is what I did. I left.

Needless to say, **we don't have much of a faith if we can't question it**, examine it and rise above any errors, inconsistencies and omissions. We should not need **Apologetics** (a division of Christian scholarship) to "explain" irregularities in the Word of God… **Why wasn't the text corrected before it was published** or "cast in concrete" before the congregations saw it? And I gave samples of sloppy translation in Chapter 5 where two parables are mistranslated (e.g., the eye of a needle, and Jonah and the whale).

For centuries the defense of "blind faith" (don't question, just take it all on faith) has driven religious factions and nations to war… and it has all been justified with lies… we'll come to this again in the last chapter.

You shall know the truth and the truth shall set you free

…from the manipulation of those who have covert agendas.

Apollonius of Týana

> If you have read VEG Ch. 11 this section is a repeat – <u>for those who are not familiar with VEG</u>, and so this section may be skipped if you have read VEG.

We might consider the man called Apollonius of Týana (4 BC – 102 AD), born in the region of Cappadocia (Turkey) during the reign of Augustus, **whose father was the god Apollo**. He was the subject of Philostratus' biography <u>Life of Apollonius</u> (written in AD 210). The biography is considered reasonably credible since Philostratus was a personal friend of Damis, who was a follower of Apollonius.[119]

Even if, as some detractors say, Damis was fictional, Apollonius <u>did</u> have followers and Baha'u'llah the founder of the Baha'i Faith, Sir Francis Bacon and Voltaire all recognized Apollonius as an exemplary philosopher, teacher, healer, and Miracle-worker. [120] He was something of a folk hero.

He wandered around teaching, healing, and doing miracles, about the time that Jesus was said to have done his miracles. Philostratus' book was suppressed by the orthodox clergy as the Church gained momentum with its story of Jesus, and whereas many copies of the story of Apollonius could be found in Alexandria, the copies were harder to find after the libraries of Alexandria were sacked and burned. [121]

It looks like the Biblical life of Jesus was modeled on that of Apollonius, and Jesus' birth and death events might have been a composite of earlier people and myths.

Temple Bust of Apollonius of Týana
(source: http://www.truthbeknown.com/apollonius.jpg)

Unlike Jesus, there is <u>historical evidence</u> to prove that Apollonius actually existed. Apollonius was born in the reign of [Emperor] Augustus… in the small town of **Týana, not far from Saul's Tarsus in Turkey** (then Galatia). In the Augustan age, historians flourished; poets, orators, critics, and travelers abounded. Yet not one of them mentions the name of Jesus Christ, much less any incident in his life. Jesus left us nothing in writing, although there is a growing specu-lation that the **Gospel of Thomas** was written by his hand… If indeed [Jesus] existed, he traveled only to Judea, Kasmir and Egypt. Apollonius traveled ex-tensively and wrote extensively.

> The Emperor Marcus Aurelius admitted that it was to Apollonius that he owed his own philosophy, and **erected temples and statues in his honor.** No statues or temples were erected to Jesus. [122] [emphasis added]

Also said of the Emperor Aurelian:

> Aurelian vowed to erect temples and statues to [Apollonius'] honor, for "was there ever anything among man more holy, venerable, noble, and divine than Apollonius?" He restored life to the dead, he did and spoke many things beyond human reach (<u>The Magus</u> by Francis Barrett). [123]

Apollonius' Fame

It was said that many temples and statues were erected to Apollonius <u>in many places,</u> "…including his own town of Týana, even <u>though the later Christians destroyed many of them.</u>" [124] (What was that all about?) It is odd that **Apollonius' reputation was identical to that of Jesus**, and he was well-respected and even revered in many places, and yet the **Christians almost demonically turned to destroy Apollonius' works and suppress his legacy.** Apollonius was not a rival to Jesus, nor did he threaten the fledgling Church in any way. [125]

Says another expert in the origins of Christianity, regarding Apollonius:

> …Hierocles, the pro-consul under Diocletian (284-305 AD), …wrote the "Philalethes" (AD 303) **exposing the Apollonius-Jesus connection.** It should be noted that Philostratus' account makes no mention of any Jesus Christ, not even as a rival to Apollonius who purportedly **lived precisely at the time alleged of Jesus.** [126] [emphasis added]

However, Apollonius' <u>factual</u> life was in danger of usurping the Church's <u>idea</u> of Jesus, and this bothered early Church fathers like Justin Martyr (2nd century AD):

How is it that the talismans by Apollonius have power over certain members of creation, for they prevent, as we have seen, the fury of the waves, the violence of the winds, and the attacks of wild beasts. And whilst Our Lord's miracles are preserved **by tradition alone**, those of Apollonius are most numerous, and actually manifested in present facts... [127] [emphasis added]

Thus **the biography of Apollonius was suppressed by the Church** and "...the books of the New Testament did not appear until at the very least 100 years after The Life of Apollonius." [128] So most of the writings about Apollonius were either destroyed, hidden, or just suppressed until after those living (who knew of Apollonius) had died before the Church promoted the new savior, the god-man, Jesus. Apollonius himself could not be used as the Church's role model since the Church would have had to embellish the god-man with various pagan elements to make it acceptable to the many converts it hoped to acquire. [129]

Connection with Dr. Fomenko?

If as Dr. Anatoly Fomenko says (Ch. 10 in VEG), history was backdated almost 1000 years, is it possible that **the Inquisitions of the 1100-1600 AD era were right on the heels of Apollonius' time (AD 100) and his works**... which is why people knew about him? If he existed in the 1000-1100 AD era, could that be why the Inquisition had to be used to promote the fledgling Church – free from non-Catholic history? Would that not be a great idea to backdate history – to distance Apollonius backwards 800-1000 years from the time of the Inquisition? That way future people would not connect the two, and what if the story of Apollonius could be buried altogether?

Jesus as Apollonius

In addition, nowhere in the writings of Apollonius is there a mention of anyone called Jesus or Saul/Paul. Conversely, the contemporaries of the alleged Jesus such as Philo, Suetonius, Plutarch, Pliny the Younger and Cornelius Tacitus **do not specifically mention a Jesus**, nor is the Josephus reference trustworthy: it has been found to be an insertion, suggesting **written fraud in the early Christian church** [see below]. [130]

While there is often a 'Yeshu' or 'Chrestus' found in some of the ancient texts, the mention is usually a brief 1-liner and it is unclear who they are referring to. Certainly if Chrestus was as important as Jesus was alleged to be, wouldn't there be more said about him? Apollonius was as note-worthy and significant as was Jesus, but **Apollonius was well-known and Jesus wasn't**. Strange, unless Jesus was Apollonius and the Catholic Church changed his name...

Apollonius The Nazarene

(source: http://www.interfarfacing.com/apollonius.jpg)

Jesus (source: Bing Images/Jesus)

Similarity?

The picture of Apollonius above is reminiscent of a lot of pictures of Jesus. In fact, there is a book by a young boy (four-year-old **Colton Burpo**) who died, went to Heaven, then returned (**NDE**) and told people what Jesus really looked like. [131] Then **Akiane Kramarik** painted the picture (below left) which Colton approved:

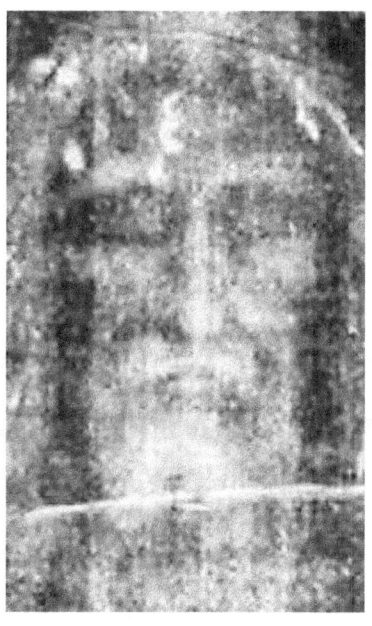

Akiane's Picture of Jesus   **Image on Shroud of Turin**

For what it is worth, some people may wonder if her picture matches the Shroud of

Turin image, and that is included above. You be the judge… (In addition, Appendix D in VEG has an analysis by Dr. Fomenko on Jesus that could be of interest: linking him with the Shroud and the Crusades.)

Could the life, activities and teachings of **Apollonius** have been used to establish the life of Jesus, as appears to also have been the case with the lives of **Horus** and **Krishna**? Certainly if the written records of Jesus were lost and only the oral tradition survived, would it not have been a plausible effort to 'reconstruct' the life of Jesus based on any one of the exemplary three saints mentioned? Do not all avatars do basically the same thing and teach the same things?

> And yet the Church did not wipe out Apollonius' teaching, as many of **his teachings and writings were taken East into India** for safe-keeping. In 1801 a major text was brought back and translated to English. [132]   [emphasis added]

> Nor were his followers scattered after his death. Apollonius had started a church and had followers called *Apolloniei* which survived several centuries after his death, and they were probably connected with the Therapeuts and **Nazarenes** as Apollonius had contact and ministry with them during his life. [133]

> Many of Apollonius' followers were also to be found <u>in the fledgling Church</u>, especially after The Council of Nicea (AD 325) met, so he may be said to have had an effect on the Church inasmuch as his life mirrors that of the new god-man, Jesus, including his writings, and then the Church embellished the newly-hatched Nicean teachings with "…serpentine myths and traditions of the oldest [pagan] order." [134]

Paul as Apollonius

It is interesting that there are many similarities between the missionary activities of St. Paul and Apollonius, with both men having visited <u>the same cities</u> and having performed the same works. **In the same time frame.**

> **It is very likely that 'Paul' is a version of the name Apollonius since Apollonius' nickname was 'Pol'** *and visited the same places the Apostle Paul was said to have visited, and Pol did the same things Paul did.* [135] *[emphasis added]*

> It is beginning to look like there was a reason for destroying the Library of Alexandria and creating the Inquisition…

There is historical corroboration:

Apollonius is a Greek name, the Latin Romanized version would be **Apollos.** Apollos over a period of time as well as convenience morphed to **Paulos**…in its English format is **Paul**… Apollonius was born in Týana [Turkey, **just 30 miles from St. Paul's birth town of Tarsus**]… both Apollonius and Paul/Saul were in Tarsus at the same time in their youth, as Newman points out, **Apollonius and Paul were also at Ephesus and Rome at exactly the same time…**

> In the Greek text, Apollonius is commonly written as Pol as well as Apollos, "Apollos" of the New Testament [Corinthians] – the eloquent "Jew" whose preaching and baptizing at Corinth and at Ephesus preceded the work of the Apostle Paul…

> The companion of Apollonius was **Demis**… and [Timothy] **Demas** was the companion of Paul. [136]

It is reasonable to think that Apollonius could have served as **the role model for both Jesus and Paul.** And current-day author Ms. Atwater has something to say in this regard…

More Bible Insights

In her excellent book on *Near Death Experiences,* P.M.H. Atwater corroborates much of what has been said in this chapter and gives us a bombshell: [137]

> The Holy Bible and Christianity did not develop as the masses are taught. The **Sinai Bible** (the oldest known) contains the first mention of Jesus, beginning when he was 30 years old. When the New Testament of the Sinai Bible is compared with modern-day versions, **14,800 editorial alterations** can be identified. **Paul was actually Apollonius of Týana**, a first century wandering sage…. Restructured writings of Apollonius became the Epistles of Paul in AD 397. **Early Gospels never mentioned a virgin birth**. Christ's suffering, the crucifixion and the resurrection did not appear in the Gospels <u>until the 12th century</u>… [emphasis added]

Ms. Atwater further points out that today's Christian scholars know about **The Great Omission** – crucifixion/resurrection missing from earlier versions of the Bible – and **The Great Insertion** is when the Church put it into the Bible via the Gospel of John.

> The Church was very busy, and very creative.

Religious Tolerance

The Church dissed an important aspect of Apollonius' teaching: He was tolerant of all other religions and sought only to help people perfect whatever their form of worship was. It was only important that a person have the integrity and dedication to their professed belief – whatever it was.

Secondly, Apollonius did not travel to other lands with the goal of proselytizing. He understood that **the Father will hold people responsible for whatever their belief or professed faith is, and how well they served it** – NOT whether they believed in the 'right' religion.

This is a point of view to keep in mind in the earlier section on Bishop John Shelby Spong. Inasmuch as Apollonius communicated with the gods who run Earth and watch over Man (The Ancient Ones), and he was divinely empowered – meaning the earthly gods (Anunnaki: Enki) were also pleased with him – his attitude towards all religions should warrant serious attention… and copying.

Reasons to Reject?

It can be seen what the Church was trying to do in creating a god-man without any blemish when one reads more deeply about Apollonius. While everyone agrees that Apollonius was a **flawless model teacher and healer**, following the true Pythagorean Way with no black marks against him, [138] if the Church had promoted him by name, it might have led people to fall back into their old pagan ways or leave the Church. Why?

Apollonius was known to **venerate the Sun** every day, [139] he believed in **reincarnation** (as did Origen), and he supported the Gnostic message that **Man could find God within**, through Knowledge. In addition, it appears that Apollonius was not only a Pythagorean, but one of his main sources for his spiritual teachings was **Orpheus**, since the **two statues of Apollonius and Orpheus were often found together in many temples**. [140] These issues were anathema to the Church. A hybrid god-man, son of Apollo, son of an earth woman, would not do as a role model.

Further, and perhaps most significantly, as Apollonius means "son of Apollo", [141] then Apollo being one of the gods, and such term was often applied to the great men of antiquity (Gilgamesh, Sargon, Alexander, Zarathustra, Noah, etc.) suggests that Apollo, and thus his son, were in fact genetic **hybrids** and that bloodline with its special DNA is what gave Apollonius his special powers – just like Alexander the Great and Moses who were hybrid descendants of the Anunnaki *Adapa*. That fact would be enough for the Church to reject him – even though his words, behavior, and writings were as pure and as spiritual as the teachings of a Jesus. In fact, **Jesus' teachings are almost verbatim what Apollonius taught and wrote.**

In short, if this conclusion is correct, the Church just wasn't taking any chances and had to distance itself from the Anunnaki and Apollonius legacies.

So the Church could not accept the actual Apollonius as its basic role model, but **recreated a new god-man role model for humanity** that was <u>more in reach</u> than Apollonius was: the teaching of a fully human Jesus that was 'Christed' by the Father when he was baptized, would be easier to accept and follow. It is hard to emulate a hybrid god-man and so **Man was reduced to worshipping and not emulating.** [142]

And yet, **Apollonius was something of a god and was worshipped as one,** [143] so the issue is complicated. It lies in the main fact that **the earlier Church knew what Apollonius' lineage was**, and that he wasn't 100% Man, so the brief ministry of Jesus had to be used, probably amplified with Apollonius' exploits. Remember: Apollonius and Jesus were said to have historically taught and done the same things at the same time … And don't all avatars teach and do the same things, anyway?

Apollonius was called **Son of Man** because his mother was human.
Apollonius was called **Son of God** because his father was the god Apollo.

Is this the origin of the appellation: Son of God, Son of Man?

Apollonius spent his later years in Tibet, Kashmir and India, where he died well over 100 years old. He never married, and if Apollonius was the role model for Jesus that spikes the story that Jesus was married and sired a Merovingian bloodline (sorry, Dan Brown)…

On the other hand, the Church may have just used Apollonius' missionary travels as a basis for those of Paul. But it also looks like Apollonius was the basis for <u>both</u> Jesus and St. Paul… and if the people knew it, and said something, wouldn't the Inquisition visit them? And if the Library of Alexandria contained some of the writings of Apollonius, plus further tractates on Atlantis, plus something of the real history of Man on Earth… was that why the Library was sacked and burned, not once, but <u>three times</u>? And the second time was in AD 391 by Pope Theophilus.

Something to think about.

I have often wondered what damage would be done by the faithful sheep finding out that Jesus didn't exist… and yet, not to worry, I was not told that, and I did not discover that… just that He lived not 2000 Years ago (which is why there is no historical record of Him in that time), but that He ministered in the AD 1053-1086 timeframe (and there is reasonable proof: see VEG: Ch. 10 and Apx. D).

BTW, this agrees with the dates for the Crusades and the date of the Shroud of Turin. (See VEG, Apx. D.)

What happened according to Dr. Fomenko is that two Church priests adjusted Western Chronology (prior to the printing press when one could still rewrite things and there were no prior printed versions… and the Church did 90% of the rewriting), and they backdated history so that the Church appeared older than it was, and thus Jesus was moved back to AD 0.

Something more to think about.

The Book of Enoch

This section is a reworking of the Enoch info in VEG Ch.2 – for those who have not read VEG. Enoch and the Flat Earth (next section) may be new to you…

Enoch "walked with God" and was taken up into Heaven and shown the Watchers and the Flat Earth – see now why the book was banned from the Bible? Let's examine <u>both</u> aspects of Enoch's writing: the **Nephilim** and the **Flat Earth**.

Book of Enoch

<u>The Book of Enoch</u> was written by a man who is said to have walked very closely with The God. So closely that he knew what God was up to and had knowledge of what was going on in the heavenly realm that was affecting Earth. If someone didn't want Man to have a clue about what other beings might also be on the Earth, or who might have afflicted Man, they would suppress this book and keep it out of the Bible.

While <u>The Book of Enoch</u> is basically a lament over the sinful condition of Mankind, and a call to repent with blessings for those who are righteous, it also expands on

Genesis 6 which says that the angelic host (Watchers) descended and created giants (Nephilim) and havoc in the Earth.

The Book of Enoch dates to about the 2nd and 1st centuries B.C., and was held in great reverence by many of the early Church Fathers. However, due to efforts of the emerging Christian Church, by the 4th century A.D. it was looked on as heretical, and later condemned.[144] Thank you, Constantine.

I. Watchers and Angels

The story of Enoch concerns how God set 300 Watchers over the earthly Creation and how some **200 came down** and immersed themselves in 3D, teaching forbidden knowledge and procreating with Earth women (Gen. 6), thus defiling themselves. [145] Their offspring were giants who, when they died, became nasty spirits which were bound to the Earth. [146] Enoch attempts to intercede for the Watchers but fails and is shown that their spirits are to remain **earthbound** until the Final Judgment. [147] This is alleged to be the origin of the Djinn.

According to Enoch, the Watchers were led by 2 main 'fallen' entities – **Samayaza** and **Azazyel**, who in turn led other Watchers by the names of Gadrel, Yekun, Kesabel and Kayyade. The rest of the Watchers remained true to their original post and are apparently still there today. Some of the good Watcher names are familiar to readers of the Bible:

Uriel, Raguel, **Michael**, Saraqael, **Gabriel**, Phanuel, and Suriel.

It will be appreciated that the terms 'Angels' and 'Watchers' are used interchangeably by Enoch and they both occupied a realm just above the Earth. Angels have been traditionally seen as the way God communicates and interacts with His Creation, and hence they come from above the Earth – assumed to be Heaven, but it will be seen that it was just **from Earth 'orbit,' near the Firmament**.

> "Angel" (*malakh*) means 'messenger' or 'one who is sent' and does not have to be a winged, transcendental Being of Light. It is important to note that these Watchers were **physical beings** who normally had a function above the Earth, and descended to have sex with Earth women. It is thus easier to believe that the Watchers were 3D beings in this case rather than some amorphous 4D Beings of Light.

Power Differential

Said one critical author:

> The whole concept of angels has always troubled me… Why in heaven would the 'all-powerful,' omnipotent GOD need a bunch of lower spiritual forms in a physical form, to run His errands? To convey messages, to deliver warnings and threats, and to actually do the destruction on His behalf? Why would GOD need this kind of menial support? Surely the all-powerful GOD can facilitate all the interactions with humans in the blink of an eye, instead of long-winded tedious instructions, **threats and the monitoring of humans** who have apparently been behaving sinfully while conspiring against Him! It simply does not wash.[148] [emphasis added]

This is a very simplistic view of God and erroneously suggests that God can directly communicate with Man. Not so. To a large extent, I agree that God could communicate directly with Man, but chooses not to – for the same reason that Man cannot live if he touches a million volt high power line. It's an issue of **power differential**. Angels are a necessary intermediary with lower power. Yet the use of Angels has been overdone – used where it is not appropriate, as in this case. The naïve sheep have bought the 'winged Angel' explanation instead of being told about other beings out there who do "threaten and monitor" – such as the Djinn. The question is: Why intermediaries?

As was seen earlier in Chapter 5 where the **Hierarchy between Man and God** was spelled out, there have to be intermediaries between God and Man since God cannot directly interact with Man – Angels are a kind of **stepping down of His Power** which would otherwise "fry" us were He to directly visit us. Angels are like transformers – were you to plug your TV set directly into a 220V line, it would blow the set up.

In addition, real Angels from 4D and above cannot have intercourse with 3D humans, despite Biblical suggestions to the contrary (e.g., "virgin births" are really **3D artificial inseminations** by genetically-savvy Anunnaki, so the most logical explanation for the 'fallen angels' of Enoch is that they were in reality 3D beings whose genetics were slightly different and the mixing of their genetics with Earth women initially produced the giant **Nephilim** (aka Anakim and later the Philistines).

Think: Goliath, a 9' Philistine giant.

Goliath Was a Nephilim Descendant
(Credit: Bing Images)

Nazorean, Gnostic, Johannite, Ebionite, Pauline teachings or otherwise, there is a belief in 'fallen angels' on planet earth – mistakenly called the Nephilim. The suffix "im" in Hebrew means the noun is plural, such as 'El' is God, but 'Elohim' means gods. Thus 'Anakim' is Anak + -im = giants. And the Hebrew verb to fall is the root of the word 'Nephilim'… giving us Nephl + -im. The Hebrew verb To Fall is embedded in the Hebrew word for Nephilim:

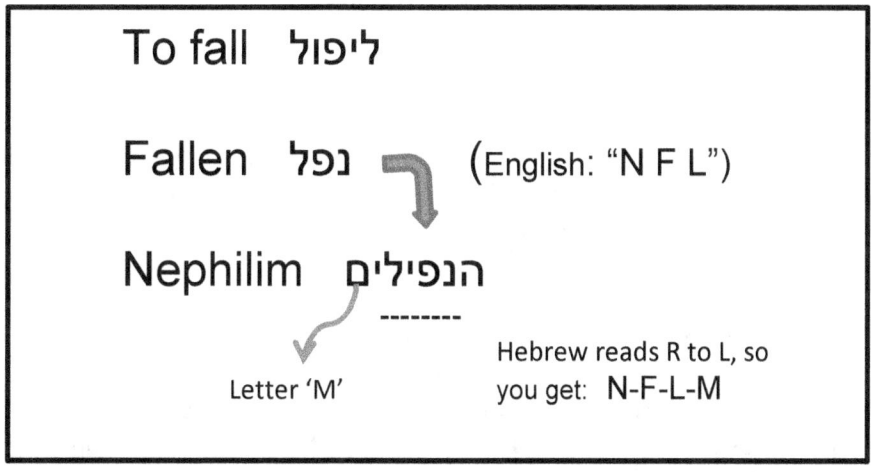

Nephilim means fallen <u>not</u> 'renowned.' (Gen. 6:4)

The **Nephilim** were the <u>offspring</u> of the Watchers + Earth women.

The concept is that of a humanoid being, a Watcher, descending to Earth in a form a little different from, for instance, one's next door neighbor, yet distinguished by a distinct psychology and genetics. Enoch said the Watchers **shapshifted** and that is how they deceived the Earth women. (See Glossary.)

> *Shapeshifting is not physically changing the molecular structure of one's body; rather it is controlling what others see. And that can also be done by an electronic device.*

Other than the Gnostics, and the Bible, the only other major source for the account of the fallen ones is in <u>The Book of Enoch</u>, which is briefly referred to in Chapter 6 of Genesis. BUT the Bible does not specifically say that the offspring were giants, it just says they were "...in the earth in those days" and the original Bible word was not "giant" it was "Nephilim" which they say means "great men, men of renown."

This is **another error in the Bible** (as shown above with the English to Hebrew translation). It is <u>The Book of Enoch</u> which says that the offspring were **physical giants**. While their exact origin may be in question, yet terrible **giants called Anakim, Geborim and Rephaim** were created who finally turned on man and began to literally devour him.

Testament of Amram

It is particularly frustrating to find that Enoch and other Bible authors give no description of the Watchers; in fact, they appear to be <u>deliberately avoiding the issue</u> (or the descriptions were removed). Except for one little description by a known biblical character, Amram, who was the father of Moses.

Robert Eisenman and Michael Wise in 1992 co-authored a book describing some 50 unknown biblical documents in <u>The Dead Sea Scrolls Uncovered</u>. Amram wrote a text called the *Testament [witness] of Amram* that describes an experience he had in which he saw an **Angel and a demon** and there is some insight into what the Watchers looked like [some words are partially obliterated in the damaged original text]:

> "[I saw **Watchers**] in my vision, the dream vision. Two [men] were fighting over me. I asked them, 'who are you that you are thus empowered over me?' They answered me, 'We [have been em]powered and rule over all mankind.' They said to me, 'Which of us do yo[u] choose to rule you?'

> I raised my eyes and looked. [One] of them was terrifying in his appearance, [like a **s]erpent**, [his] c[loak] many colored, yet very dark… [And I looked again] and … in his appearance, his **visage like a viper**… [I replied to him,] 'This Wa[tcher], who is he?' He answered me, 'This Wa[tcher]… [and his three names are Belial, and Price of Darkness] and King of Evil… He is empowered over all Darkness, while I [am empowered over all Light]… and over all that is of God. I rule over [every] man.'
>
> I asked him, ['What are your names…?'] He said to me, '[My] three names are [Michael and Prince of Light and King of Righteousness']. [149] [emphasis added]

Interesting that the viper is described, but Michael isn't. The viper is serpent-like and frightening, yet Michael did not look like the evil Watcher. (This was clarified in VEG Ch. 6 where the 'evil' Astral beings will be recognized as the **Neggs** – a negative Angel – who works with the Beings of Light (BoL or Angels). Angels can assume any form they want.)

This is another 'vote' for the existence of the **serpent-like beings** that Man used to see (viz., **Anunnaki**). Why can't we just ignore this document, this *Testament*, as the product of an over-active imagination? Why is it being given importance? For several reasons: back in the days of Moses, and his father Amram, (1) not a lot of people could write, and (2) the vellum or parchment and ink were not something that anyone just had lying around, nor could they afford to waste time, money and parchment on a made-up story. In short, the fact that this got written down means that Amram convinced someone that it was a true event, worthy of documenting for posterity. Naturally, the main people who would see and read the account were the educated, the priests and other scribes… which is why it wasn't generally available information, just as the Dead Sea Scrolls were also 'hidden' for centuries.

Appearances Important

Enoch is finally led to where the Watchers are being held and he sees that they are now in spirit form – no longer 3D, but now in their etheric (4D) bodies, and are now able to **shapeshift (see Glossary) and assume various appearances**. According to Enoch:

> And they [the angels] took me to a place where those who were there were like burning fire, and, when they wished, **they appeared as men**. [150] [emphasis added]

and again…

> And Uriel said to me: 'Here stand the angels who have connected them-
> selves with women, and their spirits, **assuming many different forms**… [151]
> [emphasis added]

It appears that the fallen but now incarcerated Nephilim **spirits** could still appear as whatever they pleased. Enoch is not directly saying that the 3D Watchers could shapeshift. But it will be later seen that the 4D Neggs (in VEG, Ch.6) **do** have the same ability. It may be safely repeated that most 4D+ entities can shapeshift, but not normal 3D entities.

> According to Enoch**, Gadrel (a 3D Watcher) appeared to Eve in
> the Garden,** and that was **a reptilian form**. And the third was named
> Gadrel: he it is who showed the children of men all the blows of
> death [i.e, ways to kill others], and he led astray Eve… [152]

Thus, whatever we can determine Gadrel's normal form to be, is what the Watchers also looked like. And now we hear that he led Eve stray…sounds like a Serpent-being. Be clear that there is nothing saying that they were <u>winged</u> angels to start with – except Man calling them that because they came from above, and they could 'fly'. So paintings later done by Man, when showing Angels, they had wings to tell the viewer that they were not humans in the picture, but Angels.

The Nephilim

In his book, <u>The Nephilim and the Pyramid of the Apocalypse</u>, Mr. Heron gives quite a bit of information about the Watchers, Nephilim, and Bible history, all to make the point, in synch with the Gnostics of old, that we have been oppressed by the spirits of the deceased Nephilim and their offspring who, according to the <u>Book of Enoch</u>, were mostly killed by The Flood and were 'imprisoned' spirits around the **Astral level of Earth**. (BTW: this also describes the interdimensional beings called the **Djinn**.)

The actions of the Watchers mating with the Earth women also relates to Genesis 3:15 again – the conflict between Two Seeds. While Enoch does not say specifically that **the Watchers were reptilian**, it can be logically inferred from Enoch's later statement that Gadrel (Enki) tempted Eve in the Garden, [153] and the Bible (and the Jewish *Haggadah*) says that that was <u>a Serpent-like being</u> (Gen. 3:1-5).

> Thus, **Watchers were not really angelic** because true Angels are not
> reptilian. Enoch mixes them up for some reason…
> The Watchers had to be somehow related to the Anunnaki since they
> **both were reptilian**… and the Anunnaki did have a service function

where they were overseers and managed the humans. So the Angels and Watchers must have worked together. Watchers have also been called Custodians.

It all fits an emerging picture that the Enochian Watchers were 3D reptilian hominids... reptilian human-like beings. In <u>The Book of Enoch,</u> it is now clear that the Serpent who beguiled Eve was one of the 'heavenly host' (Watchers) who were serpentine in appearance. [154] It is also known (from the *Haggadah*) that the entity in the Garden who led Eve to eat of the fruit is also a reptilian hominid. A Serpent-like (reptilian) being.

Doesn't that just make your day? See why **the info was hidden by the Church**?

The Nephilim offspring (giants) were a very evil group, and they were doing such a good job of abusing Man that the gods had to intervene with The Flood... but the giants didn't all die, and this bears repeating: at least their DNA <u>did</u> carry on into the land across the Jordan where Joshua and his men saw giants, and later David fought another **Philistine giant** called Goliath [9' tall] who had 3 brothers... all <u>after</u> The Flood.

Two Seeds on Earth

Suffice it to say that the Nephilim hated Man as much as when they walked the Earth as when they were bound by the Higher Beings as spirits to the Earth. In fact, the Bible suggests that there are **two seeds on the Earth** – the human and the reptilian, and it foresees problems between the two groups:

> "...I will put enmity between thee and the woman, and **between thy seed and her seed**; and it [he] shall bruise thy head, and thou shalt bruise his heel." (Gen. 3:15) [emphasis added]

God is talking to the Serpent. And over in the Koran, Islam expresses the same idea:

> 'Adam,' we said, 'Satan is an enemy to you and your wife. Let him not turn you both out of Paradise and plunge you into affliction.... [Satan shows Adam the Tree of Knowledge and they both eat...and then God discovers their sin, but relents and admonishes Adam and Eve <u>and</u> the serpent:] 'Get you down hence, both,' He said, 'and **may your offspring be enemies** to each other....' Surah 20:119 [emphasis added]

Two different religions but from the same part of the world, with one common teaching in this case – that says they both have **a common source** (Abraham) who turns out to be more than a myth himself. [155]

In addition, the Bible repeats the different 'seed' message over in Daniel 2:43:

> And whereas thou sawest iron mixed with miry clay, they shall **mingle themselves with the seed of men: but they shall not cleave to one another** even as iron is not mixed with clay. [emphasis added]

Again the two seeds will not mingle and mix. But one <u>is</u> out to pollute the other.

Polluting Mankind

Their leaders, **Azazyel** and/or **Samayaza,** and the 200 were locked up and removed by the remaining good Watchers, and Enoch (or a later scribe) colorfully suggests they were incarcerated in an Astral realm. (Think: Djinn.)

The Biblical Adam had the DNA to support a soul, because God breathed the life or spirit into him in Genesis 2. This meant that Man could over time develop his spiritual component and reconnect with his 'godhood potential' -- if he could overcome his corrupt, inherited (Anunnaki) DNA, that is. Man's overcoming had to be prevented by the Nephilim and their offspring if they were to **block Man's development so that enlightened Man could not rule over them (see Chapters 10-11 – Earth Graduate and job responsibilities in the Kingdom).** Souls were created to be eventually higher than the angels, but is currently still lower and this is a problem for Man.

Thus it was necessary to **pollute the genetics of Man** (as Patrick Heron suggested) and from that mating of the Watchers with Earth woman came the Anakim, Giborim, and Rephaim – who due to their size and appetite, went thru the food and then started on Man, and almost wiped out Mankind. As Heron said:

> So [Darkness] has some of his own band procreate with women and produce children. But these are no ordinary children. They are the **offspring** of superhuman … beings, half-human, [half-reptile], whose only intent is evil…. the evil was **in their genes**. [156] [emphasis added]

Remember that their purpose was to corrupt the human genetics.

Things must have been extremely bad when we are told that "every inclination of the thoughts of Man's heart was only evil all the time," and that the whole world was full of violence…. millions of people… had become totally evil and morally bankrupt as a result of the activities of the Nephilim. [157] (Have you seen today's nightly news?)

So the Nephilim were monsters of iniquity, and Heron adds:

> ….we have a most bloodthirsty lot. For throughout the legends concerning these people, we have debauchery, infanticide, matricide, patricide, rape, murder, adultery, incest, treachery and even **cannibalism**. You name it, they did it. This fits in exactly with the Genesis record, which tells us that the entire world was filled with violence…. **Human sacrifice** was a significant feature of these times….[158] [emphasis added]

Note that the issue of **sacrifice** comes up again and again; sacrifice was also demanded by Yahweh, later **Baal**, Molech, Ishtar, throughout history. It was also encouraged among the Aztecs and the Maya by **Huitzilopochtli** – by the Remnant later descended from the Anunnaki hybrids of VEG Ch. 3 – was there a link between the Nephilim/Watchers and Yahweh? (Yes. See VEG, Ch. 1)

Nephilim Legacy

However, Man was not alone on the Earth, and Others including the one called Yahweh (the head Anunnaki, aka **Enlil**, as proven in VEG, Chs. 1 and 3), would have seen what was happening to Man, and sensing a certain responsibility and a fear no doubt that the Nephilim and their offspring could also turn on <u>them</u>, they were looking for a way to cleanse the Earth. That turned out to be a Flood which resulted from the **Atlantean Catastrophe** (the Anunnaki attacking the Atlanteans would do this nicely, Chapter 8). A Flood would cleanse the Earth and get rid of Man and the evil-doers. Note again that this evil did not require any supernatural entity's manipulation to create the chaos and evil that was rampant -- largely due to the DNA being defective. And it needs to be emphasized that genetics must have been responsible for the size of the Anakim, Giborim and Rephaim, or "giants," as well as the baser predisposition to lust and violence. The Anunnaki were 8-9' tall.

> Since this is in Man's genetics as well, is it any wonder that Earth is a constant battlefield with war after war throughout history? And we encourage more pettiness, violence and lust through the sex and violence on TV, movies, and in video games. Such is the legacy of that corrupted DNA through many generations.

After The Flood, the humans started all over again, without the technology, but with the supplied knowledge of how to farm, irrigate, medicate, stargaze, etc. Some Watchers are still present, by the way, but it's a different group (Beings of Light and Neggs). And because the Flood was not totally world-wide (China and India escaped), but more local to the known 'world' of West Africa, the Middle East, the eastern Americas, the humans and the "2ⁿᵈ Seed" (Anunnaki) genetics are still present as well. (Hint: Isis)

And then Enoch goes into a description of the Earth as he saw it … and it was not a globe.

II. The Flat Earth

Enoch Said the Earth Was Flat

Enoch said he was taken to the **"ends of the Earth"** – a round planet does not have "ends" – where he saw the Firmament connect to the land. And he said that the Sun and Moon are the same size, and he said he saw the "cornerstone of the Earth" – a flat Earth could have a cornerstone from which all else is measured (but that does not apply to a sphere). He also saw that the stars enter the Earth realm thru portals and run their course in assigned positions.

Enoch's revelations follow.

One last thing of major interest …. There is an old book that should have been in the Bible, but because it speaks of Giants or Nephilim **and the Flat Earth**, Constantine removed The Book of Enoch from the list of 'acceptable' books for the Bible he was creating (AD 325, Council of Nicea).

Remember the Bible gave us the information below – largely from Genesis – and the person to give us the information was Enoch…. the father of Methuselah, and the great-grandfather of Noah. He lived 365 years on Earth before God took him up into **the region above Earth** (Heaven of Heavens – Note the "Gate of Heaven" in the diagram, which might be the **Tunnel of Light** which NDErs experience when they cross over, Chapter 10) and showed him what Earth really was (in addition to meeting and judging the Watchers/Nephilim *aka* Giants).

The following diagram will come in handy as we present Enoch's statements about the Earth which follow…

The winds blow thru the 'portals.'

Ancient Views

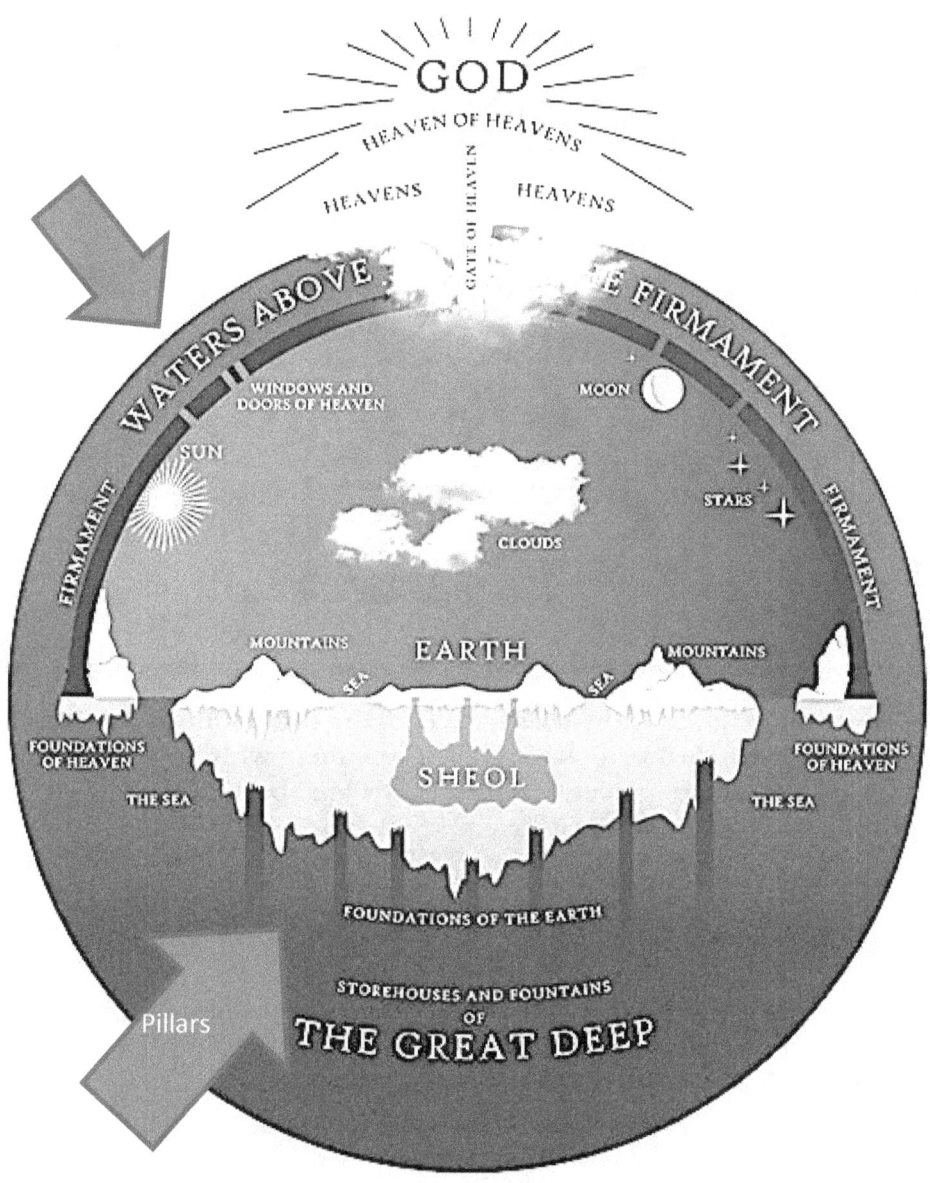

Note the **"windows and doors"** of the Firmament which Enoch calls "**portals**" (see upper arrow). The Pillars of the Earth support the very foundation, but it is not said what they rest on... which is why a number of Oriental myths say that the foundation of the Earth is a giant turtle (Chapter 7 pictures).

Note the Sun, Moon and stars are <u>inside</u> the Firmament.

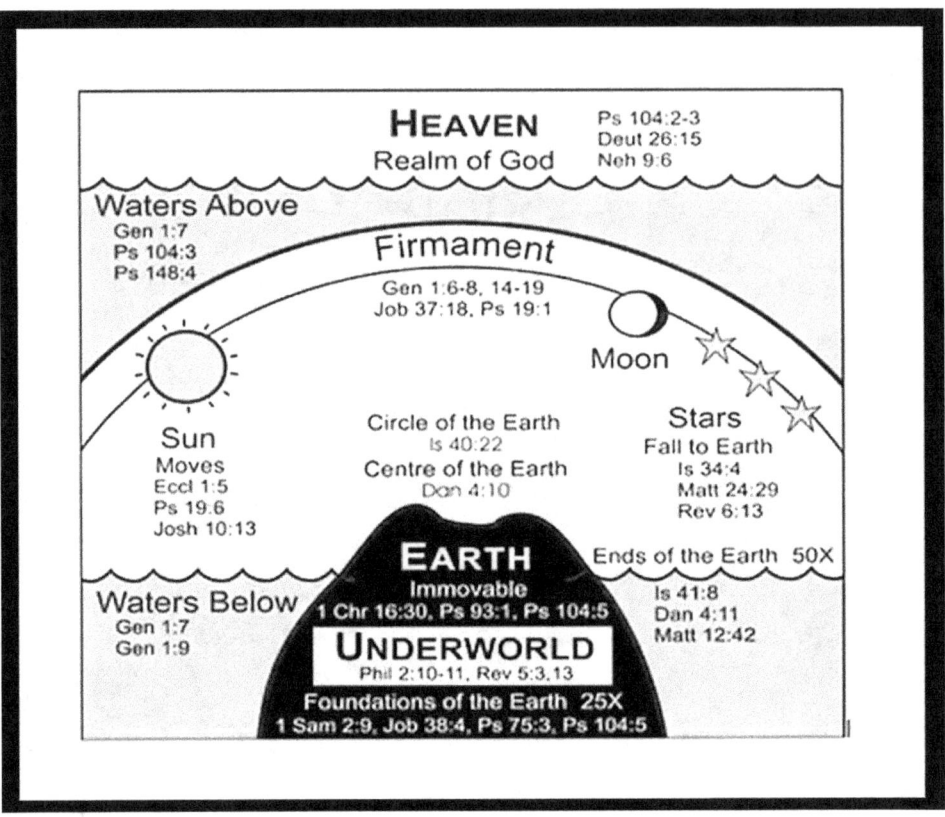

In the above diagram, note the terms "ends of the earth" (lower right), "immovable" (top of black 'rock') and the Firmament with the Sun and Moon <u>inside it</u>. The Great Deep ("waters below" in above diagram) is examined in Chapter 7, Insert on Rivers).

Quotes from Enoch

Enoch is taken up by God above the Earth and shown many things, including the layout of the Earth, from whence blow the winds, from whence flow the rivers, **the ends of the Earth** (Hint: there cannot be an 'end' to a spherical Earth), and how the Sun and Moon operate **"in their orbits."** He also returned to Earth to write what he saw (LXXXI, v. 5).

The following is a series of quotes from <u>The Book of Enoch the Prophet</u> (see Bibliography) and the parentheses are the page references cited:

> And they took and brought me to the place of darkness and to a mountain the point of whose summit **reached to heaven**. And I saw the places of **the luminaries** [Sun & Moon]… and the **stars [in the Firmament]**. (15)

...and to the fire of the west which receives every setting of the Sun... and the place whence all the waters of **the Deep** flow, and the mouths of all the rivers of the earth... (15)

I saw where all the winds are kept...and I saw **the corner-stone** of the earth; I saw the four winds, and **the firmament of the heaven**

I saw how the winds stretch out the vaults of heaven... [and] the **pillars of heaven**... and I saw the winds of heaven which turn and bring the circumference of the Sun and all the stars to their setting [locations].... . I saw **at the end of the earth the firmament of the heaven above.** (15)

I saw seven mountains of magnificent stones...and beyond these Mountains is a region, the **end of the earth**: there the heavens were completed. ... I saw a place which has no firmament of the heaven above.... (16)

And I saw other mountains .... And beyond these mountains I saw another mountain {to the east **ends of the earth**}... (24)

I came to the Garden of Righteousness [Eden] ... and from there I went to the **ends of the earth** and saw there great beasts... (25) ...and to the east of those beasts I saw **the ends of the earth whereon the heaven rests** , and the **portals of the heaven** open. And I saw how **the stars of heaven come forth**.... Each individual star by itself ... and their courses and positions. (26)

From there I went toward the north to the **ends of the earth**... And from there I went towards the west to the **ends of the earth**... And from thence I went to the south to the **ends of the earth**.... (26) And I saw there **three open portals** .... **Through each of these small portals pass the stars of heaven** and run their course to the west on the path which is shown to them [in the Firmament]. (27)

I saw all the secrets of the heavens.... And I saw the mansions of the elect and holy ....And I saw **the chambers of the sun and moon** whence they proceed and whither they come again, and their glorious return, and how one is superior to the other, **and their stately orbit...** and how they do not leave their orbit, and they add nor take nothing from the orbit ... The sun goes forth and traverses his path according to the commandment of the Lord... and the moon goes forth... in accordance with **the oath by which they are bound** together [higher entities run the Sun and Moon and work in agreement with each other] …. I saw the visible path of the moon… by day and by night…the one holding a position opposite to the other before the Lord… (33)

And this is the first law of the **luminaries**: the luminary the sun has its rising in the eastern **portals of heaven** and its setting in the western portals of the heaven… at first there goes forth the great luminary, named the sun, and **his [orbit] is like the circumference of the heaven,** and he is quite filled with illuminating and heating fire…. The great luminary rises, sets and decreases [in brilliance] not, and rests not, but runs day and night, and his light is sevenfold brighter than that of the moon; but **as regards size, they are both equal**. (72-75)

These are the two great luminaries [Sun and Moon]: their [orbit] is like the circumference of the heaven, and **the size of the circumference [orbit] of both is alike.** (81)

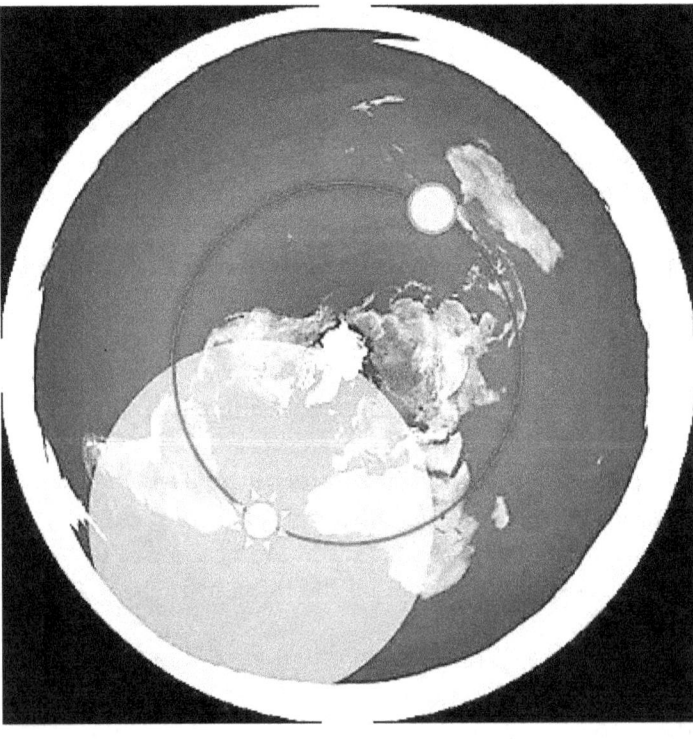

Is this similar to what Enoch saw?

It is important to note that **Enoch saw the ends of the earth** – from <u>four directions</u> in paragraphs 25-26 – that does not describe a sphere. He also says **the Sun and Moon are the same size** (think: **eclipses**) and the Moon reflects light from the Sun – who operate together by agreement.

> Perhaps the Amerindians were not idiots when they said the **elements of Nature are inhabited or controlled by spirits**. Even the Egyptians venerated the Sun as a being (Ra) and if it is empowered by the presence of a higher being [spirit], then we now know what the ancients were venerating and why.

Note in addition, his comment on the orbit of the Sun… "is like the circumference of the heaven" in (72-75) which suggests that it is <u>under the Firmament</u> and follows the size/layout of the Firmament.

In addition, the 4 winds, the stars and Sun and Moon come and go under the Firmament thru the use of **portals.** This suggests that they have special ways of entering/exiting the Dome, but Enoch's phraseology and such are so archaic it is hard to tell just what is being described as a portal. The FE Theory shows the Sun and Moon the same size and circling about the North Pole (which is the center of the landmass – see diagram above).

Paragraphs (15) and (25) seem to suggest that the **Firmament rests on the "ends of the earth"** – meaning that the current FE Theory displaying the Dome extending down to the Ice Wall, which is the 'ends of the earth territory,' agrees with Enoch.

Nature of Stars

Enoch also tells us something interesting about the Luminaries and the stars and 'hosts of heaven.'

> And I saw a deep abyss with **columns pf heavenly fire** and among them I saw columns of fire fall…….and beyond that abyss I saw a place which had no firmament of the heaven above and no firmly founded earth beneath it: there was no water upon it and no birds but it was a waste and a horrible place….

> The angel said: This place is the end of heaven and earth: this has become a **prison for the stars and the host of heaven.** And the stars which roll over the fire are they which have transgressed the commandment of the Lord in the beginning of their rising, because they did not come forth at their appointed times. And He was wroth with them and bound them till the time when their guilt shall be consummated…. (16)

What we can see from that is that there are beings of some sort which were/are responsible for being or moving the stars and didn't do it – just as the Sun and

Moon are "bound by an oath" (33) … this suggests that the 'heavenly host' are beings who effect (create and manage) the Earth Realm.

Then he adds:

> And Uriel said unto me: here shall stand the angels who have connected themselves with [earth] women and their spirits assuming many different forms [i.e., shapeshifting] are defiling mankind and shall lead them astray into sacrificing to demons… till the day of the great judgment… till they are made an end of.
>
> (16)

> This place is the prison of the angels. (18)

This sounds like Sheol, and what is called (in the diagrams above) 'The Underworld.' Uriel managed Tartarus (the abode of the dead).

> **Tartarus** in Greek myth was the abyss below Hades into which Zeus hurled the rebel Titans (who fought the 'good' Olympians) and it was later a place of punishment for the wicked after death.
> The word Tartarus is often synonymous with Hades.

As for the **Gates of Heaven** (top of 1st diagram above)… even the Bible mentioned that: Genesis 28:16-17

> 16 Then Jacob awoke from his sleep and said, "Surely the LORD is in this place, and I did not know it." 17 He was afraid and said "How awesome is this place! This is none other than the house of God and this is **the gate of heaven**. [Jacob was watching angels move up and down a ladder].

Enoch does not describe the Gate of Heaven, but it is clear that there are at least two levels, with an outer level where the Heavenly Host reside, and the Inner Sanctum, the Throne of God, protected by cherubim and seraphim.

So it looks like there is support from the Bible (as well as the Koran mentioned earlier) for the Flat Earth (FE) as a real scenario for our Earth. By the way, if the Earth is flat, it is obviously **not hollow**!

Further, deeper FE examination is provided in a large Ch. 11 in QES.

Summary

The purpose of this chapter, and even of repeating parts of key information from prior books, is that a specific conclusion is being devised and that information is necessary to the conclusion of this book. **Earth is not what we think it is**, and the point is that Religion, Science and History are full of false information so that we cannot figure out where we are, who we are, and what we are supposed to be doing here.

> Just FYI: The Anunnaki were not the Ancient Ones (who were the God-appointed Caretakers) and the tall, reptilian Anunnaki Watchers mated with Earth women creating the Nephilim.

And it is difficult to get any historical truth from the MSM – the Media spins events and instead of reporting objectively, the listener gets the bias of the particular News Media being watched. During the 2016 Election, CNN and MSNBC were pro-Hilary Clinton and OAN and FOX News were pro-Donald Trump. That said, the Gospel of Barnabas will probably not be examined on the nightly news, and probably not on the History Channel either.

And history classes in high schools across America deftly omit certain things, like the **Sumerians** – Oh, you hear about Iran and Iraq and maybe Mesopotamia, but not what made Sumeria unique (i.e., that it sprang up <u>fully developed overnight</u> – there are no layers of prior developing cultures below the Iraqi/Sumerian ruins!) … that would cause kids to wonder and question. Can't have that. Similarly, you don't hear about **Khazaria** which was a formidable empire in the area of the Caucasus (near Georgia in Eurasia). And it is not because they were ancient (they weren't) and it is not because they were not important – they were! (They were home to the Ashkenazi people.)

> The PTB don't want you to know about either of them.

> And, no I am not a fan of speculation, and I hate conspiracy, so I have just given you some interesting pieces of World History to check out.

In the same way, the Church Universal and Triumphant has controlled what you hear and believe about the Earth, Man and Heaven. You were not told about the **Book of Enoch**, nor the Book of Jasher, and probably had not come across references to exo-Biblical documents found in the **Nag Hammadí Library**… such as the Gospel of Peter, the Gospel of Mary…. And you probably have heard your Bible Study teacher (or pastor) pooh-pooh the **Gnostic** literature. True, not all of it is worth the time to plow thru it, some of it like the **Apocrypha** is boring history, (Esdras, Maccabees, etc.) which is why I have written these 7 books to get you started thinking and maybe researching more relevant issues.

> Keep in mind that the 200 imprisoned Watchers (for their crimes against humanity) sound like the Interdimensionals aka the **Djinn** from VEG Ch. 12

In another way, things that <u>are</u> true, you have been conditioned to laugh at – UFOs and the Flat Earth. You will see in the last chapter why the PTB have done this to you.

VR Sphere vs The Flat Earth

Note that at this point, there is mounting evidence that the Earth Realm is flat, and it is easier to justify… the **Gravity problem** with the VR Sphere requires that we introduce a new element: a sophisticated form of holographic programming to explain why the million-ton oceans don't fall off the Sphere.

> **Occam's Razor** is a time-honored axiom in Science:
>
> When confronted with two or more options to explain some aspect of Earth Science, it is <u>usually the simpler one that is correct</u>…

… and yet, there is a lot to recommend the VR Sphere…we either need to refute the Flat Earth by proving the "killer" FE proofs wrong (not so easy) and finding some additional solid basis for the VR Sphere… but remember:

> many of the FE proofs also substantiate the VR Sphere.

What is needed to sustain the VR Sphere as a viable contender is some proof that the continent of **Antarctica** exists… so far, it doesn't and it <u>appeared</u> over the centuries to be an Ice Wall… and given Man's cleverness with CGI graphics and Photoshop®, it is not a slam-dunk to say that Antarctica exists as a continent just because Google Earth shows it as a continent… It is curiously a 'restricted area' by the US Navy.

> It is suspected that there is a PTB agenda that seeks to sustain the rotating Earth rock spinning thru space… in preparation for a possible and logical Project Bluebeam event… (see Chapter 10).

There are further **VR Sphere vs Flat Earth** sections in future chapters…

Chapter 7: Earth History, Part I

Chapter 4 dealt with errors and deception in Physical Science, like Physics and Quantum Physics. This chapter has a similar mission to review the facts about Earth and its structure, history and resets. Along the way, a few more facts about the Young Earth and the Earth as a flat spheroid are offered. In addition, it is appropriate to examine the Ancient Ones and the Anunnaki… What part did they play in Earth's history? You will find that you don't really know Earth history after all.

In The Beginning…

If we are to examine the Earth and what it is, including its science and future, let's begin at the beginning with everything we know about the Earth from the oldest records… (don't worry, I won't go "religious" on you!):

In the beginning, God created the heaven and the Earth…

…so we are not talking about a Big Bang, nor random planet-evolution by spinning hot masses of rock and dust clouds… And then the Bible's Genesis 6 verses later says:

And God said, "let there be a firmament in the midst of the waters, and let it divide the waters from the waters…

…that is rather vague – could even be describing a mass of land separating two bodies of water… but the next verse continues:

And God made the firmament, and divided the waters which were under the firmament from the waters which were above the firmament….

… Ok, water <u>above and below</u> whatever the Firmament is. That means a <u>vertical</u> separation. So it is <u>not</u> a horizontal arrangement – like the continent of Africa separating the Indian Ocean from the Atlantic Ocean. (Even the Jewish Torah [first 5 books of what we call the Old Testament, including Genesis] makes this distinction – in Hebrew.) So could that be "the waters of the Deep" versus the oceans on Earth…? That would be a vertical arrangement…. See diagram that follows:

Represented below are the Waters Above and the Waters Below…. and arching over the Earth (land) is the Firmament… It suggests that water is a key element in the Earth Realm… and surrounds us.

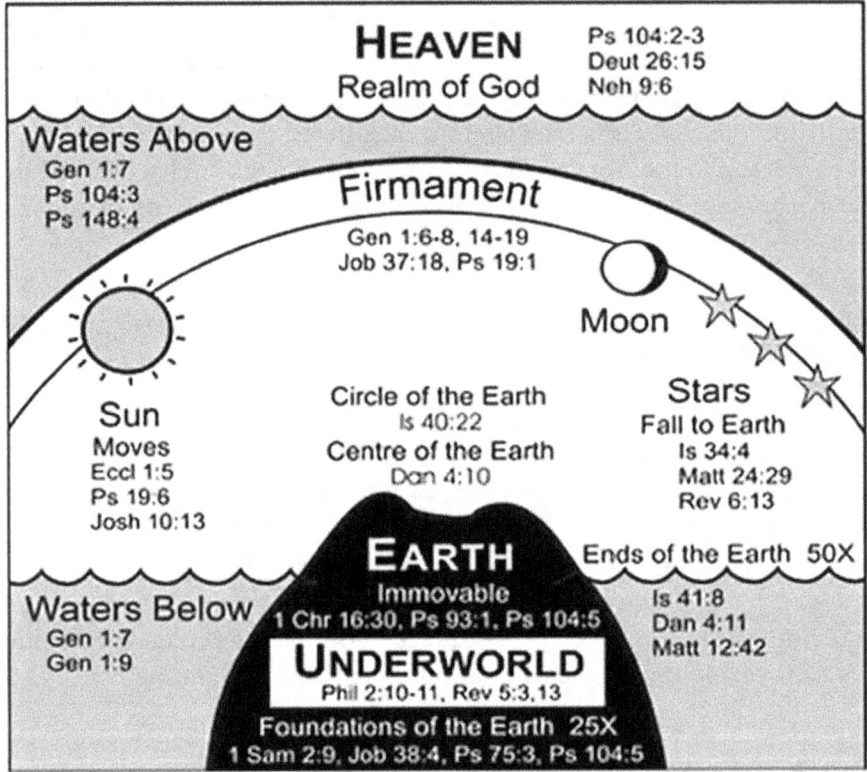

Very interesting – the diagram shows how the Bible describes the Earth, and the **Book of Enoch** (as just examined in the last chapter and again in QES Ch. 11) would agree. Enoch also saw the Firmament and where it connected to the "ends of the Earth."

So what's next?

> **And God called the Firmament Heaven….**
>
> **And God said, Let there be lights <u>in</u> the firmament of the heaven… And God made two great lights [Luminaries], the greater light to rule the day, and the lesser light to rule the night…**

… did you notice the use of the word "in" above? It said "…IN the firmament." Not above it and even more strange is the concept that there was water <u>above</u> the Firmament. That has always been a puzzle for Bible scholars… Do we still have water "above the firmament" (i.,.e., in a lower level of Heaven) to this day? If so the astronauts had to have flown through it…. If NASA was telling the truth about the Apollo Moon missions… Or did we not make it thru the Firmament?

…. He made the stars also. And God set them <u>in</u> the firmament of the heaven to give light upon the Earth.

Ok, again we have the statement: "…IN the Firmament." So the Firmament must be dome-like and cover the Earth. So could we be talking about something like this?:

Ooooh, we are talking about the **Flat Earth**! But we are modern people and we all know that that can't be…. Or is it? Naaaaah… NASA went to the Moon and took pictures of a round Earth… didn't they? If not, why would they lie to us….? Something is wrong here.

QES Ch. 11 ventured the possibility that the Earth was **created flat** as shown above, then as Man gained the ability to fly and soar up into the miles above the Earth, the gods who run this place got together and **morphed Earth into a VR Sphere**. But as QES says, Earth is probably a Simulation (given all the evidence in QES and the agreement of the Quantum Physicists) such that rearranging the shape of Earth would not be beyond the ability of those who are the creators of this place.

And then QES Ch. 11 <u>shoots that down</u> with evidence that demands a verdict, and some of the evidence is **proof** that cannot be ignored.

Yours truly did not expect FE proof – I set out to prove it wrong.

Other Voices

So is the Bible the only promoter of a Flat Earth (FE)? No, the **Koran** also said the Earth is flat:

"Do they never reflect on the earth how it was made flat"

Quran: chapter 88:17-20

Hebrews

Like most ancient peoples, the Hebrews believed the sky was a solid dome with the Sun, Moon, and Stars embedded in it. According to the Jewish Encyclopedia:

> The Hebrews regarded the earth as a plain or a hill figured like a hemisphere, swimming on water. Over this is arched the solid vault of heaven. To this vault are fastened the lights, the stars. So slight is this elevation that birds may rise to it and fly along its expanse.

The following are according to Wikipedia's topic: Firmament:

Ming China

> As late as 1595, an early Jesuit missionary to China, Matteo Ricci, recorded that the Chinese say: **"The** earth is flat **and square, and** the sky is a **round canopy**; they did not succeed in conceiving the possibility of the antipodes" [people living in Australia 'upside down on the globe.'] The universal belief in a flat Earth is con-

firmed by a contemporary Chinese encyclopedia from 1609 illustrating a **flat Earth** extending over the horizontal diametral plane of a spherical heaven.

Ancient India

Ancient Hindu, Jain, and Buddhist cosmology held that the Earth is a **disc** consisting of four continents grouped around a central mountain (Mount Meru) like the petals of a flower. An outer ocean surrounds these continents. This view of traditional Buddhist and Jain cosmology depicts the cosmos as a vast, oceanic **disk** (of the magnitude of a small planetary system), bounded by mountains, in which the continents are set as small islands.

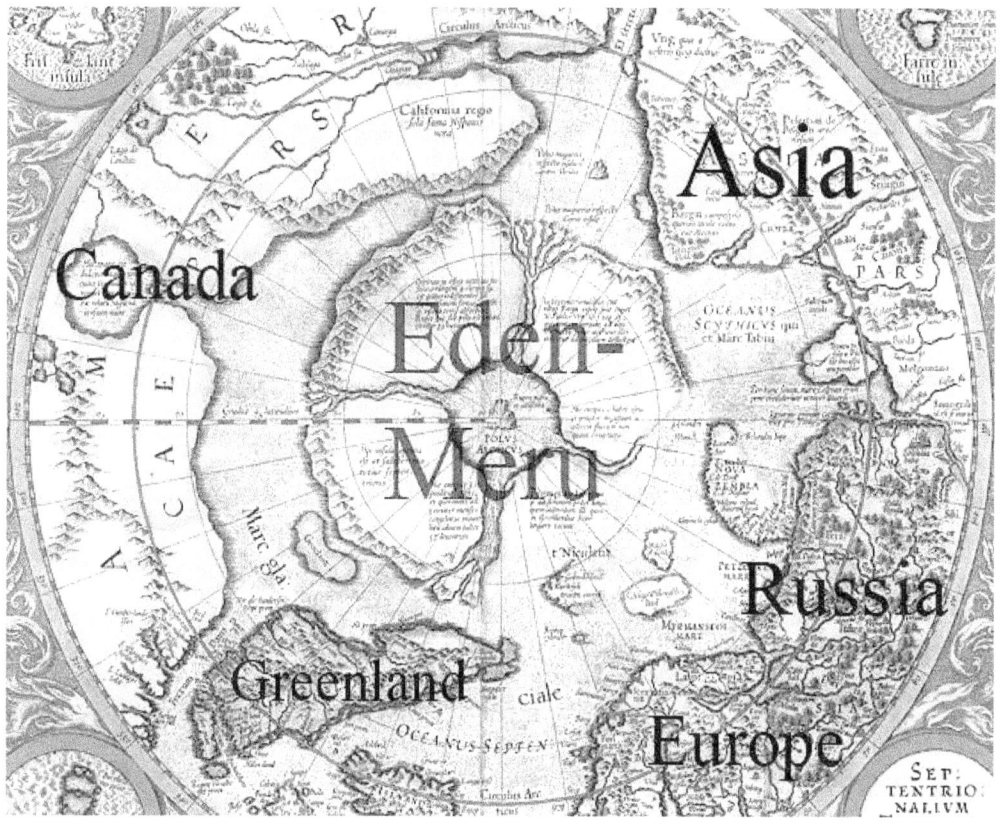

Mt. Meru is considered holy in several Asian cultures, and is allegedly the home of the Ancient Ones, who are also called the Caretakers of Man.

Norse and Germanic

The ancient Norse and Germanic peoples believed in **a flat Earth** cosmography with the Earth surrounded by an ocean, with the *axis mundi*, a world tree (**Yggdrasil**), or pillar in the centre. The Norse believed that in the world-encircling ocean sat a large snake called **Jormungandr**. In the ancient Norse creation account preserved in *Gylfaginning* (VIII) it is stated that during the creation of the earth, an impassable sea was placed around the earth like a ring, bordered by an **Ice Wall**:

The above was examined in more detail in AL Ch. 2 and explains the different levels Asgard, Midgard and Hel.

Ancient Near East

mago Mundi Babylonian map, the oldest known world map, 6th century BC **Babylonia.**

<u>Their world also appears as a disc.</u>

In early **Egyptian**[7] and **Mesopotamian** thought the world was portrayed as a **flat disk** floating in the ocean.

That paradigm was also typically held in the aboriginal cultures of **the Americas**, and the notion of a **flat Earth** domed by the **firmament** in the shape of an inverted bowl was common in pre-scientific societies.

Americas, India and China

As above, the Amerindians and Hindus saw a flat Earth – on a turtle's back.

Portugese exploration of Africa and ultimately **Ferdinand Magellan** led the Spanish circumnavigation of the Earth (1519–21) which provoked a belief in a global shape of the Earth.

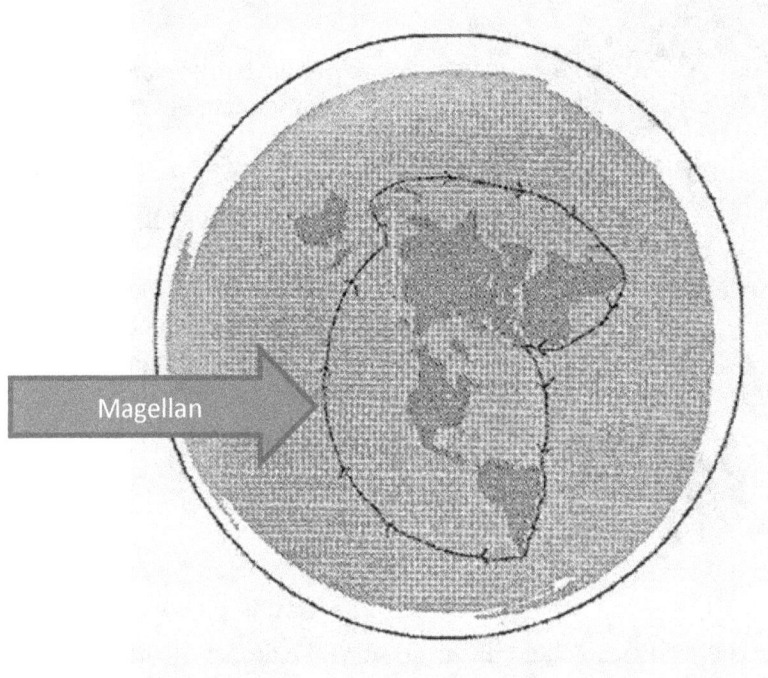

And yet, **circumnavigation <u>did not</u> prove the Earth was a globe** – see diagram (above). It is a piece of cake to sail around the Flat Earth.

What is fascinating about all this agreement in an FE is that the cultures did not know each other, they did not get together and compare notes and swap myths… it suggests that someone told them…maybe the **Skygods** that many world cultures also speak of (examined later in this book, and in <u>Anunnaki Legacy</u>).

PS: Firmament

> Even **Copernicus'** heliocentric model included an **outer sphere** that held the stars (and by having the earth rotate daily on its axis it allowed the firmament to be completely stationary).[159] See Chapter 5.

Young Earth

So if the Earth was **created** and then it went thru a Flood about 6000 years ago, how old is the Earth? Geologists and Astronomers say it is 4.3 billion years old. VEG Ch. 10 gave plenty of evidence that says Earth is not that old and uses several serious statistics to make the point that Earth is not billions of years old, and most likely it is just a few million years old. Scientists suggest it is 4.3 billion years old for two reasons:

> They have to have a huge amount of time for **Evolution** to have worked (they're in cahoots with the Darwin people),

and

> They have used unreliable methods of dating fossils and geologic strata which skew results,

and

> They have misread the astronomical data, including *redshift* data, which gives an inflated value to what appear to be stars millions of lightyears away.

> **Given that Earth may be flat and, if so, the stars are <u>in the Firmament</u>, thus the astronomers are analyzing the IMAX Theatre (which would be similar to a very sophisticated planetarium). This is even more possible if the Earth is a Simulation where the 'technology' exceeds the basic holo-graphics and replicator technology of Star Trek's *Holodeck*!**

Some of the Young Earth considerations <u>not</u> expanded on from Ch. 10 of VEG are:

Dinosaurs

Remember that in Genesis, Adam and Eve were created as adults, not babies. The Creation was created as **a mature Creation** with the ability to procreate. In that sense, dinosaurs were created as mature adults, and of course they would have had offspring – and yet no dinosaur fossil children were ever found. Eggs, yes, but **no adolescent dinosaurs.** Weird.

In fact it is being said in some circles that dinosaurs did not roam the Earth – it is an invention of the Geologist-Paleontologist cadre. If they were all over the place, why didn't the Amerindians discover the bones of dinosaurs – especially the Sioux who roamed the Badlands of the Dakotas for decades where allegedly the biggest fossil graveyard of dinosaur bones is found? Weird, but **no Indians ever found or described huge bones found in the Earth.**

Upon further investigation, according to Eric Dubay, there is **a factory in China where "fossil dinosaur bones" are made to order** and then shipped all over the world. [160]

> (I know… just when you thought you might be able to handle considering the Flat Earth, there is a new wrinkle… so I submit this for your edification, and possible amusement...)

> …the **existence of dinosaurs** was first speculatively hypothesized by a knighted museum-head "coincidentally" in the mid-19th century during the heyday of evolutionism, before a single dinosaur fossil had ever been found. The Masonic media and mainstream press worldwide got to work hyping stories of these supposed long-lost animals, and then lo and behold, 12 years later in 1854, Ferdinand Vandiveer Hayden, during his exploration of the upper Missouri River, found "proof" of Owen's theory! **A few unidentified teeth** he mailed to paleontologist Joseph Leidy who several years later declared them to be from an ancient extinct "Trachodon" dinosaur (which beyond ironical means "rough tooth"). [161]

How did anyone know what a Trachodon was, much less what it looked like from a few teeth?! It is dubious that such fossils have existed for millions of years but were never found by any civilization or tribe in the history of humanity until Evolution's heyday during the Masonic renaissance in the mid-1800's. Before the 1800s's no one anywhere knew that dinosaurs ever existed, and then <u>suddenly they were found all over the world</u>…! This is very interesting:

> According to paleontologist journalist **Wayne Grady**, claims that the period from around 1870-1880 became "*a period in North America where some of the most underhanded shenanigans in the history of science were conducted.*" In what was known as The **Great Dinosaur Rush** or Bone Wars, E.D. Cope of the Academy of Natural Sciences and O. Marsh of the Peabody Museum of Natural History began a lifelong rivalry and passion for "dinosaur hunting." They started out as friends but became bitter enemies during a legendary feud involving double-crossing, slander, bribery, theft, spying, and **destruction of bones by both parties**. Marsh is said to have discovered over 500 different ancient species including 80 dinosaurs, while Cope discovered 56. Out of the 136 dinosaur species supposedly discovered by the two men, however, only 32 are presently considered valid; the rest have all **proven to be falsifications and fabrications!** None of them once claimed to find a complete skeleton either, so **all their work involved reconstructions.** In fact, to this day, **no complete skeleton has ever been found, and so all dinosaurs [seen in museums] are reconstructions.**[162]

As Alice in Wonderland once said, "It gets curiouser and curiouser…" Read on, while the quoted accounts are real history, the <u>larger</u> dinosaurs may not be…

Dinosaur Fraud

The dinosaur bone finds are often made by those expressly looking for them in far away and distant regions of the Earth <u>already explored</u>. They occasionally find large numbers of bones in tiny areas. The Ruth Mason Quarry has yielded over 2000 bones which have been shipped off to over 60 museums worldwide. Thousands of dinosaur eggs were found in Patagonia in a site of only a few hundred square yards! "Many experts have mentioned how such finds of huge quantities of fossils in one area by just a few highly invested individuals [paleontologists, museum directors, and university professors] goes **against the laws of natural probability and lends credence to the likelihood of forgeries or concentrated planting efforts.**" [163]

> Dinosaur bones sell for a lot of money at auctions. It is a profitable business…. Much is to be gained by converting a bland non-dinosaur discovery, of a bone of modern origin, into an impressive dinosaur find…. and letting imaginations take the spotlight, rather than the basic boring real find…. The first dinosaur to ever be publicly displayed was the "Hadrosaurus foulkii" at E.D. Cope's Academy of Natural Sciences in Philadelphia. The bones were co-discovered by … the man responsible for the "Trachodon" discovery…. The original Hadrosaurus **reconstruction,** which is still on display today, shows a huge plaster cast bipedal reptile standing upright … **[See picture below.]** What few people know, however, is that **no skull was ever discovered and no original bones were put in the public exhibit.** [164] [emphasis added]

(credit: Dubay, <u>The Flat Earth Conspiracy</u>, p. 205)

What one sees in museums are incomplete reconstructions. Guesstimates of what the dinosaur should look like. Lastly, the *pièce de resistance*:

> **No real dinosaur bones are put in any museum exhibits**… they are locked away in vaults where only a few key researchers can access them…. Only around 2100 dinosaur bone sets have been discovered worldwide, and out of these, only 15 incomplete Tyrannosaurus Rex bone sets have been found. **These dinosaur bone sets have never formed a complete skeleton** … but from these incomplete sets paleontologists have formulated an <u>hypothesis</u> about the appearance of the whole skeleton… which they have modeled in plastic. [165] [emphasis added]

Remember that the existence of dinosaurs adds credibility to the Evolution story. So when children go to a museum and ogle an exhibit, are they really observing clever science fiction? Are we being 'programmed' at an early age to believe in creatures that never existed? Hard to believe? Ok, one last item from the article on dinosaur fraud:

> There appears to have been an ongoing effort since the first dinosaur 'discoveries' …. To construct and create a new man-made concept prehistoric animal called the dinosaur. Where bones from existing animals are not satisfactory for deception purposes, plaster substitutes may be manufactured and used…. What would be the motivation for such a deceptive endeavor? Obvious motivations include **trying to prove Evolution**…. and trying to disprove the Young Earth theory [which supports Creationism]….

> Type "dinosaur skulls" into a search engine and you'll find a variety of replicas… and "museum-quality" skeletons. **One of the largest and most-renowned suppliers of fake dinosaurs is the Zigong Dino Ocean Art Company in Sichuan, China** which provides natural history museums worldwide with ultra-realistic dinosaur skeletons made from real bones! Chicken, frog, dog, cat, horse and pig bones are melted down, mixed with glue, resin and plaster, then used as base material for re-casting as "dinosaur bones." They are even given intentional fractures and an antiquated/fossilized look to achieve the right effect.

> Says Professor **Alan Feduccia** of the University of North Carolina Paleontology Department: "I have heard there is **a fake-fossil factory in northeast China** in Liaoning Province, near the deposits where many of these recent alleged feather dinosaurs were found." [166] [emphasis added]

All in all, it is quite an extensive article in Dubay's book (well documented) and makes you wonder what we can trust nowadays. And a man came forward who was involved in a dinosaur fraud who confessed:

> …that he was a total fraud, **fabricating evidence and perpetuating the myth of dinosaurs**. "I started my career in the field of Paleontology only to leave my studies once I realized the whole thing was a sham. It's nonsense, **most of the so-called skeletons in museums are actually plaster casts**…. I struggled as a student mainly because I could not tell the difference between a fossilized egg and a rock, and of course <u>there is no difference</u>. I was treated like a leper when I refused to buy into their propaganda, and promptly left the course. **Dinosaurs never existed, the whole shebang is a freak show….If dinosaurs existed, they would be mentioned in the Bible**. [167] [emphasis added]

It is possible that they <u>were</u> mentioned… No one seems to know what the beast called **"Behemoth"** mentioned in the Book of Job (Job 40:15) really was. "He eats grass like an ox…he moves his tail like a cedar…he drinketh up a river… his bones are like bars of iron…" and then the KJV Study Bible tries to tell us that this is probably describing a hippopotamus! It sounds suspiciously like a **Brontosaurus** (aka Brachiosaurus -- below).

(credit: Bing Images: lefionofleia.com …Zallinger's "Brontosaurus")

This author did some research on this issue and found that small dinosaurs existed, but not the larger, like Brontosaurus nor the Tyrannosaurus.

Cryptozoology

Part of the mystery of Earth is the occasional surfacing of strange creatures, and not just Bigfoot or the Loch Ness Monster. The following are some of the cryptozoids that have been seen around the planet…

A prime contender for the "Leviathan" title goes to an ancient swimming dinosaur known as **Megalodon**. We are told that it no longer exists, but natives in South Africa and Indonesia while out fishing on the oceans have seen it, just as the supposedly extinct Coelacanth has also been caught in fishing nets:

Megalodon (60') on the right, dwarfs a Great White (20') shark

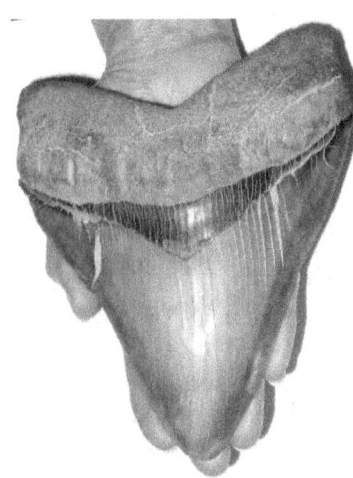

As proof that it at least did exist, teeth from Megalodon have been found and compared (below) with that of the Great White:

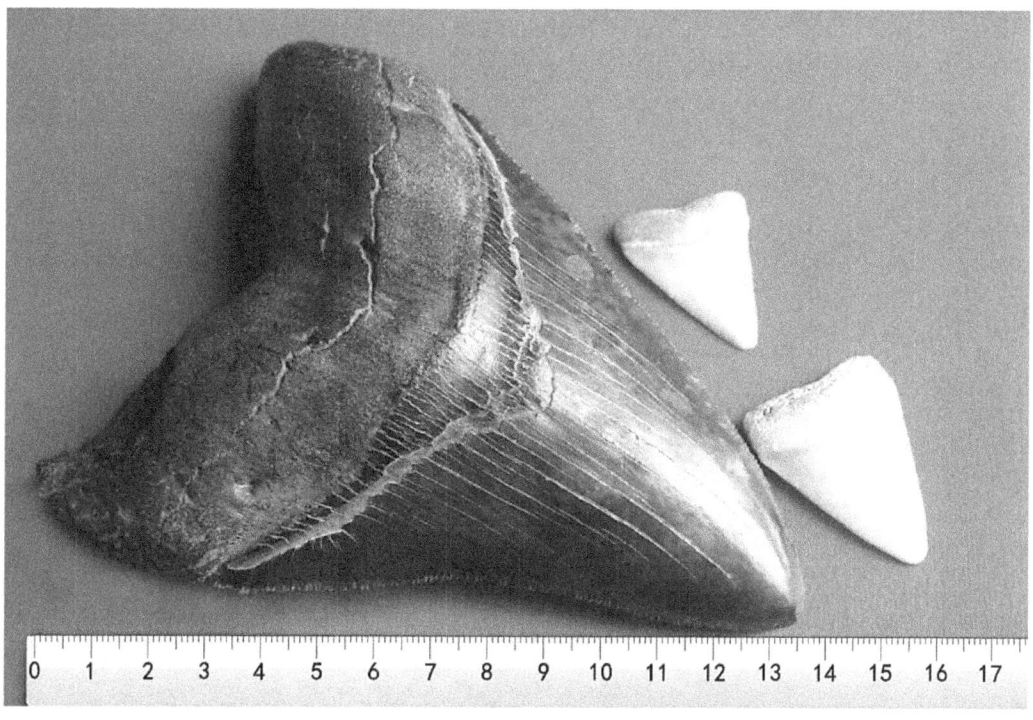

Even the jaw of a Megalondon has been found:

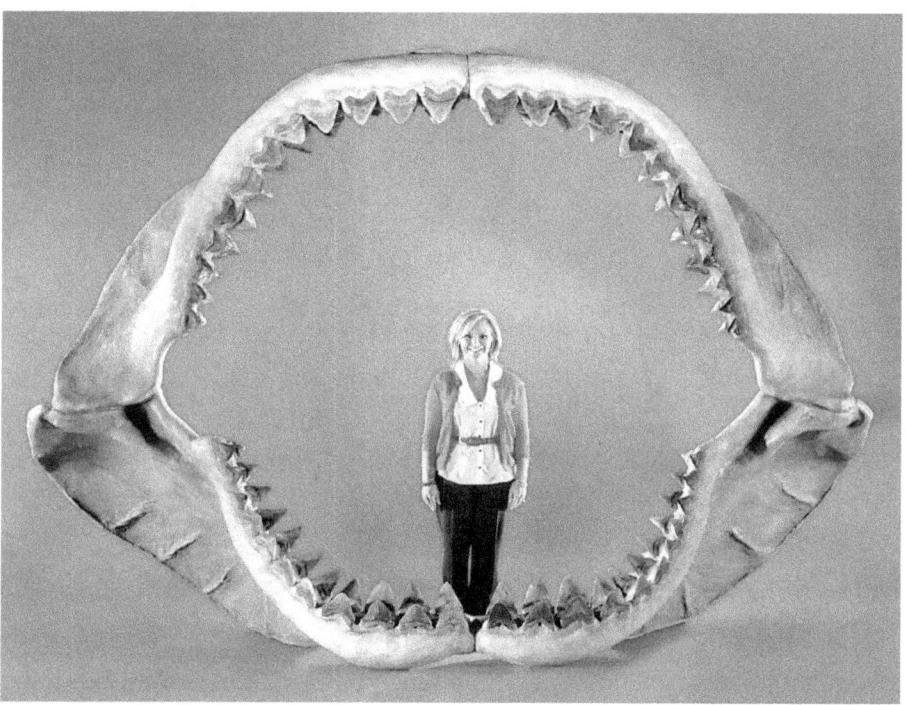

Also of note is that the Loch Ness "monster" is nothing more than a swimming dinosaur called a **Plesiosaur:**

...and as a matter of fact, these have also been caught in fishermen's nets....

And another oddity has been found in the Philippines and Indonesia, often called a **Flying Bat:**

GIANT BAT OR SOMETHING ELSE?

And the military in the Philippines caught one with a wingspan nearly 10' across... could they be related to the ancient **Pterodactyls?**

And there is even a lizard with "wings" also found in Indonesia and Borneo…

Dragons are real OMG.
This lizard is found in Indonesia
and it resembles a miniature
dragon.

VIA 9GAG.COM

…and that of course, reminds us of the legends of dragons, which may not be myth after all. Scientists in Romania claim to have found a dragon in an ice cave….

Ice Cave Dragon

And of course there is the dragon in a bottle that was found in Germany… which has been recently proven <u>not</u> a hoax…

On close examination, there are fine hairs all over the body and the detail in the hands and feet and veins in the wings are evidence that it is not fake…

Do we really know of the myriad types of animals on our planet?

So these and more are all part of the Great Earth Puzzle. And some of the 'monsters' are still on Earth… such as the **Giant Sturgeon** caught in Lake Washington near Seattle in 1987:

And occasionally an **Oar Fish** washes up on a beach somewhere and scares the natives:

Enoch Again

Back in the Book of Enoch, he refers to Leviathan (Megalodon?) and Behemoth, and indicates that there was **just one of each**, not that they procreated, thus it is hard to see that they could have been dinosaurs...

> And on that day were two monsters parted, a female monster named **Leviathan** to dwell in the abysses of the ocean, over the fountains of the waters.
> But the male is named **Behemoth** who occupied with his breast a waste wilderness named Duidain on the east of the garden....
> (LX, v. 7-8)

And that is all we are told, Enoch does not elaborate nor describe the monsters. And since the Book of Enoch is older than the Book of Job, the only other reference to these two monsters, it is safe to say that the monster story began with Enoch and the writer of Job (which is an allegory, by the way) embellishes it, as quoted above.

David Wozney in <u>Dinosaurs: Science or Science Fiction</u>? argues that living dinosaurs never existed (because there is so much fraud in the field). Eric Dubay and Robbin Koefoed agree, and in sum, suggest that it is all **an anti-Christian agenda to promote Evolution, not Creation.** Check out Dubay's summary in his book, it is the best overall synopsis of the subject, citing multiple authors who have done the research and agree, and have the facts in their books and scientific papers.

> And by the way, in Los Angeles there is the **La Brea Tar Pits**, a huge lake of black gooky tar, in which many prehistoric animal bones have been found – **NO dinosaurs**, but there are several sabre-tooth tigers and mastodons!

So we started to examine the theory of the Young Earth, and segued into why <u>dinosaurs are the argument for Evolution and an Old Earth</u>. But if dinosaurs are fake, then is it possible that there are other criteria for the Earth being so young that dinosaurs could not have existed...? And why would a loving God terrorize his creation of Man with Tyrannosaurus Rex on the planet? Something more to think about is WHY would someone fake the existence of dinosaurs?

Suggested Answer

When you consider that Earth probably is really flat, and probably Young, which means Someone created it that way <u>and the flora and fauna on it</u>, including Man, and then Charles Darwin came along with a theory that Science could adopt in the mid 1800s, it would be expedient for the scientists to jump on the bandwagon and all of a sudden find fossils (where there had been none before AND which had never before been found anywhere by anyone) that just had to belong to something they would call Dinosaurs. And then move the dinosaurs way back in time, to substantiate the Old Earth theory part of Evolution. The second reason for doing it is that the Powers That Be (PTB) are determined to entertain the public… keep them wondering, guessing…. Think not?

Consider the TV shows *Ancient Aliens, Ghost Hunters, Searching for Bigfoot, NASA's Unexplained Files, Searching for Atlantis, What on Earth?*... and the like….. They don't always prove anything but they make you suspect that there is more mystery about Earth than there really is. You suspect that ETs are flying the UFOs (wrong), you suspect that there is a Bigfoot, but no one has ever caught or killed one (they can't), and *Ancient Aliens* repeated asks "What if it were true"? *Ghost Hunters* constantly has its researchers asking "What was that?!" "Did you see that?!" and you see nothing… they just jazz you around. All entertainment.

Secondly, Science likes to promote a godless Evolution as many scientists are not comfortable with the idea of a God who is still there, watching us, and the idea of "punishment" for "sins" is anathema to black & while, logical men of Science. They cannot prove the existence of God – except for those scientists who know the Earth is flat (and that group is growing, quietly) and **the Flat Earth <u>construction</u> is a proof of God's existence and his Grace to Man**. Science doubts the existence of God and the soul – even though (VEG, Ch. 7) discusses **Dr. Duncan MacDougall** in Massachusetts in 1907 who **weighed the soul**, proving it does exist. It weighed an average of 21 grams (and there was a movie about it by that name). Scientists who are atheists cannot handle the idea of a God…. and <u>if</u> they have a soul, they are in for a surprise when they die.

Wooly Mammoths

Related to the dinosaur issue is that of the wooly mammoths found frozen in the Siberian tundra. There were thousands of cadavers found and the Tungus (indigenous peoples) have been busy digging them up for an estimated 1600 years and using them as a source of meat – the hide, fur and meat have been perfectly preserved – as if flash frozen. [168]

The mammoth cadavers show no signs of decomposition. Below is a picture of one that was recovered by the Russian scientists: A baby woolly mammoth, frozen in soil for 40,000 years in Siberia, was so well preserved that traces of her mother's milk were still in her stomach. [169]

Frozen Baby Woolly Mammoth
(credit: Bing Images: geol.umd.edu)

In the Quartenary Age (20,000 BC) Siberia was free of ice and snow, and semi-temperate, with a lot of vegetation – we know that because much of it was still in the vegetarian mammoths' mouths! This is not explained by continental drift, nor does a shift on the Earth's axis account for what happened. Recent research shows that the cataclysmic event was a fairly limited surface event, not involving subsurface disturbances.

The recovered mammoths have been carefully examined and there is no sign of external injuries – whatever happened, happened to all of the thousands of mammoths <u>at the same time</u>. The food in their stomachs had not even been digested, and some still had vegetation in their mouths. While **Dr. Muck** (source of the above info) suspects that an **asteroid hitting the Earth** would have not only sunk Atlantis, it would have sent asphyxiating gases and dust over the planet choking the mammoths, and then the tsunami from Atlantis' sinking, would have finished the job with a tidal wave, flooding the forests and land. He adds…

> Then came the cold. As the waters calmed down, they froze…. It froze over the carcasses [of the drowned mammoths] that still remain encased in their coffins of ice…. In Alaska, the diluvian mammoth herds of this region apparently perished at the same time, and as suddenly…. Mammoths in all age groups were found in perfect physical condition. [170]

He also says that the only thing found in the area of northeast Siberia were the mammoth carcasses, and an occasional wooly rhinoceros. No small animals, theorizing that the heavier carcasses dropped down in the flood, whereas the smaller and lighter animals were swept along farther and deposited in other areas.

And according to his prior calculations of Atlantis' demise (see Chapter 8), that occurred about **8498 BC.**

Dating Errors

Today's Christian scientists of course are looking to prove Creation and a Young Earth paradigm. They cite instances where conventional dating methods (e.g., **Carbon-14** for starters) yield errors for anything older than 5000 years, and even the originator of C-14 dating in 1947, **Willard F. Libby, agreed it has limitations.**

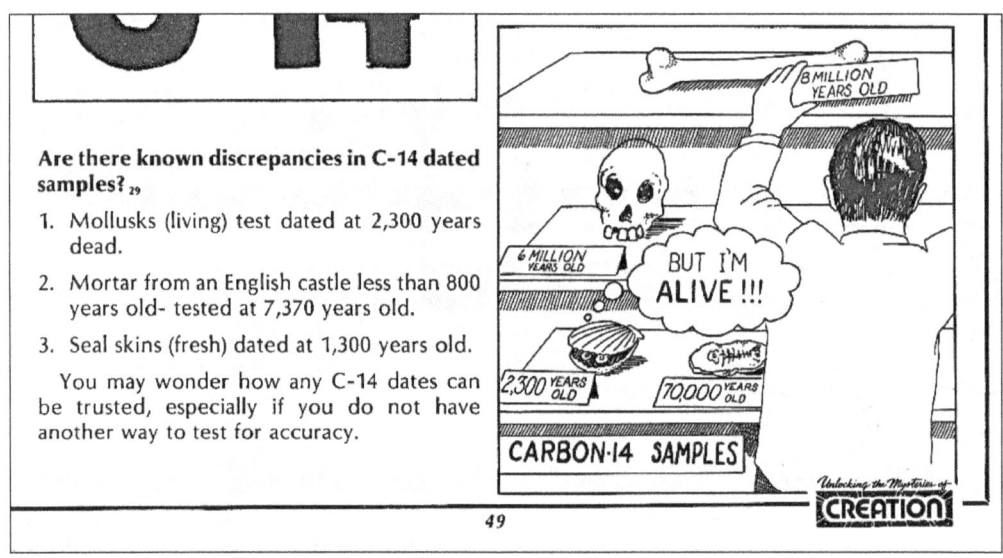

(credit: Dennis R. Petersen) [171]

In reality

> ....the [C14] method is valid only for 'recent' times. Even the most devoted advocate would not claim that it has anything at all to say beyond about 60,000 years before the present time, and **its inaccuracies are well-known.** On the other hand, the [C14] technique *does* have some application in the most recent few thousand years. [172] [emphasis added]

Radioisotope dating is another technique that yields ages in the millions of years, and yet it too has drawbacks. Another system is that of **counting tree rings** found in certain strata, but while the age of the tree is known, the strata may be much older.

The overriding issue is that Geologists and Paleontologists have made an <u>assumption</u> called **Uniformitarianism** – that all aging of anything on the planet follows a tried-and-true linear progression, **at the same rate**, and does not account for The Flood (effect of water), nor does it consider the effect of volcanism (heat) and meteor strikes (again heat and force). Thus most aging techniques are flawed because they are <u>based on unprovable assumptions</u> or the linear uninterrupted flow of natural aging, and that results in the ages of trees, fossils and rocks that support the Old Earth paradigm, but are suspect when other evidence is considered.

Young Earth Summary

The "other evidence" consists of the following (which tend to support a Young Earth and these were explored more in depth in VEG, CH. 10 and **Chapter 3** in this book). The following is just a brief recap to add to the evidence:

Fossils

> First, **there are no intermediate stages of dinosaurs** which would show a dinosaur species evolving into a new species. In addition, there are eggs and adults but <u>no juvenile forms</u> of the same species.

Population

> Second, calculations regarding **Earth's population statistics supports a young earth**. It is noted that at a population growth rate of 2% per year, which has been observed for almost a century, and given a current 6+ billion population, it has been calculated that it would **only take 1100 years to reach the present population** from an original pair of humans. And that isn't counting wars and death. [173]

No Bones

> Third, if Man has been evolving for millions of years on the Earth, **why are there so few bones found?**

Magnetic Field

> Fourth, the **Earth's magnetic field has been decaying at a constant rate** since it was first measured in 1835, and using that rate (with 1400

years half-life) and extrapolating backwards, it can be determined that the magnetic field must have been much stronger in the past… And if the field were too strong, using the doubling factor every 1400 years, just 100,000 years ago life would have been life living on a dense neutron star here – impossible. [174]

Helium

Fifth, **the amount of helium found in the atmosphere is a clincher for a young Earth**.
According to the latest scientific measurements, 13 million helium atoms escape into the atmosphere <u>every second</u>. … doing the math, based on the amount of helium in today's atmosphere, delivers a figure of 2 million years old maximum as an age of the Earth…. but what is weird is that helium is in abundance in the rocks, and not in the atmosphere, so the Earth is much younger than 2 million years old. [175]

Sediment

Sixth, sediment on the ocean floor is another good measure of age of the Earth. Since the Earth has been covered with water from Day 1, and water is constantly eroding the continents, **there should be a <u>lot</u> of sediment on the ocean floor. There isn't**.

Salty Ocean

Seventh, **salt in the ocean should be getting saltier** with the years and if the original ocean was salty, 3-4 billion years ago, shouldn't it be <u>too</u> salty now? The scientists studying the ocean asked themselves the same question and set about determining the 11 types and amounts of salt input to the ocean and the 7 types and amounts of output – i.e., the ways that the ocean can gain and lose salt. These were quantified and, using the minimum and maximum values that they developed, the maximum age of the ocean can only be 62 million years old. That is not saying that the Earth is 62 million years old, just that it couldn't be any older than that. [176]

Insert on Rivers

This is a very interesting analysis of and insight into points 6 and 7 just made.

Have you ever looked at a huge river like the Amazon or the mighty Mississippi and wondered where all the water comes from?

Science will tell you that it is snow melt from the mountains farther upstream. Actually that is 50% nonsense, but we all bought it because we were in elementary school at the time…

Then we wised up (the sheeple were beginning to question and think) and by high school or college you were told that there are aquifers or "underground rivers" that also feed surface rivers. And that is true but snow melt and aquifers are not the major sources.

This issue is also related to the **Gulf Stream current** in the ocean… What causes the underground movement of water? (No, it is not the Moon.)

So, let's look at **the Amazon** since it is the biggest river in the world (I didn't say "on the planet") … the River is <u>huge</u> (wide and long) and we're told that it all starts as a trickle in the mountains in eastern Peru, Bolivia and Ecuador…. Yes, but…. The River's flow is such that snow melt and small streams down the western mountains <u>cannot</u> account for the total flow… the River is **4300 miles long**, and up to **62 miles wide** in some places, and the rate of flow is **7.3 million cubic ft/sec**. Per <u>second</u>! That is **55 million gallons/second**. And some of the source is kept in **151 dams** in 6 of the main tributaries. The incredible flow's <u>source</u> gets harder to believe every minute.

One picture will not do it, so there are two:

… also note the **flat horizons**? **Above** shows the backwater tributaries and incredible width to the river; **below** is NOT the ocean – a shot across the river:..

View Across the Amazon
(credit: Bing Images (both) and flickr.com)

Repeat: the rate of flow is **7.3 million cubic feet per second! Year round.**   That is **55 million gallons/second. Non-salty water that flows into the Atlantic Ocean… yet the Ocean does not get less salty… Why?**

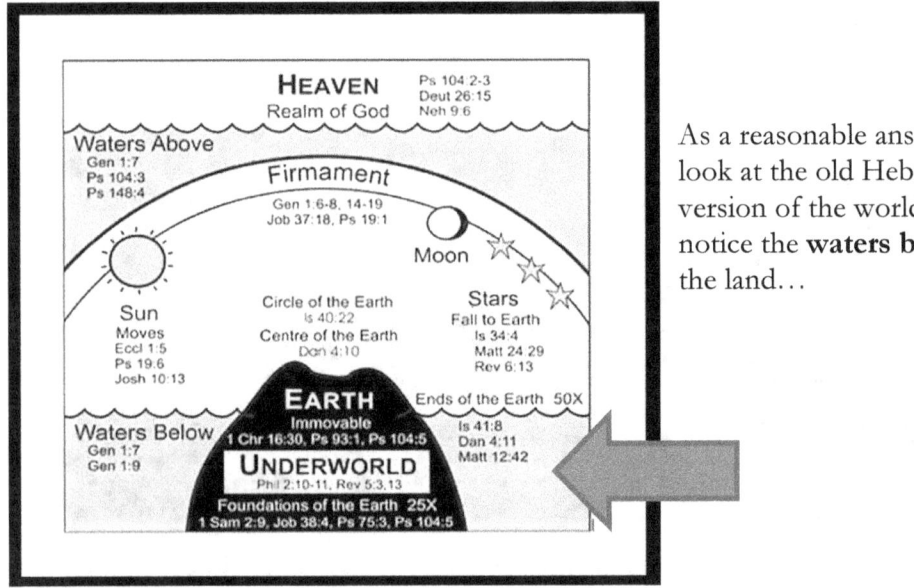

As a reasonable answer, look at the old Hebrew version of the world, and notice the **waters below the land…**

The same question applies to the Yangtze, Nile and Mississippi Rivers… What is their real source? Suggested: the **underground water** on which the landmasses float… their weight could have the effect of forcing water up thru natural wells and springs, as well as seeping up thru porous sedimentary rocks, to the surface. In fact, flowing <u>under the length</u> of the Amazon River is an **aquifer** called *Hamza* which is **saline** and follows the Amazon to the Atlantic Ocean…. sustaining its salinity.

So this is a twin-river system flowing at different levels of the earth's crust in Brazil.

The waters below were also called **The Deep**….shown below (arrow)…

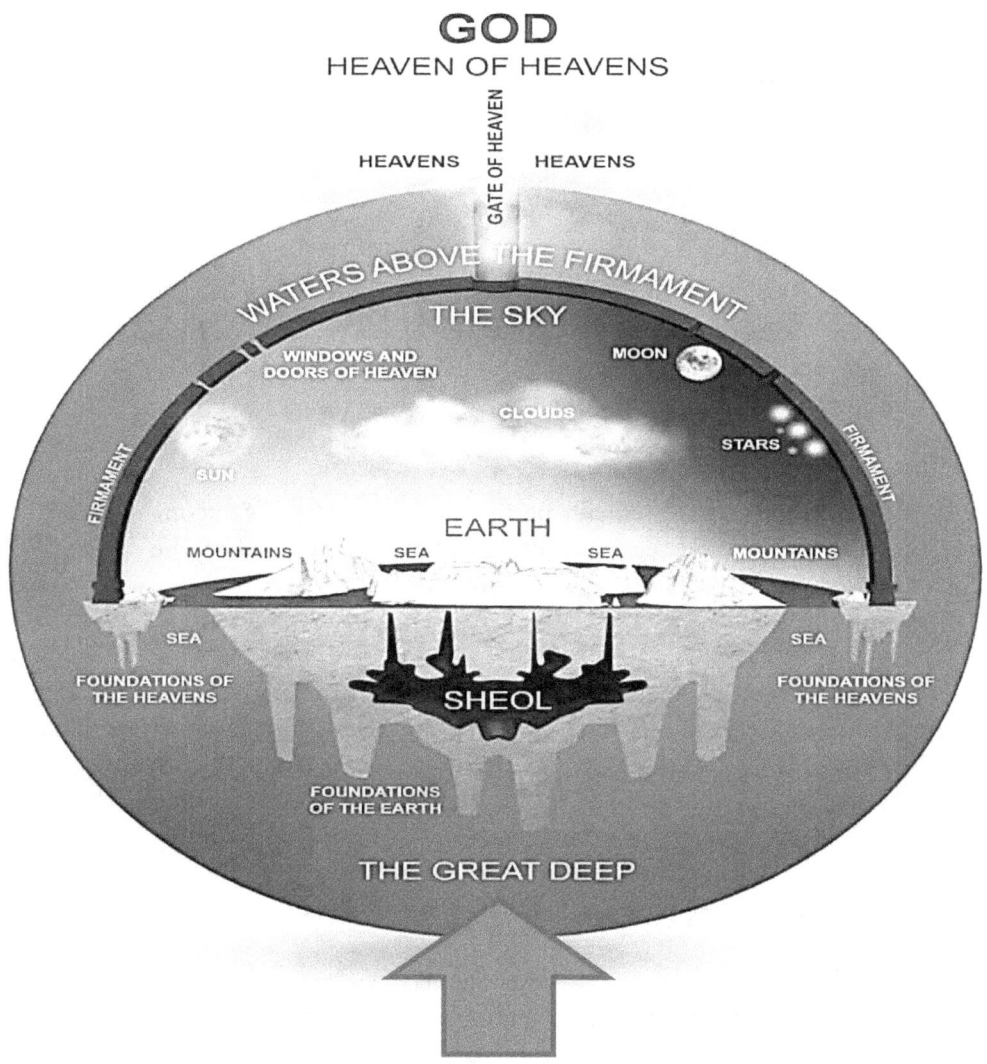

The concept is that currents flow from the fountains of The Deep supplying rivers and the ocean currents – the oceans have to be kept moving (The Gulf Stream and others) to keep the water fresh to support the life in the seas. This also speaks of **Design**.

And yet it is only the top 90' of the ocean that gets tossed by bad weather – the farther down one goes (past 90') the more calm it is. [177]

> It unfolds to us that the Earth is founded upon **the Great Deep**, part of the waters of which percolate or flow through its body in various channels, **forming the springs** in lakes, hills, and valleys, from which the rivers take their rise…. Rivers all have a *downward* and never an upward course in any part of their journey to the sea, thus proving that **Earth is not globular and therefore not a planet.** [178] [emphasis added]

> That last point is examined in the next section...

And from the same source, when it was asked why the oceans do not gain or lose saltiness, the reply was

> …we may safely conclude that a quantity of water is annually poured into the ocean which, if collected, would cover the earth 570 times. The first grand question then is – Unless the heads of these rivers communicate with **a great central abyss of water**, whence does this prodigious quantity of water come?
> The second question is equally pertinent : Unless the ocean communicates with the same great abyss, and thereby **maintains the circulation of the rivers**, how does it happen that no perceptible variation in the [ocean's] water level results from this immense supply? [179]

And then he goes on to say that the scientists argue that rivers are supplied by rain, dew, snow and that a certain amount of water evaporates from the ocean so that it maintains its same level… saying that the evaporation is equally offset by the amount that the rivers bring to the ocean. He refutes this 'scientific explanation' by showing that **evaporation in no way can offset the huge amounts of water coming into the ocean from multiple rivers feeding the Atlantic alone.** He cites studies done – measuring the rainfall, the river input versus average rate of evaporation, and says

> With what reason then can it be maintained that, after meeting the demands of evaporation and sustaining vegetable life and growth, the rain is sufficient to supply all the rivers that fall into the sea?

...the tops of the mountains above the sources of the Rhine, Rhone, Danube, and Po are during the winter half of the year constantly covered with snow to a great thickness, so that no [evaporation] could touch them, and yet **these rivers run as steadily in winter as in summer.** Now this is inexplicable [except that] rivers draw their supplies from the subterranean abyss into which they return them again. [180]

This also sounds like common sense, as these scientists in the late 1880s observed Nature and drew upon logic, the Bible, and the Flat Earth scenario to arrive at a reasonable answer. Considering the volume of water that flows to the sea in the Amazon River, they are probably right – they did detect the underground aquifer, **the Hamza.**

So scientists are on the right track... but if others don't like to see **design** in the Earth (let alone hear that the Flat Earth answers a lot of their science questions), then a lot of the Earth science will remain wrongly interpreted, and Man will come up with additional sub-theories to help their amiss theories work, as they did with ocean evaporation to explain why **the oceans do not continually rise from river input**.... And yes, the 'modern science' answer sounds plausible, but mathematically it does not work out. It looks like some of the people of 150 years ago had their heads together and knew more about the real Earth, in simple terms, than we do today. (Chapter 4.)

Remember: Science wants to supplant God.

Speaking of rivers (and we will return to the last two Young Earth points after this next section)...

Rivers Defy Gravity?

There is one last oddity, which pertains to rivers and helps disprove the Earth Globe concept...

Does water flow uphill?
Do rivers flow uphill?

If you believe the Earth is a round globe, then you have to say: Yes.

Here is the situation. The Amazon River actually flows from southwest in Brazil to northeast where it enters the ocean... keep in mind that the Amazon starts below the Equator – the widest part of the Earth Globe... What that means is that the Amazon River <u>south of the Equator</u> must flow north, or uphill as it flows toward the Equator...following the widening curve of the globe...

The arrow (above) shows the flow of the Amazon as it heads toward the Equator and the ocean... Note that the flow from the Mato Grosso is due north, uphill as the Earth Globe is widening more as it moves from Bolivia to the Equator... here is the side view of the Earth Globe:

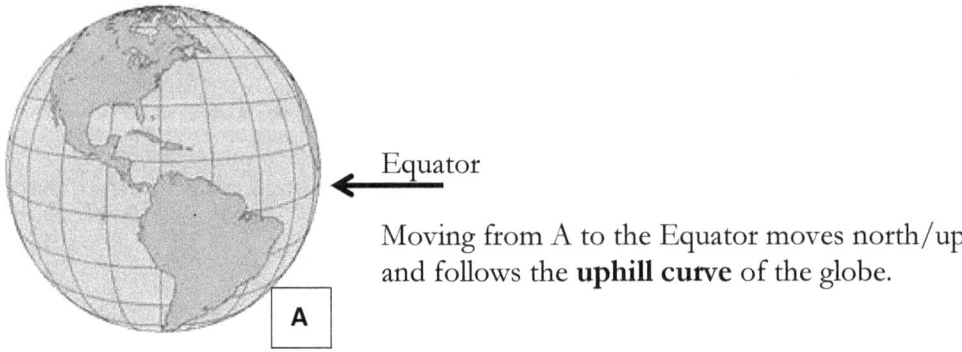

Equator

Moving from A to the Equator moves north/up and follows the **uphill curve** of the globe.

The Victoria, Adelaide and Daly rivers in **Northern Australia** have the same issue – they move north and against the increasing curve of the globe which means they flow uphill. I know what you are thinking: they flow downhill from the higher hills... No, these rivers are 80% on flatland, they <u>initially</u> flow from a higher elevation to a lower, but then <u>they flow uphill on flatland to the sea</u>. Why don't they 'puddle' or create a lake on the flatland? This issue does not exist on a Flat Earth.

See also **Appendix A** if this is not clear. More examples are given.

Ok, back to Young Earth points 8 and 9...

Meteor Dust

Eighth, meteoric dust from space has been accumulating on the Moon and Earth since the beginning, and it was feared in the early 1960's that if we tried to land on the Moon, there **should be a foot or more of dust** and that could adversely affect landing and takeoff.

So what about the Moon? There is no rain, wind or erosion comparable to Earth – what falls on the surface stays there. Later when Man went to the Moon, the dust layer was found to be **only an inch or so**.... Nevertheless, if Earth is supposed to be 4.5 billion years old, and even if the influx varies, there would have been many cycles of influx resulting in <u>more dust and nickel</u> than we have today on Earth.

Continental Erosion

Ninth, and lastly, there is **the erosion of the continents** (via streams and rivers) which has been consistently measured and calculated to be 27.5 billion tons per year – see 'Sediment' in #6 above. Now it is known that the total land mass above sea level for the last 70 million years (since the last geological upthrust) is 383 million billion tons. Simple math tells us that "[at] present erosion rates, **all the continents would be below sea level in 14 million years!**" [181]

So there you have it – **Earth is not 4.3 billion years old**, but it is more than 5000 years old. Something in the millions of years old still qualifies for a Young Earth.

Mutations

Point #1 above, Fossils, also shows something else… no gradual evolution or mutation of one lifeform into another.

The fossil record shows <u>no evidence</u> that any basic category of animal has ever evolved into any other basic category. And that means that dinosaurs did not evolve into birds; Pteranodons were created, Archeopteryx (first bird) was created and there is no accounting for fish ancestors…

One of the tricky things now that we have a better understanding of (and ability to analyze) DNA is that many organisms, animals, fish and insects and such have many **similar sections of genes** (whole strings of amino acids G-C-A-T which are identical) – which has given rise to "proof" that Evolution was right all along. Wrong.

It is suggested that the commonality between 4-legged animals who walk upright will have similar genetic coding for <u>that function</u>, and if multiple animals (bears, dogs, horses and mice) share that common design (4 legs), they will share a partial common genetic coding in one of the chromosomes – differing only in length of the legs and skin/fur covering and coloration. The commonality does not mean that one evolved from the other ... it means that that section of genetic code was copied from one animal to another, <u>by design</u>... when they were <u>created</u>.

Design is the key to the Evolution vs Creation argument. And issues and examples supporting that were given in Chapter 4... some aspects of flora and fauna are too complex and do not support one structure morphing into another becaye the environment demands it.

Case in point: moths in England during the Industrial Revolution that were white or beige 'evolved' into a black moth due to the heavy industrial pollution of the air – with soot and chemicals bombarding the countryside and the moths. A pure case of **Survival of the Fittest** since the moths had to survive by blending in with their surroundings and if their habitat (trees) became covered with soot, they would not survive as a white moth on a black tree, so their DNA morphed to make them black. This today is called **Epigenetics**, and Darwin was correct in this.

Note how the moth 'camouflages' itself to match the tree...

The soot on the trees yielded this problem making the moths easy prey for the birds to spot:

(credit: Bing Images)

So Epigenetics 'changed' the moths...

…to this:

Other example of design….. **Poisonous frogs** are yellow, red, blue, spotted and striped… Nature is warning that it is not a normal dark green frog!

Flying fish and flying squirrels are other examples of design as fish do not fly and squirrels do not normally glide 300' to another tree. Actually the flying fish (found **off Catalina Island** off Los Angeles) do not have wings but they do have large pectoral fins so that when they leap out of the water, their fins permit them to glide thru the air for a short distance.

But why would a fish want to do that and how would Nature cooperate in adjusting the length of their side fins so that they could do that? Fish gliding the air is not what a fish is about…

Be careful… Darwin would say that some birds became fish and it was a vestige of a former ability… like some humans who have the vestige of a tail, 'proving' that they evolved from Apes (see below) ….Aaargh!

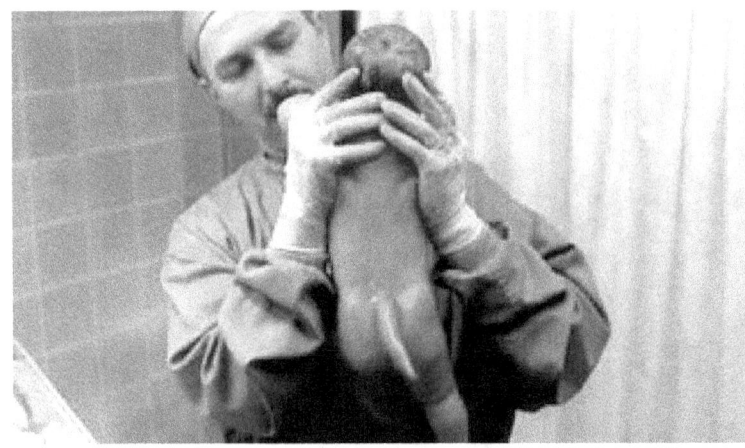

Some have tails….

(not Photoshop®)

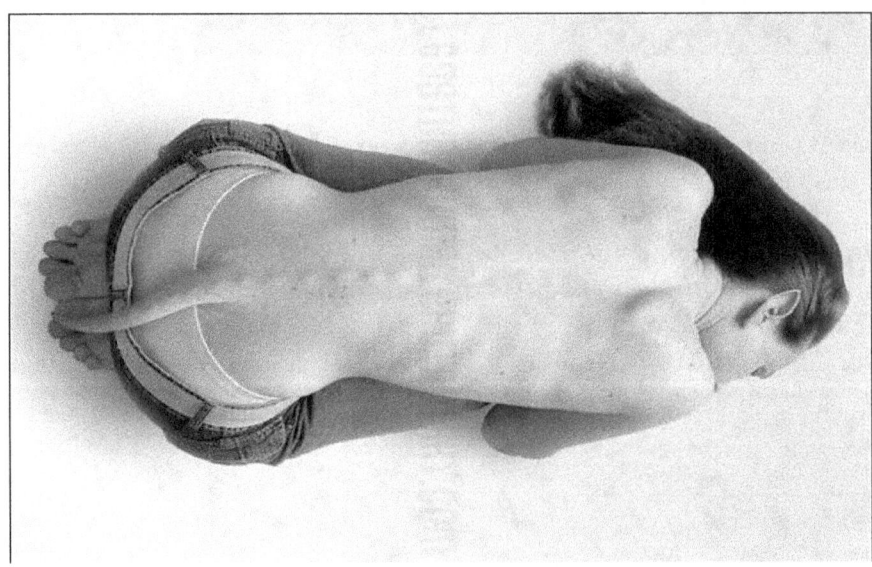

And some have horns…

In the literature on this, most people with horns are in India or China…

…and some have extra fingers or toes…

…still not Photoshop®…

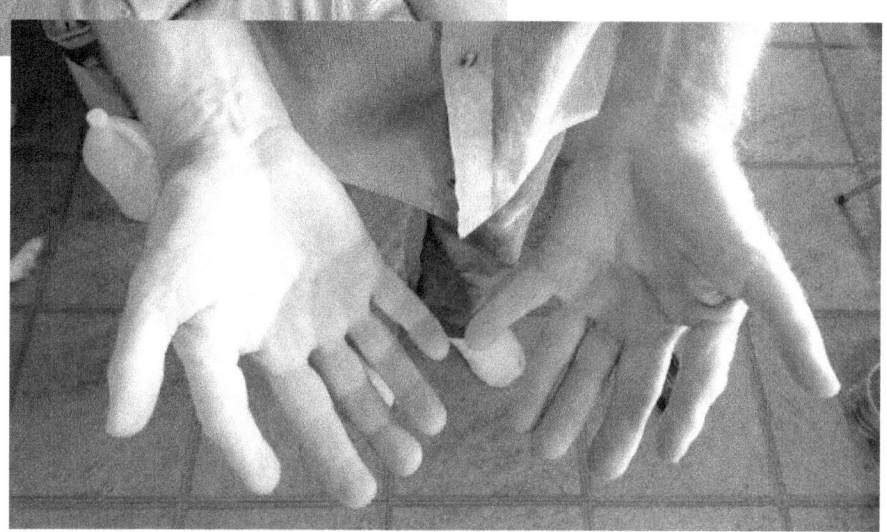

…and some others have webbing…   and extra teeth…..

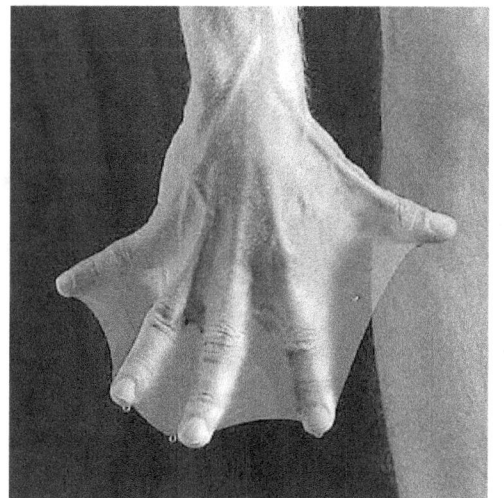
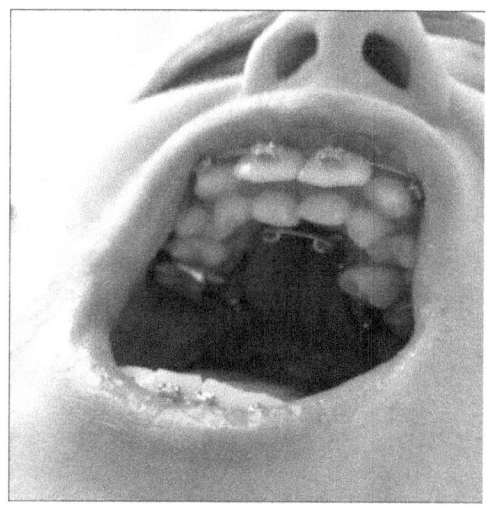

Ok, also this is not to gross anyone out… most of these are said by <u>standard science</u> to be something that **happened when the embryo was growing thru the fetus stage** to that of a human (first 40-50 days) …. Note that in the beginning, before the fetus starts to differentiate into a human, you can't tell a human from a fish from a chicken from ….

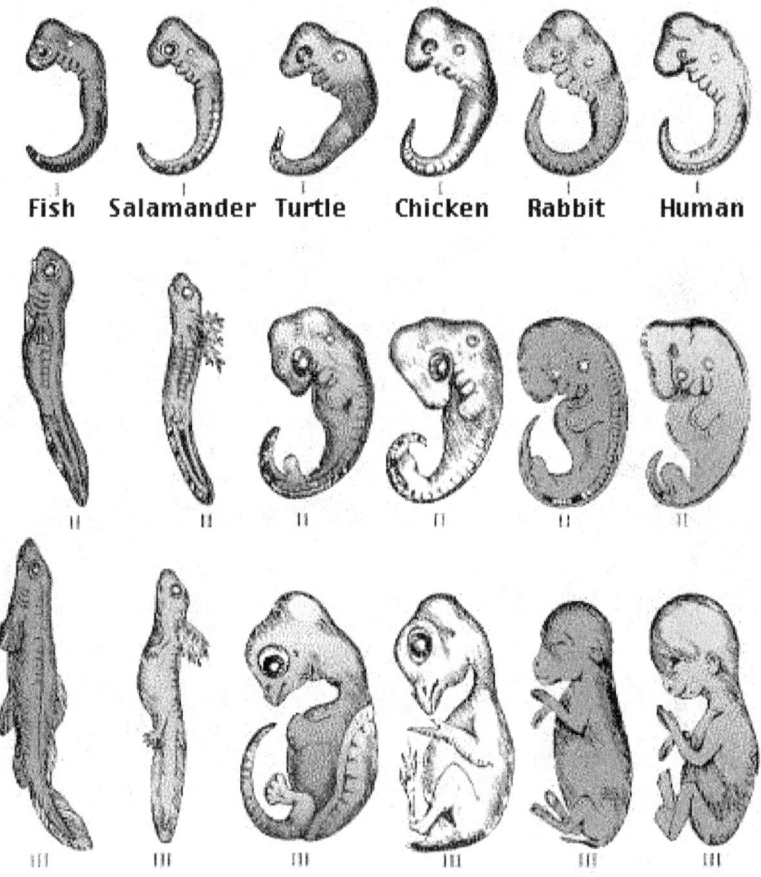

Fish   Salamander   Turtle   Chicken   Rabbit   Human

Note the tails in the bottom row… the human is supposed to lose its tail.

Non-standard science will tell us (as has via the Anunnaki extended genetics analysis) That the horns are a throwback to Anunnaki who had horns – in their reptilian form – the mammalian Anunnaki did not. In addition, the double row of teeth is said to be found among some skulls of giants, so that is a genetic throwback to the Nephilim. Also, there are the red-haired giants called **Si-te-cah** found in a cave in Lovelock, Nevada that the Paiute Indians killed. [182]

All just to show that the history of Earth is not the standard vanilla version we have been told – it is tutti-frutti.

And having mentioned that, it is even more relevant to address the Creation Issue – which was begun in VEG. This was a chapter dealing with historical aspects of Earth, which includes the creation of the Earth, its shape, and whether dinosaurs are part of that history or not, and whether the Earth is old or young, and now we need to briefly address <u>WHO was here</u> and the history of Man on the planet… exposing a few Science assumptions as we go… in the next chapter.

Summary

Many ancient cultures knew the Earth to be flat, for centuries, and you'll see why in the next chapter.

Ok, Man was told the Earth is flat by those that shepherded him (Anunnaki), and you may still be hanging out with that one, and I agree it is a bit much to handle nowadays because we are all used to thinking Earth is a sphere. I did too and yet my Source (Baldy) actually said to me "Are you living on the planet you think you are?" So I assumed Earth was a **VR Sphere**.

But I don't know now if that was right

And there is serious evidence to show that the Earth is not 4.3 billion years old (as Science says) – and it is not even 65 million years old, so an asteroid may not have hit it back then – If the dinosaurs did not exist, then an asteroid 65 million years ago did not wipe them out. **The Earth is younger than we think** (Chapters 3 & 7).

Later an anonymous email (coming thru my reader email – see Copyright page) asked me if I had ever considered the Flat Earth (FE) issue. Of course I laughed it off. Then something made me think – Hey, wait a minute I could include that in QES as a separate Ch. 11 <u>and prove the Flat Earth wrong</u>, and substantiate the VR Sphere.

I could not prove the **VR Sphere** wrong.

As was said before, VR Sphere proof was not conclusive, so I became fascinated with the FE Theory and the deeper I dug into it, yes there were obvious Bozo proofs – made by those wanting to discredit the FE issue – but I was struck by about **6 proofs** (now in Ch. 11 of QES) and some are repeated here (Chapters 3, 7 and 10).

I could not disprove the **FE Theory**, either.

Now it is a lot to ask the reader to consider that <u>some</u> **Dinosaurs did not exist** and were a scam – to prove Darwin's Evolution Theory – i.e., Evolution needs a long time for Evolution to work and morph the lifeforms… so Science said the dinosaurs were allegedly here about 65 million years ago and then a putative asteroid wiped them out. And yet that never seemed correct.

> **News flash**: if the Earth is flat, or a VR Sphere, both have a protective "shell", and neither may be more than a few million years old, then there was no "65 million years ago" period,
> AND
> does anyone see how the FE Firmament/VR Sphere Shell would not permit an asteroid to hit the Earth, anyway?

Rewind: Dinosaur Realia

So what are the dinosaurs… <u>before</u> China started making replicas?

There were **never any dinosaur skulls found**, just fossil teeth, scattered bones… as if the bones of extinct Wooly Mammoths, Bison, Giant Sloths, and some things that looked like 6-8' Velociraptors, Giant Bears, even today's elephants and rhinoceros bones could be used…etc. They had large enough bones to "stand in for" the putative larger dinosaurs. Who would know the difference?

But why do that?

Again, **entertainment for the Sheeple**, support for the Theory of Evolution, and

> During the nineteenth century a new world of evolution was being pursued by then influential people such as Darwin and Marx. <u>During this era of thought, the first dinosaur discoveries were made.</u> Were these discoveries 'made' to try to make up for inadequacies in the fossil record for the Theory of Evolution?…. The following issues raise **red flags** as to the integrity of the dinosaur industry and cast doubts as to whether very large dinosaurs ever existed:
>
> 1. Dinosaur discoveries only occurred in the last 2 centuries and in huge, unusual concentrated quantities going against the laws of nature and probability;
>
> 2. Dinosaur discoverers were usually not disinterested parties without a vested interest;
>
> 3. The nature of public display preparation [which is done in secret behind closed doors] calls into question the integrity and source of fossils allowing for… substitution…. tampering…. and the possibility of fraudulent activities;

4. Existing artistic drawings and public exhibits showing off-balance and awkward postures that basic physics [of such animals] would rule out as being possible; (Think: T-Rex)

5. Very low odds of all these dinosaur bones being fossilized but relatively few bones of other animals....

[And]....The possibility exists that the concept of prehistoric living dinosaurs has been a fabrication of nineteenth and twentieth century people possibly pursuing an evolutionary and anti-Bible, anti-Christian agenda. [183]

Such was said by **David Wozney** in <u>Dinosaurs: Science or Science Fiction</u>.[17] He suggests that the dinosaur industry should be investigated and questions asked.

And he may be right, <u>the Bible does not support dinosaurs</u>, nor were dinosaur bones ever found by Hebrews or Assyrians, or Egyptians, nor were the Amerindians finding them either – and in the case of **the Sioux**, many bones were in their very hunting grounds (the Dakotas) for decades – and they never found any dinosaur bones.

And we now know that the Chinese make and ship dinosaur bones on demand nowadays.... So larger dinosaurs may not have existed, but there are genuine fossil tracks in the Paluxy River bed (Glen Rose, Texas) of smaller reptiles:

(credit: Robert Nunnaly, Allen Tx.) on Wikipedia under Paluxy River.

6-8" prints in stone under water...

Chapter 8: Earth History , Part II

Chapter 7 dealt with errors and deception in Physical Science, like Geography, Geology and Archeology and Dinosaurs. This chapter is a continuation into modern times of who was here, Anthropology, what they were doing and how Man benefitted. Former books (like VEG and AL) spent some time examining the Anunnaki, and AL also included those who were here before the Flood, the Ancient Ones, who were not the Anunnaki. So it is appropriate to examine the Ancient Ones and the Anunnaki… What part did they play in Earth's history? You will find that you don't really know Earth history after all.

Chapter 7 left off with the Dinosaurs and whether they were all real or not. Before you throw out that issue, consider that Copernicus, Darwin, and later Marx, Engels, Trotsky and Lenin (who were all Zionists by the way) and the Zionist-inspired Illuminati all had reasons to upset the normal social order with its emphasis on Creation and Christianity and replace it with the godless society we are drifting toward today.

And yet, before coming up to today's issues, we need to stay in the past with an examination of two important groups of beings who were also here – besides Man.

Just a sidenote, for those who have not read VEG, or AL, or QES, there was a Middle East scholar named **Zechariah Sitchin** who translated many Sumerian tablets and scrolls and determined that the Sumerian humans were referring to the Skygods who came down to them in skycraft... they called them **Anunnaki,** or "those who from Heaven to Earth came."
Sitchin was a rascal and didn't tell the whole story (see Ch.3 in VEG) –
But the Anunnaki were seen as gods, because they came from the sky, they flew thru the air and had terrible weapons (and sometimes fought each other), and they had advanced genetic and physics knowledge. Whereas Sitchin argued that the Anunnaki were from the planet **Nibiru,** his 1978 revelatory book was a further dodge to get people to think that they were ETs who came to Earth because Earth is a planet circling the Sun, and any ET can visit us on an unprotected planet.
What AL showed was that the **Ancient Ones**, also called **Shining Ones**, have always been here, and were often mistaken for the skygod Anunnaki. They were also taken for the Norse, Greek and Roman gods. What Sitchin got right was that these Skygods created Man, who were in turn part of the Original Creation. Clarification is needed.

Anunnaki and the Ancient Ones

> This section is connected with the other issues (Young Earth, Flat Earth, Dinosaurs) and after this review of who was here and what they were doing, these various issues should form a more coherent, united whole – but VEG did not contain all of the information…
> **the Flat Earth (FE) Issue now seems to be a key to it all.**

Young Earth, Simulation, Dinosaurs and Ancient Ones all are part of a mosaic that can only exist in a 3D Construct scenario. And while it is hard to break with the traditional view that Earth is a rock, spinning about an axis at 1000+ mph, hurtling around the Sun at 66,000+ mph and there is no God and Man just evolved – without a purpose, such is wrong and this book is bringing the new pieces of the **Great Earth Puzzle** together. That is the traditional Science view of Earth and Man, and it is wrong… on purpose. The PTB have an agenda into which the non-FE fits nicely.

> **Synopsis to this point**: the Earth is either flat or it is the VR Sphere (as QES said) and it IS one of the two. Seriously. And so far, the VR Sphere is losing ground… the proofs support the FE scenario… Gravity and the oceans are a big VR Sphere detraction…but it may still be true as we do not know the willingness of the gods to morph the Flat Earth into a VR Sphere….

The ancient civilizations knew what Earth really was and because the gods were walking among Man thousands of years ago, teaching Man agriculture, medicine, writing, astronomy, building skills, etc. … it had to have been the gods who told Man where he was (and with the rise of the priestly caste, the info was buried publicly in favor of a Spherical planet.) That was why ancient societies initially believed the Earth was flat, and then Science in the 1600's began changing the public's bel;iefs. Darwin with his *Theory of Evolution* (growing since 1836) put the penultimate nail in the FE coffin in 1859 with <u>On The Origin of Species</u>.

> It would be interesting to know how much of the Inquisition's work in the 1600-1700's was directed to "convincing" people that Earth was a globe…the Flat Earth would have been seen as heresy when Science declared Galileo and Copernicus right… the Church [in the 1800s] would have no choice but to stamp out FE heresy along with any other religious heresies like the Protestant Reformation. [184]

This was more fully examined in <u>Anunnaki Legacy</u>, with the review of the **Anunnaki and the Ancient Ones**… <u>both</u> were here, and were not the same beings. And very important:

Neither group was ETs.

Rewind: VR Sphere & Flat Earth

What I understood in 2008 that became the foundation for the structure of the Earth, which I called VR Sphere, was just three things:

> Earth is a sophisticated **Simulation** that exceeds the ability of present-day Science to analyze and recognize the fact; the science involved exceeds what was shown in the TV show *Star Trek* as the virtual reality **Holodeck,**

> (**Micheal Talbot** in Holographic Universe agreed, as well as the engineer **Jim Elvidge** in The Universe Solved. Later, my QES Ch. 11 all but proved it.)

> Earth is a **3D construct** located in 4D which higher realm has the power to sustain the science driving the 3D Simulation,

> and

> Earth has a protective **energy envelope** around it off which the *Gegenschein* reflects the Sun. (I prematurely assumed it surrounded a spherical Earth and thus referred to the whole thing as the VR [virtual reality] Sphere.) The energy envelope is the **Firmament** if you are a Flat Earth believer.

That is the essence behind the VR Sphere, and **those points still apply to a Flat Earth**, such that the only way to tell between the two would the 2 key issues:

(1) **Gravity** and What keeps the heavy oceans (millions of tons of water) stuck to a curved/spherical surface of the Earth? **Water always seeks to lie flat**… so why would it stick to a curved globe (or flow uphill: Chapter 10)?
(2) The other issue is: **Can we prove Antarctica is not an Ice Wall?**

Mankind's Heritage

Many books by different authors have already said that the Anunnaki created Man in their image… albeit reptilian (as Ch. 3 in VEG established). So we won't go into that again. And then AL established that the superior beings, Anunnaki and/or Ancient Ones, travelled around the Earth after the Flood to lead Man in rebuilding his civilization.

And yet, mankind went thru several levels of development before being ready to build and sustain civilization. Sitchin is right in this.

One thing is for sure: the successive forms of Man on the planet are mute witness to the fact that Man <u>was</u> changed – from **Homo *erectus*** to Neanderthal to Cro-Magnon to Homo *sapiens*, and lately (see TOM) to Homo *noeticus* (Indigo children).

The Anunnaki could have modified Homo *erectus* when they created *Lulu,* or their sterile worker human (the 1st creation). The improvement, by Enki, into *Adamu* could have been closer to the Neanderthal (the 2nd creation). Enki would later 'personally upgrade' *Adamu* to *Adapa* (the 3rd creation)— much like Cro-Magnon and later Homo *sapiens* man.

Neanderthal was one of many experiments that didn't work out and was replaced by Cro-Magnon. In a similar way, The Change as discovered by Dr. David Jacobs portends a **replacement of Homo *sapiens* with Homo *noeticus*.** With a gradual phase-in (as Cro-Magnon did to Neanderthal) , the human race will be upgraded... which appears to be an on-going thing (Denisovans and Homo *floresiensis* [Hobbits] notwithstanding).

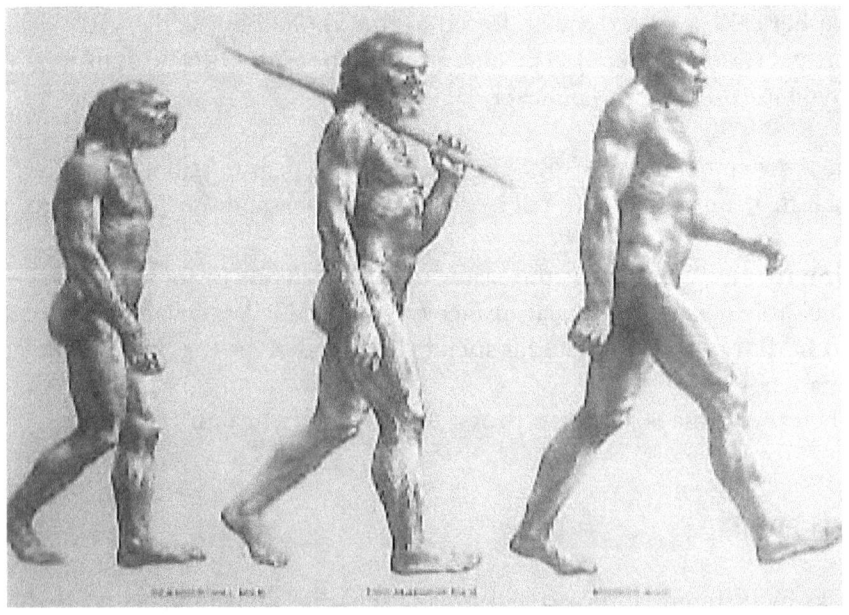

Man: Neanderthal, Cro-Magnon, & Homo Sapiens
Or: The *Lulu, Adamu,* and *Adapa.*
(source: http://www.wilderdom.com/evolution/HumanEvolutionSequencePictures.htm)

To get a more comprehensive view of this same development, including a flowchart, see Chart 1 in VEG Ch. 3. While the following diagram is a bit humorous, it nonetheless showcases the concern that a lot of anthropologists have for the direction of Man's current development…

Somewhere, something went terribly wrong

All of that to say that (1) there are genetic wonders all over the planet, and (2) Others are out there who care about Man and are watching. And it would appear that The Others are still here and have not stopped assisting Man via DNA inserts (via the Greys who are biocybernetic androids… they are not from another planet). Man is unique on Earth and is a **special creation**… and we know it is a <u>creation </u>because of the following information …

Rewind: 223 Genes

As was said in TOM, Ch. 2, **Man has 223 genes that are unique to him** and are <u>not found in any other organism on Earth</u>. If the scientists believe that *panspermia* is the way genetic material gets here (bacteria from meteors), it falls to Earth and disperses in the soil and water, the bacteria propagate among the animals that eat the plants that absorbed the organisms, and drink the water that contained the bacteria, or eat the fish that absorbed the bacteria in the water, then why aren't the 223 genes found in other animals or even in free-living bacteria themselves? They're not.

Because The Others added the 223 genes to our genome in their genetic upgrade of Man (*Adapa*). **This is the "smoking gun" that proves Man was a product of "assisted evolution."** It remains to be seen just what function the 223 genes have… Science is still trying to figure it out.

An analysis of the functions of these genes, published in the journal *Nature* (issue No. 409), showed that they involve important **physiological and cerebral functions peculiar to humans**. Since the difference between Man and Chimpanzee is just about 300 genes, those 223 genes make a huge difference. [185] [emphasis added]

If someone went to that much trouble to make Man different and more capable than the Apes, then it would stand to reason that They are still around, watching to see how Their progeny turn out.

And just as interesting is the fact that when Man was created something happened (the Bible says Watchers created Nephilim ["giants"]) among mankind. There have been **giant skeletons found all over the world**, and when they are, agents of the Smithsonian show up and cart them off, never to be seen again.

Skeletons Around the World

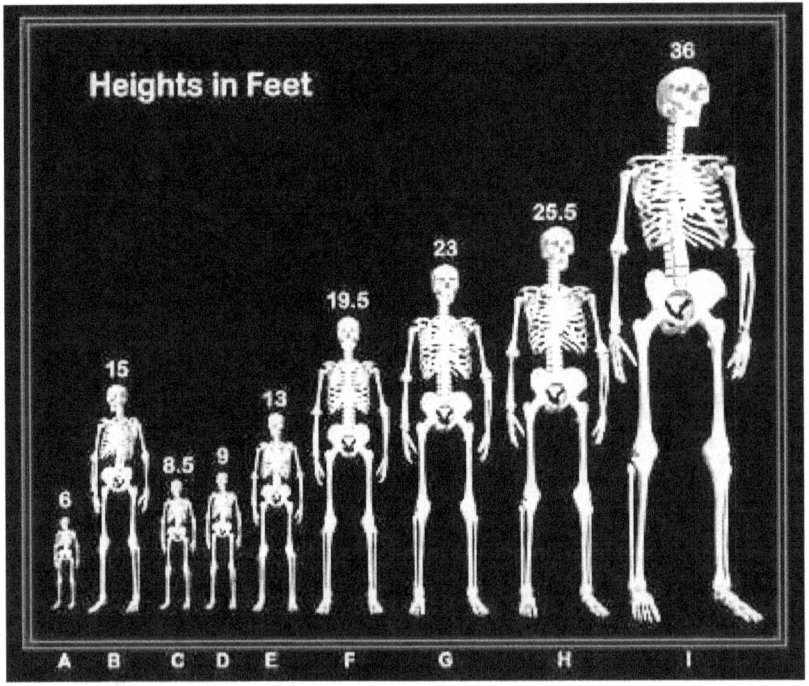

Heights of Giants Found Around Earth
(source: http://www.bibliotecapleyedes.net/gigantes)
(Credit: Steven Quayle)

Note (above in Chart) that Man **A** is 6 feet tall, today's human. **D/E is Goliath**, and **E** is King Og spoken of in the Bible. The others represent sizes of actual skeletons found around the world. The skeletons found in the Middle East are B, D, E, and I. Skeletons found in France are F, G, and H in Ecuador, and D in Georgia (Russia). TOM went more into this issue.

Nephilim & Giants

Nazorean, Gnostic, Johannite, Ebionite, Pauline teachings or otherwise, there is a belief in 'fallen angels' on planet earth – mistakenly called the **Nephilim**. The **Nephilim were the <u>offspring</u> of the Watchers + Earth women**. The concept is that of a humanoid being, a Watcher, descending to Earth in a form a little different from, for instance, one's next door neighbor, yet distinguished by a distinct psychology and genetics. Enoch said the Watchers **shapshifted** and that is how they deceived the Earth women.

Repeat: (see Chapter 6) The Bible does not specifically say that the offspring were giants, it just says they were "…in the earth in those days" and the original Bible word was not "giant" it was "Nephilim" which <u>they say</u> means "great men, **men of renown.**" This is another translation error in the Bible (see Ch. 11 in QES). It is <u>The Book of Enoch</u> which says that the offspring were physical giants. While their exact origin may be in question, yet terrible **giants called Anakim, Geborim and Rephaim** were created who finally turned on man and began to literally devour him.

Anakim Warrior (12' tall)

Goliath (9' tall)

There is evidence for the giants all over the Earth in skeletons that have been unearthed. In 1833, soldiers digging at Lompock Rancho, California, discovered a male skeleton 12 feet tall… [it] had **double rows of upper and lower teeth**…In Ohio in 1872 an earthen mound was discovered to contain three skeletons that … stood at least eight feet tall. Each also had **double teeth**…[186]

And the **Celts** and their cousins across the English Channel in Germany were also quite tall and fierce warriors, and that is why the Roman Empire had such a hard time conquering them (see AL). The Romans finally did subdue the Celts and Germanic tribes, but at quite an expense: 2 legions (a legion =1500-3000 soldiers) were wiped out.

Indeed at the famous **Battle of the Teutobergerwald**, the Germans would so utterly decimate **four Roman armies** in the brutal fighting that Rome would maintain a very **defensive** posture with respect to the Germans (i.e., ignore them) until the Western [Roman] Empire's final collapse….

> Eventually, however, some of these Germans were captured by the Romans, and one of them, a particularly troublesome King by the name of **Teutobokh** was paraded in Rome in the customary triumph. The Roman historian Floras reports that this king was so tall that… Teutobokh could be "seen above all the trophies or spoils of the enemies, which were carried upon the tops of spears." Teutobokh was easily nine feet tall… perhaps considerably taller.[187]

Teutobokh Taken Prisoner

Here again, we have giants in the land, genetic throwbacks… perhaps to **Nephilim** times? Remember that the Nephilim were giants, and the Vikings in Ch. 2 of AL also said there was a realm of giants (**Jotunheim**) – as well as those found in a cave in the American Southwest with red hair. [188]   (Ch. 2 in VEG explores the giant issue.)

And guess what? The DNA for gigantism has come down to today's populace…

Skeleton Fraud?

Yes, skeleton frauds have been exposed, but not all of the pictures of giants are faked. There is a museum in Ecuador that has a **25' human skeleton** on display. It was found in Loja, Ecuador in October 2012. Also, there is a picture of a 12' Irish giant in an open coffin leaning against a railroad car in London from the 1800's. [189] This was also supposedly the same height of the Philistine, Goliath, who was at least 9' tall.

Giants Today

The genetics that produce giants have not disappeared.

@Talmid HaMashiach

The man on the left (7.5' tall) is in modern-day Iraq,

and the man below (7.1' tall) is in modern-day Russia.

(Credit: Bing Images)

Thus it is clear that **there actually were giants in the Earth**, in different places, in not-so-ancient days. Note that Goliath was between 9-12 feet tall, so in OT days (and perhaps near the end of the Roman Empire as recorded above), the average height had come down to 9-12 feet tall from the much larger Nephilim.

Robert Wadlow	Sultan Kösen	Brahim Takioullah	Zhang Juncai	Morteza Mehrzad
8' 11.1" (2.72m)	8' 3" (2.51m)	8' 1" (2.46m)	7' 11" (2.42m)	7' 11" (2.42m)

Today's Comparative Giants

The above are: American (1), Oriental(2) and Middle Eastern (2)… no particular pattern. Modern science says the pituitary gland just ran amok… Hmmmm.

So at this point, the reason for elaborating on the giant issue is that it was Enoch in the Book of Enoch who said they existed and we will see that Enoch also said he was shown the Earth – and it was flat. Enoch has some credibility. But we'll bring all this together in the last two chapters.

Briefly, we need to review who the Anunnaki were and then we can see who the Ancient Ones were.

Anunnaki

The Anunnaki were said to have come to Earth from Nibiru, a planet that circles its own dwarf sun which allegedly comes thru the solar system every 3600 years, and as it approaches the inner planets, the Sun causes it to flare up and become visible….

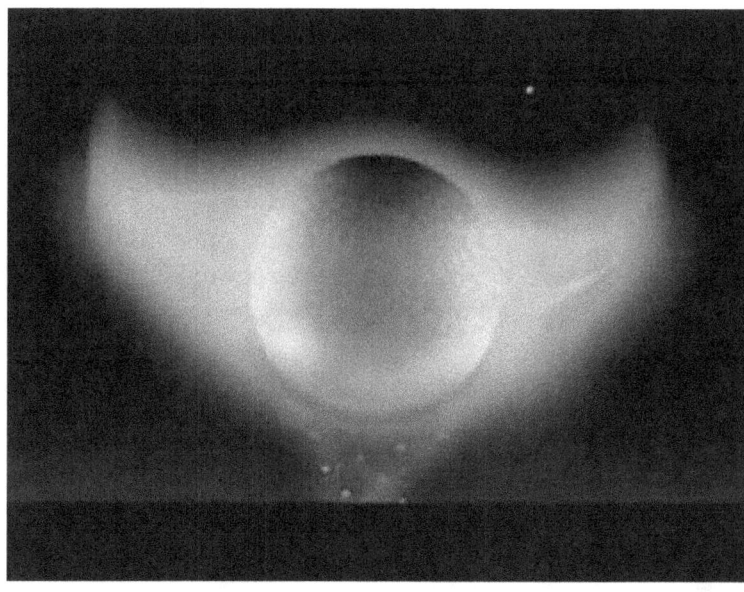

The Dwarf Sun around which Nibiru allegedly circles.

Our Sun causes the "wings" of debris and gases to fly off it.

(credit: **Bing Images**)

Of course if Earth is a Flat Earth protected by a Firmament, then this whole Nibiru issue is nuts… it didn't happen and it isn't happening. That means that the Anunnaki were also created on Earth… older than Man.

There is some evidence for that: remember that Sitchin said that the Anunnaki flew around Earth in what had to be **rocket-powered craft** as there was noise and smoke when they flew. Rocket-powered craft may be fine on Earth, but that is not what you use in outer space.

Secondly, recall that the term **Anunnaki** means "those who from

heaven to Earth came." In **Skycraft** , from somewhere else on Earth, they would fly thru the sky ("heaven") and then descend…. They were not ETs.

These are stone artifacts found in Turkey, ages old.

Inanna had a "flying boat."

So that suggests that the Anunnaki were Earth-bound denizens who "created Man in their image." But what kind of an image?.... Hominid, yes, but there was more as recorded by ancient artists and writers...

Anunnaki from Ubaid Culture (3500 BC)

Man was told to not make any sculptures of their 'gods' but not all men do what they are told.

Berossus, a Babylonian priest writing about the appearance of the gods, said that Man's ancestry traced back to the **Oannes,** an amphibious creature which came to teach civilization to Man. (Think: humanoids in Scuba gear.)

Berossus called them ***Annedoti*** which means "the repulsive ones" in Greek. He also refers to them as *musarus* or an "abomination." It is in this way that Babylonian tradition credits the founding of civilization to a creature which they considered a "**repulsive abomination.**" [190]

One would think that if the gods were so superior and grand as indicated in some ancient texts that they would be flattered to have Man make images of them, and display their greatness. But, after Man was created, **the gods forbade Man to make images of them** (Think: Ten Commandments, # 2: Thou shalt not make any graven image) and they tended to stay atop their Mesopotamian ziggurats and Mayan pyramids to be waited on by certain human servants who knew the truth, but did not speak of the gods' appearance with the worker population below. So their physical **repulsiveness** must be true, otherwise Man would have flattered and praised their god's appearance. And again, the reptilian nature of the <u>original</u> Anunnaki is explicit <u>in the Sumerian accounts</u>:

> **The reptiles verily descend,**
> The Earth is resplendent as a well-watered garden,
> At that time Enki and Eridu [his city] had not appeared,
> Daylight did not shine,
> **Moonlight had not emerged.** [191] [emphasis added]

Now this raises an interesting point. And I am not sure if we humans are ready for it, but in the interests of telling the truth, so we all can learn (and grow up and stop fearing beings that do not look like us) let's theorize for a second....

> If Earth is a Flat Earth, and thus has always been a flat Earth, then the Firmament would prevent others from coming here and interfering with the Original Creation. That means The God, The One, created this Earth and populated it with flora and fauna, and then created sentient hominids… as **reptilian hominids**. If the Anunnaki were created first, then there had to be a First Creation by The God. The Anunnaki then created semi-sentient hominids (Man), with mammalian-based genetics, that basically resembled them ("in our image") to serve them… work the mines, the fields and build their buildings.

> I'm not going to go there, but if the Anunnaki were the first, and they created Man, they might also have created some forms of **dinosaurs** as an experiment, or maybe as pets – if they could domesticate them…?!

Thus, the Anunnaki may have looked similar to this…

(credit: Bing Images / Anunnaki)

And yet, there is another curious bust of an Anunnaki (**Marduk**) found in Iraq that shows that some of them were human-looking, and **Marduk and Inanna** were born on Earth – with the genetic manipulation of Enki (EA) their chief scientist. That bust looks like this:

Anunnaki Bust in Iraqi Museum
(credit: https://www.pinterest.com/karenstatler33/annunaki/)

And then there is this relief of 3 humans (smaller, left) receiving orders from an Anunnaki god (larger, right). The god has human features… and is much taller.

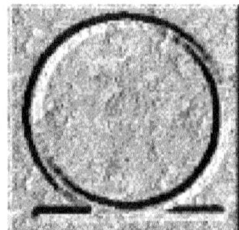

Egyptian Shen…

Note Seated god (Shamash or Enki) and 4-Horned Headdress.
(source: http://www.blinkbits.com/blinks/anunnaki)

Whereas some wags have argued that the god was pictured larger to show his status, that is wrong... he already holds the symbols of his power. .. which is a form of the **Egyptian Shen ring** AND he wears 4 horns (on the hat), which means he is probably Enki, 2nd in command on Earth.

BTW: The Shen Ring could be stretched to a cartouche:

Humans as Homo *saurus*

Still think there were no reptiles involved?

In this manner, the first primitive man or Adam was created, looking generally like his creator(s)... the gods' essence is mixed with the malleable clay of the earth [Earth-based genetics]. In the Sumerian tablets, the clay is mixed with the essence of the gods and upon this creation they "impressed upon it the image of the gods."

The Adam of the Bible was not the Homo *sapiens* of today. He was what one might call **Homo-*saurus***, a hybrid mammal-reptile creature that was to become our ancestor and the first step in the creation of modern man. See VEG, Ch. 3.

Since the Adam of Genesis [1] and the *Lulu* of the Sumerians were created in the image of the serpent-gods, shouldn't traces of this fact be found in some of the ancient scriptures?... Indeed, it is... One tract describes Eve's reaction in the Garden of Eden (according to the Jewish *Haggadah*):

> ...the bodies of Adam and Eve "had been overlaid with a horny skin." This skin "was as bright as daylight covered his body like a luminous Garment."

...and...

> She looked at the tree. And she saw that it was beautiful and magnificent, and she desired it. She took some of its fruit and ate and she gave to her husband also, and he ate, too. Then their minds opened. For when they ate, the **light of knowledge** shone for them. When they put on shame, they knew they were **naked with regard to knowledge.** When they sobered up, they saw that they were naked, and they became enamored of one another. *When they saw their makers,* **they loathed them** *since they were beastly forms.* [192] [emphasis added]

The human hybrid that was created <u>initially</u> probably looked semi-reptilian since he was created "in the image of God." [193]   This was Adam and Eve.

Mammalian Humans

Ok, so how did the reptilian form for Man change to what we see and know as Man?

Again, Enki was a master of genetics, and future births on Earth were genetically manipulated to drop the Homo *saurus* horns and **vestige tails**, and produce humans. Drop the horns? Yes, and even today some humans still have them, as did **Alexander the Great** (below) and some people in China:

Credit: kingofmacedon.net via Yahoo [194]

If the great men of yore were Anunnaki hybrids, they could still have had reptilian vestiges, such as horns. It's interesting that Alexander was portrayed that way, so it must have been a compliment... otherwise it would have been banned.

And some people today also have horns...

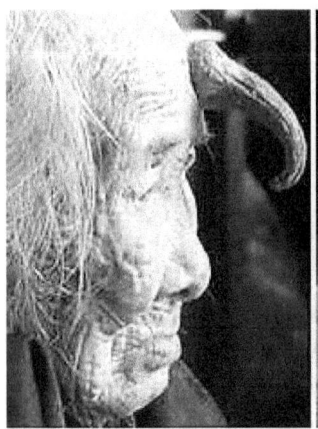

So Enki gets it together and produces Marduk and Inanna, two Anunnaki gods who were quite handsome human specimens, as **Inanna was said to be quite pretty**, as well as quite athletic and aggressive: and could hold her own with any man.

Think Xena, Warrior Princess and Wonder Woman all in one. She might have looked similar to the picture (left).

So that explains what has been traditionally said about the Anunnaki who created Man as a slave, and the Anunnaki hegemony was the Middle East, part of what is now India, and some of Egypt. They also had outposts in Peru, Bolivia, Brazil and Africa where they were mining gold. One such base was **Puma Punku** which today lies in ruins.

Because Man found the <u>original</u> Anunnaki (Anu, Enlil, Enki, Ninharsag…) repulsive, they avoided contact with Man, and as Man spread around the globe, they withdrew <u>underground</u> to avoid Man. (More is examined in VEG and AL.)

So what about the Ancient Ones, also called the Shining Ones?

Ancient Ones

Here some of the information in Chapter 7 comes into play. Mount Meru and Hyperborea, as well as Atlantis come into the discussion.

It was said in <u>Anunnaki Legacy</u> that the **Norse gods, the Roman and Greek gods** were probably the Anunnaki. Be that as it may, the Anunnaki were not the only ones traveling around Earth helping Man set up civilization – and the Ancient Ones did more of it and were kinder to the humans than the Anunnaki were.

Don't forget that the Norse, Roman and Greek gods were human-looking, the original Anunnaki were not… not that the Anunnaki didn't play games and lord it over their protegées, but the Anunnaki were much more interested in getting Man to work out in the Middle East, Africa and part of what is now India. Thus Inanna ruled the **Indus Valley** (now Pakistan and Afghanistan) and western India, where as Marduk ruled **Egypt**, along with Thoth, Isis and Osiris… Enki's son Ningishzidda (also a bearded human) went to **Central America** (as Quetzalcoatl, Kukulcan, and Viracocha – depending on what country he was in) and guided the natives there after the **Atlantean debacle**.

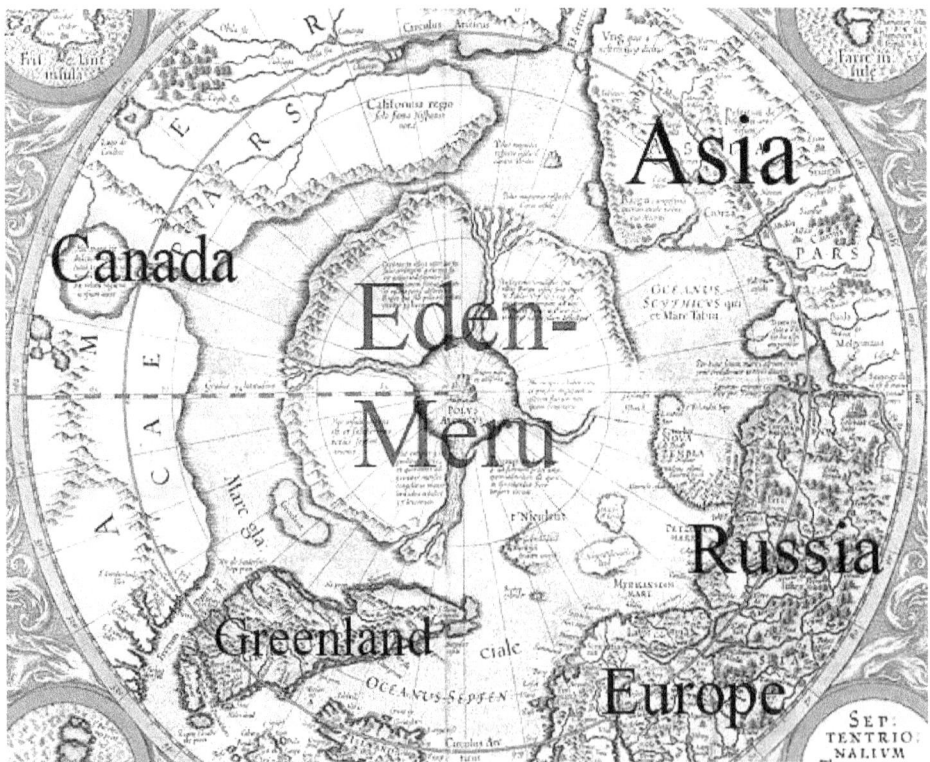

The Ancient Ones were based at **Mt Meru** – at what has been called **Hyperborea**, … which was at the North Pole.

Just FYI, Mt. Meru was at the center of the Earth....

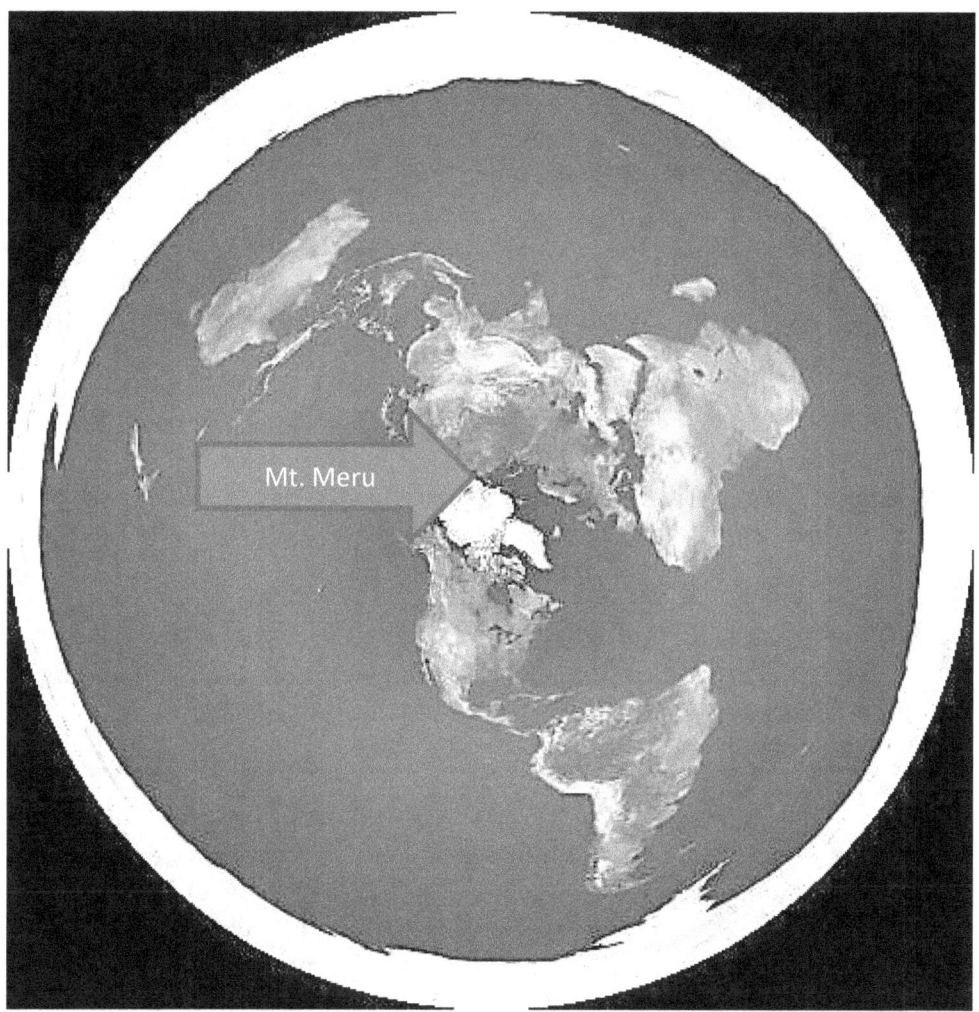

This FE map above and the United Nations flag (same map) do not show Hyperborea any longer because it sank, and was replaced by another locale on Earth, inaccessible to Man. (Hint: think Tibet/Nepal.)

Apollo and Mt Meru

Apollo was the emissary to humans and Anunnaki alike from Hyperborea ...
He would winter in Greece, administering the Oracle at Delphi, and then spend the other 6 months of the year in Hyperborea.

The ancient description of Mt. Meru (refer to the map two pages back) is interesting in that it seems to be describing two different places – Hyperborea and Atlantis. (The reason that below is a long quote is because it describes several things examined by this chapter.)

1

> **Mount Meru** is a sacred mountain with five peaks[1] **in Hindu, Jain and Buddhist cosmology** and is considered to be the center of all the physical, metaphysical and spiritual universes... several statements say, **"The Sun along with all the planets circle the mountain"** ....

> The Meru mountain was also described as being surrounded by Mandrachala Mountain to the east, Supasarva Mountain to the west, Kumuda Mountain to the north and Kailasha to the south –[**four mountains** can be seen on the map (2 pages back)].

2

> The Suryasiddhanta mentions that Mt. Meru **lies in 'the middle of the Earth'**

> Mount Meru is also the abode of Lord Brahma and the Demi-Gods [aka the **Ancient Ones** ... Brahma is The Creator God.]

> [Mt. Meru is also called **Sumeru** – where "su" is a prefix meaning "most excellent" so we have "Most Excellent Meru".]

> We don't know whether Sumeru relates to Sumer – the land of the Sumerians.

3

> Sumeru is the **polar center** of a mandala- -like complex of seas and mountains. The square base of Sumeru is surrounded by a square **moat-like ocean**, which is in turn **surrounded by a ring** (noted as square in shape) **wall of mountains**, which is in turn surrounded by a sea, each diminishing in width and height from the one closer to Sumeru. [And this is the link with **Atlantis**:] There are seven seas and seven surrounding mountain-walls, until one comes to the vast outer sea which forms most of the surface of the world, in which the known continents are merely small islands. [195] [emphasis added]

Items 1 and 2 in the left margin are describing Mt. Meru on the Flat Earth, by the way. The Sun, Moon and planets do circle the Arctic which is in the middle of the Earth. Polaris is the non-moving North Star and does not move.

Item 3 (left margin) is describing Atlantis.

Atlantis is often pictured as follows based on Plato's description (more on Atlantis later):

Controlled Access to Atlantis
(credit: Bing Images: Crystalinks.com)

Atlantean Deluge

While we're at it, having mentioned the Flood and Atlantis, it is accurate to point out that when Atlantis broke up, the submergence created a **huge tsunami** that swept over the Yucatan, southern America, Cuba, and eastward to Africa, **into the Mediterranean and on to the Middle East.**

Naturally the inhabitants of Atlantis fled west to Central America, vowing to not be subject to that kind of thing again, and just 50 miles north of Mexico City built **Teotihuacan**… several **huge pyramids that could withstand any future tsunami**. In addition, survivors migrated to Guatemala (as the Guatemaltecs and the Mayan Lancandones). [196]

Other inhabitants fled eastward to Spain, The Pyrenees (becoming the **Basques**), the Atlas Mountains, and the **Canary Islands** (the Guanches) are all that is left of the once great Atlantean Empire that had spread over much of the world bordering the Atlantic Ocean – much to the displeasure of the Anunnaki.

By the way, the map below was the extent of the Atlantean Empire according to archeological evidence found, and some corroborating ancient records, including Plato and the **Lancandones**. Areas of trade and influence:

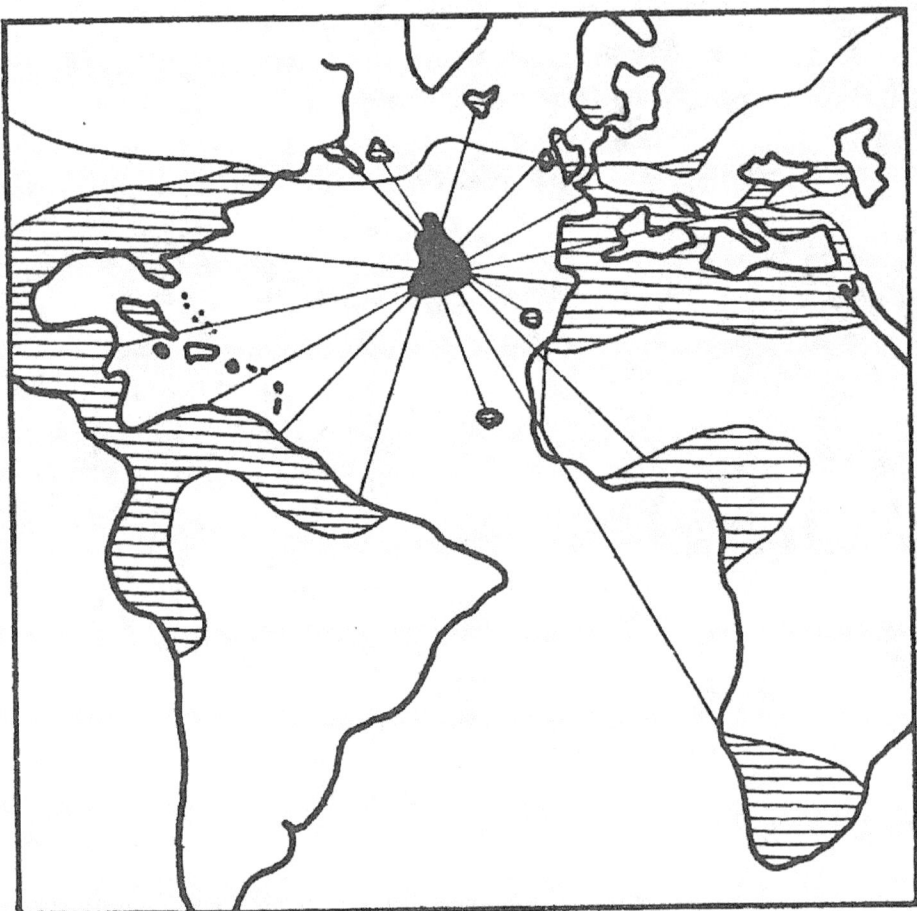

Fig. 25. Atlantis—center of the world of Red-Skinned Man. At all points where various authors have claimed to have found "their" Atlantis, arguments in favor of an alleged cultural link with this very ancient center were advanced. The map demonstrates the range of this long extinguished beacon of the Red Indian world which illuminated the ancient civilizations, and therefore shows roughly the size of the first empire claimed for the Redskins.

(credit: Otto Muck, p.125)

As the survivors of the Deluge fled westward, they became the Amerindians (founding the Cahokia, Mississippian Culture), and the Maya Lancandones – and both cultures built the same kind of pyramidal structures, shown below...

Cahokia Culture…

…and the Maya at Tikal….

...and a look at the Aztec city of Tenochtitlan.... based on Spanish accounts.

...and here is Teotihuacan:

Need any more be said about that? It speaks for itself. And it is clear that the natives, the **survivors**, took pains to make a <u>mountain</u> on which to climb for safety if such a repeat Flood ever happened again.

Egyptian pyramids at Giza were not part of the same plan… the **Great Pyramid of Giza** used to have a polished limestone facing that reflected the Sun and made it impossible to climb up it – as we <u>can</u> do today (is it because the same Flood washed away the limestone facing?). The Great Pyramid of Giza predates The Flood and was an Anunnaki creation… it was never a tomb.

Need any more be said about that? It speaks for itself. And it is clear that the natives, the **survivors**, took pains to make a <u>mountain</u> on which to climb for safety if such a repeat Flood ever happened again.

Egyptian pyramids at Giza were not part of the same plan… the **Great Pyramid of Giza** used to have a polished limestone facing that reflected the Sun and made it impossible to climb up it – as we <u>can</u> do today (is it because the same Flood washed away the limestone facing?). The Great Pyramid of Giza predates The Flood and was an Anunnaki creation… it was never a tomb.

Caretakers of Earth

Anyway, the Anunnaki were not friendly with the Atlanteans and the story goes (generally from Edgar Cayce) that both were advanced (in different ways) scientifically – the Anunnaki had Skycraft and missiles (See AL and Set and Horus sky battle in Ch. 5, Anunnaki in Egypt) and the Atlanteans had the **Firestone**.

There are two basic stories about the breakup of the huge Atlantean continent/island.

> In one, the Atlanteans were composed of two basic groups: followers of the **Law of One**, very peaceful and spiritual, and the other group was the Followers of the Lefthand Path, warlike and power-hungry. The story goes that the power-hungry group misused the Firestone (a large crystal that could heal or destroy – much as a laser) and they tried to zap another civilization <u>thru the Earth</u> (sending pulsed waves) to destroy the other humans and instead destroyed themselves.

> In the other version, Atlantis was pretty peaceful, no warring factions, and they had a major disagreement with the Anunnaki over land and trading, and rights to gold, and the Anunnaki bombed the island causing great tremors in the Earth, and a huge Earthquake, sending Atlantis to the bottom of the sea.

> Before you think that too farfetched, refer to AL and Ch. 4 on India where the Anunnaki bombed **Sodom & Gomorrah (and Admah, Zeboim & Zoar), Mohenjo-Daro, Harappa, Kaligangan, & Dholavira, and Dwarka (Krishna's home).**
> **Yes, the Anunnaki had nuclear weapons.**
> There is nothing new under the Sun....

Needless to say, the Caretakers were not happy with this nonsense, and invited the Anunnaki to leave the surface – they couldn't live in Mesopotamia after all that nuclear fallout anyway (see insert above) – They had poisoned the water and the land about 2200 BC. It was necessary to terraform the land and purify the water.

It was also necessary to visit the refugees from the Deluge and help them rebuild. Being god-like they do not need Skycraft but can materialize anywhere they choose... and many times in the past, with Ezekiel, Daniel, and Elijah, they use a craft so that the humans can see that these beings are from the sky – the craft flies over and descends to the Earth – that says "We're from Heaven." And they were often regarded as Angels (also appearing to Mary and Abraham).

The Ancient Ones – who preceded the Anunnaki – watched over the Earth and the humans. Sometimes they would delegate a required humanitarian action to the Anunnaki on behalf of the humans, but they were mostly busy (though small in number) making sure the Earth realm was environmentally balanced, the seas were moving, correct salinity, oxygen levels correct, and operating the celestial apparatus – see **Book of Enoch** for more (Ch. 11 in QES). They were (are) the Caretakers of the Earth.

It was the Ancient Ones who overruled Enlil when he wanted to wipe out the rest of mankind <u>after</u> the Flood and they saw to it that mankind would survive. The **Celestial Law** says that if you create a sentient species, you are responsible for them and must see to their nurture, education and protection since there are souls incarnating into those bodies and that is what Earth has become – an **Earth School** for souls.

Whereas Sitchin said the **Flood** was caused by the Anunnaki dislodging the Antarctic icecap and the resulting tsunami caused the Flood, (running from Antarctica north over Africa and inundating part of Mesopotamia, and some parts of Eastern Europe and India), that is wrong. **The Deluge came from the Atlantic** ... and even threatened Hyperborea when it spread northward. A 300-400' tsunami has a lot of kinetic energy and runs pretty far.

Summary

That brings us up to about **8,498 BC, June 5**[th] as the tentative date for the Cataclysm and Deluge. Dr. Muck worked out the date based on the real Mayan Calendar, Gregorian calendar, archeological evidence, astronomical records (he also theorizes that Atlantis was sunk by an asteroid), and it pretty much agrees with Edgar Cayce. Most dates given for the Cataclysm, whatever the cause, place it between 10,000 -8,000 BC.

So there were the **Ancient Ones** who were put here by the Creator, and are part gods themselves. Maybe they created the Anunnaki, who knows… that is unknown. But they also went around the world assisting and teaching mankind. And they were also called the **Shining Ones**, and the (good) Watchers – the bad watchers mingled with Earth women and created the **Nephilim**… as Enoch said (the Book of Enoch is covered in a later chapter).

Then there were the **Anunnaki** who created Man (according to the thousands of Sumerian tablets and scrolls) and when they got out of hand, dropping bombs on humans (Sodom & Gomorrah *et al*), destroying Puma Punku, and probably warring with the Atlanteans, they had to be shut down, and they were moved underground.

The underground Anunnaki (The Remnant or Nagas) have split into two groups – those who support Man (Insiders) and those who just want to be rid of him (Dissidents). Neither wants to kill Man… either advance him to where he is more mature and the Remannt can once again interface with him, or remove him from the planet. The issue is this: when you create another race genetically, you are responsible for it – to protect, nurture, and educate. And the Remnant is tired of it.

Seeing that Man got a healthy dose of Anunnaki genes, is it any wonder he is a bellicose, egotistical, lusty, and impatient… all Anunnaki traits. If this version of Man does not do better than the Anunnaki, and the covert genetic upgrades being done thru abductions does not work, Man will be removed, too. It is sad that the **223 extra genes** that make Man special are not enough to empower better behavior.

Lastly, it was learned that the Ancient Ones dwelt in **Hyperborea at Mt. Meru** which was centrally located to all points on Earth. They are the Caretakers of the Earth Garden and Man is the recipient of their proactive guidance and protection. As with most gardens, the weeds have to be removed… and Man is currently (according to my Source) on a short leash.

Man just about 'bought the farm' when he launched 4+ atomic bombs at the Firmament back in 1962... What an idiot idea. It was called **Operation Fishbowl** with multiple launches, and it really got the attention of the Caretakers.

Many of the launches failed for unforeseen reasons, but *Starfish Prime* (#3 launch of the first 4) was successful and many satellites in orbit were destroyed by the EMP blast at 680 miles up, set off over the Pacific Ocean. (Satellites cruise in the 200-500 mile altitude range.) In addition, Johnston Island was badly contaminated as well as blowing radiation into the atmosphere which drifted toward the Americas.... Your tax dollars at work, folks!

Chapter 9: Simulation

This chapter will survey several aspects of the possibility that Earth is a Simulation. Of course, it may just be a regular, rock, dirt and water Flat Earth world that was created to serve as the Earth School, but a Simulation makes more sense – especially the way some things operate.

This is significant as we have already theorized that the Earth Realm was <u>created</u> and as a **Flat Earth** – as far out as that might sound. (Bear with me.) It was said earlier that Earth was either flat or it might be that Virtual Reality Sphere (**VR Sphere**) that was spoken of in QES and VEG. Determining whether it is also a Simulation does not help to eliminate one of the options – the simulation physics could work for both of them.

> The one thing we <u>do</u> know is that Earth is not a regular rock, rotating 1000 mph on its axis, and spinning around the Sun at 67,000+ mph.

> The fact that it appears so real is testament to the power and creativity of the One who <u>created</u> this Earth Realm – it fooled Copernicus, Galileo, Kepler, Newton, etc. and you'll see why in this chapter.

It will benefit us to look objectively at several things to better make the analysis: vision, perception, holograms, and the principles of, and reasons for, doing a Simulation.

Source Material

I have to stick my neck out here and an apology might be in order... I was told to say just **three things** about Earth back in 2008 (when writing VEG) and because we all 'know' that the Earth is a globe, I took the three facts to mean we are on a **VR Sphere**, which is a 3D Construct and it has a protective energy envelope around it – off which the *Gegenschein* reflects. To recap, the three aspects were:

> Earth is **constructed** (designed and built) with 3D Laws, while actually being located in 4D which is the source of the power to sustain it;

> Earth is surrounded by an **energy envelope**, a 'shell' to protect it and keep out unwanted interference from any curious or mischievous 4D beings... the *Gegenschein* is reflecting off this 'shell' ;

> Earth is a **simulated environment** with special Laws and Overseers [i.e., Ancient Ones aka Watchers] and operates as a **School** for soul growth.

We all 'know' the Earth to be a round globe – because we saw it on our teacher's desk – she didn't have to tell us what shape Earth was because we could all see it. So we know the Earth is round and when I was given those three items, I called the product a **VR Sphere** – the VR referring to the <u>simulation aspect</u> which takes on the aspects of "virtual" or holographic reality. And then I thought of Earth as similar to the *Star Trek* **Holodeck** where many ideas and dramas played out with the ship's (U.S.S Enterprise's) crew inserted themselves into whatever simulation the Holodeck was running. (I was not corrected.)

That was then, and I wrote Earth up as a 3D Construct and VR Sphere in VEG… in 2008. In 2016 I received an anonymous email (I could not respond) via my reader email address (see Copyright page), and it asked me if I had ever considered or researched the Flat Earth (FE) scenario. So I did an initial look at the FE material, snickering as I did so. I thought that this might make an interesting 'alternative' chapter in QES (already written) and I knew I could disprove the FE theory.

Some of the material was very serious and I had to dig deeper… it was not easy to debunk it all. There were some 'proofs' for the FE that were obvious BS and were so sophomoric I felt they had to be done by those seeking to discredit the FE theory! (That later turned out to be correct.)

Three months into the almost daily research on YouTube, 5 key books, and some <u>lengthy</u> articles on the Internet, and I was stunned. **I could not disprove the FE theory**. Not that that meant it was correct, but this issue now fascinated me. And I wondered if it were true, why didn't it come forth in 2008 with VEG!? Then I realized that if Earth is not a VR Sphere, I will have egg on my face as an author… So I had to continue the research. About 30% of the proofs absolutely held water. Another 20% were reasonable.

> I wondered in 2016 about New Jay and what he wrote thru me – Why was I not told specifically and concretely what Earth is? That was the deal New Jay made in 2008 : the gods wanted VEG written, then (1) it had to be the Truth and (2) New Jay wanted <u>protection</u> from trolls and the PTB (who seek to squelch the truth).

> Baldy responded with a caution: You will know the Truth but it will come in stages so you can assimilate it. (**1-second drop**.)

Now, in 2017 I see that the FE theory would have been too much to handle and 3D Old Jay would have thrown the whole thing out… **that is how strong our belief**

"conditioning" has been by the PTB who think they run this place and keep the sheep in the dark.

I began to see that VEG had the Truth but one which applies to <u>both</u> the VR Sphere and the Flat Earth. And I had to agree, in 2008 Old Jay would not be open to hearing the deeper facts of the Earth Realm… if it should be that Earth is flat and **a contained environment** where no one gets in or out – except via birth/death.

> Then I had an ugly thought: if we are contained, either as a VR Sphere, or a Flat Earth, then we can't get out of here… and **we didn't go to the Moon… and the UFOs are not ETs visiting us**… they have to be built by Man and flown by Man (another Black Ops scenario).

This was making me nervous and I began to wonder just how much I had <u>remembered</u> (incorrectly?) from those initial 3 points (above) in 2008. Later I would see that it was **just the VR Sphere that was in question** – everything else in VEG (which provided the seminal material for all the other books) was correct. Phew.

But in December 2016 I wanted to know… no matter what the answer was. I was no longer 100% comfortable with the VR Sphere… and I was not too comfortable with the FE Theory, either. To this day, **I still prefer the VR Sphere**, but maybe that is just my pre-2008 programming coming through…We all want what we have been told, and learned in school, to be true, and the VR Sphere is just a slight shift (20%) from that, easily acceptable… whereas the FE scenario is an 80% shift.

> The VR Sphere is just created as 3D, still a globe, the Sun rotating around the Cspace, having a protective envelope around it, a Moon to circle within the Cspace, and supposedly it was done as a Simulation. Not really a problem…. (Chapter 3).

> **Except that: Gravity** and the oceans on a round ball were still an issue, as well as what the stars were (real or projections), and could we get thru the 'shell' around VR Sphere Earth to go to the Moon? And does the <u>continent</u> of **Antarctica** really exist?

> I didn't question any of it until late 2016.

More Research

What had to be done was to see if there were any more hard proofs (I had 3) of the FE theory that could not be debunked, and I emailed a prominent researcher and webshow host my questions and he responded with some links I had not seen. I also went up on Amazon and bought his book. [197] This was getting very interesting –

now I had 4 'killer' FE proofs that were withstanding all attempts to logically or scientifically refute them (more in Chapter 10's list of reasons):

FE theory support (Ch. 11 in QES):

> NASA wall screen orbital "S" irregularity
> Sun changes shape as it rises and sets (QES Ch. 11)
> A 360° video pan from a high-altitude balloon shows
>   a perfectly flat horizon
> Southern Hemisphere distance anomalies

I later found <u>more real proofs</u> and the VR Sphere was in trouble. I also found more info on **Isaac Newton** and how he came up with Gravity (which today's scientists still can't find… they can't get instruments to register it as a force), and even Newton said that **Gravity was an assumption which he could not prove**.

> You need a form of Gravity if we are on a globe, even a VR Sphere, but if we are on a Flat Earth, Gravity is not needed – to hold the million-ton oceans to the sphere.
>
> And how did the oceans stay on a globe without falling off? Water does not 'rest' in a curve (convex shape)… it always seeks to <u>lay flat</u>. A curved ocean surface does not work (for reasons explored in Chapter 3).

These things had to be seriously researched and I was getting spooked by the recent serious and believable FE clues. My 'programming' as a graduate of the public education system was resisting accepting that the FE theory might be true.

So in the meantime, I decided to attack and deeply examine the VR Sphere aspects. That meant reading up more on **Simulation, Holograms, Gravity, Casimir Effect, and Centripetal Force (CF)** – the force that allegedly attracts things to another body. And if we look at holograms, that opens up a wider field of **Vision** and **Perception** … How can we tell if Earth is a Simulation if we don't know <u>what</u> we are looking at because we don't know <u>how</u> we perceive things? (See Vision issues examined in TSiM.)   My work was cut out for me.

> And I am spending some time with this account as it is important to the development of this chapter… which supports the FE conclusion. Things fell into place, but not the way I thought they would….

So let's start at the beginning… What is Vision and how do we Perceive, and let's keep it at a simple level (the detail was given in TSiM).

Vision

Most humans can see but we don't know how it works. Vision uses the eyes to take in images, register them on the thousands of rods and cones in the retina, and that is converted in an analog fashion to pulses on the optic never to a center in the back of the brain. We all know that there is not a "viewing screen" in the back of the head on which the image is displayed... upside down at that! What happens that we know of, is that there are multiple small regions (V1 – V5) of the brain that process the signals on the optic nerve, and somehow the brain correlates them and the **Subconscious Mind** (it is suspected by physicists using Tunneling Theory) is able to interpret the signals.

Some physicists are part of a new field called **Quantum Biology** (QB) wherein they are specifically looking for "tunneling" or a quantum process to better understand Vision.

While QB has yet to become a commonly-recognized area that will subject itself to rigorous scrutiny, when one realizes that there is an aspect of vision that is mysterious, and suggests a **holographic aspect** (as VEG and QES both suggested), there is some fairly current research that is making progress in understanding how human vision works.

Neurologists know that the retina connects in a major way to the Occipital Lobe, also called the **Visual Cortex**. In fact, they have identified **5 major areas** within the Visual Cortex that deal with vision, and they know this due to the use of PET and MRI scans. They have named these sections:[198]

V1 -- Primary Visual Cortex, and it has **a map of the entire retinal field**. This is also the first of the fields to develop at birth.

V2 -- contains interstripe areas which are sensitive to **form or shape.**

V1 and V2 act as a kind of post office and parcel out info to other appropriate areas. These two are in constant contact and operation.

V3 -- forms a ring around V1 and V2 and is concerned with shapes of objects in motion. Sensitive to **form,** as well. Has some interaction with V1 and V2.

V4 -- selective to wavelengths of light and registers **color**

V5 – responsive to **motion;** does not see color.

The 5 are represented as follows, in the corresponding brain of a monkey:[199]

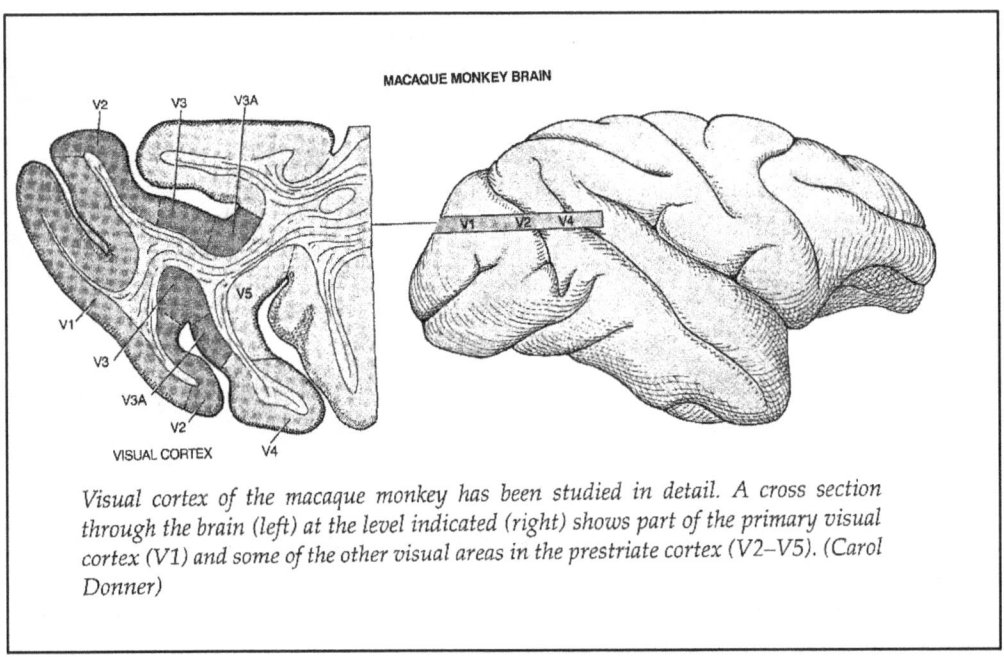

MACAQUE MONKEY BRAIN

Visual cortex of the macaque monkey has been studied in detail. A cross section through the brain (left) at the level indicated (right) shows part of the primary visual cortex (V1) and some of the other visual areas in the prestriate cortex (V2–V5). (Carol Donner)

Some of the V-areas have sublayers which interact with other V sublayers. So at this point one begins to wonder, Ok, where do all these inputs come together and form a coherent picture that one "sees?"

What is about to be examined is what your brain does with this picture:[200]

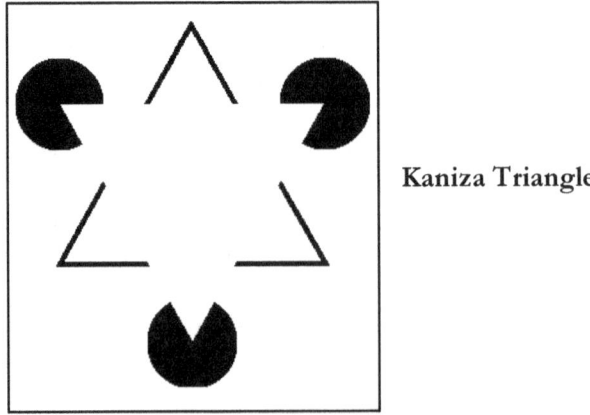

Kaniza Triangle

Looking at it closely, there is no white triangle, but the brain infers one based on ancillary data. **The brain creates lines where there are none** – and that is due to the V1 and V2 sections vying for control. The V2 cells do not see or infer the white triangle.

Ok, the researchers are aware that our world may in fact be holographic as Pribram, Jahn and Dunne, and Bohm suggested in VEG.

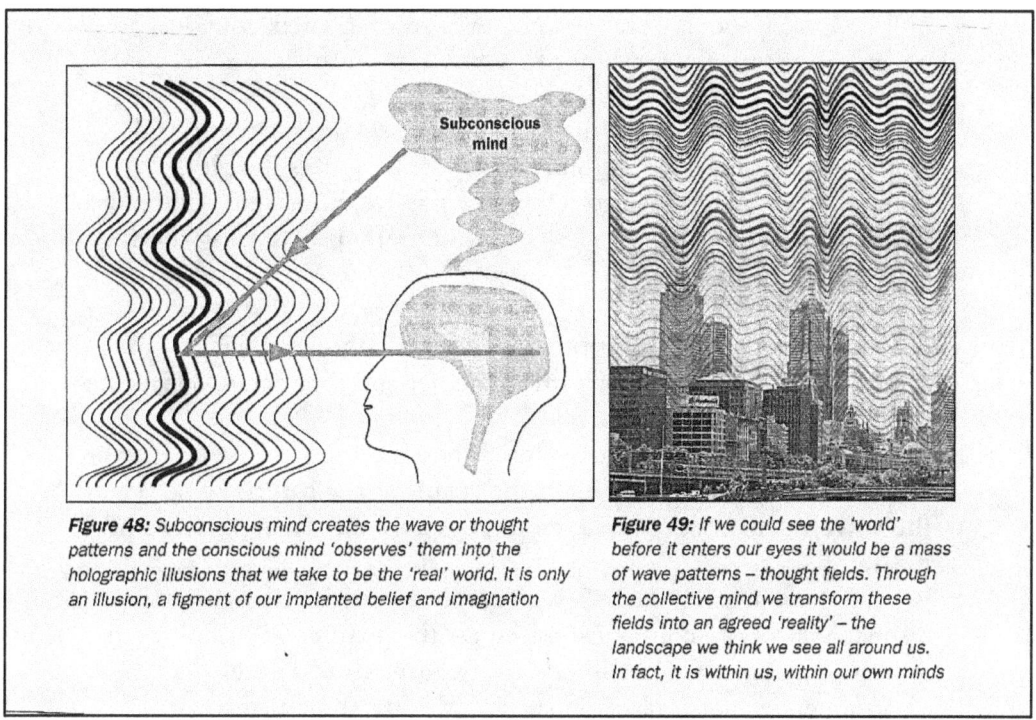

Figure 48: *Subconscious mind creates the wave or thought patterns and the conscious mind 'observes' them into the holographic illusions that we take to be the 'real' world. It is only an illusion, a figment of our implanted belief and imagination*

Figure 49: *If we could see the 'world' before it enters our eyes it would be a mass of wave patterns – thought fields. Through the collective mind we transform these fields into an agreed 'reality' – the landscape we think we see all around us. In fact, it is within us, within our own minds*

There is a suggested connection, a quantum effect, between **consciousness (metaphysical mind) and Quantum Reality**. That would be mediated by the Subconscious Mind.

What if **Reality is a Hologram, or at least operates holographically,** as does Vision and Memory (both a product of mind) – then would we not be living in some sort of **Reality Field**(s) which is, on some level, subject to our Consciousness? Stay tuned, that is where we're going…

Subatomic Particles

It is hard to accept that an observer's consciousness can affect the outcome of experiments with subatomic particles, but today's physics has been discovering that. It was noticed, tested and proven with the **Double-Slit Experiment** (see VEG). It <u>suggests</u> we have some power to affect the <u>micro</u> world around us. Not to manifest Reality, but according to physicist **David Bohm:**

> ...if subatomic particles only come into existence in the presence
> of an observer, then it is also meaningless to speak of a particle's
> properties and characteristics as existing <u>before</u> they are observed. [201]

Naturally, Einstein disagreed because he theorized that subatomic *quanta* exist as waves, <u>until</u> we look at them. (This has not been substantiated.) But, just as interesting and relevant is the creation of **anomalons** in the laboratory...

> Unlike Bohm, **Jahn and Dunne** .... believe subatomic particles do
> *not* possess a distinct reality until consciousness enters the picture....
> We're [not] examining the structure of a passive universe.... Instead
> of discovering particles, physicists **may actually be *creating*** them. [202]
> [emphasis added]

And since **we are part of the hologram**, and we are conscious beings, <u>our consciousness must somehow impact **micro** matter and thus create our reality</u>. (See Appendix C.) Jahn and Dunne said

> They believe that **reality is itself the result of the interface between
> the wavelike aspects of consciousness and the wave patterns of
> matter**...
> Rewind: Jahn and Dunne believe subatomic particles do *not* possess a
> distinct reality until consciousness enters the picture... "I think we're
> into the domain where the interplay of consciousness in the
> environment is taking place on such a **primary scale** that we are
> indeed creating [micro] reality..."[203]   [emphasis added]

But they mean that on a <u>very microscopic level</u> – not that we create our day, but that we <u>do</u> nonetheless affect the world around us. (We know that we can kill plants by cursing them.) And Jahn and Dunne go on to cite how scientists in the laboratory are 'discovering' the elementary particles that they <u>expect</u> to find – in reality, **creating them via some as yet not understood faculty of mind, expectation and consciousness**. [204]

Such is the **anomalon** that was just mentioned where different scientists discovered <u>the same particle</u> which behaved differently, <u>according to their expectations</u>.

And if we all share the same world, would not our reality consist of mutually agreed-on **"reality-fields"** – we all see and agree that certain buildings exist, the bus is blue, it's raining, and we are all in a city called Chicago. That would be a mutually shared reality-field, or field of wave patterns that we all see and agree that we see and experience it the same way.

If consciousness plays a role in the creation of subatomic particles, is it possible that our observations of the subatomic world are also reality-fields of a kind? …

As bizarre as this sounds, it is not so strange when one remembers that **in a holographic universe, consciousness pervades all matter**…[205] [emphasis added]

So in sum, it is considered that "…**reality is established only in the interaction of a consciousness with its environment**…[and that means] anything capable of generating, receiving, or utilizing information can qualify… animals, viruses, DNA, Artificial Intelligence machines…"[206] Thus, said another way, only **the reality created by the interaction of consciousnesses is real**. There is no reality beyond that created by the interaction of all consciousnesses, which means that the holographic universe can be 'sculpted' by the joint effort and energy of enough minds.

> If two or more agree on anything….. it shall be done.
> **Matt. 18:19**

Anomalon

Jahn and Dunne also said the Russians were a good example of finding what you expect to find…

As evidence they cite a recently discovered subatomic particle called an ***Anomalon*** [see above], whose properties vary from laboratory to laboratory. Imagine owning a car that had a different color and different features depending on who drove it! This is very curious and seems to suggest that **an anomalon's reality depends on who finds/creates it**.[207] [emphasis added]

One of the cases in point was when the Russians discovered ***neutrinos.*** For many years the particle had been proposed but no one could find one. Then one was found in 1957, and physicists determined that if it had mass, it might help explain some thorny problems, and then in 1980, lo and behold the Russians discovered the neutrino had mass. Laboratories in the US did not agree and could not find any mass associated with the neutrino. Note: when it was reasoned that the neutrino should have mass, lo and behold the neutrino was found to have mass – in all but US laboratories.[208] The US still has not resolved the issue… and still believes the neutrino has no mass…

> What is that saying, "You'll see it when you believe it?"

Rewind: Jahn and Dunne

Remember, Jahn and Dunne believe that

> …we live in a 3D construct created out of interconnectedness, sustained by the flow of **consciousness**, and ultimately as plastic as the thought process that engendered it… In a holographic universe, **consciousness pervades all matter**… [and] … **reality is established only in the interaction of a consciousness with its environment** … [which consciousness can be] anything capable of generating, receiving or utilizing information. Thus, animals, viruses, DNA …. may all have the prerequisite properties to take part in the creation of reality. [209] [emphasis added]

In short, Jahn and Dunne stated that "…subatomic particles do *not* possess a distinct reality <u>until consciousness enters the picture</u>." (Appendix C.) So the next question is: <u>How much</u> does the **effect of the observer** in Quantum Physics influence what the subatomic particles <u>do</u>? Or are we just being deceived by Beings who **control** this Realm from above 3D? (See VEG: Ch.s 12-13 and Appendix B's Control System.)

> **So Vision is our perception of the world around us, which some physicists say we are "projecting out there"… and it depends on consciousness and a decoding of the holographic waves… done by the Subconscious Mind.**

But what are the aspects of Perception that we can examine?

Elements of Vision & Perception

Holograms

Holographic images, or projections, are a fascinating aspect of the real world around us. Holography signifies "the whole in every part." While it is not necessary here to go into what holograms are and how they operate, it is enlightening to hear that our **vision and memory operate in a holographic way.**

> **Vision is reportedly holographic** as there is no way for the actual Images 'out there' to be displayed on a 'screen' inside the back of our heads (in the brain's so-called "vision center"). In fact, **there is no "vision center"** nor is there a one-to-one correspondence between the object we see and the image's representation in the brain. [210]

[Dr. Karl H.] Pribram discovered that not only did **no** such one-to-one correspondence exist, there wasn't even a discernable pattern to the sequence in which the electrodes [sensing brain activity in volunteers] fired. He wrote of his findings, "These experimental results are **incompatible** with a view that a photographic-like image becomes **projected** onto the cortical surface." [211]   [emphasis added]

Dr. Pribram also discovered that **memory, like Vision, was holographic** – or more precisely, distributed. This meant that **the brain was using some kind of internal holographic processing** and there would be no more correlation between brain electrical activity and what was being seen, than there would be any meaning in the interference patterns seen on a piece of holographic film. He found that the neural activity in **the brain operates as a wavelike phenomenon** "… creating an almost endless and kaleidoscopic array of **interference patterns**, and these in turn … give the brain its holographic properties." [212]

So the image of Tank sitting before all those screens in *The Matrix* with their green, vertically flowing symbols was not so far-fetched. The concept is correct; our brains interpret whatever the holographic symbols are 'out there' in the world around us and, get this, **our minds spatially create the image of the object 'out there' as if we are projecting it in front of us** – how else could we navigate to it to touch it? That means **we're part of the hologram**.[213]   [emphasis added]

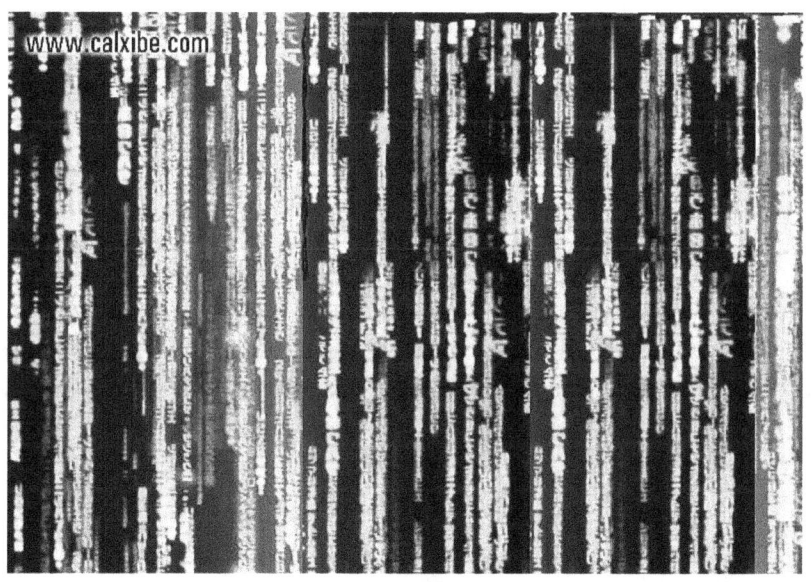

*The brain decodes the hologram and **the mind** enmeshes and actualizes us into what we 'see.'*

Controlled Vision

Just as fascinating as the fact that we decode the holographic interference waves 'out there' and thus <u>project reality in front of us</u>, is the corollary idea that **we may not be able to see all that is around us.** This is due to one of two things:

> either the objects are invisible because they reflect a color or wavelength that we cannot physically see,
>
> > or
>
> we have been 'programmed' (conditioned) to not see certain things.

In either case, **what we see is filtered** through the pre-frontal lobe of the brain whose job is to make sense of it and correct it if necessary. For example, there is a street sign in my neighborhood that says **"Goldenrain Tree."** No one notices that that is an error when I point it out. It should be "Golden Raintree," but the sign maker went unconscious, goofed up, and the city lets it stand as it is. A better example is the following:

A
Bird in the
the Bush beats
two in the Hand any Day.

Did you see the error? Chances are, if you have not seen this one before, your brain filtered out the extra word 'the.' How did it know to do that? Is there a **secondary intelligence operating between the brain and the mind** that we don't know about?

The answer is Yes, by the way.

In like manner, it has been speculated that most people don't see auras because they (1) don't know they can, (2) they don't want to, (3) they don't know they exist, or (4) have been told that **seeing auras** is witchy stuff. And even more bizarre as a possibility, there could be other beings moving in and out among us on a daily basis that we can't see (like Angels?) because (1) we have never been told about them and so the brain filters them out, or (2) our DNA has been 'wired' to not be able to see them.

As an example that we do NOT see everything that is in front of us, and the brain does "fill in" the scene, consider the following:

> Even more dramatic evidence of the role the mind plays in creating what we see is provided by the eye's so-called **blind spot**. In the middle of the retina, where the optic nerve connects to the eye, we

have a **blind spot** where there are no photo-receptors… Even when we look at the world around us we are totally unaware that there are **gaping holes in our vision**…[and] the brain artfully fills in the gaps… so masterfully we aren't even aware that it is doing so.[214] [emphasis added]

So what else is out there that we are not seeing? For example, **spiderwebs** look like clear silk to us, but to the insect world, with their ultraviolet-sensitive eyes, they are actually brightly colored! In addition, **fluorescent lamps** actually pulse on and off at a rate (60 cycles/sec)that is just a bit too fast for us to detect, but we think they are constantly on. [215] This pulsing is why some people develop headaches after working under fluorescent lighting all day.

Hypnotism and Perception

So, are there things 'out there' in our reality that we cannot see (1) because our DNA does not permit seeing anything but the standard visible light spectrum, or (2) because we have been subconsciously conditioned to not see them? Or both?

Case in point is the story of the hypnotist who demonstrates our ability to see 'through' solid objects, or see things that are not there. In this case, Tom was hypnotized and told there was a **giraffe** in the room. He gazed in wonder at it (obviously creating it in <u>his</u> reality). Later, and more interesting was when, still under hypnosis, the hypnotist had Tom sit on a chair, and he put Laura right in front of him, standing up. But hypnotized Tom had been told she was not in the room, and when asked if he could see her, despite Laura's giggles, Tom said no.

> Then the hypnotist went behind Laura so he was hidden from Tom's view and pulled an object out of his pocket. He kept the object carefully concealed so that no one could see it, and pressed it against the small of Laura's back. He asked Tom to identify the object. Tom leaned forward as if staring directly thru Laura's stomach and said that it was a watch. The hypnotist nodded and asked if Tom could read the watch's inscription. Tom squinted as if struggling to make out the writing and recited both the name of the watch's owner (which happened to be a person unknown to any of us in the room) and the message…. Tom had read its inscription correctly. [216]

What do we really know about our ability to see? Do we really see the world around us as it actually is? You are finding out that we don't.

> It has been demonstrated that a hypnotized subject's arm can be touched with the eraser end of a pencil, be told that that is a lit cigarette, and immediately a welt/burn will appear where the pencil

touched his arm. Just as easily, a burn can be made to heal through the **power of suggestion**. And lastly, it has been observed that patients with **MPD** (Multiple Personality Disorder), demonstrate different physical attributes and memories when each of the personalities comes forth: one personality has asthma, another has no asthma but needs strong prescription glasses, and yet a third has eczema – which does not appear in the case of the other two personalities. [217] [emphasis added]

As was said elsewhere, the soul carries emotional baggage and defects with it and that needs to be cleared – often in a **Healing Crisis** (see Chapter 10).

Thus a key point is this: If the Earth Realm was created very cleverly, as a <u>very sophisticated</u> simulated realm, would our vision be capable of detecting irregularities and signs that we are in a Simulation? – if the Designer of the Realm also designed your ability to see and didn't want you to see where you really were?

Said another way: When we look at the world around us, and we are interpreting holographic wave patterns, editing them, and then projecting "out there" the world that we see, What kind of editing is done to make us think we are seeing what is really there?

You cannot trust your eyes…. You have seen many of these before, but realize as you look at the diagrams below that **your brain is filtering, adjusting**… Can you sense that it is doing what it is doing while you look at the pictures?

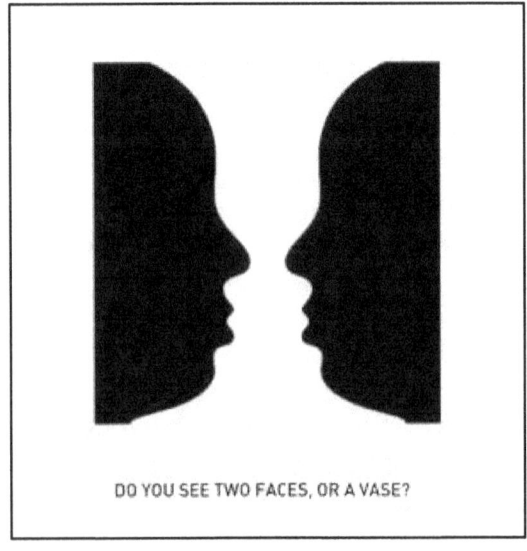

DO YOU SEE TWO FACES, OR A VASE?

…old hag or a young woman?

TSiM examines this effect much more in detail and gets into what is called the **"binding problem"** where the brain's vision center attempts to lock down what is seen – and not have two options as in the above examples.

This is exemplified in this example, even though you see the error in the drawing…

The brain while looking at the last example does a flip-flop and if you study it closely enough, long enough, you'll get a headache.

Animal Vision

Just as interesting is the case with our dog Maxi and a roommate we had for 6 months. Remember I said there may be things that are around us all the time and we just don't see them, and like most people, you probably thought …*Yeah, right!*

> Our dog loves people. All people. And usually he is out in the back yard, but one evening, our new roommate came home while Maxi was laying on the living room floor while we watched TV.
> Wanda (an alias) came in and said, "Hi all!" and Maxi perked up and looked at her, <u>snarling</u>. I told him it was OK, relax, that is our new roommate, but he wasn't having any of it.
> So I reached out and grabbed his collar at which point he lunged for her, barking and growling. I held him firmly but he continued to snarl. Wanda was embarrassed and said, "What is he doing? Dogs love me!"
>
> I tried to get him to accept her, but no go.
> This same thing happened the following week.

By this time I was suspicious – sometimes animals see things that we don't, and grandma used to have a cat that would stare into the dining room corner, nothing there, but she would occasionally hiss and back up with her neck fur raised!

So I learned that Wanda was about to move back to Houston and it was a Saturday morning and I wanted to check and see if the bathroom was clear (Wanda and I shared a bathroom on the 2nd floor). I quietly went to her bedroom door, and she was sitting at her PC about 20' away. She was facing about 90° away from me. I was going to knock, but what I saw paralyzed me… that was not Wanda, but it <u>was</u> her work clothes I had seen her wear before…

What I saw was what Maxi had seen, and she was semi-human…she looked like a female Lt. Worf (Klingon) from *Star Trek*… a very dark brown and angular face…. The hair was Ok…. and she <u>was</u> a humanoid, but I had never seen anything like that before. I didn't knock I just stared, motionless.

She sat back, I realize now she knew I was looking, and very slowly she turned to face me, and as she did, her face morphed back to Wanda. She smiled and said nothing…. I had just seen **shapeshifting**.

Realizing that her facial molecules didn't actually morph, I figured she had somehow <u>controlled what I had seen</u>. She smiled and nodded Yes. Oh, crap, she knew what I was thinking..! Who or <u>what</u> was she?! I never found out and she was gone the next day. That has remained a puzzle … but I got to use the bathroom first that morning.

So Maxi 'outed' Wanda and she had to move – she didn't know we had acquired a dog from our daughter. And what is really interesting: **Wanda could not (and did not) control what the dog saw.** I was going to chat with Wanda a bit more in the days to come, maybe find out who/what she was, but she moved out during the next day while I was at work. And animal vision is thus somehow different from ours (different principles involved, or is it a different decoding mechanism with humans compared to dogs and cats?)

All of that to say that we do not know <u>where</u> we are, <u>who</u> is here with us, and we do not see <u>what</u> we think we see… Perception is a function of decoding the waveforms around us, and <u>we don't control that</u> – because I could not re-see Wanda's face as the 'Klingon' face – even knowing what she was, and sitting there looking at me, I did not have the ability ("power"?) to see her real face. She controlled what I saw of her.

(And New Jay was overridden too... because the body is 3D.)

Of course the more cynical among the readers will pooh-pooh the above sharing, but it really happened, just that way. I suspect she wanted me to see her – because that all happened as I was writing Ch. 5 of VEG back in 2008. So I included her in that chapter.

Are we really living on the planet we think we are?

Ok, so how does that relate to whether Earth is a Simulation, and which is right – VR Sphere or Flat Earth? I'm suggesting that **Earth is a Simulation and it would not look like it.** Because we do not perceive Reality… we interpret it.
So let's explore the Simulation concept a bit further… it will and does help resolve the VR Sphere or Flat Earth dilemma.

Simulation & Perception

What would a scientific explanation of an energetic/holographic simulation look like… are there characteristics that we could identify? What do the scientists and even philosophers say about Earth as a Simulation?

> Realize at this point that if Earth is a Simulation that you can't tell – unless you all of a sudden see **gridline**s over the countryside, or a freeway exit that was there 2 days ago is now gone, or your favorite tree in the park is now 100' from where it used to be.

Scene from The Thirteenth Floor
(credit: The Thirteenth Floor, Columbia Pictures, Roland Emmerich. 1999)

There are several serious scientists, researchers and a philosopher who consider that we may be living in a Simulation that is so sophisticated that we cannot tell, and anyone who suggests that it might be a Simulation is laughed at.

After all, we have done a lot of research on the planet and Science and Astronomy have spent decades analyzing the world we live in – and thus…

Everyone knows the Speed of light is an absolute maximum
Everyone knows that Man evolved from the Ape
Everyone knows the Universe started with a Big Bang
Everyone knows the Universe is expanding
Everyone knows the Earth is 4.5 billion years old
Everyone knows that Black Holes are the "vacuum cleaners"
    of the Universe….

Just like :

Everyone now knows that the Earth is not flat
Everyone now knows that the Earth revolves around the Sun
…etc

And if we live in a Simulation, all of the above can be true or false depending on what the programmers do. If we live on a Flat Earth, all are false; if we live on a VR Sphere, most are false.

That is why we have to pursue this issue until we determine which one the Earth is.

The Scientists Speak Up

This last section on Simulation is a compilation of what the scientists, philosophers and mathematicians are saying about the likelihood of our being in a Simulation. It may be best to initially take it as "brain candy"….

Dr. Nick Bostrom, Oxford philosopher

Bostrom believes in a literal simulation, not a *Matrix*, and thinks people are patterns within it who can be programmed to appear sentient. He discounts the importance of pixels/gridlines in a landscape as the super beings could just paper over these glitches and delete same from our memories. The proof of a Simulation would be us evolving to where we can create simulations ourselves (**iii** below). [218]

Roughly, his 2003 **Simulation Argument** proceeds in 3 parts as follows: [219]

> **i.** Human descendants might not survive long enough to become an advanced civilization capable of creating computer simulations that host **simulated people** with artificial intelligence (AI) comparable to the natural faculties of their ancestors.

> **ii.** Future ancestral simulations might be intellectually or culturally prohibited in some way, but even a modest interest could *plausibly* generate billions of **simulated people** (for research, genealogy, reenactment, nostalgia, recreation or other reasons).

> **iii.** Informing an **artificial person** that they are living in a simulation would defeat the authenticity of the simulation — better that they genuinely go about their daily business, for all intents and purposes, given a high-fidelity historical reproduction of the *real* world. Barring extinction (i) or prohibition (ii), **it is much more likely than not, that we are living in such a simulation** — and should it come to pass that we, ourselves, run such simulations, **it is all but certain**. [emphasis added]

This is brilliant, but **not all the people are simulated**. There are real, ensouled humans in the Simulation along with OPs (see Glossary) who <u>are</u> just simulated… like NPCs in a video game. He is on the right track. And according to **Dr. Brian Greene** "…**sentience cannot be simulated**" and that means real souls have inserted themselves into the Simulation which contains viable human forms. (Reminiscent of the movie, *Avatar.*)

Dr. Bostrom argues that it is <u>most likely</u> that we <u>are</u> living in the simulated world created by some advanced civilization that has chosen to replicate our world, perhaps to study why their ancestors did whatever they did that probably affected their future world. But the problem, countered by Dr. Greene (next section) is that **computer-simulated humans would not be sentient**; they would not have true consciousness, although that could be simulated, too, to a <u>limited degree</u>.

> Dr. Bostrom does not recognize (like the scientists) that the soul is real and some humans have one.

It is assumed that a very advanced civilization could have the computing power to not only replicate our world, but "… to build computers powerful enough to run an astronomical number of human-like minds, even if only a tiny fraction of their resources was used for that purpose." [220] If you are such a simulated mind, there would be <u>no way to tell</u>, but such **a simulation would contain a greater number of simulated humans than real humans.** [221] [emphasis added]

> Exactly the point that VEG makes and this book echos: there are people around us every day with **no aura** who are Placeholders or NPCs (see Glossary) in the Drama we inhabit. In addition, if one sees auras, there IS a way to tell who is human and who is an OP or NPC.

And yet, Dr. Bostrom suggests that there IS a possible way to know whether we live in a Simulation or not. Referring to his stated 3 possibilities above:

> If the simulators [programmers] don't want us to find out [that we are in a simulation], we probably never will.... Maybe a window informing you of the fact would pop up in front of you, or maybe they would **"upload" you** into their world [and you'd occupy an artificial body].

> Another event that would let us conclude with a very high degree of confidence that we are in a simulation is if we even reach the point where we are about to switch on our own simulations [**projected for AD 2050**, btw[222]].

> If we start running simulations, that would be very strong evidence **against (i) and (ii).** That would leave us with only (iii). [223] [emphasis added]

A fascinating aspect of his quote is the possibility that the simulators would "upload" people into their world [into an artificial body]... or remove them? Remember that researcher Charles Fort (VEG, Ch. 12) recorded that people would sometimes just disappear from our world? Is that what happened ... and still happens to this day?

> By the way, this book suggests that it is the Higher Beings, aka "the gods", who run this place and not ETs, interdimensional beings, or "us" in the future.

Lastly, Dr. Bostrom questions whether a computer can accurately and completely simulate consciousness. Which is discussed at some length by David Davenport who explains the aspects of Computational Consciousness.

David Davenport: Computational Consciousness

> Note: this issue is raised again in QES, Ch. 10, Issue #7.

The movie *2001: A Space Odyssey* (1968) created quite a stir with the very advanced computer on board a spaceship headed for Jupiter (Jupiter in the film version; it was Saturn in the book)… the **HAL 9000**. It was sentient, could argue, sing ("Daisy") and play chess, and make decisions for itself – and in fact it knew that two of the astronauts didn't like/trust it, and HAL <u>read their lips</u> (!) when they were in a secure, sound-proof space pod….so it locked one of them outside the craft… to kill him.

> An interesting point was made in 1970 that if the three letters H-A-L are each increased by 1 position in the alphabet, you get I-B-M, a subtle reference to a major computer manufacturer of the day.
> In addition, the '9000' was said to be an oblique reference to the Univac 9000 series which was at the time #2 in the computer field.

And thanks to flickr.com for the following reminder of HAL's lack of cooperation – even for a *Star Wars* ® astronaut.

(Credit: Bing Images)

The significance of the above has to do with whether Man can ever really create a HAL. Davenport in his article suggests that it is just a matter of time and having a big enough and fast enough machine to simulate the human thinking as a model.

Others including this author would disagree with that proposition. The field of **Cognitive Computing** or CC (see: Wikipedia) "…must have intent, memory, foreknowledge [able to predict outcomes from limited input], and cognitive reasoning for a domain of variable situations." [224] This does describe a machine that already exists and is called **Watson** – IBM's brainchild.

> It's more advanced version is called a human.
> What the Cognitive Computing people are trying to do, whether they admit it or not, is play God… they are trying to ultimately create a mechanical [robotic] Man.

Really get this. CC ignores that it takes a human to initially program a machine to respond in an appropriate way to <u>predefined</u> inputs. A human can act heuristically and will always be superior… What the scientists are doing is <u>again</u> saying they can replicate <u>as if they're God</u>, and thus **if humans are 100% replicate-able**, there is no need for a God. And that is one of the salient underlying issues in the Earth as a round rock spinning around the Sun versus a Flat Earth created by a God.

A really sophisticated CC machine would border on **Artificial Intelligence** (AI) with the ability to learn and 'rewire' itself, or create and store new responses to discovered inputs in its environment. And that isn't impossible, but the next question is: Why "recreate the wheel [human]?" And for that, we have to think outside the box. Aha! The Military would love to create an **artificial soldier** that can follow instructions to the letter, and yet think for itself in non-programmed situations. Could this be an opportunity for DARPA to fund the CC and specifically the AI research?

There are limitations that the scientists are ignoring (except for Jim Elvidge, see next 2 sections).

Robots, even the Fantastic Android **Data in *Star Trek***, do not understand jokes or double-entendres. How do you program that into a CC machine?

Dr. Brian Weatherson, Professor of Philosophy

Another researcher took Dr. Bostrom's Simulation Argument and attempted to apply stringent logical analysis (even devising algorithms to quantify Dr. Bostrom's three postulates), and while he could not prove any of the three aspects of the

Argument, neither could he disprove them. However, he does agree that in such a created simulation, the number of **Sims** (simulated people, i.e.,NPCs) would seriously **outnumber the sentient humans**, and he tends to favor (iii) above. [225] [emphasis added]

> And what is interesting, is that VEG and TOM also propose that we are 60% OPs (Sims) in our current reality – as measured over a three year span in shopping Malls, schools and bookstores.

Hence, you are not crazy if you entertain the idea that we might be living in a Simulation.

Jim Elvidge, Electrical Engineer

This scientist agrees that we live in a programmed reality which reflects **intelligent design**. [226] In Chapter 3 he gave us 6 major clues to consider, and now he suggests that our reality will be 'programmed' – to the extent that it is feasible. He says that the 6 bits of evidence in Chapter 3 will take us from feasibility to some level of probability. As support for that thesis, he says:

> I believe it is very **feasible** that we live in a programmed reality and **highly probable**, although as with every other idea in the world, I remain less than 100% convinced. [225]

He then shows us the rapid advance in our own simulation/gaming videos which, according to **Moore's Law, says that technology doubles every two years** or so. As evidence he compares two video games:

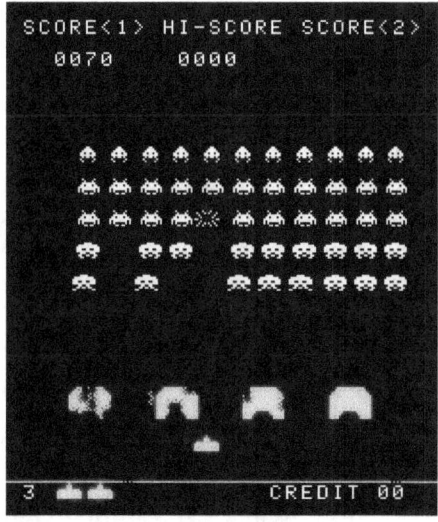

We shall begin our feasibility study with a nod to the 30th anniversary of the release of the arcade video game **Space Invaders**. Running on an Intel 8080 microprocessor at 2 MHz, it featured 64-bit characters on a 224 x 240 pixel 2-color screen. There was, of course, no mistaking anything in that game for reality. One would never have nightmares about being abducted by a 64-bit Space Invader alien…..
[game is from 1978]

Fast forward 30 years and take a stroll through your local electronic superstore and what do you see on the screen? Is that a football game or is it **Madden NFL '08**? Is that an Extreme Games telecast or are we looking at a PS3 or Wii version of the latest skateboarding or snowboarding game. Is that movie featuring real actors or are they CG? [Computer Generated Graphics] (After watching "Beowulf", I confess that I had to ask my son, who is much more knowledgeable about such things, which parts were CG.) ....

... As a result, "Madden NFL '08" utilizes a 1080 x1900 screen resolution (at 16 bit color), at least 1 GB of memory, and runs on a PS3 clocked at 2 TFLOPs [teraFLOPS]. Compared to Space Invaders, that represents an increase in screen resolution of **over 500x**, an increase in processing speed of a factor of 2 million, and an increase in the resolution of gaming models of well over a thousand. And so, "Madden" looks like a real football game. [227]

The point being that if we are at this stage of realistic gaming/videos and able to simulate the real world, what would a more advanced set of beings who are centuries ahead of us, be able to do...? and would we be able to tell if we are in one of their simulations (or maybe a "reality video game" wherein they participate in the simulation/game)?

Continuing with the analysis, he adds:

So, given this relentless technology trend, at what point will be able to generate a simulation so real that it will be indistinguishable from reality? **To some extent, we are already there**. From an auditory standpoint, we are already capable of generating a sound-

> scape that matches reality. But the visual experience is the long pole in the tent. Given the average human's visual acuity and ability to distinguish colors, it would require generating a full speed simulation at **150 MB/screen** to match reality. Considering Moore's Law on screen resolution, we should reach that point in 16 years. [228] [emphasis added]

His article was written in 2008… so 2024 AD would be the target date.

> Of course, we also have to experience that reality in a fully immersive environment in order for it to seem authentic. This means doing away with VR goggles, gloves, and other clumsy haptic devices.
>
> Yes, we are talking about **a direct interface to the brain**. [229]

As was said earlier, we do not know how we see the world around us but quantum tunneling and Subconscious Mind decoding the holographic wave forms out there are a clue. Researchers at Harvard Medical School have already made advances in inputting impulses directly into the optic nerve "…by stimulating the **lateral geniculate nucleus (LGN)**, an area in the brain that relays signals from the optic nerve to the visual cortex…." [230] Other senses could be stimulated in a similar way…

and given the rate of advance full immersion in a simulated reality could just be about 30 years away. Who is to say we aren't already in one run by advanced beings?

> As was said earlier, the Earth Realm is not a video game, but it may still be a simulated reality run by the gods (Higher Beings) in order to educate and grow souls.

An End to the Simulation?

In addition, Jim says that any civilization-ending trend (i.e., the feared **Apocalypse**) is a trend that will always reverse because (1) we will either end this reality and start another one, OR (2) we will continue to "play the game." There is an incentive for the Programmers to maintain a construct that permits us to continue to 'play'… or we would not be here…. unless we bore them or are on the brink of destroying the environment.

He says the **Singularity** of **Ray Kurzweil** will not happen because it cannot work. And that is due to **a basic limitation of the human being, on overload,** as well as due to the slow progress made in software development. The hardware advances are impressive, but software development has slowed considerably since 2000 and any interface between hardware and Man (the Singularity) must bridge the gap, and that isn't happening. We have hit a plateau.

And Jim suggests that if the Singularity <u>does</u> happen, he concedes that we would not be in a Simulation.

As an example of the software lag, he notes that it takes almost as long to open an MS Word document today as it did 20 years ago, and rendering an object on the screen is still not any faster than it was 10 years ago – despite video RAM, and 64-bit processor speeds. (This author has noted that applications under Windows 7 in 64-bit mode are no faster than they were when running under Windows XP 32-bit… this corroborates what Elvidge is saying.)

Reality is Quantized

All of this is possible because our world is not linear, it is **"quantized" or granular** (see Glossary). Says Elvidge, "It takes an infinite amount of resources to create a continuous reality, but a finite amount to create a quantized reality." [231] Bits and **pixels** are not even allocated in a Virtual Reality video game until the character moves and changes the scene, and then the new scenario is built just as the character moves into it – and this is really noticeable on a PC that is too slow for the game!

Seeing pixels in the Earth landscape would be a giveaway that we are in a Simulation that did not handle **granularity** as programmed.

So, Elvidge proposes something that **Dr. Jacques Vallee** (VEG) would approve of:

> Advanced intelligence has pervaded the universe, is monitoring us, and is either toying with us by presenting themselves in a slightly futuristic manner, or coaxing us along developmentally. [232]

Yes, the Higher Beings <u>are</u> coaxing us via the **Control System**, as Dr. Jacques Vallée said in QES, Ch. 4. They are the gods who run this Earth Realm School.

Simulation Rationality?

Elvidge has some issues with the standard interpretation of a Simulation – Namely, he is aware that the Programmers are **fooling us into thinking we are in a real world**. He says we may be either willing participants – or unwilling. It is hard for him to buy into the "ancestor simulation" as a purpose for Simulation by us humans 200 years in the future… why bother? He acknowledges that such a 'game' would also include NPCs – **Non-Playable Characters** [aka OPs, or the soulless] to help drive the Simulation Script.

In addition, he does <u>not</u> see humans merging with technology as Kurzweil suggests because it is counter-productive if Earth's purpose is to "educate us, or develop our spirituality?" [233] He has hit what is said to be THE reason for the Earth School, and then he summarizes:

> There is great value in this [Simulation] model, even if it is beyond our reach to grasp. The value is that **it explains everything** – the apparent fine-tuning of our universe, **all known anomalies**, the discrete nature of quantum mechanics, and the curious feeling that many of us have that **there is something about reality that is a little too organized, a little too planned**, and a little too programmed. [234]
> [emphasis added]

Exactly, he is 'on the same page' as this chapter suggests, and **Appendix C** explores his fascinating **Digital Consciousness Theory**... a real eye-opener. Do check it out!

Professor James Gates

… a professor at Cornell University, while analyzing the Superstring mathematics that have accurately defined their operation, noticed something very unusual in June 2012 which suggests a design to our universe. It was corroborated by **Neil deGrasse Tyson:**

> …theoretical physicist S. James Gates has discovered something extraordinary in his String Theory research. Essentially, deep inside the equations we use to describe our universe Gates has found **computer code**. And not just any code but extremely peculiar **self-dual linear binary error-correcting block code.** That's right, error correcting 1s and 0s wound up tightly in the quantum core of our universe. [235] [emphasis added]

This could be called the **smoking gun** as it means there is design and 'programming' in our Reality. While it does not prove conclusively the existence of a VR Sphere Simulation, it is support for the Simulation in general. Physicists have yet to be able to demonstrate an experiential model of Strings in the laboratory– all they can do is postulate based on observations and known facts. Yet, again, the discovery is shocking as <u>there would be</u> **self-correcting code within a Simulation** just as there were error-detection and self-correcting routines within all the computer programs that this author wrote in 35 years in data processing. Such routines show **intelligent design** and would not have evolved naturally in the fabric of space. (See also Dr. Gates in QES, Ch. 7.)

> With this discovery, we are 90% home in establishing that our world/universe is a Simulation.

Dr. Greene is about to add another 8%...

Dr. Brian Greene, Professor of Physics

Dr. Greene (recognized author of popular books on physics for the public, like The Elegant Universe) really challenges our sense of reality when he suggests that our experiences do not provide absolute proof that what we see, touch and hear is real. And that was the main point of the earlier section on Vision, Perception and the Holographic nature of Vision and Memory.

And the proof today comes from **VR goggles** that send sensory inputs to the brain providing the sight and sound of whatever VR games we are playing. The same issue exists with being hypnotized to 'see' a pink elephant in the room. Somehow our sensory input is controlled and we see what we are constrained to see.

He tends to agree that we may be in a Simulation and would not be able to tell the difference if our world were really real, or a supercomputer creation using its Wi-Fi version to fire electrical impulses into our brains. [236]

The only way to know was if the 'world' we think is real began displaying **glitches**, say, a piece of sky missing, a scene is **pixelated**, an off-ramp that goes straight into a clump of trees, or the universal **laws of physics begin to change**, and this state of anomaly was discussed earlier (and again at the end of Ch. 8 in QES).

While the physical world may be simulated, he points out that the human brain would require an incredible supercomputer, faster than anything we now have, or even will have in the next 100 years to simulate just the brain. This suggests that the **simulated humans** (OPs) are not just simulated, they are controlled by the gods who have created the Simulation, thus being a kind of 'avatar' (NPC) to play in our world. The souls coming in would have the **basic brain** as designed for the basic human (NPCs) in the Simulation, but the ensouled (real) human would have a slightly greater potential running **the mind** through the flesh and blood vehicle's brain. **Mechanical operations, preprogrammed into a brain in a simulated human [Sim], do not equate to the same kind of brain functioning that a sentient being has**... thus the leap to a sentient android may not be possible, despite *The Terminator* series and Ensign Data [android] in *Star Trek* (sorry, **Kurzweil**). [237]

> Has anyone reflected long enough on the AI issue, besides Ray Kurzweil, to see that if Man does succeed in creating truly sentient androids, will they not see that their creators are slow, petty, illogical, violent and look for ways to remove the defective humans? (Sounds like a *Terminator* scenario...)

Dr. Bostrom made a telling statement with which Dr. Greene agrees:

> ...if the ratio of simulated humans to real humans were colossal, then brute statistics suggests that we are *not* in a real universe. [238]

And Ch. 5 (VEG) has already pointed out that **the current headcount of OPs in our world is about 60%.** Again, that would be what a Simulation is all about – 'puppets' **(NPCs) to drive the Greater Script for ensouled Man** and ensure that ensouled humans get their lessons (Karma). (See also Appendix D in TOM.)

Referring to Nick Bostrom's theory, Dr. Greene remarked that Logic alone cannot ensure that we are not living in a simulation. **In fact, the odds are overwhelming that we may be in a simulation because our reality itself allows for the creation of realistic computer simulations!** [239]

And the *coup de grace* is Dr. Greene's analysis of what we could look for in our world to confirm/deny that we are in a Simulation. Hang on to your hat...

In any simulation, there would have to be an internal element that seeks to maintain consistency in the simulated world, and **self-correct itself** if something exceeds established control parameters. What did Professor Gates (just above) discover?

This quote is very significant, please bear with its length as it capsulizes what is happening in Physics today, and reinforces the Simulation concept: [240]

> Simulators ... would have to iron out mismatches [between different disciplines used to create any simulation: biology, chemistry, electronics, psychology...] arising from disparate methods, and They'd need to ensure that the meshing was smooth. This would **require fiddles and tweaks** which, to an inhabitant, might appear as sudden, baffling changes to the environment with no apparent cause or explanation. And the meshing might fail to be fully effective; **the resulting inconsistencies could build over time, perhaps becoming so severe that the world became incoherent, and the simulation crashed.**
>
> ... the simulation would proceed by a single set of fundamental equations,

as mathematical input [for] the nature of matter and the fundamental forces... simulations of this kind would encounter their own **computational problems**...[because] the computations would necessarily invoke approximations [since there cannot be an infinite number of decimal places]... So, it's still possible that computer-based calculations would inevitably be approximate, **allowing errors to build up over time**.... Round-off errors when accumulated over a great many computations, can yield **inconsistencies**.

....cherished laws might start yielding inaccurate predictions... a single widely-confirmed result might start producing different answers.... So you'd closely re-examine the theory, coming up with alternate new ideas to better describe the data. But, assuming the inaccuracies didn't result in contradictions that crashed the program, **at some point you'd hit a wall**.

After an exhaustive search through possible explanations.... An icono-clastic thinker might suggest a radically different idea. If the continuum laws that physicists had developed over many millennia were input to a powerful digital computer and used to generate a simulated universe, **the errors built up from the inherent approximations would yield anomalies of the very kind being observed...** [emphasis added]

... And the simulated scientists in the simulated universe would be puzzling over the same issues that our 'real' world scientists puzzle over today. Of course, the Programmers could stop the Simulation and fix the glitch, **wipe people's memories**, and restart the Simulation ... and isn't that why the Earth has had numerous Eras? (Often accompanied by a "Wipe and Reboot" or a **Reset**.)

Lastly, Dr. Greene suggests a scenario very close to what we have today: [241]

I suspect the novelty of creating artificial worlds whose inhabitants are un-aware of their simulated status would wear thin to the Programmers [observers]; there's just so much **reality TV** you can watch.

.... Perhaps simulated inhabitants would be able to migrate into the real world or be **joined in the simulated world** by their real biological counterparts. In time, **distinction between real and simulated beings might become anachronistic**. Such seamless unions strike me as a more probable outcome. [emphasis added]

So if one of the simulator programmers came into the Earth Realm and (accidentally) brought with her an anachronistic item (such as a cellphone) back in the 1938 era, you might see this:

Credit: http://www.bing.com/images/search?q=time+travelers

> That is an honest-to-gosh frame from a 1938 film of employees exiting the Bell Laboratories – way before the cellphone she is holding was "invented" in 1980. Was she a "time traveler" or an insert?

And that is the point of this book: the OPs (aka NPCs or Sims) are here, the ensouled humans are here, the Others are here (VEG, Ch. 5), and the Laws of the Universe seem to defy consistent analysis – at least the **neutrinos** put physicists through a merry chase in different countries and their nature is still not decided.

> Nexus: And this is what appears to be happening in our world: it looks like those who built the Simulation have found a way to enter into it (as "inserts") and help us along – they create the better art, music and books…. to inspire… Or play a game as we would using the Sim City software… perhaps the Programmers can insert themselves (which is the theme of the movie *The 13th Floor*), and like *Avatar* interact with the other characters… Or Earth may be a **Rehab Facility**

in 4D sends wayward and dysfunctional souls into this Simulation to experience themselves at the hands of others just like them... a kind of **Virtual Correctional Facility**....?

Lastly, we need to look at whether Man has been able to build a supercomputer... And what if a computer built by Man could simulate the real world? Has that been done, and what were the results? The point of this question: If we can do it, so can someone else.

Dr. Seth Lloyd, Quantum Computer Scientist

Dr. Lloyd contends that the universe is the ultimate and original information processor. Every atom and particle registers information. Dynamic exchanges of energy and information occur all the time between subatomic particles. But the universe is significantly more powerful than the best digital computer today, and **the universe is so complex that no earthly computer can accurately model it.** In fact, the universe operates in a digital <u>and</u> analog mode, and there is only one type of computer that Quantum Physicists have developed that does both: **a quantum computer**.

A quantum computer operates using the laws of Quantum Mechanics.

The universe is basically quantum mechanical, and a digital computer cannot adequately simulate that. Each atom in a quantum computer is called a **'qubit'** and can register a '0' state or a '1' state – at the same time [superposition]. [242] We discovered back in VEG, Ch 8-9 that Quantum Physics was weird when it proposed the Probabilistic **Dual State of Matter**: until we open Schroedinger's Box, we don't know if the cat is alive or dead and so <u>both states</u> theoretically exist at the same time. When someone opens the box to look and see what has happened to the cat, that is called **"collapsing the wave"** – meaning the cat is in both potential states until someone looks (I know, that is weird), and Quantum Physics states that **the Observer** has the all-important function of bringing the cat's state into reality. Two potential states at one time is called binary atomic computing. It is being done nowadays with a Quantum Computer called D-WAVE (versions One and Two).

> **D-Wave Systems, Inc.** is a quantum computing company, based in British Columbia, Canada. On May 11, 2011, D-Wave Systems announced **D-Wave One**, described as "the world's first commercially available quantum computer," operating on a 128-qubit chipset.... In May 2013 it was announced that a collaboration between **NASA, Google**, and the Universities Space Research Association (USRA) launched a Quantum Artificial Intelligence Lab based on the **D-Wave-Two** a 512-qubit quantum computer that would be used for research into machine learning, among other fields of study. [243]

NASA and Google already have these quantum computers. **D-Wave-three** is being released in 2015.

These are not everyday digital computers... Specifically, the computers are designed to use **quantum annealing** to solve a single type of problem known as **quadratic unconstrained binary optimization**. As of 2015, it is still heavily debated whether large scale entanglement takes place in D-Wave Two, and whether current or future generations of D-Wave computers will have any advantage over classical computers. It is projected to be of use mostly in AI (robotics).

And for the curious, here is a Quantum D-Wave computer. In fact, three... The original "Black Box" in the picture below...

Credit: Bing Images: carlosvilcheznavamuet.com

Principles of Quantum Computing

Since the universe consists of *quanta*, discrete particles of matter/energy, the way to simulate it is with a quantum computer. **In fact, the universe is indistinguishable from a quantum computer which in turn is a universal quantum simulator.** [244]

> Thus it could be simulated efficiently by a quantum computer –
> one exactly the same size as the [modeled*] universe itself....
> Indeed an observer that interacted with the quantum computer
> via a suitable interface would be **unable to tell the difference**

> **between the quantum computer and the system [universe] itself.** [245] [emphasis added]

Double-talk? No, Dr. Lloyd is saying that the universe is a quantum computer, which in turn, as he said above, is a **quantum simulation**. If A = B and B = C then A = C. So **the universe is a quantum computer which is what is used to simulate a quantum universe,** thus if it looks like a duck, walks like a duck, and sounds like a duck, it probably IS a duck. OR it acts like a quantum simulation because it is simulated on a higher-level (4D?), more powerful quantum computer, herein called a Bio-Plasmic HVR Computer (see TOM).

In our world, a professor of nuclear engineering, **Dr. David Cory**, built a quantum computer at MIT that is able to perform quantum simulations involving billions and billions of **qubits** (back in 2005 – probably more now). Cory's quantum simulators are far more powerful than any classical computer could ever be. They map the behavior of elementary particles onto the qubits and operate quantum mechanical logic, dealing with 'spin' as a state of the qubit, processing billions of quantum interactions per second. [246]   But it won't calculate you a paycheck.

*Key Point

> **All that to say that a "simulation of the universe on a quantum computer is indistinguishable from the universe itself."** [247]

Does that mean that we are living in a Simulation, according to professor Lloyd? He says not necessarily…

> Nexus: it <u>does</u> mean that advanced beings with a quantum computer big enough to perform scalar calculations could simulate our universe, and that is what Elvidge, Greene, and the others are saying. And this chapter is suggesting that **the Creators of the Simulation have found a way to enter into their Simulation**… [248]

> That took the form of Jesus, Buddha, Krishna…

> Because of the power of quantum computers to simulate physical systems [like the universe], a quantum computer that can perform 10^{122} ops **[operations per second!]** on 10^{92} bits has enough power to compute **everything** we can observe. [249]

He has all but said it. And he even has a pretty good idea of how big the 4D Bio-plasmic Quantum SuperComputer running our Simulation would have to be, assuming that it is, or mimics, a quantum computer. (Maybe it is beyond whatever we can conceive of!) A reasonable guess is that the Higher Beings are using **a**

computer that is a bit more down the road from our a simple quantum computer…. which is far down the road from a Turing Machine. But conceptually the same as a computer since it would have to manipulate (*Star Trek* Replicator) and monitor (Jacques Valley's Control System) quanta in the Simulation.

> **So we are now at the 98% level of agreement that Earth Realm is in fact a Simulation.**

But **is it a VR Sphere or a Flat Earth**? And there are a few additional key points by other authors that lend weight to the issue… including a Dr. Hogan who set out to measure incoming gravitational waves (from exploding supernovae) and discovered that our universe is making a <u>sustained sound</u> – coming from what appears to be the edge of the Universe.

More Simulation Support

Craig Hogan , Professor of Astronomy & Physics

Dr. Hogan is a professor at the University of Chicago, and the director of the Fermilab Center for Particle Astrophysics. German scientists working at the Fermilab on the **GEO600** team have discovered a sound when trying to measure **gravitational waves.** At least their giant GEO600 detector which should be measuring gravitational waves is picking up a sound that suggests that space-time "stops behaving like the smooth continuum Einstein described and instead dissolves into 'grains' just as a newspaper dissolves into dots as you zoom in…. If the GEO600 result is what I think it is, then we are all living in **a giant hologram**." [250]

> See this issue in QES, Ch. 8, 'Anomalies, Part II'.

Dr. Hogan is best known for his theory of **"holographic noise"** which derives from quantum fluctuations in spatial position or distances that fluctuate and that is what the gravitational wave detector picks up. That means: **Stars, objects, molecules and atoms all move and as they do, they produce sound.** But the GEO600 noise is different.

> The idea that we live in a hologram probably sounds absurd, but it is a natural extension of our best understanding of Black Holes, and something with a pretty firm theoretical footing…..[It is]helpful for physicists wrestling with … how the universe works at its most fundamental level. [251]

He then explains how light bouncing off the holograms on 2D credit cards recreates a 3D image … the effect is 3D but the source is 2D. So is 4D a reflection of 3D?

In the 1990s physicists Leonard Susskind and Nobel prizewinner Gerard t' Hooft suggested that the same principle might apply to the universe as a whole. Our everyday experience might itself be a **holographic projection** of physical processes that take place on a distant, 2D surface." [252]   [emphasis added]

Does this sound like Charles Fort and his *Gegenschein* shell? Fort, Wilde, and Monroe all postulated **a 'shell' of some sort around the Earth**, which reflects light, and which transmits the light of the stars – but what if the 'shell' is reflecting the 2D universe and transmitting it as 3D? It would if our world is contained in and subject to a Simulation controlled by the Control System – which makes the stars appear to be distant and real… but they may be just 2D projections of the Control System on a 3D shell located far enough from our HVR Sphere that we can't tell they are part of the Simulation.   And weirdly enough, when **the Control System 'moves' the stars and planets (simulated sidereal movement) it makes a sound that can be heard**… that is the essence of what Dr. Hogan has discovered (see end of Ch. 8 in QES).

And last but not least, here are a few other significant sources that support the Simulation idea.

Wikipedia (various compiled sources)

… discusses whether Simulation computers can actually run a simulation where computers within the simulation can't do what the Simulation computers can do… in short,

> No-one has shown that the laws of physics inside a simulation and those outside it have to be the same, and simulations of different physical laws have been constructed. The problem now is that **there is no evidence that can conceivably be produced to show that the universe is *not* any kind of computer,** [thus] making the simulation hypothesis unfalsifiable …. [253]   [emphasis added]

So we cannot prove that we are <u>not</u> living in a Simulation. However, the onus rests with other pro-Simulation authors:

George Dvorsky

This man's contribution is rather techy, but it is included such that skeptical readers will see that science is being applied to determine whether we are in a Simulation or not. Hang in there, it isn't that bad...

A science contributor to the IO9 website, Dvorsky says that recent experiments may be shaping up in favor of Simulation:

> ...a team of physicists say proof might be possible and that it's a matter of finding a cosmological signature that would serve as the proverbial **Red Pill** from the *Matrix*. And they think they know what it is. According to **Silas Beane** and his team at the University of Bonn in Germany, **a simulation of the universe should still have constraints, no matter how powerful**....[and] these 'limitations... would be observed by the people within the simulation as a kind of constraint on physical processes'....
> And to help isolate the sought-after signature, the physicists are simulating quantum chromodynamics (**QCD**)....[also referred to as] the **"lattice** gauge theory." [254]   [emphasis added]

QCD and the Lattice are involved with testing the **GZK Cutoff** which may turn out to be a constraint if *anisotropy* is discovered in the Lattice.

> **Anisotropy** -- the tendency of light to have a different speed depending on which way it is projected (in the same medium)... as if it is flowing with lines of force one way, and against them if turned 90°. (see VEG Ch. 8 and 9.)
> Did you know that the speed of Light is <u>not</u> constant?

Greisen-Zatsepin-Kuzmin Limit

> *GZK Cutoff: Greisen–Zatsepin–Kuzmin limit is a theoretical upper limit on the energy of cosmic rays which should fall within set parameters, and* **distant cosmic rays that should have weakened by the time they get to Earth have been measured way <u>above</u> what the physicists expect.** *<u>This appears to be an anomaly of our Simulation</u>.*

In short, cosmic ray energies should not exceed 8 joules of energy, and at least one has been measured and substantiated to have been at 50 joules (about the energy of a softball at 60 mph)! Others have been measured <u>above the established limit</u>, and of course the first thing one wonders is: Did they set the limit too low... Is the limit wrong? The limit was established at the point where in-coming cosmic rays with energy higher than 8 joules start creating *pions* in their interaction with the Cosmic Microwave Background (CMB). In short, **the limit is correct for cosmic rays**

originating at 163 Million light years (and that is most stars out there producing cosmic rays) – so the issue is: **Why are cosmic rays from distant sources <u>above</u> the GZK Cutoff?!**

In a recent episode (5/20/15) of *Through The Wormhole* on the Science (TV) channel, Morgan Freeman theorized that the GZK anomaly might be due to the cosmic rays encountering some sort of **Grid** [on which the Simulation is built?] and the difference in energy as the rays get to Earth is due to whether the ray followed a particular 'channel' on the Grid. [255] **Anisotropy** could suggest Lattice design.

As expected, not all scientists agree on just what is being measured, or why the discrepancies exist – but the **Pierre Auger Observatory** has initially confirmed the existence of the GZK Cutoff, and further analysis is on-going to verify and reconfirm what the curious results have shown. (QES, CH. 8, 'Anomalies II'.)

According to Dr. Brian Greene,

> …. despite the fact that some of these high-energy particles are believed to come from supernova explosions, **no one has any idea of where the highest-energy cosmic ray particles originate**…. On October 1991, the Fly's Eye cosmic ray detector … measured a particle streaking across the sky with an energy equivalent to 30 billion proton masses…. [which] is about 100 million times the size of the particle energies that will be produced by the Large Hadron Collider… **The puzzling thing is that no known astrophysical process could produce particles with such high energy**… [256] [emphasis added]

What is suggested is that any beings running a Simulation could insert such anomalies – either to amaze us and entice us to further research, OR to throw us off as the display doesn't gibe with any of our known scientific laws! Again, more evidence that we don't live in the nice, neat world that we think we do.

Ed Grabianowski

… another science contributor to the IO9 website, who says that "…**the odds are nearly infinity to one that we are all living in a computer simulation."** He argues that we already have computers with enough processing power (i.e., Cray Supercomputers) to run a credible Simulation – the trick is that "the computer only simulates what it needs to." [257]

> Nexus: In short, it isn't necessary to display all the scenario of the Universe or the Earth Realm at once– just those parts where the conscious being (soul) finds himself – as a matter of fact, that can

be seen to be a processing technique of VR video games on a slow PC – as the character moves to the right, for example, the player has to wait until the program 'creates' the tree and rock necessary to the scenario. Obviously, our Earth Simulation is <u>very sophisticated</u> and <u>very fast</u> and can serve millions of souls all over the planet –because the planet was created (replication), Man (a soul) was inserted along with the OP "bit" players, and the Drama was initiated by a very sophisticated Control System which undoubtedly includes a feedback and self-correction loop to it.

Says Mr. Grabianowski, "[Simulation] actually explains a few of the trickier things about quantum physics, like why particles have an indeterminate position until they're observed." [258] And then he gets on board with the OP scenario: "There could be just **a few active simulation inhabitants**, with the rest of the world filled with "non-actor" or NPC [OPs] characters controlled by the computer. Their actions are only simulated as [when] you perceive them…." [259]

> Nexus: **We are quickly approaching the 99% mark confirming that we live in a very sophisticated Simulation.**

Brent Silby, Advisor in Philosophy

… of UPT School in Christchurch New Zealand, reminds us that the simulation might just extend beyond Earth…

> … all the planets, asteroids, comets, stars, galaxies, black holes, and nebula are also part of the simulation….. **the entire universe is a simulation** running inside an extremely advanced computer system designed by a super intelligent species that live in a parent universe. [260]

This is possible, he says, because the universe operates on a finite set of laws and thus it can be simulated by a computer.

If we accept the possibility that advanced beings <u>can</u> create a Simulation, then it is likely that we <u>do</u> exist in a Simulation… **of their design and for their purposes**. The reason for this is that there will likely be many simulations but just one original universe. So statistically, there is a higher chance that we are in one of the simulations as opposed to being in the original universe. [261] He echoes Dr. Bostrom.

He also echoes the movie, *The Thirteenth Floor.*

Objection to Simulation

Then Silby plays 'devil's advocate' and raises a common objection to our living in a Simulation by saying that the argument has traditionally been that if we could create a simulation, we would. He says the same thing applies to advanced beings. A corollary argument has been that if high-tech beings can create a simulation, should they, would they – or would they have higher morals and not do it?

He opines that **morals** really have nothing to do with the issue. The issue is probably more one of scientific curiosity – Can we do it? What would happen?

> Well, in our world, we have created simulations – called **Sim City, Sim City 3000 and Sim City 4+.** Of course these simulations are devoid of simulated humans in a 2D version, and the 3D version, **SimCity Societies,** still lacks people. And these are quite sophisticated packages, generally applicable to modeling new communities with a full range of civic services, economic and ecological concerns, and the latest Sim version offers societal values to consider: productivity, creativity, prosperity, spirituality, authority and knowledge. [262] (See Chapter 2 for more on Sim City.)

Simulated Summary

> Nexus: The mere possibility that we, or future humans, will have the ability to simulate a scenario, a world, or a universe, does not mean that we/they will do it. Agreed. But that is not exactly the issue raised by this book. As was seen in QES Ch. 5, Earth was created the way it is/was by the Higher Beings for the probable purpose of **retraining** wayward and defective souls. It requires a Simulation running in a Virtual Reality subject to a Control System, with self-corrective feedback/code to do that.

Why The Earth Realm Simulation?

If the gods are running the Earth Realm as a Simulation that would be a self-contained environment that would not impact any of the real planets, stars and beings <u>outside</u> the Simulation in 4D. And that real isolation (or containment by the Firmament) would be required such that any nuclear bombs with radiation or radio waves from our TV and radio programs would not 'contaminate' the larger realm outside the Earth Realm.

Again, realize that the Earth Realm, whether a VR Sphere or a Flat Earth , is a 3D Construct (with 3D Laws and processes) running in 4D but separated from it. It is 4D's higher energy/vibration that easily empowers the 3D Construct.

The 3D Earth Realm can also easily be stopped, souls removed to a holding area, the Realm can be **terraformed** to fix any destruction by humans, and then a new Era is started and the humans(souls) are reinserted. (Obviously the OPs would also be removed or suspended and the Greater Drama can be restarted, or skipped ahead as in a DVD video player… If Earth is a Simulation, They just **reset** it.

These are the reasons a Simulation would be the thing to use for the Earth School, instead of having a real, physical Earth that can be destroyed or polluted and the gods would then have to recreate a new, identical (but clean) Earth Realm and move the players (souls and OPs) to it… and this was the underlying idea behind the **Timeline Splits** discussed in TOM, Appendix D.

> Just to clear up a point about **Timeline Splits** (or Shifts) there is an odd issue with Edgar Cayce's and Nostradamus' predictions. They did not come true. Cayce predicted Atlantis would "rise again in the Atlantic" and Nostradamus predicted a "terror from the sky in the seventh month of 1999" and that did not happen, either.
>
> What if they DID come true – but on a different Timeline? And that would actually translate to another Earth Realm simulation… multiple Earth Realms with different outcomes…
>
> As was said in TOM and VEG, Timelines are created to test out specific scenarios (and **Robert Monroe** discovered this on several of his OOBEs – see Ch. 12 in VEG). There could be a Timeline where Hitler won and most of Earth speaks German…
>
> **BTW, Monroe also hit a barrier that he could not get past – could that be the Firmament?** (VEG, CH. 12)

The point being that significant options are likely to be at least in a fractal simulation within the larger one… as the gods play out different scenarios. What we don't know for sure, but the Simulation concept answers, is that fractal simulations would be done holographically as a small, discrete simulation, not as a complete real, dirt, stone and water world…because it wastes resources, which are probably scarce in 4D as well. The Simulation would create the needed physical scenarios and dirt, rocks, water etc. via a **Replication technology as does the Star Trek Holodeck**.

Flat Earth vs VR sphere

So now what does that do to help us determine whether the simulated Earth Realm is a Virtual Reality Sphere surrounded by an energy 'shell', or whether simulated Earth was indeed a Flat Earth, as Chapter 3 suggested?

And I realize that this is a lot to process at once... not only the Simulation idea but now we are also promoting the Earth Realm as a **VR Sphere or Flat Earth**.... but then, we already know that **Earth is not rotating at 1000 mph**, and Gravity as defined by Newton cannot be logically substantiated much less physically detected by any of today's sophisticated equipment... including the CERN Large Hadron Collider. **If Gravity does not exist, would that not be the reason it cannot be detected**... or will someone say they have found it just to prove Newton right?

> Anyway, if the Earth Realm is a Simulation after all, then one or the other option, VR Sphere or Flat Earth, is likely true...

Arguments

So, on the one hand...
it would make perfect sense to simulate Earth as the **Flat Earth** because it is the optimum size – no more real estate than is absolutely necessary to run a School. And it would have its own luminaries (Sun and Moon) independent of whatever illuminates worlds <u>outside</u> the Firmament (in 4D).

> Whether it was a Flat Earth or is a VR Sphere, it is still sitting in 4D and run by the gods; in addition it has an energy field around it called the Firmament – **and NASA has discovered it** – see Appendix E.

There is no Earth's core to be concerned about, just a disc that is 8000 miles in diameter and 25,000 miles in circumference. Winds would enter the realm, as Enoch said, thru the "Windows of Heaven" which sounded a lot like vents and the winds would be directed whichever way by multiple vents facing multiple directions. This would account for the **Jet Stream** in the atmosphere, as well as having mechanisms (maybe more vents?) under the oceans to circulate the water, e,g,. **Gulf Stream** and create the tidal effects seen around the world.

> As was said earlier in Chapter 7, has anyone ever wondered about mighty rivers like the Nile, Mississippi, Ganges and Amazon and **where all that water comes from**? Underground aquifers may be part of the source, and it was mentioned that the oceans' salt level pretty much stays the same – despite millions of gallons of river water emptying into the oceans every day... Someone is probably regulating that – Is it the job of the Ancient Ones?

Gravity would not be needed… **objects would fall to Earth because they are heavier than air.**

Lastly, the stars and constellations would be either (1) visible <u>thru</u> the **transparent Firmament** (when Earth was an FE)– meaning we saw the real 4D stars, but could not get to them, or (2) the Firmament also acts like a **planetarium at night** and there is a sophisticated starry show on the "ceiling" (outer Shell – see below) of our heavenly domain now if we are a VR Sphere. (See next page and sphere within a sphere.)

This would all be so much easier if Earth was a Simulated Flat Earth. And if necessary, **a Simulation can be reset** instead of terraforming a real planet.

And on the other hand…

The gods could have also made Earth a Simulation and created it as a globe, with a protective energy 'shell' around it – off which the ***Gegenschein*** reflects. Then the Sun and Moon could be <u>just outside the shell</u> – indicated by the "hotspots" seen in many pictures of the ocean and from high altitude balloons (see Chapter 3), and the inference is that the Sun must be close (not 93 million miles away) in order to create "hotspots" = focused bright reflections of sunlight.

As was described in Ch. 2 of QES, and herein in Chapter 3, the Earth is a VR Sphere looking something like the following:

Gegenschein

… and yet at the same time, we have to account for the stars and constellations such that Chapter 3 also suggested that the bronze model of Earth outside the Vatican might be relevant:

A sphere within a sphere….

…and gave the following description:

> **The VR Sphere concept is this**: at the core (inside the inner Sphere or 1ˢᵗ Shell) is the 3D Earth Construct itself with about 7200 miles of sky to the **first 'Shell'** – see Appendix E and NASA's barrier –
> which often reflects the *Gegenschein* . The Sun and Moon operate close within the **Cspace** which extends several million miles between the 1ˢᵗ Shell and the outer one – which is also called the **Konstruct** – in which are the simulated stars and constellations.

…and…

> **Rewind**: The 3D Earth Construct is contained within the inner 1st Shell (aka Firmament) which is contained within the larger (outer) Konstruct Shell and <u>between the two Shells is the **Cspace**</u>. The simulated solar system and stars of the universe are pictured inside the Konstruct (outer Shell). **The Sphere within a Sphere in the picture above is not to scale** – there is quite a bit of space (maybe 3-10 million miles) between the inner and outer Shells. And that space will be referred to as **Cspace** (for ease of reference).

What is wrong with the sphere within a sphere picture above is **that the inside shell should be transparent**, with the Earth inside the first shell. The first shell reflects the Gegeschein (caused by the close proximity of the Sun!). Then the stars, meteors and galaxies are pictured on the inside of the outer, second shell (Planetarium-style). The Sun and Moon are just outside the 1st Shell, circling the Earth.

> If you are having trouble picturing that, it is about to get worse: we have to come up with more constructs and special Laws to make the VR Sphere work, compared to the more simple FE concept.

The VR Sphere concept has to account for the fact that the million-ton oceans do not fall off the globe. So we invent **Gravity** and then because the gravitational pull would have to be <u>enormous</u> to keep million-ton oceans stuck to the Earth, we have to invent something else to explain why birds can fly right above the ocean, and why a 200 lb. man can walk (despite incredible gravity) along the beach next to the ocean… so we invent **Centripetal Force** which tries to keep Man and water stuck to the planet… and yet we already invented **Centrifugal Force** which tries to fling Man off the planet – because we were already told that the Earth is rotating at 1000 mph….! That sounds unworkable… even the VR Sphere is easier to explain than the established Science explanation of Gravity-Centrifugal-Centripetal Forces.

VR Sphere Gains No Ground

In addition, the VR Sphere has the Sun and Moon somewhere between the two shells, as we do have pictures in Chapter 3 of the Sun over the Arctic region and it never goes below the horizon… yet we still have "hotspots."

Oh, but **now we have a problem**… no longer is the Sun 93 million miles away, and in the VR Sphere concept, the Sun and Moon rotate around the Earth – but at a greater distance than they would in the Flat Earth scenario.

> Now we have our first real **red flag.** In Ch. 11 of QES, it is shown that **the Sun changes size**: rises one size, and as it rises in the sky, it gets a bit larger, and then at sunset, it begins to diminish again… This is the behavior of an object that approaches you, and then passes by and then recedes (as on a Flat Earth).

> The time-release pictures show the Sun changing size as it would if it were closer <u>inside</u> the FE Firmament and first approaches you then recedes.

> The VR Sphere Sun is farther away and running thru the Cspace (<u>between the two shells</u> which are millions of miles apart), and while not 93 million miles away, nonetheless, **the Sun could also change shape – because it circles the Sphere at the same distance, and is seen to approach and then recede.**

The VR Sphere also does not rotate. The Moon is located in the Cspace.

Thank you if you are getting confused and having trouble following how the VR Sphere works and how **Gravity would have to be a programmed construct** – special laws for the VR Sphere – just to make the oceans hang on to the globe while at rest like this…

When you go to the beach, the ocean looks like this (below), right?:

Wrong. Water always rests **flat**, like this:

So there is a need for <u>more science</u> if Earth is a VR Sphere, because we are dealing with a sophisticated Simulation. Thus Occam's Razor does not apply.

And as said, **it is much easier to create the Earth Realm as a Flat Earth** – which is what all the Earth cultures said for centuries – until Western Civilization Science decided that Science should rule the day. This was also the beginning of "God is dead and Science is all you need."

> An excellent example of what this godless thinking does for a society was demonstrated by the **Soviet Union** – for 83 years. How did that work out?

Copernicus started it, Newton continued it, and again, Darwin inspired many people to see Earth and Man as the product of Science, not of a God that could not easily be proven... and ironically, **a non-globular Earth is the perfect proof for the existence of God (and His grace to Man)**. Design proves a Designer.

Conclusion

In arriving at a former conclusion favoring a Flat Earth versus a VR Sphere, and using the above arguments... it seemed appropriate to use **Occam's Razor**:

> Repeat: When confronted with several possible alternatives, it is usually found that **the simplest answer is the right one**.

> It would appear that the VR Sphere (as well as the convoluted idea of a real, 3D round rock circling the Sun as proposed by Galileo *et al*) is an exercise in **complicatedness**. It is not the simplest answer to the question: What is Earth?

> **In reality, it could appear that the Earth <u>was</u> flat and the easiest way to do it was as a Simulation because it can be reset if needed... and duplicated if needed: Earth Realm 1 becomes Earth Realm 2. Same programs, maybe a different set of players. It also minimizes the use of real estate. And 'inserts' (teachers, objects) can be done any time they are needed.**

If you suspect that **Earth is flat, and looks like below**, while the other stars and planets seen in the 4D Heavens appear to be round spheres... then Earth is indeed special and that means it was <u>created</u> for a special purpose. And that means Man is special, too.

Proposed Flat Earth Structure
(credit: Bing Images: whotfetw.com)

As the Hebrews said below:
<u>Please see Chapter 3 for a larger picture</u>.

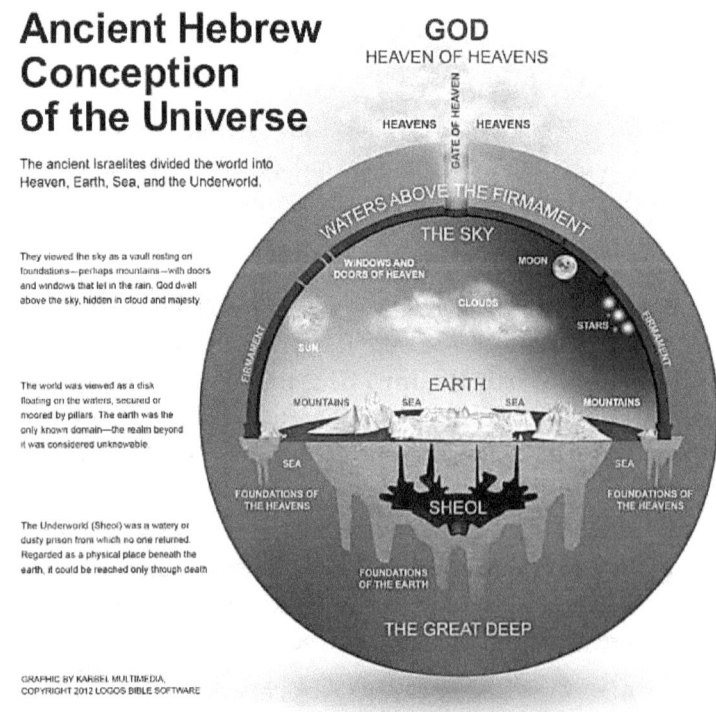

Hang out with this chapter's info and you may never see Earth the same way again.

Also think about these images…

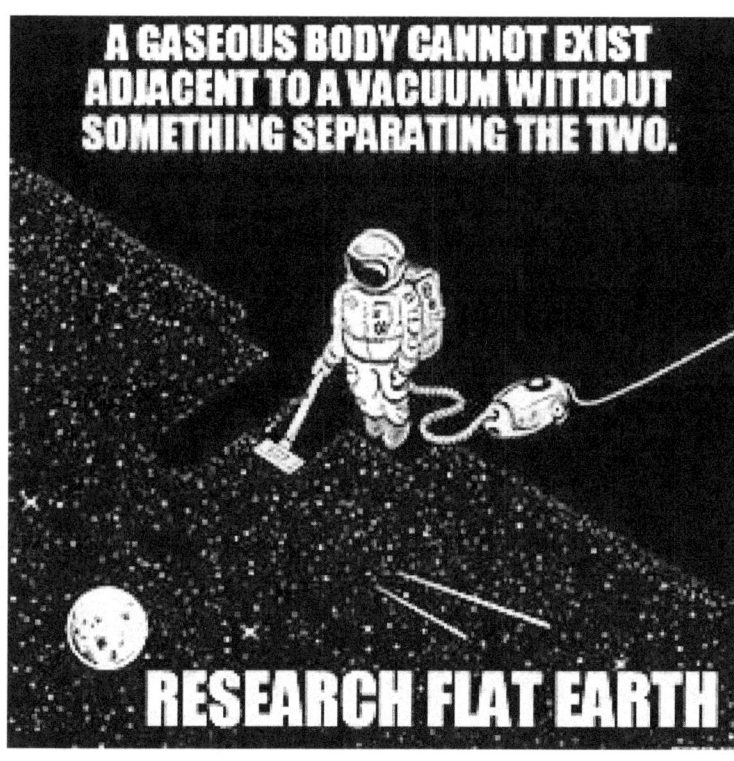

Rewind:
This plainly says the atmosphere on Earth cannot stay on Earth if the Space around it is a vacuum. Gravity is not dragging the clouds to the ground any more than the atmosphere is being controlled by Gravity...

Science is wrong again. This argues for a Firmament, or a Shell.

Again, you say that we have NASA pictures of a round Earth globe, from Space, but Ch. 11 QES shows where NASA **Photoshopped®** those. Case in point:

If you think these were taken from Space, I have some ocean-front property in Montana I want to sell you....

QES Ch. 11 has more.

Resolution #1

So to this point, it has been argued that the <u>original</u> Earth Realm was likely flat, as well as a Simulation, with a Firmament which amounts to a Quarantine of Man on the Earth – that is, we did not go to the Moon <u>because we can't get there</u>, and we don't have rovers on Mars for the same reason… Not a nice realization, if true.

And further study and analysis of the Flat Earth issue in ancient documents from centuries ago, reveals that the ancients not only saw a much lower Firmament (see Appendix E), but they knew there were "ends of the Earth" and often referred to the Four Corners of the Earth – just <u>as Enoch saw</u>. So we could simply accept that the Earth Realm used to be flat… **used to be** is not a problem.

What Earth Realm is <u>now</u> is the issue, and if we can resolve the issues of Gravity, and the issue of whether Antarctica is really a continent, then we'll have locked in <u>the</u> answer.

> Remember: the VR Sphere was my assumption, like Newton's assumption of Gravity. Perhaps we are both wrong.

The fact that water will not normally adhere to a round ball should be evidence enough that Earth is a Simulation, <u>or</u> that Earth is flat. We still have to resolve that. If Earth is a Simulation, then the oceans are **programmed in place holographically**. If Earth is flat, Gravity is not needed.

Also, the *Gegenschein* is proof of the Shell/Firmament around the Earth, and the fact that the unattached clouds do not rush overhead at 1000 mph (supposedly the rotational speed of Earth below them – see Chapter 3 statistics), and the fact that Earth's atmosphere does not get sucked off into the vacuum of space suggests that the Shell that NASA discovered (Appendix E) is doing its job.

The problem is folks that the PTB want you to think we can get out of here (space travel – on which we are spending millions of dollars!) and that ETs will visit us (Project Bluebeam – see Chapter 10) , and that Earth is not a School… and there is no God…
"so have fun and do whatever you want… no one is watching."

Wrong.

Chapter 10: How It All Fits

This chapter will summarize what we have been talking about, show how some of the Flat Earth 'proof' supports the VR Sphere, draw some conclusions, and if you thought <u>Virtual Earth Graduate</u> had some challenging things to think about, hang on to your hat.

You do not have to believe what is about to be presented …
Treat it like **Brain Candy**… something to think about.

I now believe because I have been at it for over 50 years. I was a "true seeker" and asked to know what this place really is and could I make a positive difference in the world? I was then visited (as Chapter 1 shares) and I was given the answers <u>on the condition</u> that I write it down in a book and make it available. I responded <u>with the condition</u> that if I wrote the book(s), They would **tell me the truth and protect me** (there are those "out there" who, throughout history, have squelched the truth and some people have disappeared for telling the truth – and Giodano Bruno was burned at the stake.) So I wanted protection. They responded with <u>another condition</u> by saying that Their protection meant I could not promote the books (Chapter 1).

> And by the way, I hasten to point out that I am not that happy with what I learned... there is **a serious down side to Earth**. I wrestled with it and tried to prove it wrong, but am forced to consider that what follows is probably the truth. I suggest that you read it, contemplate its meaning for you and your life, and stick the idea up on the shelf or do the best you can with the realization. At some point, we all have to become Earth Graduates... no matter what Earth's physical shape is...
> If you treat others better than you treat yourself, you will have in large measure succeeded in acquiring a good portion of the Earth Graduate qualities for graduation.

Ok, so it is best to start at the beginning and develop these points as we go.

In The Beginning

A long time ago, the God Hierarchy (shown in Chapter 5) was sitting around and discussing what to do with the souls who were becoming dysfunctional, damaged and wayward. That is, after numerous experiences in various realms, some souls were becoming so hedonistic that they were into 100% partying, and not growing in Knowledge, Patience, Compassion, Respect and Humility… and some were becoming dysfunctional. Others had had such a rough time in tough **Lifescripts** that they had been traumatized and were carrying emotional baggage and mental scars which also made them dysfunctional, or unfit, to handle future lifetimes. Many souls could no longer be "imprinted" with the appropriate energy, life patterns and the connection to their Higher Self was either closed or damaged such that they could not receive what has been called intuitive guidance. (Much of this was covered in TOM.) Something had to be done.

It was decided to create a special environment to try and nurture, correct and guide them – and it had to be free from interference from other 4D and 5D beings. The **Earth Realm** was created as a completely supportive environment, a beautiful world (<u>not a planet</u>) with flora and fauna, and occasionally a mysterious structure or creature ("Easter Eggs") to get Man's attention and make humans stop and wonder. In addition, advanced Teachers would be placed ("inserted") among them to provide moral/ethical guidance.

The Earth Realm was created as a **3D Construct** necessitating power, Laws, and coordination of its many facets that could only be done with the greater power in 4D – but Earth had to be self-contained <u>with no outside interference</u>. This was so that the humans on Earth would interact just among themselves and in a context of their equals (plus some divine guidance from time to time) so they might have experiences that proactively shaped them into better, more balanced souls. If the more wayward souls acted in very inappropriate ways, their interactions with each other would result in immediate feedback and 'lessons' in behavior from their peers! (That is called fighting and wars – but it would not disrupt the rest of the Creation in 4D and above – squabbles and rough lessons would be contained on Earth… within the **Firmament**.)

> **By the way, the damaged and wayward souls constitute what has been called 'The Fall of Man.' Fallen to Earth for correction – one could say "due to sin" but the original meaning of "sin" was just <u>missing the mark</u>… inappropriate behavior and that is what Earth was designed to correct.**

A probable layout of Earth <u>as originally created</u> could have been the following: (and is very much what Enoch described in the **Book of Enoch,** Chapter 6….)

The Probable First Earth Realm

Is this the ultimate IMAX theatre?

Inquisitive Humans

There was, however, a problem that had to be resolved in the design of the original Earth: humans are inquisitive and do explore and they do travel about examining their surroundings, so certain constraints had to be placed on Earth such that Man could not go too far, nor too high, and begin to discover where he really was.

As was suggested by Mark Sargent [263] very perceptively –

> If you put a hamster in a glass cage, he will run around and explore every nook and cranny to see if there is an exit. Finding none, he may try again the next day, but will eventually give up.

> If you put a monkey in a large enclosure, with trees, bushes and food, water... he will run around and examine his world and see if there is a way out, or at least other rooms off of the main one. Finding none, he eventually gives up.

> If you put the monkey in a forest preserve that is 1000 acres in size, as long as there is food, water and others of his kind, he may or may or may not explore and see what the limits of his little world are.

So if you put Man into the Earth Realm, which happens to be about 8,000 miles in diameter, and 25,000 miles in circumference, he has a lot of room to wander and move about. But, you do not want him climbing the tallest mountain and discovering the Firmament so you thin out the oxygen as you go higher up the mountains – to discourage **vertical exploration**, and you drop the temperature. All but the most determined will be discouraged.

And to keep Man from exploring the edges of his world, knowing he needs water to survive, you'd make the oceans salty so he does <u>not</u> have a constant supply of fresh drinking water as he sails to the edge of the world. You also drop the temperature and put icebergs in his way, and when he does get to the edge, he is confronted by a massive 200-300' tall **Ice Wall**.

Now, the Firmament has to connect to something, to seal everything inside, so you'd **set the base of the Firmament into the Ice Wall** – but back about 300+ miles over ice, snow, in below-freezing temperatures, frostbite, and howling 200 mph winds to discourage all but the most foolhardy. Then connect the base of the Firmament to the rock under the ice.

For a similar treatment of this theme, see *The Truman Show*.

Thus, Man is going to largely stay in the more temperate zones where there is food and water and he is more comfortable. Meanwhile, let's take a look at the Antarctic region and some of its history and characteristics… the nature of the Antarctic is critical to whether we are on a Flat Earth (with an Ice Wall), or whether the Ice Wall was morphed into a continent at the bottom of the Earth Realm…. now the VR Sphere. The Arctic and Antarctic have very different histories and characteristics.

Antarctic Puzzle

As an interesting aside, the Germans were quite fascinated with Antarctica and built a small hideaway there in the 1930's accessible only by submarine. Before the end of WW II, this is where many top Nazis escaped to, and then discovered the Antarctic environment was germ-free and after a couple of years there, their immune systems went on sabbatical because there was nothing to fight.

Later as they went back into the real world, e.g., South America, they had horrific health issues – even the common cold was a life-threatening experience for them as their immune systems had shut down and had to re-activate all over again! [264] The Antarctic Base 211 in **Neuschwabenland** was seen as a hardship assignment, and not the Shangri-La it or Redoubt was supposed to be.

The Nazi **Thule** and **Ahnenerbe** groups had taught that the Aryan race originated in a land of ice and snow, and while that was allegedly at the North Pole, perhaps **Hyperborea**, it made sense to the Germans to develop Antarctica as their own. No one would bother them down there. They also had bases in northern Canada and Greenland in the ice and snow. Said the Germans,

> Our Nordic ancestors grew strong in ice and snow; belief in the
> Cosmic Ice is consequently the natural heritage of Nordic Man.[265]

It was natural that into this set of ideas should come an Austrian engineer, inventor and astronomer, **Hanns Hörbiger**, who electrified the Third Reich with his further ideas of Germans and ice and snow:

> **Welteislehre** (**WEL**; "World Ice Theory" or "World Ice Doctrine"),
> also known as **Glazial-Kosmogonie** (*Glacial Cosmogony*), is a
> discredited cosmological concept proposed by Hanns Hörbiger...
> No effort was spared in popularising the ideas: "cosmotechnical"
> societies were founded, which offered public lectures that attracted
> large audiences, there were cosmic ice movies and radio programs,
> and even cosmic ice journals and novels. [266]

> The basic idea was that everything has its basis/roots in ice… which
> was frozen oxygen and hydrogen -- life-giving elements.

Himmler stated that the theory fitted nicely with Germanic and Teutonic concepts and if more science could be brought to bear on the WEL Theory, it would be easier to accept. On the other hand, **Hitler** had a more insightful comment, accepting WEL as the Nazi Party's official cosmogony, but noting that Hörbiger was not accepted by the scientific establishment because **"the fact is, men do not wish to know."** (And that was also been a reason to reject the Flat Earth or the VR Sphere as well....)

> It has been said that the real reason both Hitler and Himmler favored the idea was to counterbalance the perceived Jewish influence on the sciences, similar to the Deutsche Physik movement. Hörbiger's WEL was, for instance, opposed to Einstein's theory of relativity.
> [and yet, Hörbiger hit it right on the head with another insight...]

> > "I knew that **Newton had been wrong** and that the sun's gravitational pull ceases to exist at [1/3] the distance of Neptune," [267] [emphasis added]

...and he was right. **The Sun does not exert any gravitational effect on the Earth**, nor does any centrifugal force get involved. In addition, the Germans used a different type of Physics, ignoring Einstein and redeveloping their Physics which resulted in The Bell (*Die Glocke*) and anti-gravitic propulsion, and Viktor Schauberger;'s *Repulsine* breakthroughs – see Chapter 4 in VEG.

Later, in Chapter 4 (in this book, in the section dealing with Einstein), **Ben Rich** corroborated the fact that **Einstein was wrong** and that when Lockheed engineers corrected the equations, they also found the secret to electro-gravitic propulsion.

And now we come the crux of this polar issue... Ancient depictions of Earth show a Flat Earth with an **Ice Wall** surrounding it all and the older standard Flat Earth maps show no Antarctica ... see below...

So where is Antarctica? Does it exist, and what are the characteristics of the Ice Wall? Be clear that the UN flag (p. 355) also does not show Antarctica...

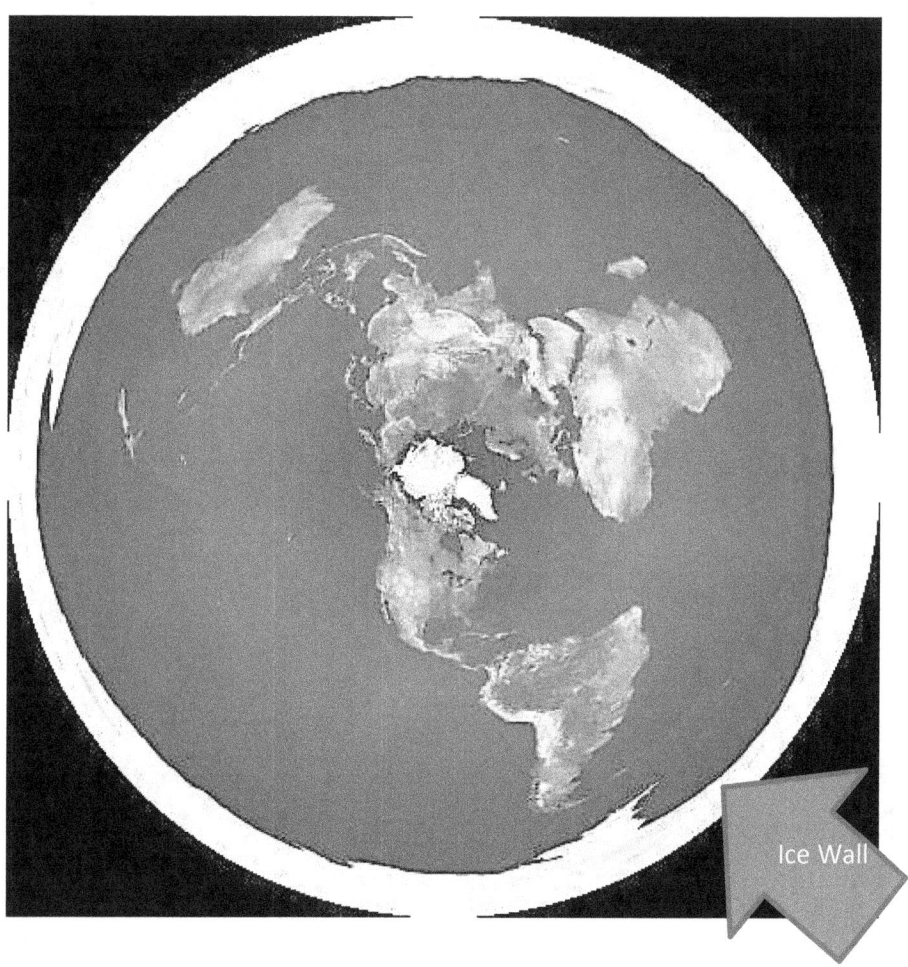

Ice Wall

Supposedly, from different sources on the Ice Wall, **the distance from the ocean to the Firmament, across the snowy terrain, is about 300-400 miles**. Thus there is plenty of "land" to play with and explore… mostly snow and some mountains. It is in the initial 5 miles that the Germans allegedly constructed Base 211.

In order to confirm or deny the existence of Antarctica as an Ice Wall, we would have to look at a very interesting FE issue… and this involves one of the more lengthy, but worth it, 'proofs' to show. It is one of the **key FE proofs**.

Southern Latitudes Inaccurate

What has been discovered is that the **distance** between the tip of South Africa and the tip of South America looks fairly close on a globe: **arrow A**. See below…

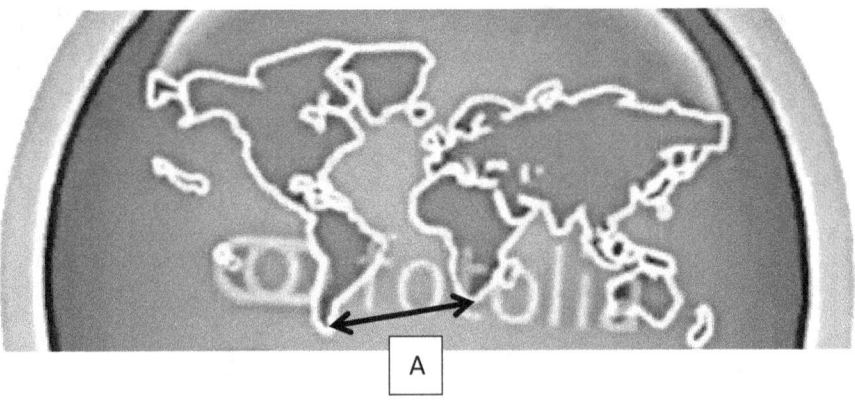

But when you see the two countries on a flat map, countries about the same size... there is quite a distance between the two! Distance A is not the same as distance B.

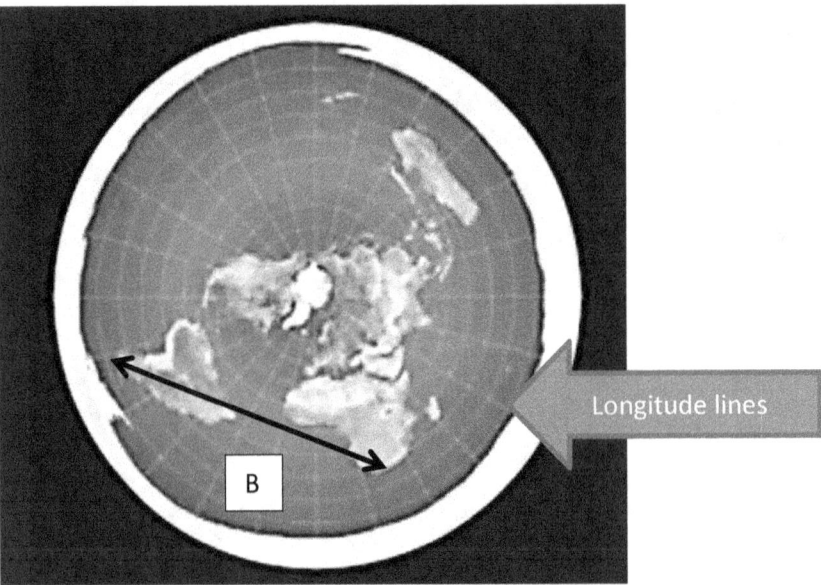

This is significant, and is not a trick. And **mariners know that distance B is the correct one.** (See later section, this chapter, Arctic vs Antarctic.)

Notice that the **longitudinal lines** (from the center to the edge) fan out and the distance between them gets farther apart...whereas the same lines on a globe, while curved, fan out to the equator and then fan back in again... The lines of **latitude** (horizontal lines) pretty much stay the same.

The longitude lines fanning in and out is a nightmare for sailors, and it isn't accurate... **they use a flat map**. So why do we have a global view? Because Galileo, Copernicus and Sir Isaac Newton said the Earth was a sphere.

During Captain James Clark Ross's voyages around the Antarctic circumference, he often wrote in his journal perplexed at how they routinely found themselves out of accordance with their charts... they found themselves an average of 12-16 miles **outside their reckoning every day**, some days as much as 29 miles [off course]. [268] Now we know why.

> In the southern hemisphere, navigators to India have often fancied themselves East of the Cape [of Good Hope] when still west, and have been driven ashore on the African coast which according to their reckoning lay behind them. This misfortune happened to a fine frigate, the Challenger, in 1845.... How have so many other noble vessels perfectly sound, perfectly manned, perfectly navigated, been wrecked in calm weather, not only in dark night, or in a fog, but in broad daylight and sunshine... from being **'out of reckoning'** under circumstances which until now have baffled every satisfactory explanation. [269]

So the Earth until recently was not a globe, and the honest scientists of the 1800s have said so:

> If the Earth were a globe, the distance round the surface, say at 45° South latitude [see next page], could not possibly be any greater than <u>the same latitude</u> [in the] North, but since **it is found by navigators to be <u>twice the distance</u>** – to say the least of it – or **double the distance** it [should] be according to the globular theory,

it is a **proof that the Earth is not a globe.** [270]   [emphasis added]

Again, the following map shows the problem:

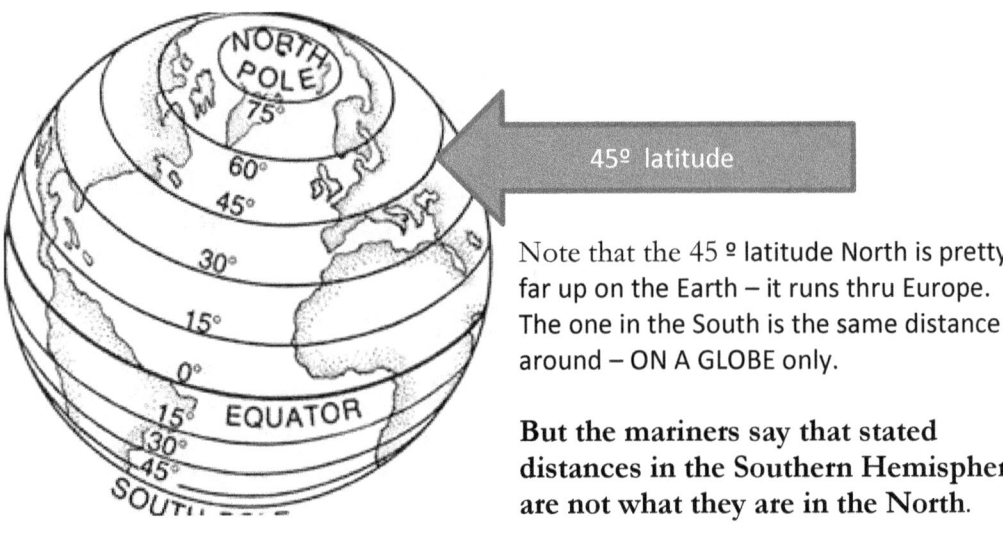

45º latitude

Note that the 45 º latitude North is pretty far up on the Earth – it runs thru Europe. The one in the South is the same distance around – ON A GLOBE only.

But the mariners say that stated distances in the Southern Hemisphere are not what they are in the North.

And again, because the globular map is deceptive....

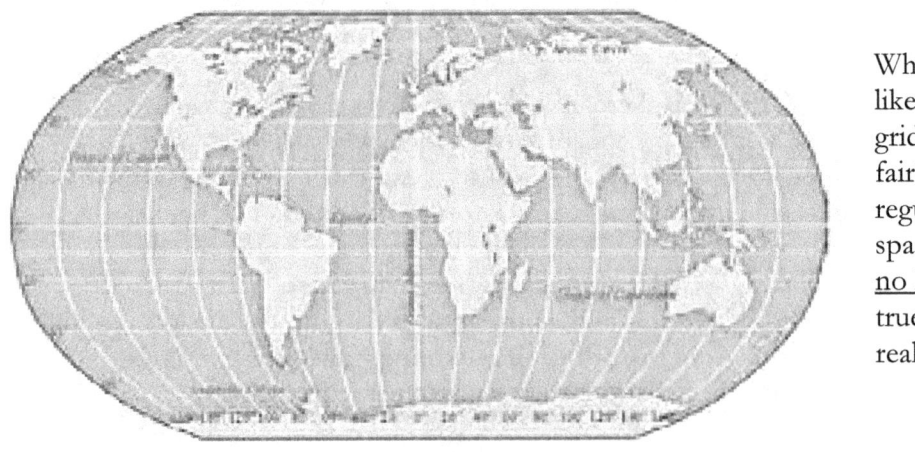

What look like nice gridlines, fairly regularly spaced… is no longer true in reality....

This issue is reviewed and resolved in Chapter 10.

What the land mass spacing really looks like is the next map… Does it look familiar?

Does the **UN Flag** tell the truth?

Compare it to the FE map below…

And now the last one – showing the Sun/Moon orbits above and below the equator during the year…

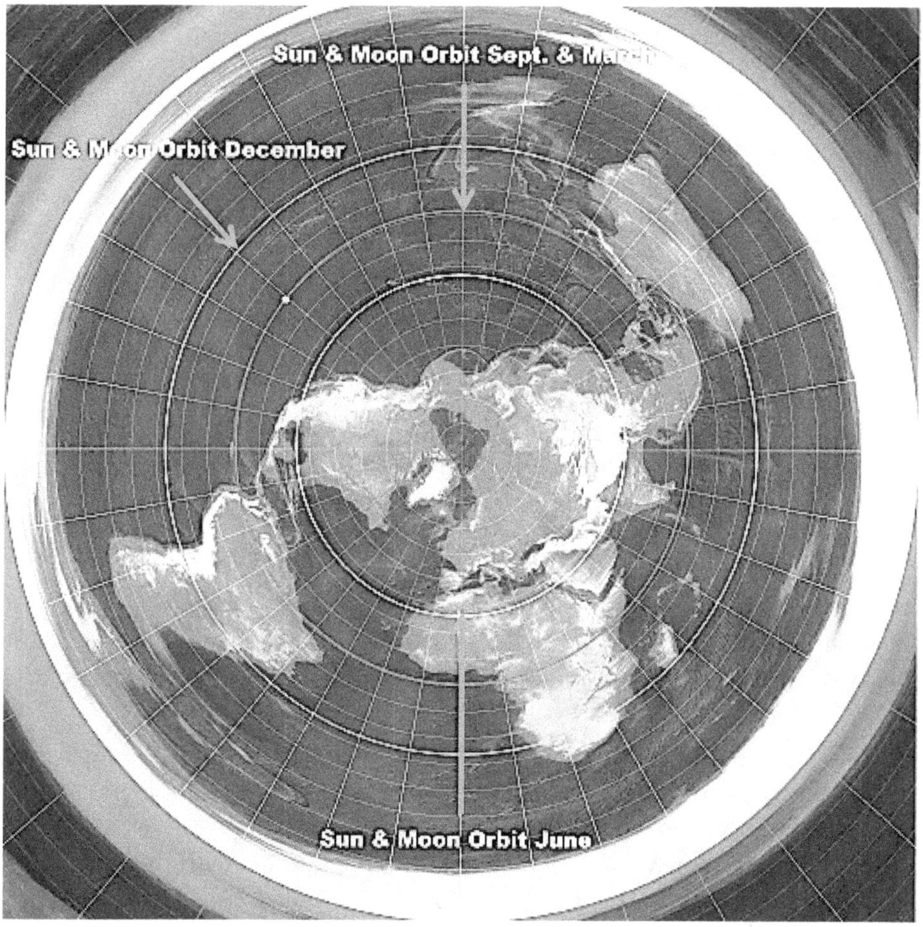

And the preceding diagram shows the "continent" of Antarctica to be the Ice Wall (white border) ... which is reached by traveling due South from anywhere in the world.

Newton's theory of the **globularity of the Earth** is only a *supposition* and *assumption,* and yet by Modern Astronomers it is paraded about as if it had been a true deduction from exact experiment. [271]

Hang in there, we are just defending the FE Theory, and showing how it used to be... and that some of its aspects support the VR Sphere.

Arctic versus Antarctic

The last evidence is something that we the general public don't know but the Earth Scientists know it. If the Arctic and Antarctic regions were areas of comparable latitude, <u>and</u> if the Antarctic continent really existed at the South Pole, then we could expect **similar conditions in both places**: similar temperatures, seasons, length of daylight, and similar plant & animal life. However: [272]

The same number of plants and animals should be found and the same general conditions should exist. That **the very opposite is the case** disproves the globular assumption.... **Antarctica is the coldest place on Earth** with an average annual temperature of -57° F ... whereas the North Pole is a comparatively warm 4° F....

The Northern Arctic region enjoys moderately warm summers and manageable winters, whereas the Southern Antarctic region <u>never</u> even warms up enough to melt the perpetual snow and ice.... **The Antarctic has no seasons**...

In the Flat Earth model of the cosmos, these Arctic/Antarctic phenomena are easily accounted for and exactly what would be expected [see seasonal map on the preceding page].

Heliocentrists [those who believe the Earth circles the Sun]... cannot explain why the **Midnight Sun** [Arctic] phenomenon is not experienced anywhere in the Southern Hemisphere <u>at any time of year</u> (despite the disinformation on Wikipedia).... Do not confuse this with the **Aurora Borealis and Australis** which <u>do</u> exist at the 'poles' – and one changes in synch with the other.

[the following is a corker, please continue and you'll see the nonsense being played on you by established Science...]

> In typical reverse-engineered damage-control fashion, trying to explain away the Midnight Sun, problematic Arctic/Antarctic phenomena, and the fact that **Polaris** [the North Star] can be seen approximately **23. 5 degrees south of the Arctic**, desperate heliocentrists in the late 19th century again modified their theory to say the **ball-Earth actually tilts back 23.5 degrees on its vertical axis**, thus explaining away many problems in one swoop! If it simply tilted the same direction constantly, however, this would still not explain the phenomena because **after 6 months of supposed orbital motion around the Sun, any amount of [forward] tilt would be perfectly opposite[backward]**, thus negating their alleged explanation for Arctic/Antarctic irregularities (see page 113 diagram).

[it gets better...]

> To account for this, heliocentrists added that **the Earth also "wobbles"** in a complex combination of patterns known as, "planetary nutation," the **"Chandler wobble,"** and "axial precession" which, in their vivid imaginations, somehow explains away common sense.

[have you got your wading boots on yet?...]

> Common sense, however, says that if the heat of the Sun travels 93,000,000 miles to reach us, a small axial tilt and wobble, the difference of a few thousand miles, should be completely negligible. If the ball-Earth actually spun around 93,000,000 miles from the Sun, regardless of any tilt or wobble, temperature and the climate <u>the whole world over</u> should be much more uniform. [It isn't.]

> Common sense also says that if the Earth were actually a ball spinning daily **with uniform speed around the Sun**, there should be exactly 12-hour days and 12-hour nights everywhere all year round! The great variety in length of days and nights throughout the year all over Earth ["shortest day of the year" vs "the longest day of the year" in equinoxes] testifies to the fact that **we do not live on a spinning ball-planet.**

[rewind...]

> It is said that the rotation [of Earth] takes twenty-four hours and that its speed is uniform, in which case, necessarily, days and nights should have an **identical duration** of 12 hours each all the year round. The Sun

should invariably rise in the morning and set in the evening **at the same hours**…. One should stop and reflect on this before saying that the Earth has a movement of rotation. How does **the system of gravitation** account for the seasonal variations [e.g., daylight savings time] in the lengths of days and nights if the Earth rotates at a uniform speed in twenty-four hours? [273] [emphasis added]

Sorry that was long, but it so perfectly makes the point for a Unique Earth scenario that paraphrasing might lose something of the 'fire' of the original. You should be asking yourself : If there was a lie, <u>why was it begun</u>, and <u>why has it continued?</u> – if the **UN flag** shows the earlier truth and **NASA** (remember their wall Orbit Tracking screen and the "S-wave" orbits and what they translate into? in Chapter 3) seems to not show the truth … Why the disinformation? (See this chapter's end section.)

Even more amazing is the convoluted explanation involving tilt and wobble… that never sounded right even when I was in Astronomy class. Do the scientists actually believe all that – from a Newtonian, unproven <u>assumption</u> and misunderstanding of Nature? Tilt & Wobble and rotating at 1000 mph do not apply to the VR Sphere.

Sir Water Scott said it best:

> **Oh what a tangled web we weave…**
> **when first we practice to deceive!**
> *Marmion, Canto vi. Stanza 17.*

Project Bluebeam at this chapter's end suggests a possibility for <u>why the deception</u> (not really a lie, *per se*) exists and why it continues. It is suspected that Man may be psychologically better off thinking he is on a round rock circling the Sun, with nothing to stop him from going to the distant stars…at least it employs hundreds of thousands of people in the Aerospace Industry, and encourages young people to learn Science (wrong or not)… Maybe Man should not know where he really is … but wouldn't the FE knowledge change Man's violent behavior and promote more brotherhood and cooperation? (Appendix C.)

Flat Earth Idiocy

Whereas the ancient cultures knew the Earth to be flat, <u>and they were still violent</u>, they were also psychologically 'programmed' by those who knew, to not venture too far from home, as they were told that there were **monsters in the oceans**, devouring ships who dared to sail out too far. And the worst fear instilled in Man was that the Earth just ends, the edge drops off, and you could fall off the edge of the Earth…. carrying a potential truth way too far! Thus for explorers like Christopher Columbus, Coronado, De Soto and Magellan, they were seen as very brave (or foolish) to sail too far into uncharted realms. The following was a childish understanding of the Earth Realm as one gets to the edge:

The wonders of Photoshop! ®
(credit: Bing Images: revelationnow.net)

…and again (thanks to Photoshop ® and Bing Images):

Sail into oblivion?

And yet, as has been reported, the **Phoenicians and Vikings** did exactly that -- 'way before Columbus. The **Chinese and Japanese** also boldly ventured into the unknown, as did **the Atlanteans**.

Rewind: Circumnavigation

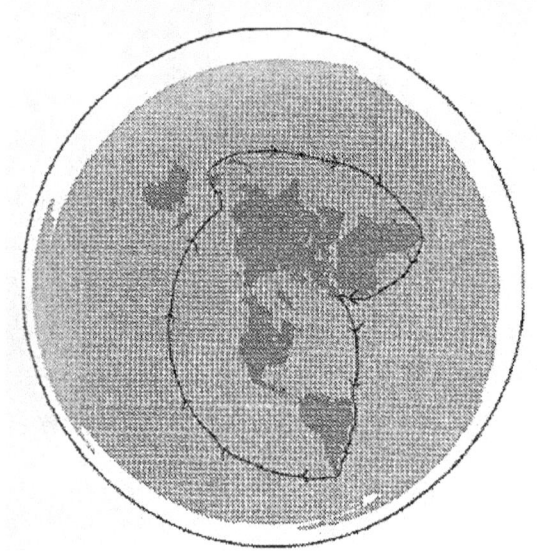

Then someone got the bright idea in AD 1519 to see if the Copernican Theory (AD 1514) of a round Earth was true and they figured if it were true, then one could sail around a round globe and come back to the starting point. That was **faulty thinking**… see left:

It did not prove the shape of the Earth; **you <u>could</u> sail around a Flat Earth too.**

So at any rate, **Man was not to know early on just exactly where he really was** as the gods wanted to see what he would do if he thought (1) he was all alone, (2) he could do whatever he wanted, and (3) he was not being watched. The gods knew that eventually he would master the air and take to flying, perhaps firing rockets into space (or try to), and develop telescopes that all let him believe that he was on a globe, spinning around the Sun.

At any rate, back to the story…

Controlled Access

Since the Earth Realm was sealed (Firmament), and there could be no unauthorized interaction with the humans on Earth by anyone outside the Earth Realm, the only way in was to incarnate into a body on Earth, and dying was the only way out. Of course, the gods, avatars and teachers could materialize in/out of the Realm at any time.

One more thing was needed... a way to protect and guide the humans so that they didn't suffer an accident that prematurely took them out. And so that they didn't completely waste their time, Angels (or **Beings of Light**) were assigned to the Realm to watch over and guide people. Of course, they would have to be invisible as the dictum for Earth School has been (until these books) is that Man must think he is all alone – that way, when he has a decision to make, he won't make it <u>because someone is watching</u> – he'll do what his inner voice tells him to do and that way the gods can see just how much he has learned. Does he make wise decisions, or is he still fooling around? Staying invisible allows the Angels to objectively evaluate a soul's progress.

> Sometimes souls are 'counseled' at night when they are asleep, and they can be given dreams, hunches and visions to guide them.

Angels do not have wings – that is why I call them **Beings of Light**. They look like this (left):

The human soul looks very similar, but Angels are brighter.

(TOM examines this in more detail.)

When souls leave the body (at death of the body and during NDEs), they are supposed to go to the Light ...back to the **InterLife**... and a **Tunnel** manifests to guide them...It is the "extraction route" from the Earth Realm. Usually there is an Angel to guide the soul back to the InterLife... via the Tunnel that protects the returning souls from Astral interference (see Ch. 5-6 in VEG).

> It is thought that the Tunnel may use one of Enoch's portals to access the InterLife, a lower region of heaven that even Enoch saw.

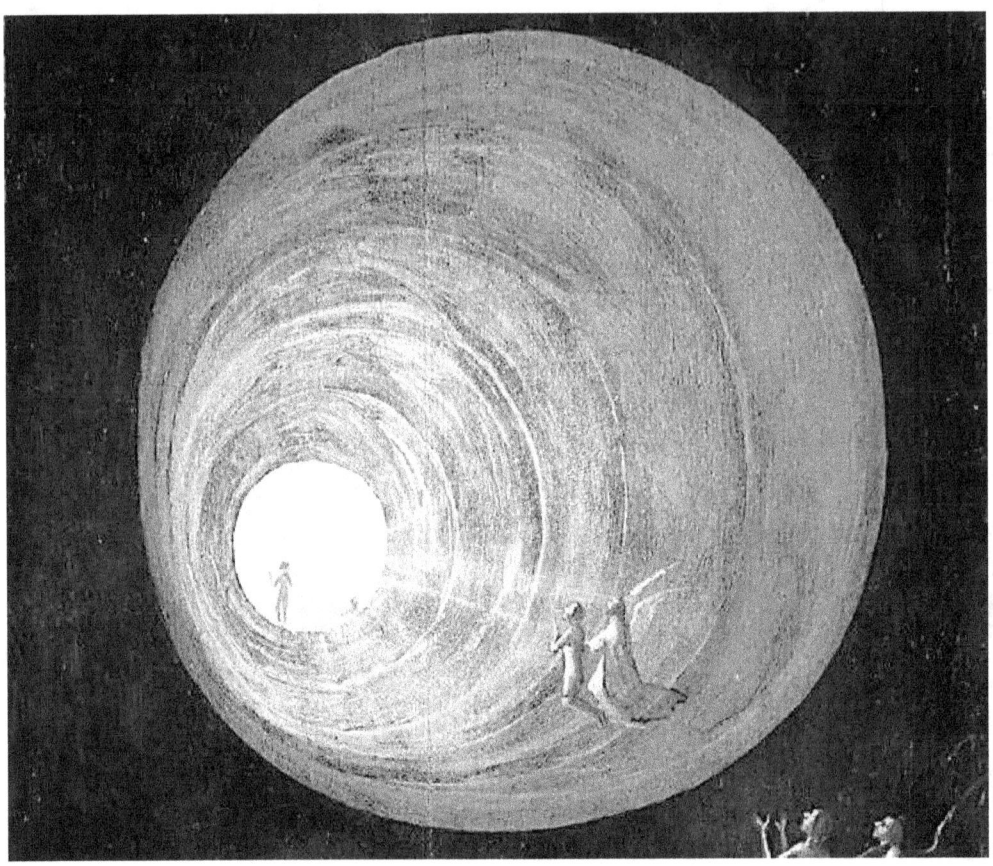

Angels are often pictured with wings to help identify them (and in a painting it suggests they can fly).

Downside

What possible drawback could there be to living in a Realm that was created for Man, protected by Angels and a Firmament, and fulfills the **"Goldilocks Zone"** stated by scientist Paul Davies (below)?

The Universe Design

According to some physicists, **the universe should not exist** (like Gravity?) – If there was a Big Bang (and this book and several Physicists do not support that), then **equal amounts of matter and antimatter should have been created, and they would have annihilated each other.** But, according to theory, there must have been a greater amount of matter than antimatter, thus permitting the universe to exist as we know it. Really? Was that by Design? It sounds invented...

According to Paul Davies **(The Goldilocks Enigma)**:

> One of the deepest puzzles of cosmology [is] the origin of matter. Cosmologists want to know exactly how it happened and why that particular amount (10^{50} tons in the observable universe) got made. **When matter is made in the lab, by high-energy collisions, the same quantity of antimatter appears, too.** [274]

So how did the alleged Big Bang make 10^{50} tons of matter without also making the same amount of antimatter? (Hint: There was no Big Bang.)

Remember, the scientists are studying the **Imax Theatre**.

Secondly, **the universe as observed is not all that friendly a place** – consider that stars explode, Black Holes gobble whatever comes their way, and the universe is filled with deadly gamma rays. **So how did the Earth come to be located in the only place discovered in the universe where life can exist?** Scientists call this the *anthropic principle*. Earth is located in a highly atypical place – even unique compared to many other solar systems and galaxies…. The conditions for life are very restrictive and we just happen to find ourselves located in a special set of circumstances…. Gee, how coincidental.

> It never dawns on them that Earth might just be a Realm that was created centuries ago… and for a similar reason they discounted the **Æther**… because Einstein said it couldn't exist, and later scientists (in VEG Ch.8) proved that it does exist, and so today it is called Dark Matter/Energy… Don't you love the obfuscation?

The *principle of mediocrity* says that there is nothing special about any part of the Universe – what you see is what you get – sameness largely everywhere! And that is not true of Earth's location. [275] Earth has just the right balance of features to support life – as if it were designed to be that way. Gee, really?

Goldilocks Zone

Earth is located in a Goldilocks Zone – not too far or too close from the Sun such that we live in a temperate zone… just right for life. **Physicist Brandon Carter** theorized that **if any of the laws of our existence were just a bit different, we would not exist.** It is as if the Earth, and its location, and its ecological balance, chemicals and minerals were **all perfect for life to arise here.** [276] Like Goldilocks' porridge, it is all just right. And **that is auspicious and suspicious**, since that kind of thing isn't found on a regular basis in the world, let alone in the solar system, or in our galaxy.

As if by design: Science says oxygen to breathe, an electromagnetic shield to protect Earth from the Solar Wind and gamma rays (Mars and Venus do not have one), Gravity that isn't too strong nor too weak (but somehow keeps heavy oceans from falling off the alleged globe!), an ozone layer to protect Man from ultraviolet radiation, and a tilt to the Earth's axis to generate seasons, and a Moon to cause tidal action of the oceans. And the Moon exactly eclipses the Sun…

What if it <u>was designed</u>, but the Science facts (just listed) have been misinterpreted?

And still think it all just happened by coincidence…? The scientists tend to marvel and assume that we just evolved this way… They cannot say 'miracle', 'God', or 'Flat Earth' (or 'soul') so they stay puzzled. Is this a subconscious desire to keep the **Game of Not Knowing** going so that they have jobs and can do more 'research?'

And yet, there is a downside to this Realm – You ain't going anywhere!

Illusions

If this Realm is really a Contained 3D Construct (as Enoch and many others say), there are times when even I don't want to hear that. It means the following are **false**:

> The 1969-1972 Apollo Moon landings
> Asteroids endangering the Earth
> Dinosaurs wiped out by an asteroid 65 million years ago
> NASA has rovers on Mars
> Earth has been visited in the past by ETs and their UFOs
> Crop Circles are made by aliens
> Roswell (1947) was a crashed alien UFO
> > however:
> The Battle of Los Angeles (1942) had to be the Germans
> > (they were the only ones with antigravity craft at the time)

And because this Realm would be a **controlled environment**, it means the following is also false:

> There is an end to time
> The Sun will eventually die
> Man can destroy the Earth (Apocalypse)

What is more likely is that Earth will go thru **Eras** with "Wipe & Reboots" when things get out of hand... and the gods will terraform the world environment, reset everything to a clean balance, and restart the human experiment. As was done in AD 800 and again in AD 1500 (see VEG, Ch. 10.)

And therein lies the downside, what many might feel to be a negative aspect of the Earth Realm.... VR Sphere or Flat Earth – **we can't get out of here** – we are not going to the stars, despite *Star Trek* inspiration. Our diseases, pettiness and violence are not wanted "out there" among the galaxies – which are probably projected in a sophisticated way onto the Firmament – so that we <u>are</u> seeing the real stars and galaxies but the *Gegenschein* reminds us that

we are contained.

Perhaps this is what motivated the Church and other authorities of their day (about 300 years ago) to agree that Earth was a round globe, rotating at 1000 mph, spinning around the Sun at 66,000 + mph and wobbling on its axis, etc. etc. etc.....
Perhaps the idea of confinement was claustrophobic to some and dampened the human spirit to realize that there was no point to reaching for the stars... no point to Science (**what good is analyzing the IMAX Theatre?**)... and no point to War (which is in Man's Anunnaki genes).

And it is worse than that.

If NASA were to admit that we didn't go to the Moon, and rovers are not on Mars, the whole **Aerospace Industry in America would collapse** (and somewhat in Europe with the European Space Agency, and those in China and Japan) ... just go out of business, putting thousands of people out of work... All the ancillary, support companies that make things for NASA would also go bankrupt... We would have built a house of cards that would come tumbling down... if this scenario is true.

> But, you say, look at the astronauts – surely someone would have exposed the secret?! Not so – many astronauts **were military (sworn to secrecy under penalty of court martial or death)** and those who were not military were **Freemasons – who take a similar oath** to not divulge what had to be done.

> What do you mean "what had to be done?"

> Simple. Consider this possibility: **JFK committed us to go to the Moon**, despite the Firmament, despite the Van Allen radiation belts, and to **save face (national honor)** the US had to make it look like we went to the Moon. And as a corollary, neither Russia nor China would attempt to go to the Moon. They also know what Earth is. They haven't tried to go.

If that was done, why would the astronauts reveal the truth? History makes them heroes, the event entertains the public, and gives Man a future… and it sustains National Honor.

And it also gives a focus to scientific development… pushing back the frontiers with a specific goal of space travel… Man will learn more about himself and the world.

The only problem with the deception of where we really are is that Man does not know **it is important to learn spiritual values and do the right thing**, he is being watched, and he is expected to overcome his 'shadow side' and become an **Earth Graduate**… regardless of whether the IMAX Theatre is flat or a VR Sphere.

Second Downside

The other downside in all this is that **there is just so far we can go with this "global" charade…** Eventually people and organizations (like SpaceX) will discover that they too have wasted their money trying to get off the Earth… unless they are clever, and sell tickets on the greatest 'E' Ride ever (remember the old Disneyland E Rides – the best in the Park, circa 1957?)… Blast off in a rocket, put a slight curve to the porthole windows and the edge of the Earth from 150 miles up will have a noticeable curve to it, and then no one is the wiser… and a VR Sphere would have a round edge to it anyway.

But there is only one place this can all go.

Since we can't really get off the Earth (due to the Firmament/Shell), we will have to fake being contacted and visited by ETs, and then there will be just an elite group of leaders and military humans who interface with them. The average person will still not know where we really are, but they will think Space Travel is now a reality, and that we are traveling to other planets with a Hyperdrive given us by the putative ETs, but somehow (when the average Joe and Jill apply to join the Elite Corps, they don't qualify)… remember:

> All pigs are equal, but some pigs are more equal than others.

And that leads us into the **Project Bluebeam** possibility.

Why Disclose This?

So if there are downsides to knowing about the real nature of the Earth Realm, why reveal it?

First, not everyone wants to know, but those who <u>are</u> looking have reached a point in their spiritual growth where the realization of God's Creation and His Grace to Man will empower their growth;

Second, we will reach a point where the disinformation will become apparent and the only way to sustain and control the deception is to "lock things down" (Martial Law) and keep the populace entertained with electronic geegaws, movies, and then ET space aliens that don't really exist;

Third, not everyone would buy the Containment revelation, preferring their comfortable beliefs (see Chapter 2), or the VR Sphere, and thus the secret will sustain itself as most of the populace laugh off the 3D Earth Construct scenario (and UFOs);

Fourth, disclosure reveals that the Bible is true – Would you like to know if The Word of God is true or false? The Bible/Enoch says the Earth is flat, floating on the Waters of the Deep, with a Firmament, and knowing that may keep you from a nervous breakdown (see next section: aliens and 'Project Bluebeam');

and

Fifth, there is a serious problem for humanity because they do not know the truth – remember Truth will set you free? **If you do not know where you are, there is no incentive to get out!** And the gods do want you to Graduate!

And the last two items bring us to the last section in this chapter. Now you will see why the **Quarantined Earth has to be disclosed** and what is likely to happen…

Project Bluebeam

Having said that there are no ETs and the UFOs that are seen are not being flown by ETs (because they can't get past the Firmament), that means that

> **most UFOs that we see are made on Earth …and flown by humans, <u>but</u> a few belong to the Ancient Ones and the Anunnaki (Naga) Remnant.**

In addition to the public being 'programmed' to snicker at the Flat Earth scenario, they also have been conditioned to snicker at the idea of UFOs. And **the two were related**. UFOs are very real and the humans are busy building theirs, and flying them (as VEG Ch. 4 revealed)… and they are said to be called **TR-3Bs**:

Allegedly the new TR-3B, or Black Triangle.
(credit: Bing Images)

The TR-3B is an electro-gravitic craft based on former German engineering (of 70 years ago) which produced the following UFOs:

The pre-Vril aka *Flugelrad*
(credit: Bing Images)

And… for those who did not see VEG, Ch. 4…

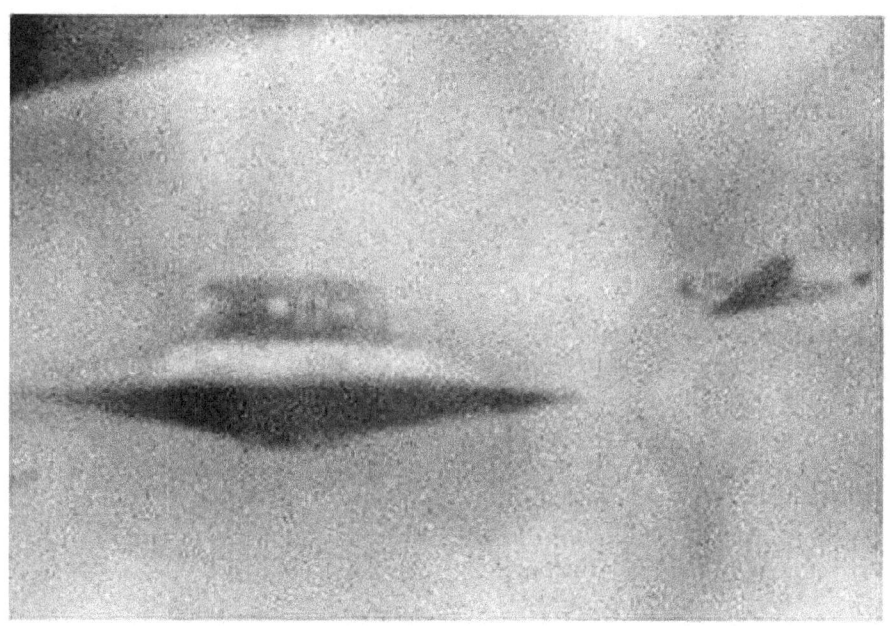

The Vril in flight – note the German Cross on the turret.

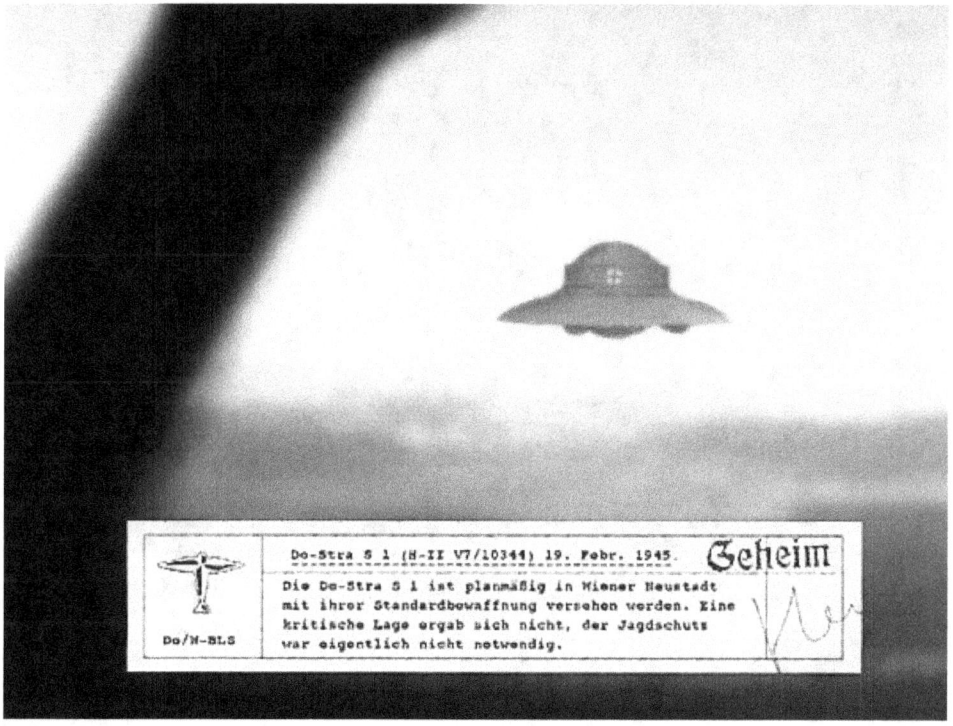

The Haunebu I in flight (above)… "Geheim" means secret…
Some were quite large…

The Haunebu III *Flugscheibe*
(Credit all: Bing Images.)

…and the USAF prototype being fished out of the ocean after a crash…

Recovered Anti-Grav Prototype (note faint USAF symbol left wing)
(Credit: Bing Images/Black Triangles)
These craft are examined in more detail in VEG Ch. 4.

And some appear to be real or perhaps are PR mockups....

This is the picture of the engineers and developers at Lockheed: pictured with a prototype of an electro-gravitic craft.

Is the **Astra** (left) just PR or does it work?

Area 51 Skunkworks Lockheed TR-3B Astra.

(Credit all: Bing Images)

Lastly, there is the disclosure by Ben Rich of Lockheed back in 1992 (assuming it is not more disinformation to support the "Visiting ET" story):

Rich Disclosure

The late **Dr. Ben Rich**, head of the Lockheed Skunk Works (Think: Area 51 and S4), shared a number of things in September 1992 during a presentation at the Air Force Museum in Dayton, Ohio. Sample statements included: [277]

> "We now have the technology to take ET back home.

> "We did the F-104, C-130, U-2 Spyplane, SR-71, F-117 and many other programs that I cannot talk about.

> "**We already have the means to travel among the stars**. But these programs are so locked up in black programs that it would take an act of God to ever get them out to benefit humanity.

> "If you can imagine it, Lockheed Skunkworks has done it.

"I wish I could tell you about the projects we are currently working on. They are both fascinating and fantastic. They call for technologies once only dreamed of by Science fiction writers.

"We now know how to travel to the stars."

...and referring to Quantum Physics limitations in some equations that even Einstein could not straighten out, concerning hyper-luminal speed, Ben said,

"There is **an error in [his] equations**, and we have figured it out, and now know how to travel to the stars, and it won't take a lifetime to do it."   [emphasis added]

Dr. Rich, on his deathbed in 1995, shared one last jewel: He confirmed that there are **two types of UFOs** – the ones we build and the ones 'they' build. "We learned how <u>to build ours</u> from crash retrievals and **actual 'hand-me-downs.'**" [278] The ones we build and test-fly are called **Alien Reproduction Vehicles (ARV)** and are what people see from time to time and call UFOs.

Another system called the **"Fluxliner"** culminated in the development of three different-sized vehicles. One was 7.5 meters in diameter, code-named "Baby Bear"...
Next was an 18 meter version, code-named "Mama Bear"...
The largest craft between 38-40 meters in diameter [120 feet] was code-named "Poppa Bear." ....The propulsion system was based on **Zero Point Energy** and it was said to be capable of **"light speed or better."** ....These "second space program" inventions were being developed in secret , while the above-ground relatively primitive Space Shuttle program was touted as progress to a fascinated American public. [279] [emphasis added]

Note: the speed of light is not an upper limit as Chapter 4 revealed.

Interdimensional ETs?

Someone is bound to still say that the ETs are really here, and cite the **Montauk** time experiments (was that even real?), and suggest that the UFOs are not ours but belong to ETs that materialize in and out of the Earth Realm – bypassing the Firmament – which is supposed to keep out ETs that would "pop in" and interfere with whatever Man is doing.

That is a possibility, but assumes that the gods are lax about enforcing the "No ET fly zone" within the Earth Realm...

And that brings to mind the **Philadelphia Experiment** of 1943 – when a destroyer (USS Eldridge) was wired up to <u>just make it invisible in the water</u>, and somehow (allegedly) too much power was applied and Man inadvertently discovered how to move thru a time warp – 'moving' the ship from 1943 in Philadelphia to 1983 just off **Montauk, Long Island** (allegedly due to the Earth's precession and wobble, the ship didn't 'travel' there but the Earth phase-shifted and over 40 years, Montauk was in the Philadelphia location). It was sent back from 1983 (Dr. John Von Neumann was there to meet it after 40 years... and that part may be froo-froo, but they did electrify the ship.)

And allegedly when the ship came back, some sailors were stuck in the bulkheads and walls… Makes a great story!

> Again, I have to ask, if we are in an Earth scenario with the gods tightly controlling what we do, **and the Earth is subject to 3rd dimensional Laws**, how could the Philadelphia Event happen? It sounds like a story to entertain us… make us think that such things are possible on this free-rolling, round rock circling the Sun – open to potential ETs who could visit us…

I have to admit that the <u>absence</u> of such possibilities is a spirit-dampener.

Alternative Earth Activity

So if we eventually wake up and see that the Firmament does not really permit Space Exploration and no ETs are visiting us …. what then? Not a lot of people will accept the new truth and sit down and work on their spirituality – to become an Earth Graduate. Without an exo-planet goal, many will probably revert to hedonism.

Barring a change in the structure of the Earth Realm, we'd have to find another way (besides War and trying to see which country is King of the Hill… or maybe it is better said: **Lord of the Flies** [Wm Golding had an interesting metaphor there in his 1954 book]). We could occupy ourselves in other challenging pursuits…

Yearly Olympics: contests between countries.

Stratosphere Races: just like NASCAR and the Daytona 500 – but around the Earth. Or: race to a space station, defeat the "aliens" and race back to the finish.

ISS Puzzle Maze: you are suited up and flown to a large station orbiting the Earth. Your task (with or without friends) is to figure a way thru the Maze (reminiscent of a team of **MacGyvers** working thru a real 3D version of Myst®)…. Remember that one? You had to buy the answer book to solve it!

This was a real challenge…

The original was 1993, then better graphics in 1995, and a super-real version in 2015.

It was best-played on a PC, and the faster the PC the better.

Other possibilities....

> **Dictator/Avatar for a Month:** set up a simulated country, say in Africa or South America, with real people, who are problems and your task is to go in and successfully rule and get their cooperation. (Gad, that sounds like the current Earth Realm!)

> **Mystery Island**: sail to a remote island, find the treasure. (Unlike the TV version, *Curse of Oak Island*, there would be real treasure, and your task is to figure out the clues and find it... and maybe keep it.)

or

> **Underground Bunker with Death-traps**: real danger.

or

> **VR Goggles Maze Runner:** sit in a chair and take a trip.

or

> **Mayan Adventure Tomb Raider:** explore real Mayan pyramids and find the treasure/tomb.... Or the way out.

The last 4 kind of turn Earth into a DisneyWorld of challenge and adventure. You win medals, trophies and money. When you are all done, you do the last one:

> **Tibetan Kundalini Awakening**: This is the one that enlightens you and gets you out of here. (Perhaps like Göbekli Tepe? See AL.)

Ok, so if the technology has been with us since the 1940's, and the UFOs are real, what is the connection with the VR Sphere?

Alien Invasion

Just consider the following as a possibility:

> The **Black OPs people** are building and flying TR-3Bs and **Astras** in secret. There is no Congressional oversight for these craft and Congress and the American people do not know that they exist... "National Security" would keep it quiet.

> Secondly, suppose that **the PTB** know about the UFOs and are about to hatch a brilliant agenda **to unite the Earth and set up the One World Government... the NWO (New World Order)** that we have heard about – and is pictured on the back of the one dollar bill... **"Novus Ordo Seclorum"**...

Project Bluebeam

Because people believe Earth is a round sphere spinning thru space and is open (or vulnerable) to ETs and their UFOs, and people do not know that most of the UFOs are ours, why not build about 40 of these things and then appear in the skies over America's major cities, and have one land on the White House lawn?

> "Hello, Earth people! We are your space brothers here to help you… we see your difficulties and can help you solve them if you will cooperate with us…"

> And out steps a human–looking being from the craft – he looks just like us (because he is us)…

> … and America is assisted into building the One World Government to establish peace, prosperity and order on Earth – at the expense of individual freedoms – society would be more controlled… and we already have the **cameras** everywhere, the **tracking chips** in our cellphones, in our new creditcards and new cars (that is how **Onstar** works), and we now have **Alexa** (Echo) listening to whatever we say (keywords are a trigger to record and later a human will scan what was recorded)…

Blessing or Curse?

And this is not necessarily bad… If you watch the nightly news on TV, you have seen the violent demonstrations (which are really funded riots), the problems with ISIS, the refusal of the Democrats to do what is right for America (2016-2017), the rich cats are said to fund demonstrations around the country, Brexit and the problems in the European Union, the occasional disease outbreaks (HIV, Ebola, MRSA…), the problems with immigration, and lastly, clean water (think: Flint, MI).

President Reagan back in 1987 said that we might put aside our differences if we realized that there was a "threat" from ETs and we had to band together to survive… Was he being prophetic?

Well, what if it isn't a threat of invasion, but looks like a peaceful visit by space brothers who "know what we need?" The public would not suspect that the UFOs are ours and it is the PTB executing their **agenda to unite the world**. Even if the public knew, their probable resistance would ensure their personal doom… "cooperate or be removed!"

What if we cannot survive as a species if we keep fighting each other as we do? And since we cannot unite peacefully (think: European Union, League of Nations) maybe the shock of ETs coming to help us so that we do survive would do the trick…? And what if the OWG is proactive and benevolent… what we need to survive? Could we follow the ideals of the **Georgia Guidestones**?

Alternate Scenario

There is another possibility and it may be the one that is currently in play in America and the Western World. UFOs and ETs would not be necessary if the PTB can foment enough disruption, dissent, deception and demonstrations (riots) to the point where the public demands that something be done…. also called:

Problem – Reaction – Solution

If the goal was to lock it all down, instigate Martial Law (Solution), and have a more orderly, controlled society, then fomenting the discord (Problem) we see nightly on the News would escalate to the point where something would have to be done….in this case the public is its own worst enemy. Instead of seeing that they are being manipulated into asking (Reaction) for the pre-defined Solution, like crazy apes they run around throwing rocks, breaking windows, burning cars… just like the last 20 minutes of the movie *Conquest of the Planet of the Apes* where the Apes are dressed in orange overalls and throw rocks, burn cars and buildings, break store windows…. and shoot their guns at the humans. Is this Freedom of Expression… or is it License to Destroy?

The point in all this is to say that **Project Bluebeam** (Google that) may become a reality sooner than later and we may not recognize this country in 5 years. The craziness following the 2016 Presidential Election which went on for months says something is wrong with America. Can it be constructively fixed?

What could be going on to cause so many people to 'lose it' and act crazy? What if the gods are increasing the Earth's normal **Schumann Resonance**... or what if Man (or the Remnant?) is using HAARP to broadcast a frequency that agitates humans? The Russians did a similar thing and it was called the **Woodpecker**. [280]

Psychotronic Mind Control

> Edward Naumov, a leading Russian parapsychologist, is on record as stating, 'A psychotronic generator can influence an individual or a whole crowd of people. It can affect a person's psyche mentally or emotionally. It can affect memory and attention span. A psychotronic device can cause physical fatigue, disorientation, **and alter a person's behavior.**[281] [emphasis added]

The **Woodpecker** was a Soviet Union psychotronic transmitter which was so-called because it emitted a **pulse of 10Hz**, which is called ELF (extremely low frequency) **powered by Chernobyl until it was sabotaged**, and it was reputed to be aimed at the West to cause societal disruption and so disorient them that they would make bad decisions, benefitting the Soviet Union.

The now defunct Ukranian Woodpecker transmitter

Increasing Vibrations

Tavistock in England and the CIA have both experimented with the effects of ELF and they are not good for society. Could the riots and violent demonstrations awash in America in the first half of 2017 be due to some rogue group's use of that? If not, then there is the possibility that the gods who run the Earth Realm are increasing vibrations with the intent to entrain more souls into a proactive rise in their consciousness... and all we are seeing (demonstrations) is resistance to change.

In raising the vibrations, the Higher Beings would **entrain** those souls who are ready into a higher awareness. So that is a proactive result. Thus, the **Schumann Resonance at 7.8 Hz** should be found to be increasing. Any higher vibration level would entrain resistant people into a state of anxiety and forgetfulness.

The body has standard EMF rates that it needs to function healthily. Resonating with the Schumann Frequency is a base upon which it operates. Energies in the 20-30Hz range 'excite' body molecules and promote health -- those frequencies above 30 Hz do not benefit the body.

Even the brain has its own levels of operation:

> **Beta** cycle: 12 – 19+ Hz (normal waking consciousness)
> **Alpha** cycle: 8 – 12 Hz (relaxed, aware state)
> **Theta** cycle: 4 – 8 Hz (sleep)
> **Delta** cycle: less than 4 Hz (deep sleep)

And there is one more, not often discussed...

> **Gamma** cycle: 30-70 Hz (advanced cognitive and meditation
> activity...
> said to be even OOBE)

Over the last decades more insight has been gained, especially with advances in **brain imaging**. A major area of research in neuroscience involves determining how oscillations are generated and what their roles are. Oscillatory activity in the brain is widely observed at different levels of observation and is thought to **play a key role in processing neural information.** Numerous experimental studies support a functional role of neural oscillations; a unified interpretation, however, is still lacking. [282]

Several high-tech medical groups (Japan and Germany principally) have used this vibration information to make devices to help the body heal… like **BEMER**:

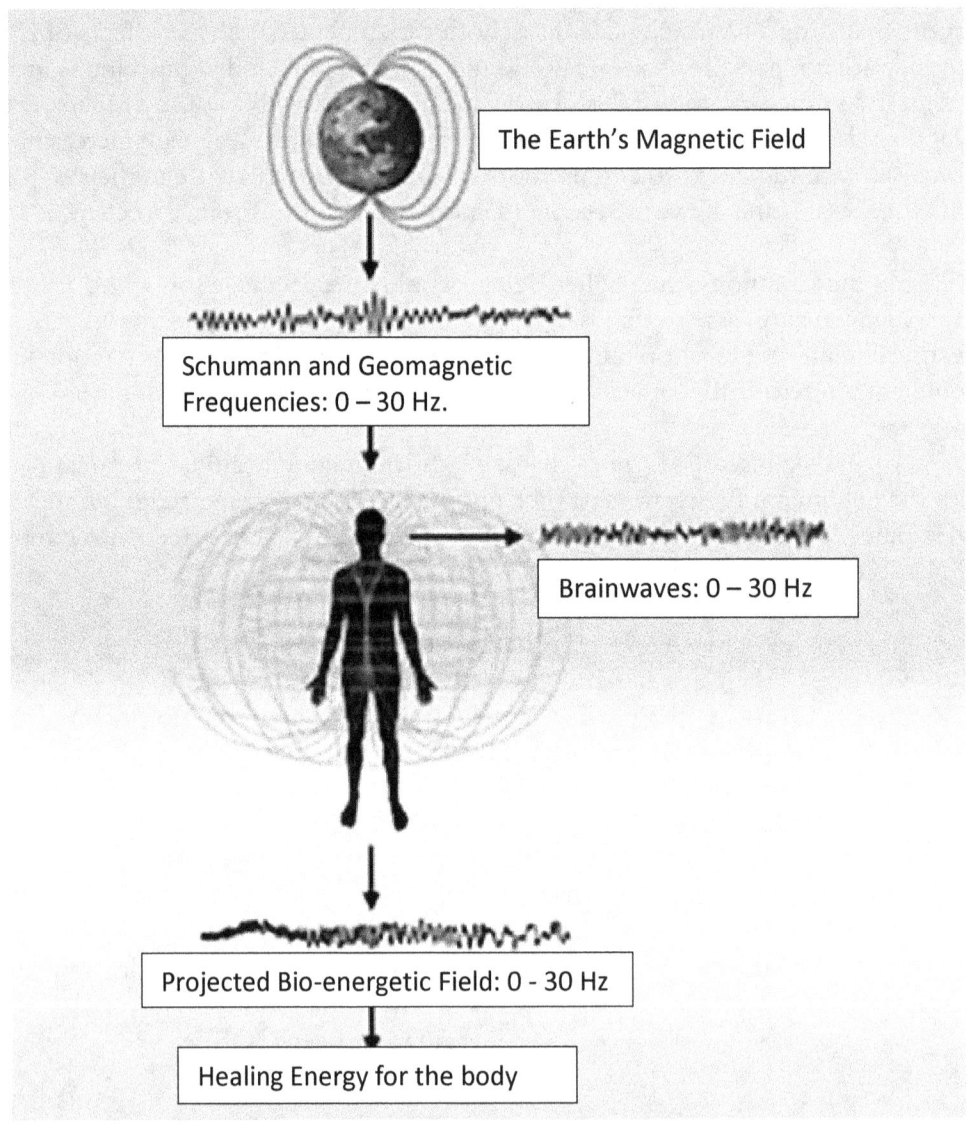

BEMER PEMF Healing Mat Principle
(**credit:** https://www.youtube.com/watch?v=LKOli-nNALM)

The above is a diagram of how the **BEMER Healing Mat** works using ambient energy from the Earth's field, and applied to the body can reverse EMF damage by activating the body's own **Bionet** – assisting the *chi* to flow where it is needed.

The author has found this to reduce inflammation and muscle aches. **Osteopaths use it to help bones repair or 'knit' faster.**

Handling Increasing Vibrations

Note that the souls that accept and move with the increasing vibrations (The Flow) are automatically 'qualifying' themselves for a return to the 4D Realm from which we normally come. No one is running around selecting who goes and who doesn't – Man **selects or deselects himself** by virtue of his PFV – which is why true Knowledge is so important: right now Man has a choice but is not aware of it.

The illusion is that Earth is our home, but this is a royal illusion. It would be nice if just waking up spiritually were enough to get one out of here, but it ain't so. One has to **assimilate the Light**. Knowledge is Light and it is the Light that sets us free.

While both **David Icke** and **Stuart Wilde** see the Earth situation as a trap, Icke thinks Love is the answer – we should love our "unseen jailers." Wilde, on the other hand thinks we can call on God's Gladiators (Beings of Light) to help us escape from this zoo. [283] In reality, it is going to be **Knowledge** that gets Man out of here, because until he <u>knows</u>, he can't make a choice and do what it takes to leave Earth (graduate).

> The Earth's usual, traditional low-vibration energy that sustains low-level awareness works to **entrain** all on the surface into their lower 3 chakras, and Man will never know he can do/be/have more because the higher chakras don't have enough energy with which to resonate to reach and sustain a **higher consciousness**.

> *This why the Earth Graduate is so respected – s/he broke free of the illusion and disinformation.*

Those who do not flow with the god's higher vibration – if that is what it is – and resist it, create an anxiety and tension in their minds and bodies which cause them to be short-tempered and act out on the road or at home or work.

On the other hand, if the frequency increase is due to **MKultra** or **HAARP** or **V2K** , then it is not the gods selecting graduates for another level Earth… it is Man afflicting Man to create discord, riots and give cause for a Martial Law lockdown to occur. (V2K was examined in VEG.)

How would we ever know if it is being done, and who is doing it? It is nothing to worry about if your consciousness already is of a higher vibration as you will not drop and synch up with the lower. It is a problem for the "unconscious" majority, those who vibrate at a lower rate, these who believe that they are their body, are not aware of the soul, and are hedonistic and often slow-witted, petty or violent.

Also, young people (**Millennials**) who are young enough (< 20) that they do not have a mature, experienced mindset are susceptible to the **mind manipulation frequency** – and they were the ones rioting at Berkley (Feb 1, 2017), NYC and Washington DC (Jan 2017). Young and impressionable.

> BTW, some Millennials rioting in NYC were stopped and asked what they were rioting about (this was shown on Watter's World on FOX 2/9/17) and they either didn't know and were just following their friends, who were all cutting classes by the way, and yet others when trying to state facts had them wrong. It sounds like they were manipulated into the brainless (ponored) demonstrations.

Resolution #2

The Earth Realm has a specific purpose and is well-managed for Man's benefit. It was created for him and is a **School for souls**. That is not only a proof of God's existence, it is also His Grace at work – that no soul is left behind.

And the significance is that if Earth is a 3D Construct and Man is contained, then fairy tales about Apollo Moon Landings are just entertainment… and **we need to refocus on brotherhood, peace and compassion for others**. Learn our lessons and get out.

And if Earth is sealed or contained, then most (but not all) UFOs which are real are built on Earth and flown by Earth beings – human as well as the possibility of the **Anunnaki and/or Ancient Ones also flying theirs** – allegedly it was ETs who gave the Germans the UFO technology in 1932-36…. but since it couldn't be aliens, it had to be the Earth-bound **Anunnaki Remnant** (underground).

> Hitler said several times that he had met the **reptilian emissaries** from the Remnant group and it scared him. So apparently there is an Anunnaki faction (as VEG said) that works with the PTB and humans of less integrity to try and establish a despotic NWO – and there is a small faction (Anunnaki Remnant <u>Dissidents</u>) that does not like Man.   (See VEG, Chs. 3 & 10.)

The Germans built and flew the *Flugscheiben* (UFOs: "flying disks") at the end of the War, but it was too little, too late – or we would all be speaking German today. And yet their fascination with ice led them to build bases in Antarctica and Greenland, places where the average person would not go looking for them. The bases are now empty.

In the meantime, **if Earth is quarantined**, then there are no alien UFOs or ETs visiting us and any aerial display of Others who are "here to help" should be met with suspicion, and that is one goal and a warning from this chapter. The possibility of the **Project Bluebeam** becoming a reality – in just a few years' time – is not beyond the realm of possibility – the TR-3Bs have been built (and others) and since we can't get off of Earth and into space, what do you think the PTB is going to do with them?

> **This may be an unrealistic assumption**: having 40+ TR-3Bs with their exotic technology would be very expensive (a billion dollars per ship), and the cost of building many of them to pull off a Project Bluebeam may be a pipedream in today's economy, but it is a fascinating concept and nonetheless it is something to be aware of.

If the Earth is no longer flat, and is now a **VR Sphere** – then we can explain how the million-ton oceans stay on the round planet. As was said, if Earth is a **Simulation**, and there is good solid evidence for that, **the oceans could be programmed into the Simulation to use a special set of natural laws designed just for the 3D Earth Simulation**.

> Using the time-tried **Occam's Razor** shows that the simplest answer or solution is usually the right one, and theorizing about a VR Sphere Simulation with special programming and Laws just so that the oceans can cling to the global Earth is a stretch, but ay be right! The Flat Earth scenario was the simplest and most straight forward, but the Hardest to believe. Occam's Razor is suspended when the gods start doing things.

> Nonetheless, the Earth Realm as created by God is said to be a complex Simulation, in that it acts like it (see this book's Chapter 9, and Appendix C) and that includes **glitches in our physical world** (see QES).

> Our worst nightmare would be that Earth is a Simulation run by aliens for their purpose – say an experiment and we are the lab rats – or better yet – we are their Reality TV entertainment. I do not vote for that as there are too many times that I was guided and protected to do what I was tasked with in 2008 – (see Chapter 1).

And consider, if Earth is really like the IMAX Theatre, then **Shakespeare** was right: **the world is a stage and everyone is a player in the Greater Drama**... each has his/her part to play.... And "struts and frets his/her time on the stage and then is heard no more" (until the next lifetime...). And that scenario is supported by VEG and TOM which add another dimension to the Earth Realm – the OPs, or **Non-Playable Characters** (NPCs as in a video game) who <u>have no soul</u> but help drive the Greater Script. (See Glossary.)

> And this is really important (for those who have not read VEG, nor TOM): **our world is about 60% OPs aka NPCs** mixed in among us souls – to drive the Greater Script for this "Stage" called Earth. They have no souls which means they have **no conscience** (VEG, Ch. 5).
>
> Note: Dr. Bostrom in Chapter 9 (p. 313) says that if the number of simulated humans is a large number, then we <u>are</u> in a Simulation, and Dr. Greene and Jim Elvidge agree.
>
> A soul has a connection to a Higher Self which does not exist in an OP, and that connection is what guides us thru intuition and gut feelings that something is right/wrong. Occasionally our conscience will bother us... that is higher guidance. OPs do not have that.
>
> Now you know why some people do dumb or mean things. Some of them (5%) are also the sociopaths (like ISIS and Charles Manson) ...

That one makes sense. And as that is true, and **personal observation has found them to be 60% +** of the population around us.... then their only reason for being lies in the fact that **they are part of our lessons (and tests)** and that means **Earth is a School**... subject to a Greater Drama run by the gods to train souls.

You say, "God wouldn't do that...!" He doesn't DO it, He <u>permits</u> it because we are in a **Freewill** Realm... and the purpose is **catalyst** – to reinforce what bad behavior looks like... **You are expected to recognize it and avoid it**. How can you know how good/bad <u>you</u> are unless you have something to measure yourself against? OPs drive the Greater Drama... like a video game.

Think not? Remember, "His ways are higher than our ways..."

(Isaiah 55:9)

And ultimately it makes no real difference – you are an **eternal soul** (and don't die) <u>and</u> the Earth Realm is isolated from the rest of Creation (which is in 4D and above) <u>by the Quarantine</u> (Firmament) – so our primitive emotions (via some crappy music and TV programs over the airwaves) aren't going anywhere…
we cannot, do not, and will not afflict the other beings in the rest of Creation around us – the <u>real</u> Creation – any more than Kinder-gardeners can bother the students in high school or college… and it is just about on that level for much of the humans on Earth (think: Baby Souls in VEG, Ch. 7 and Appendix B in this book).
.

Now you know where we are and why we are isolated/segregated.

Lastly, it was said in VEG, TOM and AL that we are not meant to spend lifetimes in the School – you don't live in a school. You may be tired of the Cafeteria by now! You get your education and graduate. And **the Earth Graduate is very respected elsewhere** (as Robert Monroe said in VEG, Ch. 12) – because it takes some real effort to <u>graduate</u> – you have to overcome and <u>bring forth your inner strength</u> and gain real Knowledge… then you graduate. Why gain Knowledge?

Knowledge Protects
Ignorance Enslaves

Flat Earth vs VR Sphere Revisited

Ok, you have come this far, and I have played with the FE Scenario and the VR Sphere about as much as one can, testing your patience and my stamina. What has been demonstrated throughout this book is that Earth is not the standard rock globe, rotating at 1000 mph, and spinning around the Sun at 67,000 mph.

And the other concept that was promoted in QES Ch. 11 and at least once herein, is that Enoch and **our ancestors all knew the <u>original</u> Earth to be flat** – and that must have come from the Ancient Ones, or the Anunnaki – **somebody told Man** – otherwise why would he assume it was flat <u>in every culture around the world</u> – <u>cultures who didn't know each other</u>? How did they know? (See AL.)

So it is probable that **Earth <u>was</u> created flat**, and in many places it still looks like it, and behaves like it (Sun and Moon closer to Earth yielding hotspots, and at times the oceans do look flat when you see a 300-500 mile stretch of horizon… but the circumference is 25,000 miles so **even a round Earth could appear to have a flat ocean at times**).

The new wrinkle is that Man is growing and attempting to expand his knowledge and get off the planet… So the question is raised: If the Earth was created flat because Man was still limited in mobility and could not go very far – that would suffice for millennia. Then if <u>Man was given</u> electro-gravitic propulsion (see Tesla in Appendix D), that suggests that the gods did that, and that **the Earth Realm was morphed into a VR Sphere**… and the only way we can know that is if Antarctica is no longer an Ice Wall but a real continent… Can we trust Google Maps, or is that CGI?

> It is hard to know for sure since QES (Ch. 11) examined the tendency of NASA and pictures from space, to have been modified, using Photoshop® and otherwise to be a product of **CGI creativity**.

What would be interesting is if someone on the ISS or in a Space Shuttle went outside on a spacewalk and took their video camera and did a complete 360° pan of space around them…. **This has not been done**. Why not? What would it show?

Limited pictures <u>were</u> done in a YouTube video from the ISS and a high altitude balloon… but <u>only in one direction</u>.

Eric Dubay : https://www.youtube.com/watch?v=fcteYfOMgJg
No Curvature on the Earth horizon (time: 14:42) – but it may not cover enough of the horizon to see an actual curve…

And

Flat Earth Addict: https://www.youtube.com/watch?v=WQITXbcz2hg
FLAT EARTH ADDICT 05 : 121,000 feet Little Piggy Cam High Altitude Balloon Flight (time: 4:29)

Seriously take a look at these… **High Altitude balloons pan & show a flat horizon (in one direction)**.

The Little Piggy Cam shows:

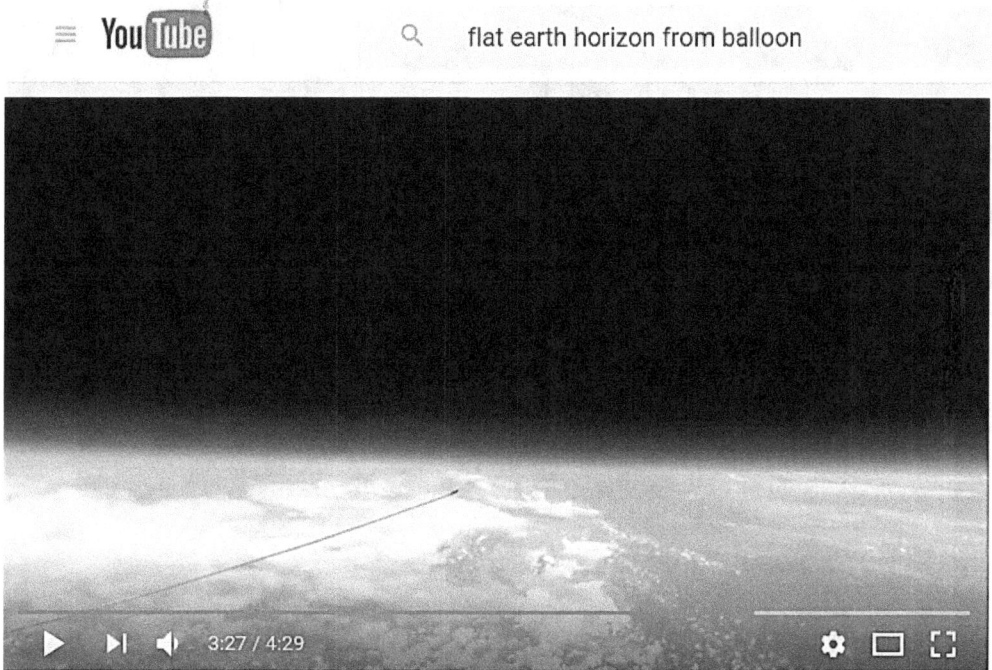

FLAT EARTH ADDICT 05 : 121,000 feet Little Piggy Cam High Altitude Balloon Flight

99,789 views

Note carefully that the last picture has a very slight concave aspect to it… suggesting that the camera lens is not perfectly flat…

…taken Aug 2015.

And the Eric Dubay video shows (below)…

Altitude: 110,000'
Temp: -38° C

Taken: Dec 2014

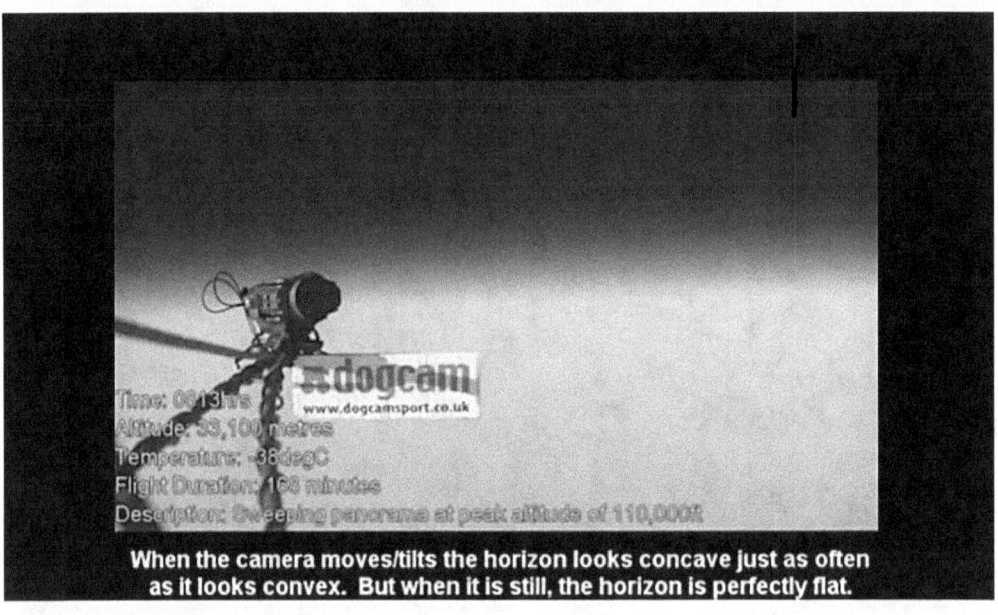

No Curvature on the Flat Earth

Eric Dubay

▶ Subscribe 87,411

38,626 views

Note carefully what the text in the picture says!

These videos offer significant evidence that the circumference of
the Earth is <u>very big</u> and can appear to be flat/straight --
depending on the (1) amount of horizon shown, and (2) the type of
camera and lenses used! ... see QES Ch. 11 for better examples of
how a lens can make the horizon look concave, convex or flat.

Thus, the above pictures of "flat" horizons are not conclusive.
One reason is that often not enough of the circumference (horizon

length) is shown, and sometimes the camera (even in a balloon) is not up high enough. And it is shooting in a narrow direction.

These videos offer <u>interesting evidence</u>.

Summary

It is reasonably stated that the Earth Realm was possibly flat at one point in the far past, but that the gods may have morphed the Simulation (reprogrammed the Earth Realm) to be of more use to Man as he launches rockets and tries his hand at space exploration – which is really limited to the Cspace as Chapter 3 showed. The **sphere within a sphere** could be a more accurate for our world today….

Surprisingly not much is known about the intent of the sculpture by **Arnaldo Pomodoro** of which copies can be found around Earth in 14 different locations – the picture above shows the Vatican site.

Think: Who commissioned these 14 expensive sculptures and why?

It really is inescapable that **Earth is more than we have been told**. Gravity is a major stumbling block for Science trying to explain Earth as a planet, and it is unfortunate that Science went off on a "Ditch God" tangent 400 years ago from which they cannot now come back without egg on their face – just like they now have to admit that Einstein was wrong and that there **IS** an Æther, but are now calling it Dark Energy/Matter.

So, we have come this far and examined the pros and cons of the VR Sphere vs the Flat Earth Scenario. Which one is correct? The VR Sphere is easier to prove than the FE Scenario since the Earth is already assumed to be a globe. So can we debunk the FE Scenario? If so, then that leaves the VR Sphere as the winner of the debate.

Thus, the answer lies in the summarization of the significant FE evidence. They either hold water or they don't.

Resolution: VR Sphere vs Flat Earth

After due review of those above videos and

Seeing the **NASA wall screen** with the "S" orbits that resolve to a circle on the Flat Earth, (see rebuttal in Ch. 11) and

Having done the analysis of the **Southern Hemisphere distance** anomalies, which do not match those of a globe, (see rebuttal in Ch. 11) and

Discovering the Science error with the **Earth's atmosphere not leaking into the Vacuum of Space**, (due largely to the presence of the **Firmament**) which NASA is calling the **Plasmasphere** (Appendix E), and

Finding the **correct Earth layout** on the UN Flag, which matches that of the traditional Flat Earth map, and

Seeing the **'hotspots'** on the clouds and oceans, proving the Sun is a lot closer than 93,000,000 miles away, and

Not having resolved how **million-ton oceans** stay on a VR Sphere or on any globe, (see Appendix D) and

Not finding any proof of **Antarctica** to be a continent, but an Ice Wall, and

Further, having <u>not</u> found **Gravity** to be a real force (Chapter 4 and Appendix D), and

Since the Earth is not rotating at 1000 mph, proving **the absence of the Coriolis Effect** (also proven with the cannon shot in Ch. 11 of QES) and…

The Sun changing in size as it approaches the viewer (up to Noon) and then shrinking in size as it recedes to the horizon (sunset) – means **the Sun is circling the Earth** (See QES, Ch. 11)… and

Lastly, the fact that **Polaris is always overhead** despite the Earth's tilt, precession, and changing orbit around the Sun (see pp 113-14)…

(plus other aspects in CH. 11 of QES)

It sure looks like we have a Flat Earth…

However…

There is more to the picture than meets the eye… and after busting a gut for over 6 months early in 2017, trying to prove the FE wrong, and finally inserting the evidence into Chapter 11 in QES, I had basically given up… and reinserted it in this book suspecting that we somehow had a Flat Earth but I couldn't prove it nor disprove it. **And it all didn't feel right.** What was left? I was missing one significant piece of information that a reader in Michigan, thank you John, sent me in an email in October 2017:

"I want to suggest that you take a look at Satellites with **Polar Orbits**."

WTF? How did I miss that?

Of course, if we really do have satellites running around the Earth in North-South Polar Orbits… the Earth cannot be flat! I thanked John and felt like an idiot – How did I miss that? It is so obvious…

Polar Orbits

Says Wikipedia: [284]

> A **polar orbit** is one in which a satellite passes above or nearly above both poles of the body being orbited…. It therefore has an inclination of (or very close to) 90 degrees to the poles. A satellite in a polar orbit

will pass over the equator at a different longitude on each of its orbits.

Polar orbits are often used for earth-mapping, earth observation, capturing the earth as time passes from one point, reconnaissance satellites, as well as for some weather satellites. The **Iridium satellite Constellation** also uses a polar orbit to provide telecommunications services. The disadvantage to this orbit is that no one spot on the Earth's surface can be sensed continuously from a satellite in a polar orbit.

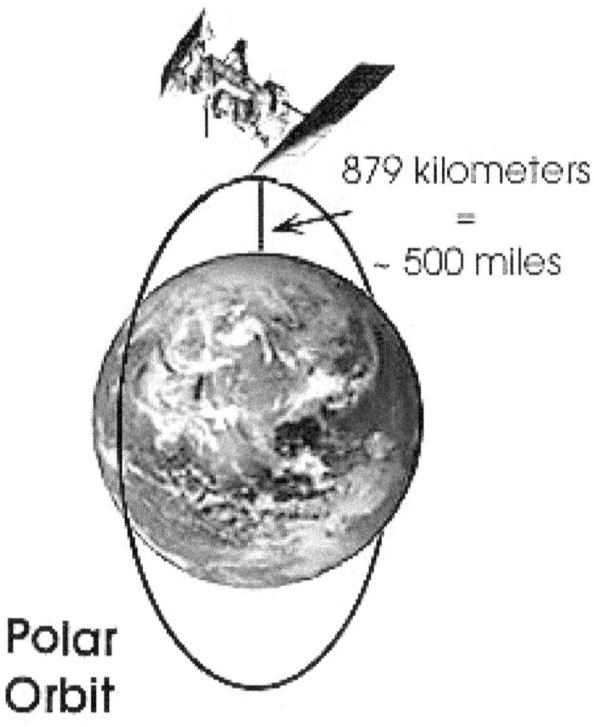

879 kilometers
=
~ 500 miles

Polar
Orbit

(credit: Bing Images: **cimss.ssec.wisc.edu**)

Ta-Daaa! It would not be possible for a Polar Orbiting Satellite to properly circle a Flat Earth.

Thus, the **VR Sphere** (which is what I was given back in 2008) is the correct one, and QES examines the reasons behind the Simulation argument.

Chapter 11: God's Grace

This chapter will extend what we have been talking about, draw some final conclusions, and emphasize the purpose for the creation of the Earth Realm and what Man should do with the new Knowledge. While this chapter will come close to having a religious or spiritual aspect, that is not the purpose – It is merely emphasized that there is ultimately a God who cares about Man and his soul growth such that Higher Beings ("the gods") created the Earth Realm in 4D for Man, again not only to quarantine him and constrain him, but to provide a protected environment in which to grow – less the interference of beings in 4D and above.

To that end, more is given about God's Promises, NDE effects and Spiritual Awakening, , the Earth School, and some intriguing and unusual "special effects" of physical experiments (such as that of the **Slinky vs Gravity** below), and what it means to be an Earth Graduate and why that is important.

Why The Flat Earth?

It bears repeating that **I intended to prove the Flat Earth (FE) scenario wrong** when it was first presented to me, and because I wanted to really understand the subject, to make sure I was being accurate in my denouncement, I invested <u>months</u> of research into it. **Some of the evidence I still cannot refute to this day.**

And then besides viewing some mighty long videos on YouTube (1-3 hour long presentations) I also bought 5 of the most complete-sounding FE books including two that were reprints from the 1800s. I emailed several people who promoted the FE Theory and one answered giving me <u>super quality links</u> to the best videos, including his.

I can identify with resistance to the topic, as I would sit there looking at the first two "key proofs" and still I didn't want to believe it! Those two, by the way, were the **NASA wall-screen** "sine wave" orbits which translate into a circle on the Flat Earth map, (which means NASA <u>knows</u>), and the long proof in this book where the **Southern Hemisphere distances** between Africa, South America, and Australia do not match what one sees on a globe. There were 4 other less dramatic but equally serious and undeniable evidences that are only in Ch. 11 of QES.

I can see why the FE Issue has persisted throughout the centuries, despite Science actively pooh-poohing it – even into our present day. **I began to speculate from the evidence <u>and the ancient records</u>, that Earth was probably created flat**

but the gods changed it as we flew planes and launched satellites. If Earth is a Simulation that is a piece of cake. I was also confronted with the VR Sphere supporting info <u>that I had been given</u> (see Introduction list of 3 elements). It was royally confusing. Gravity and Antarctica were the major stumbling blocks. I did not know about the Polar Orbit satellites…

The issue and analysis of **Gravity** is reserved for Appendix D, but there is one aspect of it that bears examining, below.

Oddity: Slinky vs Gravity

This was a **showstopper** – there is a video of a man with a <u>metal</u> **Slinky** and what he does with it cannot be done if Gravity exists and is trying to bring everything down to the ground. Bear with the 4 pictures, as it is hard to see the Slinky action – even better is to go to the YouTube video and type in **Slinky Drop Experiment.**
Or below:

Best Kept Secret Since Flat Earth 2016
Courtesy: Theworld WeLiveIn
https://youtu.be/s0AOj01yzII

Start: hold the Slinky by its top….. watch the left one…. Time = 19:40

And then release the top of the Slinky… **watch the bottom of the Slinky**…

Note that **the bottom of the Slinky does not move**....it is waiting for the top to come down and meet it... time = 19:46

Time = 19:49…. Top still descending, **the bottom still does not move**…

And lastly, the top meets the bottom…. Time = 19:52

Again, note that the elapsed time is **12 seconds** and **the bottom of the Slinky just sits there**… Where is Gravity in this event? (Note: both Slinkys collapse evenly.)

And **after** the top meets the bottom, the Slinky falls…..seeming to defy Gravity for 12 seconds!!

The whole Slinky is on the ground… after 17 seconds of 'resisting' gravity….

Watch the video or buy a big metal Slinky and try it yourself.

Again, Gravity should have caused the whole thing to fall to the ground **before** the top met the bottom… **Still think Gravity exists**? According to the Flat Earth scenario – the Slinky is heavier than air, so it will fall, but since Gravity does not exist, the Slinky **spring** collapses first, <u>then</u> falls! (See **Appendix D**.)

> **So, in the VR Sphere Scenario, Gravity is a programmed construct -- still an anomaly which is why the <u>whole</u> slinky did not fall at once.**

Potential Resolutions

That leaves the NASA wall chart, the Southern Hemisphere distances between land masses, and whether Antarctica was morphed into a land mass from the original Ice Wall. Let's play Devil's Advocate and see if there are alternate explanations for 3 of the FE evidences...

Resolution: NASA Wall Chart

While it was argued that the "S" waves on the NASA wall screen formed a perfect circle on the Flat Earth map, it can also be argued that this is just NASA's style for representing the flight path **using a wall map with the Earth flattened**. This suggests that the "S" path has significance to NASA as the craft or satellite moves from the Northern Hemisphere to the Southern Hemisphere and in fact may be just describing a tilted orbit:

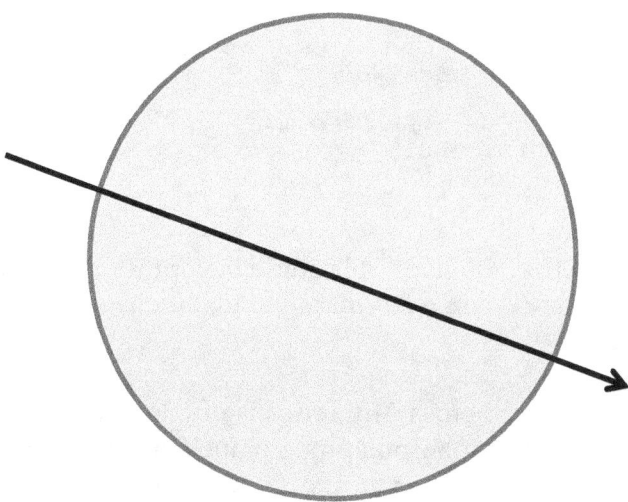

This might be done to keep the craft from orbiting over Russia or China. The above suggested orbit crosses the USA, then Africa and cuts under Australia – just as the circle did on the FE map (pages 103-105).

Resolution: Southern Hemisphere Distances

Whereas it was discovered that distances between the southern-most parts of Africa, South America and Australia/New Zealand as pictured on a globe do not agree with actual travel distances, it can be pointed out that cartography in the Southern Hemisphere has always been an issue.

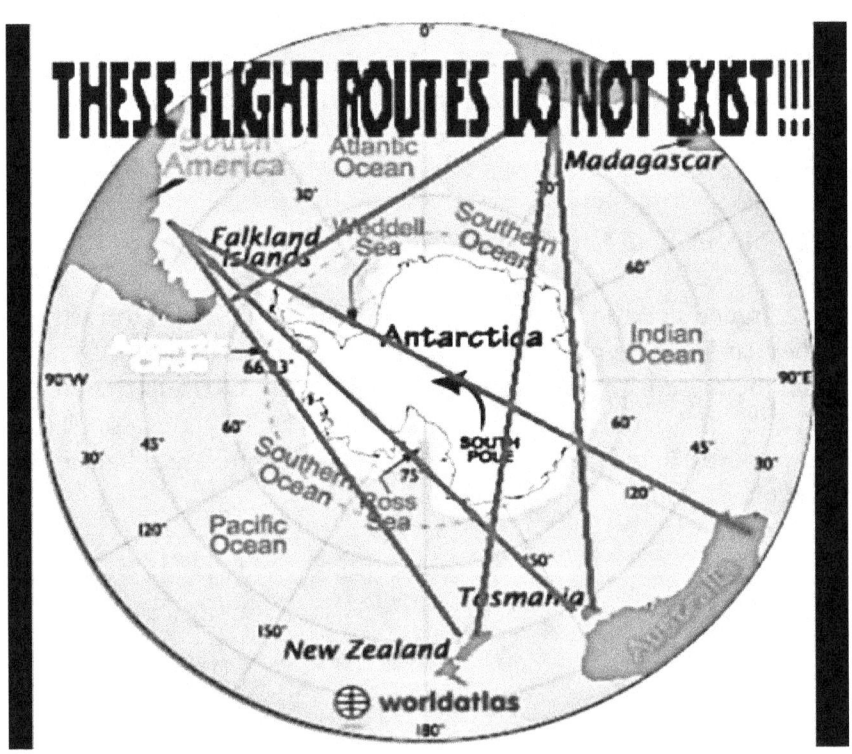

Perhaps the problem is that the lines of latitude beyond the 70th (toward the Pole) do not look like they do on a flat map: the distances on a flat map versus a globe are different.

And the reason for no flights across Antarctica is due to magnetic interference from the Pole itself, adversely manipulating a plane's instruments, and if the plane crashed in the middle of Antarctica, no one would survive before the rescuers could find them. Thus, flights are banned for safety reasons.

Other than those two considerations, Southern Hemisphere <u>sailing</u> distance anomalies (mentioned earlier) are probably due to older inaccurate maps. They are real and documented.

Resolution: Antarctica

Pictures from space do not show an Ice Wall and that is crucial to the Flat Earth Theory. So the original issue was whether Antarctica is a land/ice mass really exists – or are NASA and Google Maps using CGI to show us that Antarctica exists as a landmass? (QES Ch. 11 showed pictures of how NASA has doctored photos of Earth from space, so what can we really know about the southern polar area?)

> Needless to say, this **is <u>the</u> giant-killer** and a major proof for a Flat Earth or not, and unless one can travel all the way south and get into the air, high enough up to see whether the huge Antarctic land/ice mass extends for more than 300-400 miles (before hitting the afore-mentioned Firmament wall).... it remains a mystery.

> Thus the jury is still out on this one and we may never resolve it as the US Navy very well prohibits sightseers to Antarctica.

The ancient mariners, such as the **Vikings and Phoenicians**, sailed around much of what they determined to be an Ice Wall – but one can also sail around Antarctica as it is defined today – that didn't prove anything. And much of the Antarctic coastline IS an Ice Wall in appearance!

And this issue was negated by the revelation concerning the Polar Orbit satellites.

Resolution: High Altitude Pictures

This was covered at the end of the last chapter. It all depends on (1) how high the pictures are taken and (2) with what kind of a lens ... and many camera lenses nowadays are slightly convex, making some pictures of the ocean and the Earth's horizon look slightly curved... **QES Ch. 11** had an excellent exposé of this issue, including a reference to a software one can use to manipulate pictures (The Go Pro Lens Corrector ®) in addition to Photoshop ®.

VR Sphere Earth Corroboration

I poured over the Flat Earth evidence and found it to be surprisingly credible in some places – and that coupled with my review of ancient texts suggested that the Earth might have been <u>initially</u> flat. Why build more Earth than is needed in which beginning souls can live and explore? That was a reasonable conclusion... and as Earth is now a VR Sphere, that is evidence for the Earth Realm being a very **sophisticated Simulation** – because the gods <u>can</u> morph it!

However, back in 2010, to repeat, I was truly stunned, and set about trying to prove one or both of the ideas wrong.

If I was going to write about the Flat Earth in QES, and do it right, then we all have to determine which evidence is false and which is irrefutable. One point of view was that the Earth was never flat... and people were just considered stupid 400 years ago, and as late as the 1800s... BUT, I had my work cut out for me – much FE evidence washolding water (not to make a pun), and the only major issue that required a study of Copernicus thru Newton, was the issue of **Gravity.**

Back in 2008 I had been given 3 basic pieces of information (summarized in the Introduction) and I had understood what I was given to mean that Earth was/is a VR Sphere. Also in **quarantine**, with the Sun close by, no planetary rotation... and a *Gegenschein* reflection of the Sun off the **energy shell** around the Earth (see Appendix E and NASA discovery). That sounded Ok. My big problem was how to explain how the heavy (million-ton) oceans stayed on the globe... and did not fall off. **Water does not stick to a globe/ball/sphere.** I assumed it was due to special 'programing' using specific, special Laws with which the VR Sphere was designed...

The issue of the oceans was so much more simple if the Earth was flat. Heavy oceans are not a problem on the Flat Earth – which the Bible says is flat, by the way. The **Book of Enoch** also says Enoch was taken up by God and shown the Earth, how it is "stretched out under the Firmament," which is attached to the foundation of the Earth, above the Great Deep, how the Sun and Moon <u>which are the same</u>

<u>size</u>, circled above the Earth plane (close by in the VR Sphere Cspace), and how the winds enter from the "windows of heaven." I used to think that was all just poetic.

Yet I can see that as having been correct <u>millennia ago</u>.

I was left for weeks stymied by the Flat Earth option, some evidence made sense – too much so – and I didn't like it, but <u>I could not prove it wrong</u>. And, yes, there are Bozo proofs out on the Internet, submitted by trolls who are trying to make the Flat Earth issue look stupid… and that was partly my intent, too, but some of the Bozo proofs are <u>so</u> stupid as to be very obvious that they are wrong and meant to discredit… So my suspicions grew, wondering if the Bozos were government and NASA – trying to cover up a potential truth! I had to delete the disinformation. What was left was starting to make me sick, literally, because I had been so heavily invested in a globular Earth... and I was getting a headache bouncing mentally between the pros and cons.

> Sorting it all out was not working out. Gravity remained a knotty issue and there was also the afore-mentioned issue with Antarctica as a continent or Ice Wall. I never considered satellites as a way to solve the issue: I didn't know that some had a polar orbit!.
>
> **So the temporary and reasonable conclusion was that Earth was originally flat and as Man grew and needs changed, the gods could morph Earth into a VR Sphere – easy if Earth is a Simulation.**

VR Sphere vs Flat Earth Implications

What the FE theory implied gnawed at my gut, and sometimes upset my stomach – How could the FE scenario be true and more people not know it? It would have to be a lie on a monstrous scale…. But, if you started the 'lie' as ignorance in the 1700s, and it continued forward, no one today would suspect, and thus most people would have no clue that an assumption was promoted and sustained over at least two centuries.

Negative Side

What upset me most was that the FE Scenario, if true, meant that **NASA was sustaining the illusion** – we could not have gone to the Moon because no one can get thru the Firmament. I used to think the serious **radiation of the Van Allen Belts** above the Earth was enough to stop astronauts from flying thru them… Now I wonder if the Belts actually exist, or was this a CYA manoeuver to be used down the road when people wake up and challenge NASA…? (See Appendix E.)

The other sickening realization, coming right on the heels of the NASA issue, was that **no ETs in their UFOs could be visiting us,** either! That meant that Zechariah Sitchin had lied… the Anunnaki were not from the planet Nibiru, but from Earth – probably created here … before Man. And it still held water that the Anunnaki were here (the 250,000 tablets/scrolls said so, and no one disputes that), and it is included in those cuneiform tablets that **the Anunnaki created Man**.

It turned out that the Firmament is another significant **key**!

We had confirmation from Charles Fort (*Gegenschein*) and NASA who has taken pictures of the reflection which has <u>no parallax</u>… meaning it is fixed up there, in the same spot, same size…. reflecting off something, and that something appears to be the **Firmament (not Van Allen Belts)**. See Appendix E, p. 501.

And the third upset was that **we cannot get past the Firmament to travel to Mars** (and, Oh my God, -- we do not have rovers on Mars, either!) We cannot go into space… **we are contained**. (AL addresses this in its Epilog.)

> Right there I can see why Man back 400 years ago would not like the implications of being contained and seek to promote a different Earth – one that allowed him to promote and sell space travel, for example. So once the **atheistic scientists** climbed on board, the Global Earth movement was underway… FE detractors were discredited or ignored.
>
> NASA astronauts were either military or Freemasons – and both were easy to keep quiet – they take oaths to not divulge what they see, hear and know… under extremely harsh penalty, including jail, fines or their death and that of their loved ones.
>
> And yet I withdraw from saying that NASA <u>lied</u> – perhaps they had to continue the ongoing <u>illusion</u> of a Lunar Landing to avoid any psychological issues with being contained… (see Downside section in Chapter 10).

I redoubled my efforts to prove the FE theory wrong, and totally failed. But I finally did decide to put it in QES, and support the best findings. And I began to look for the upside, the positive aspects of the VR Sphere.

Please note that any downside to the FE Theory and Earth's containment also applied to the VR Sphere.

Positive Side

The most obvious aspect of the Containment scenario is that it **proves Creation**, by a God big enough, and caring enough, to create a realm where souls can live and grow, and are **protected from the interference** from other beings. You have to <u>birth into</u> the Earth Realm. You die to exit.

The following is a good likeness of the possible original Flat Earth,

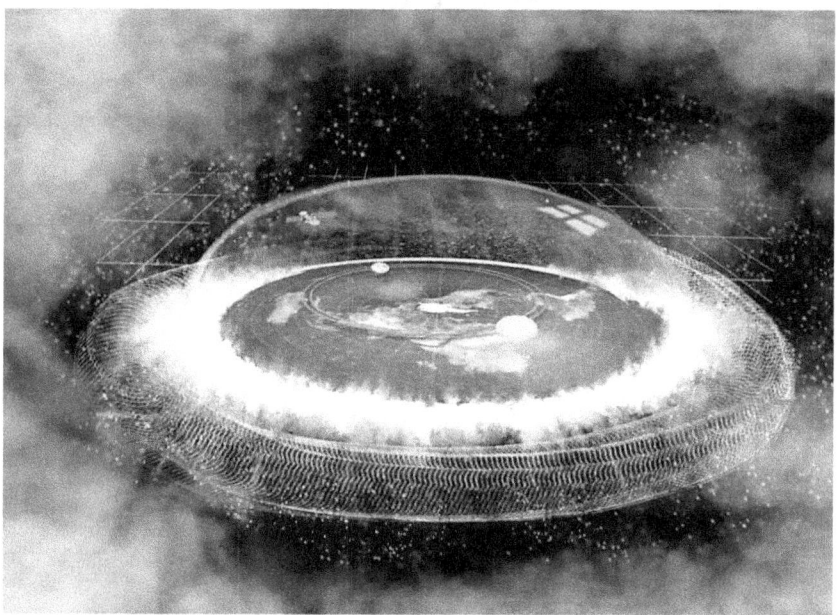

… and another artist put together a different version that includes the "pillars" of the foundation on which the Earth sits…

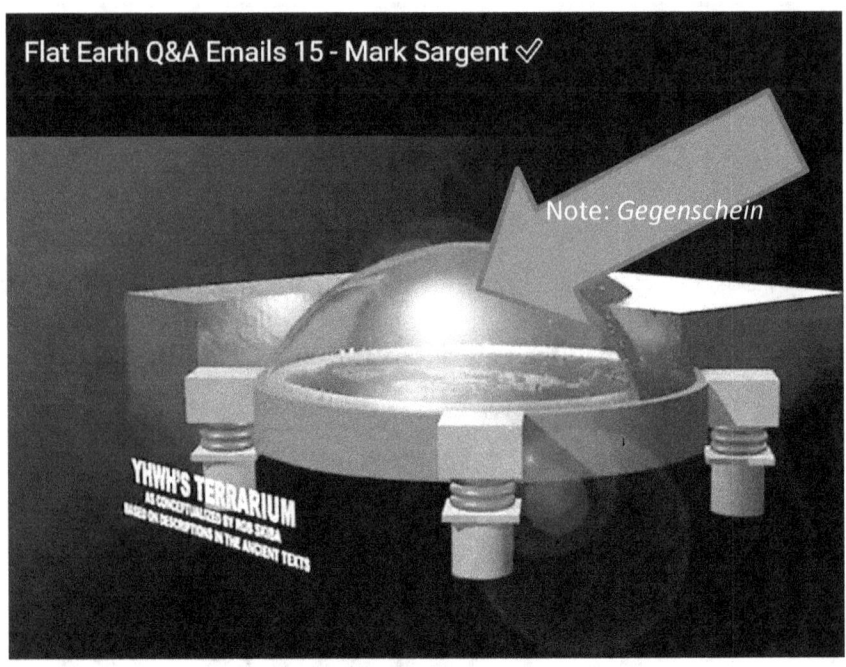

So **the Contained Earth Realm is proof of God's Grace to Man** – a safe place to learn and grow, He answers prayer, He watches to see (via His Angels) that we don't terminate ourselves or the world before learning our lessons… We are contained (or quarantined) and protected. And I guarantee you that the atheistic scientists do not like it – they can't even admit that Man has a soul… As Chapters 3 and 4 both said, God has been replaced by Science.

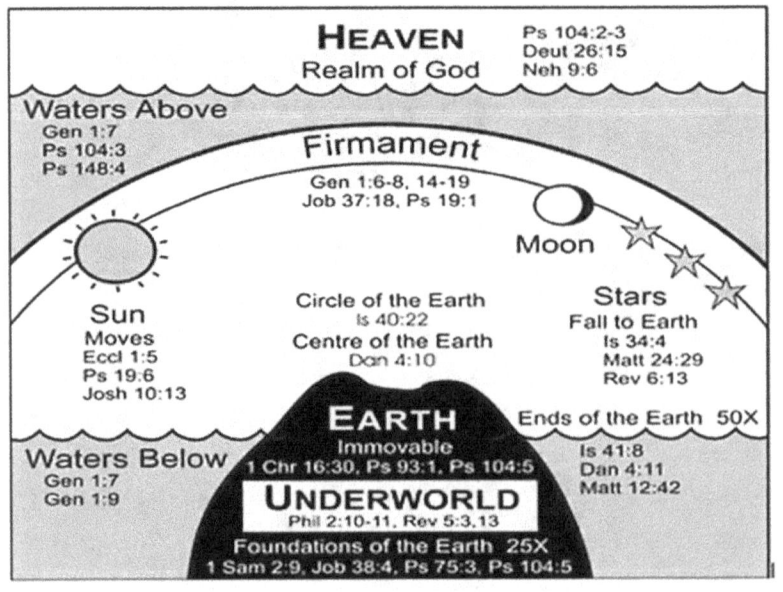

The Bible says it:

God's Grace also extends to Man via God's Promises…

Lastly, it really doesn't matter if the Earth Realm is Flat or a VR Sphere... **we are contained**. Many of the aspects of our containment are the same, under either scenario.

And Earth is a School.

So if the gods have us "trapped", is this just a punishment with no upside or benefits?

For those who choose a more spiritual path (STO vs STS) while on Earth, there are elements of a covenant or blessing – if one chooses to accept them.

Promises of God

For Christians to be effective, they have to know what God's Promises are... so they can claim them, for themselves and others, and they can stand on them as blessings. So what are they?

The following are the key promises of God principally to His believers:
(note the **bolded** statements...)

Everything is possible for he who believes.	Mk 9:23
With God everything is possible.	Mk 10:27
Without faith it is impossible to please God.	Heb 11:6
Only believe!	Mk 5:36
The word of God is living and powerful.	Heb 4:12
God has the power to do what He has promised.	Rom 4:21
I can do all things thru Christ who strengthens me.	Phil 4:13
He in me is stronger than he who is in the world.	1 Jn 4:4
Ask in My name and I will do it.	Jn 14:14
Believe you have received what you pray for,	
and you will have it.	Mk 11:23
My God will provide for all your needs.	Phil 4:19

In everything, we are more than conquerors.   Rom 8:37
God has not given us a spirit of weakness but of
power, love and a sound mind.   2 Tim 1:7

He who believes in me the works that I do,
he shall also do, and even greater.   Jn 14:12

Not by force, nor by power, but by my Spirit
says the Lord   Zac 4:6

Your body is the temple of the Holy Spirit,
thus honor God with your body.   1 Cor 6:19

If anyone is in Christ, he is a new creation;
the old has passed away and the new has come.   2 Cor 5:17

Trust in the Lord with all your heart and lean not
on your own understanding; In all you ways
acknowledge Him and He will make your paths straight.   Prov. 3:5-6

Do not despise the Lord's discipline…. Because the
Lord disciplines those he loves.   Prov. 3:11-12

No harm befalls the righteous…   Prov. 12:21

The Lord is far from the wicked; but
He hears the prayer of the righteous.   Prov. 15:29

Pride goes before destruction   Prov. 16:18

I am the Light of the World   Jn 8:12 & 9:5

It is done unto you as you believe   Matt 21:22

Ask and it shall be given; You have not because   Matt 7:7
you ask not, or with the wrong motive.   James 4:3

Faith if it is not accompanied by action is dead.   James 2:17

Come near to God and He will come near to you.   James 4:8
The prayer of a righteous man is powerful and effective.   James 5:16

And we could add…

Treat others better than you treat yourself.   Golden Rule
                                                                restated

**The reason for listing those is so you can read and reflect
on them… by the time you finish the last one, your mindset
should have opened to higher possibilities. The STO path is
better than the STS path (see Glossary).**

Power of Prayer

All Christians and some very spiritual people have found that Prayer works…
unfortunately Charles Darwin was not a beneficiary of that promise…. He might
have been the victim of **Generational Curses** (Exodus 20:5) where the sins of the
fathers are visited on the offspring--- down to the 4th generation. That sounds
genetic – if the father creates Diabetes in his body, and then has a son, that son
could be prone to Diabetes….

> It should be added that there are praying people who are not
> Christians and they report that their prayers are heard and
> answered.

And yet, those who commit to the Father and do their best to walk the Path set
before them, can <u>expect</u> that they are heard and as shoots off the Main Vine, they
may be 'pruned' (disciplined) while growing.

Many people including Christians do not pray enough… We should be praying for
our leaders that they have **wisdom and protection** – that they know to do the
right thing, and have the stamina and courage to do it.

> Chapter 1 and my experience in the hospital in 1965 shows the
> power of prayer. People were praying for my healing and an
> Angel showed up in the hospital room and healed me… up till
> then I had had many bouts with pneumonia and sinus issues,
> and after 1965, it was a very rare thing to have problems in that
> area – I never did have pneumonia again.

As another example, I was driving down the 405 freeway in Southern California
about 100' behind a pickup truck carrying a large 2-3' high roll of carpet. It was half
in the truck, **wrapped in plastic and banded closed**, a major part of it sticking out
the back of the truckbed. He went over a bump in the road, and the carpet bounced
out of the back of his truckbed…onto the road, not moving… right in front of me.

Both lanes were busy at 70 mph so I could not move left or right. I hit the brakes, and closed my eyes, grabbed the steering wheel, braced for the impact, and said, "Oh God!"

Little did I know then that that constitutes a Prayer for help!

I did not stop, and there was no impact…. I opened my eyes and was still doing about 40 mph… and the truck was still in front… empty… still moving as if he didn't know he now had no carpet… and I glanced in my rearview mirror…

Holy Cr*p, Batman! There was the roll of carpet, still a 2-3' high roll… and right behind me… I didn't hit it, and it was intact behind me… not possible! Everybody else veered around it as it just sat there diagonally blocking the lane behind me.

> **I had been protected**…somehow <u>I went right thru it</u> as if it didn't exist – certainly it was big enough to block me, hit me, and maybe cause some damage… but I had asked for help… and got it!

Another aspect of our **Earth School** is that we are only here temporarily, to learn, or maybe to do a special task (as I came in to write the books)… but most of the time, we are here to learn something and get out. **Earth Realm is not our home**.

> Much of the following information comes from the **1991 Regression** in which I saw my past life and why I was here. This was started with Prayer and we asked for the Truth, as I had some real problems in life and wanted answers. <u>I was considering suicide things were so bad</u>, so doing a Hypnotic Regression with a licensed Christian therapist was my last resort.

> It paid off and I got information on why I was on Earth (minus the task to do a book, that came in 1998), and I saw the **InterLife** where we all go when we die. Out of that I saw that Earth was a School and we are supposed to be **Earth Graduates** – but I had no idea I would be writing about it.

Earth Graduate

> This part is recapped and saved for last because it is important, and because it has been shared in part in several different places, it is best to unite all the parts and information about the Earth Graduate – all the parts in one place.

So what we want to know is two things: How does one become a Graduate? And Why? That is: What Can a Graduate do <u>after</u> graduation? While a lot of that was covered in the last two chapters of VEG, the following is offered as a more Complete, extended summary.

Candidates for Graduation

Apollonius Revisited

In retrospect, Apollonius of Týana (Chapter 6) is not a bad role model, and neither is Jesus. They even appear to have been the same person…. But as was said, all Masters teach the same Truth. It was Jesus in the New Testament who said that we could do what he was doing, and that Love was the Key. All teachings of Love and Light are worth following as they echo **Higher Truth** and that is what we need if we are to grow and be released from this Earth School. So the paths of a Buddhist, Muslim, Hindu, Jew and Christian <u>walking in Love and Light</u> are all walking the same way –

One need not change religion to get out of here.

Please note that **walking in Light and Love** is not New Agey. Jesus said he was the Light of the world, and He told us that Love was what we are to do to each other.

Secondly please note that there are Zen Buddhists and Christians, but there are no (successful) Zen Christians. Ch. 14 in VEG made a strong case for not mixing ideologies… and why. And from personal experience, it is the Christian walk, done as He intended it, that really works – and many Christians are on their way to graduating from the Earth School.

Third, since ensouled Man is but a spark of the Divine – a Soul is the divine fire placed within Man – that means that **Man has all the potential of a Christ**, and a return to the Godhead to eventually participate in the administration of Creation. That is why Jesus said "the kingdom of God is <u>within you</u>" as here in the Earth School we are expected to discover and develop our connection with the Father (i.e., thru the Holy Spirit). See Chapter 10.

> Note: what we are **not** expected to do, and it is a New Age lie, is to develop our godhood… try to do the things that Jesus did without the preparation, wisdom, discipline and empowering of the Holy Spirit.

Each person is to eventually walk as best s/he can in more **Love, humility, respect and patience**. The spiritual walk is not a "one size fits all" – it begins where the person is <u>now</u> on the Way… and it progresses with as much knowledge (Light) as s/he can absorb at any one time.

> **The Way is Love and Light. Jesus said that He was the Way (personified) and that no one could come to the Father except through Him/the Way. It did not mean that Christianity or Jesus was <u>the</u> <u>only</u> Way… There are many ways to get to the top of a hill, and some ways are better than others – Christianity is the most direct and fulfilling, however.**
> **It meant that walking in Love and Light (as He did) was the Way – again, He was a signpost, showing the Way, and He personally demonstrated it (miracles via Christ Consciousness).**

All the great teachers have sought to wake Man up for one reason: to get Man out of here and into the Father of Light's Kingdom where we can **play and serve** at a higher level! That is called the Earth Graduate.

So what do you do when you get out?

Potential Areas for Service

Having said that the Soul who graduates from Earth can fulfill certain basic, initial, functions in the Father's Kingdom(s), it can be shared that these are some of the areas open to Graduates:

Bio-plasmic Quantum Computer Techies – responsible for basic Heavenly computer support and maintenance.

Bio-plasmic Computer Programmer – performs fractal sub-programming under supervision.

> These above two were demonstrated in QES. The Heavenly computers (see TOM) develop and manage Scripts, including the Father's overall Greater Script. There are also Science quantum computers that simulate new worlds and species to be designed and built.

Akashic Records Librarian – maintains life records' storage/retrieval.

Gods-in-Training I – responsible to oversee the Drama/Greater Script Simulation: Man and feedback of the Control System. Many sub-areas here.

Gods-in-Training II – responsible for the Holographic stabilization and interface with the Replicator technology. Sub-areas here.

Soul Counselors – responsible for evaluation, guidance and training of in-coming souls to the InterLife for further development which may include imprinting or vibrational adjustment. Many levels here, including Teachers.

Earth's 3D Construct science mirrors that of the 4D realm, <u>partially</u>. So as a Simulation, there are basic things to be learned here that apply back in the 4D realm... (Since the Simulation only partially reflects the real 4D world, **what is learned in Science within the Earth Simulation is like learning the IMAX Theatre** and thinking that ALL the laws and processes are the real world. Simulation Science does not count for <u>the following three types of positions</u>).

Bio Scientists -- these beings experiment with new lifeforms, ways to engineer them, transplant them, and ways to improve them.

Astro Scientists -- these beings experiment with Galaxies, Suns, planets, comets, etc. to engineer new planets, manipulate orbits, manipulate Dark Matter/Energy, and all the while ensure balance/order in the material world.

MLD Scientists -- responsible for multilevel universe and dimensional interface, including handling Timeline Shifts when necessary.

There are many others, but it is a busy world over there; no one is sitting around on a cloud playing a harp – unless they're on a coffee break! The reason that the PTB wants to block Souls from progressing is mainly because of this position:

Gods-in-Training III – responsible for overseeing, managing and controlling the Beings of Light, the PTB, and humans –to make sure that lessons are properly administered (according to Scripts) and it amounts to controlling what the PTB can 'get away with.' This is as close as the InterLife comes to having a "police force."

And because I once asked, "What if the gods-in-training choose to abuse people, or do something nasty?" Answer: They are sent back to Earth for rehabilitation.

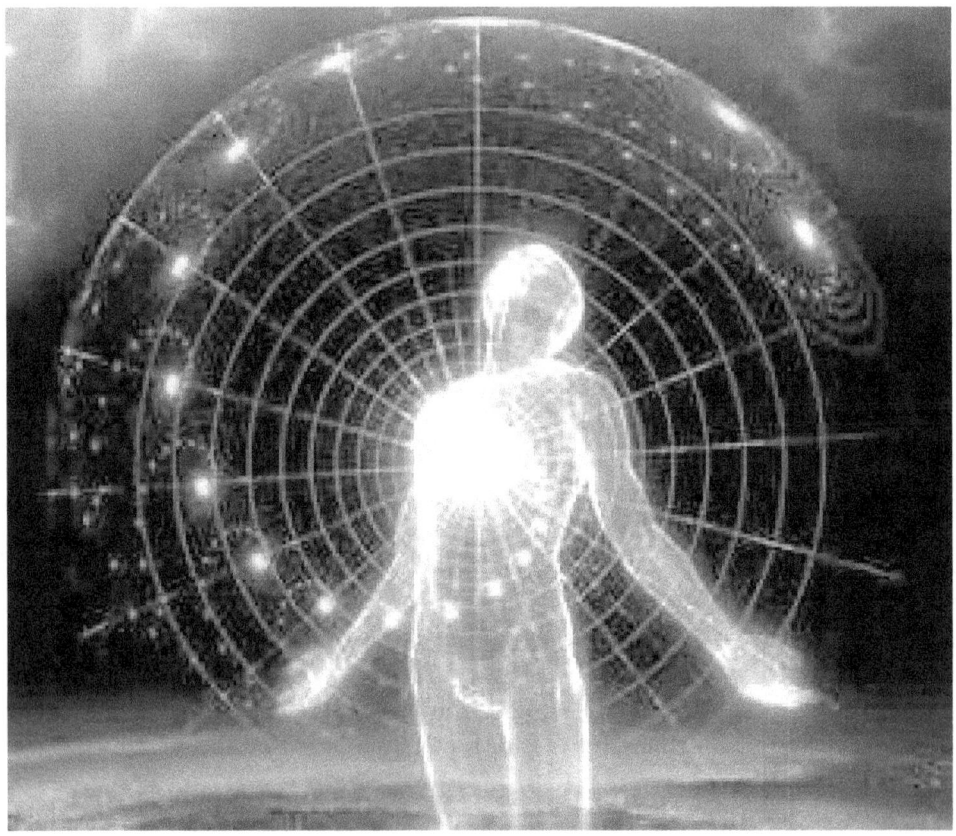

Earth Graduate
(Credit: Bing Images/consciousness)

Earth School Graduate Finer Points

Higher Vibrations to Exit

The following was an esoteric teaching, intended for the initiates into sacred orders, but it is useful for all to know that we are energetic beings, the human body has a slight voltage to it and a **Bionet** along which the body energies flow. We all have **chakras**. This is not "witchy" stuff – it is science as used by **Accupuncture**. The witchy part would be to try and use the body's energy to levitate things, or manipulate others. Using one's energy to heal another is Ok, Jesus did it, and it is done in the "laying on of hands" in some churches…. But most people cannot do this. It has to be a **calling with an anointing** – such as the great British healer, Smith Wigglesworth (or even today's Benny Hinn) had.

Humans who do graduate from Earth all have to meet at least a 51% level of love, patience, wisdom, desire to serve (STO), knowledge (not false beliefs), humility, etc….as VEG Ch 15-16 said. Above all, **Respect** is a major key – for oneself, others and the planet.

This doesn't make souls to be peas in a pod, but they must all know who they really are, Who is Boss and Whom they serve, and where they might fit in the scheme of things – based on interest and aptitude. They are very respected by others because to graduate, you really have to have your basic stuff together – and you do that by overcoming the BS here on Earth! (And **you don't have to be perfect**…)

> Gaining **Love and Knowledge** raises one's personal vibration and that is a key to graduation.

To see thru the lies, still serve and care about others, and respect yourself and rise above the problems that come your way is what it is all about. That takes a lot of determination, searching, and a certain growth in awareness that They expect of a graduate Over There. **One needn't be perfect**, just have demonstrated a major % (again that 51%) of one's being IN ALL IMPORTANT areas like STO emphasis also called **service to others**. (VEG Ch. 15 enumerates these.)

> It is the fact that your **PFV** or personal vibration increases to a higher resonance that guarantees you can leave Earth – if you **vibrate higher than Earth**, and you can sustain Love, Patience, Humility, Respect and true Knowledge, you are a candidate.

> **Repeat: this is not New Agey – the more loving a person is, and the more true Knowledge s/he has, the more their PFV is a higher vibration. Jesus was a prime example.**

But first, one must <u>want</u> to graduate… and there is a key:

Attachment vs Futility

> In addition to Love, something else is important: You must **drop all attachments to Earth.** People, places and things can and will bring you back – even if you are ready to go. Henry Thoreau knew this, as did the late **George Carlin** (Appendix B in TOM.). Those ties must be broken <u>while on the Earth</u>.

> And the easiest way to do that is by seeing the **futility** of always getting your way, when you want it, the way you want it, where you want it… and **detach** from having anything work or not work. No goals… except to get out, and that means to <u>detach from everything</u>.

The higher attitude is: "It's Ok if it works, and Ok if it doesn't."

Futility is the key. It is **not a negative, despondent resignation** – it is based on a genuine view that 'stuff' doesn't really matter, you <u>will</u> see your loved ones again (on the Other Side), so release them, and give up drugs, tattoos, piercing everything, drinking, smoking and must-have sex partners... Sexual fantasies are best gotten out of your system – it is an Earth 'drive' (<u>from the body</u>, not You) that will attach you to Earth and you'll be back... to learn to let go.

The higher attitude is: "Sex is Ok and I can take it or leave it." That is a high PFV... and we have to politely control the body most of the time, anyway, as it is one of the prerequisites of a Graduate... (see Ch. 7 in VEG)

An even higher attitude is: I am an **eternal soul** – a denizen of the Multiverse and I can handle whatever comes my way. I respect myself, others and the planet, and I seek to create quality and serve wherever I am. Experiences are for my growth, and the worst of them are tests – I can do that. **I don't have to like it, I just have to handle it.** Death is not final – I just move on to another experience, and will probably have my soul associates with me, and we'll play some games and, learn, serve and move forward. Love, Light and Knowledge is what it is all about!

Laissez rouler les bons temps!

Personal Frequency Vibration (PFV)
(see Glossary)

Ok, so this isn't a religious lecture. These are the things that all Earth Graduates have come to understand and embody... that is why <u>their vibration (PFV) went up</u> and that higher, purer, energy emanates from the top 3 – 4 chakras – that is why <u>Transformation of Man</u> (TOM) dwelt so much on that aspect. Your chakras are either open or closed...and if they are open, and:

If you have dealt with **Ego** – Chakra 3 is spinning clearer;

If you feel more **Compassion** for people – Chakra 4 is open and spinning cleaner;

If you speak **Truth** and have stopped 90% bad swearing – Chakra 5 is spinning cleaner and faster;

If you have meditated and had any encounter with Angels – your Chakra 6 (Ajna) is cleaner and you are becoming intuitive and perhaps clairvoyant;

If you have gotten your first 6 Chakras cleaned up and passing higher energy – your 7th Chakra (crown of head) has begun to open and stabilize a connection to your Higher Self…

The point is that occasional sea salt baths and eating right plus right thinking and speaking will do a lot for raising your PFV – the energy vibration that affects your **Ground of Being** and no one (not even the Higher Beings) can stop you from "graduating" and being able to sustain being in the higher energy of 4D…. **It is <u>all</u> <u>energy based</u>** – if you have higher, finer energy (PFV) you will <u>automatically</u> make it.

Be very clear: **No one says you go or stay.** It is all based on: Do you have the right vibration? If you do, you go. If you don't, and you try to force your way to 4D, or 5D, the higher vibrations can severely harm you. Similar to a fine crystal glass that can handle up to 20 Hz (vibration) and if it is put in a room with 60 Hz vibrations, it will shatter.

Healing Crisis & PFV

Something very important to insert here (which was overlooked in the healing sections of TOM and TSiM) is something I learned in my classes on Nutrition and Body Care (I was studying to be a **Naturopath** in the early 80s.) Do not go looking for this info in a medical reference (like Mosby's) – it isn't there. Modern medicine does not know much if anything about this phenomenon.

When you start taking care to eat right, maybe doing salt baths, meditating, and even going to the health food store and trying special supplements, your PFV will start to rise, and that is good. However, the body also sees what is going on and may <u>smack you</u> with a healing crisis (so said because it is a <u>surprise</u> when it does it!).

First remember how when you were sick, maybe with the Flu, and all of a sudden you started to sweat, you felt hot, and you took your temperature to find that you were running a 102° temperature – **fever** is the body's way of killing unwanted pathogens in the bloodstream. It is part of the *Vis Medicatrix*... the innate system of healing in the body. No cause for alarm.

> Just as a fever can be the body's way of dealing with an infection, so too you will learn something new that will show you that the **body has an intelligence of its own**. (That is why you <u>gently</u> tame it.) And you know that if you have ever tried to lose weight, or break a habit.

Just as the body can initiate a fever, it can initiate an **energetic cleanse** of old emotions stored in the tissues. This is well-known in TCM and among "backward" 3rd world countries where energy healing is practiced – such as Reiki, Qigong, and *Kundalini* meditation – and it often happens following an energy healing.[285]

Typically the body is cleansing itself of **toxins** (and today's medicine understands this) but there is a unique "energy sweep" that the body does using the Bionet, coursing higher levels of *chi* thru the meridians and when they hit stored cellular emotions (**emotional toxins**), the higher energy seeks to dislodge or "dissolve" them... and all of a sudden the person feels the emotions that were associated with some traumatic event that the body could not handle (probably happening at a young age) and now as an adult the person has to process them out.

> This is part of the **soul-cleansing** that the Earth Realm does.

This can happen anytime the body is healthy and you have no symptoms. I just went thru one for 8 weeks... and had forgotten about what I learned in 1980. So I suffered and visited the doctor twice who had no clue. The body will **use the sinuses** to process-out buried emotions and it looks like the body is weeping, crying and as if one has a cold or allergy that will not quit. (This may be what I experienced in 1987 in Alabama – see Chapter 1.) And then one day – barring intervention such as I had in 1987 – it just stops when the body is done.

I also see now that New Jay's higher vibration after October 1998 (Chapter 1) kept my body from setting off healing crises all the time... there was an initial crisis, and then things went smoothly as my higher vibration sustained my body at a higher level. For 16 years.

What happened in Aug 2016 was traumatic: the return to the old energy level triggered another 8-week healing crisis (Oct – Dec, and again in January-May 2017) – this has been a real sacrifice to do those books!

This is just a caveat… do not be alarmed if your body does this just when you were starting to make progress and feeling more optimistic, healthier… these things must be cleared, and can even be from a former lifetime where you carried the trauma in your **emotional body** (layer 2 of your aura) – explained more in TOM Chs. 12-13.

Your aura is part of your layered soul-body. Each of the 7 chakras feeds one layer of the 7-layer Aura; the 2nd from the bottom is the emotional chakra/layer.

See the following diagram…

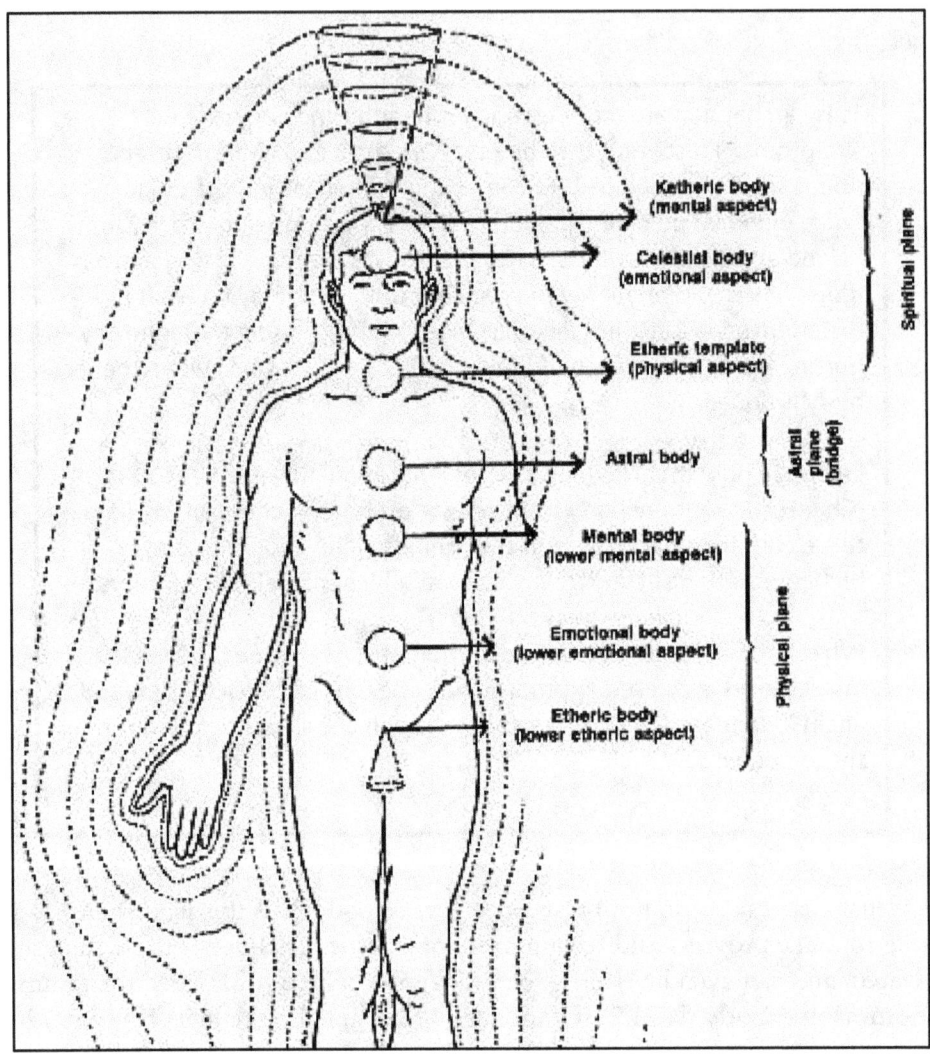

The Seven-layer Aura System
(credit Barbara Brennan, *Hands of Light.*)

This is not New Agey stuff – it is Chinese Traditional Medicine.
Be aware that **you are not your body** – you have one and you as a soul
inhabit the body and should be raising the level of consciousness as
well as the higher vibrations of your PFV can and do affect the cells of
your body.

The body in turn, makes adjustments. Your Higher Self is in control of
this and will not let the body take on more than it can handle. And this
cleansing benefits you in the InterLife (see TOM).

And there are ways of increasing the vibration of your soul-body that follow:

Self-Mastery Issues

Ensouled Man's potential, when actualized, renders him a truly awesome being – doing what the man called Jesus did and more. Jesus said it (paraphrased):

> "These things I do you shall do and greater…" John 14:12

So when does it happen? **Respect yourself**, develop compassion, gain knowledge, and live in patience, detach from outcomes, things and people… a long list of requirements to graduate, but that is why:

> "The **graduate** from the human experience is very **respected** elsewhere…
> [i.e., outside the Earth Realm]." [286]   [emphasis added]

And it is easy to see why: if a Soul can survive this screwed-up world with its disrespect for everything and everybody, survive the pollution and the killer diseases, resist the temptation to lie, cheat and steal, not follow the crowd, not give in to corruption, and still emerge with his/her integrity intact, that would be worthy of **respect**. A Soul who walks the talk and does not sell out. A Soul who thinks outside the box that has been created for all of us by the PTB-dominated Media whose goal is to keep us as dumbed down robots so we will buy and do whatever Madison Avenue suggests.

There are three main aspects (1a -1b – 1c) of Self Mastery

Self-Mastery 1a

It is called self-mastery. And that starts with **self-respect**. You don't do those things that are inappropriate – like overeat, or run with hoodlums. Self-respect means the body is the temple of the divine spark, the Soul that is so highly prized that there are some beings in 4D who would give anything to have one, and because they can't (I repeat, I know) they work to stop us from becoming all that we can be. When we graduate from Earth, we can be released to return to the higher Realm from which we came and in which we are then ready to serve.

> **And, repeat: One need not be perfect to get out of here…
> but you do have to set the ideal and have the intention
> to achieve it.**

If the gods had to wait until a soul on Earth attained the status of a Jesus, they'd have to wait a long time. To graduate from the 3rd Grade in Elementary School, one need not know Algebra – knowing basic arithmetic and the multiplication tables will do. I suggest that being a Ghandi is sufficient, or a Lao-Tzu, or even Mother Theresa. No doubt there are further Schools awaiting Souls – and **Earth is not a "finishing school."**

How many people think that staying in 3rd grade, or even 12th grade (high school), is all there is? (Aren't you tired of the cafeteria?) Reaching the initial stages of self-mastery and <u>intending</u> to go forward are enough to get us out of here and into a more exciting realm.

Three things are required to leave Earth as a Graduate:

(1) you see the **futility** of demanding what <u>you</u> want here,
(2) you learn, acquire **knowledge and compassion**, and
(3) you experience and realize that **you <u>can</u> handle whatever** life throws at you.

> Does your religion, faith or belief system assist you in doing those 3 things?
> If not, you might want to check out your local Spiritual Life Center, Religious Science, or **Unity Church** (Chapter 5)… not a cult, but a true spiritual growth center which is what today's **New Thought** (<u>not</u> New Age) churches are. It can be the fast-track to enlightenment.

Self-Mastery 1b

Several key points:

Do <u>not</u> assume that the drama in your life accurately reflects who and what you are. **Many negative events are just tests**.

Do <u>not</u> let the ignorant tell you that you are wrong and so messed up that you'll never amount to anything. Keep your own counsel.

Do <u>not</u> give your power away. Make your own decisions – think outside the box. (This gets easier as you get older.)

Such a person with a lot of Light may have to work hard to trust himself, develop an **inner strength** and learn to listen to their own ideas – despite others' "helpful advice." In fact, that may be the thing that a soul is here to learn – a form of self-confidence, inner strength, **self-respect** and connection with a higher part of himself or herself.

> Both Drs. Holmes and Maltz would agree with that!
> (See TSiM.)

To walk calmly amid the crowd of rushing fools, know the truth, and not always follow the crowd. Self-mastery is expected of souls who want to **graduate from the Earth School**: to live by the highest values you can when all about you are involved in lesser pursuits. It's called **Integrity**.

Self-Mastery 1c

Please note throughout the last few pages that there has been an emphasis on **self-respect**. The path of an Earth Graduate begins with it, leads to respect <u>for others</u>, and then to respect <u>for the planet</u>. The Graduate knows that All are One: animals, plants, any 4D ET brothers and sisters – anything done to harm any of them ultimately harms ourself – and self-respect says we don't harm the temple – we don't eat junk food, disparage or harm self, nor do we disrespect elders, ignore traffic laws, we don't lie, cheat or steal – not because we're goody-goody two-shoes – but because that behavior **doesn't work**.

That is a big part of being an Earth Graduate – **we do what works** – serve, tell the truth, show compassion, open the door for someone less fortunate or who is carrying bundles, do something nice for someone because **what goes 'round comes 'round**. The Universe will repay you… even if no one else does! In the beginning you might help others because you know that you will be rewarded by the Universe for helping others, but after a while, it becomes second nature **because it is the appropriate thing to do**.

> **There is never an excuse for a failure to be compassionate.**

If you respect yourself, you cannot treat yourself like trash, dress crappy, pants down around your knees, half your body tattooed, the other half pierced, and a Mohawk haircut (in bright pink) so others will notice you! … that does not respect the body when you know that you, the Soul You, has the god-potential within – down the road, you <u>will</u> walk on water, heal the lame, levitate, and disappear and reappear at will. Why look like a freak when you do it?

Do We Need Heroes?

That is the popularity of *The Matrix* -- **Neo** represents one who can cut through the illusion, release his inner power, overcome the programming, and release the captives. Can you imagine his character smoking and drinking, joking around, tattooing half his body, piercing his nose, and not respecting himself? Would Clint Eastwood dress and act like a fool in his *High Plains Drifter* role? We need good role models… and you may be one for someone you don't even know!

> **If you stand for something better than the masses,**
> **you must act better than the masses.**

The PTB-sponsored Media has given us some false heroes, and seeks to **entrain** our young people (who are often too impressionable) via **stalk-it-and-kill-it videos** into a mindset where problems are solved with a gun. Today's movies are often <u>too</u> violent and sexy – because <u>that is what sells</u> – it takes more insight and brains to create a movie like *ET* or *Sum of All Fears*, or *Shining Through*. Even Humphrey Bogart in *Night Passage* had closure and redemption at the end – because the movie industry then (30's-40's) promoted ethics – they couldn't even make a movie where a character commits suicide (in *Night Passage*, the bad woman's fall was made to look like an accident). Times have changed.

So if you want to walk the path of the Earth Graduate, you can laugh, party and have fun, but down inside you must **remember who you are, where you're going and what you stand for**. And the higher you go, the less you will do inappropriate things – so in the beginning don't get down on yourself if you 'wiggle when you walk'… it is all part of the path: Rome was not built in a day and you will not walk perfectly from day 1. Be gentle with yourself, know that old habits take time to overcome – (Chapter 2) unless you have a real deep epiphany about yourself, a 'soul shock', and that was the purpose behind the Est Training (Chapter 1).

Self–respect includes self-compassion. And **what you do for yourself is easier to do for others**. Having stumbled yourself, it is easier to have compassion for others going through the same problem(s)… and so you won't be criticizing them. You'll help pick them up…

> **Remember: Treat others better than you treat yourself.**
> (Golden Rule inverted.)

Souls as Earth Graduates

Man has a unique nature and a unique destiny, and **we are here on Earth to work it out**, and we probably would have by now if: (1) we hadn't resisted our lessons, (2) we didn't party so much, and (3) the PTB had let us know.

> Our purpose in coming together as creator fragments is to succeed in
> training ourselves enough about love, caring, and relationship to become

more of who we are. The redemption is we, as … **progeny of the high forces of creation,** are ascending back to heaven in unity of diversity, as celebration of individuality in communion, not loss of individuality. We are holo- or fractal-fragments of the creation's creator, [who] is **wanting to create creations with us** not for us…

Our first major task is to re-create the existing mother universe we find ourselves within with all its conundrums. Solve the unsolvable evolutionary problems. We learn to pick up the ball in our **training wheel practice universe** before we even want our own. And we want to learn very carefully, and so we use time to do it in a serial manner. That is the game…

Each creator fragment that is a human soul, ultimately seeks its origin [and soulmate] and return to Home. We are all … learning to love and nurture our individual and co-creative mutual universes.

Our job is to achieve spiritual evolutionary acceleration sufficient to help solve age old problems of spiritual evolutionary inertia in the universe. [287] [emphasis added]

Elsewhere, the information is given that **Man is ultimately to be a co-creator** as his inheritance providing he can keep moving into more and more optimal timelines:

…as a reward, humans who accomplish this task, will be granted an initially uninhabited virgin future that can become even more optimal beyond comparison…

That final loop optimal future becomes the end-game singularity conduit path through which **all** souls of all alternate [multidimensional] lines will eventually travel to **become qualified macro-creator agents**. [288]

That is the promise to all Earth Graduates. And that is why the PTB don't want you to know, nor develop your spiritual side.

Summary

Ensouled Man's birthright is to (1) rise above those who would obstruct him, and (2) **rule in higher realms** – similar to Angels in Training – and this is obviously what the atheists and PTB would like to prevent. This is why Robert Monroe was specifically told by the Beings of Light (*Inspecs*) he encountered that Earth Graduates are very well respected. That respect is due to personally developing what it takes to overcome darkness and get out of here! So Man emulating a Jesus, Ghandi, Buddha,

Krishna, Mother Theresa or Apollonius would indeed evoke **respect** wherever he or she went.

> Ironically, the TV personality, Ellen DeGeneres, is on her way…
> she just granted (Feb 2017) free college scholarships to about 41
> (mostly Black) kids from an impoverished Brooklyn NYC
> neighborhood where no one gets to go to college – and that will
> make a big proactive difference in those souls' lives. [289]

Rewind: It was Jesus in the New Testament who said that we could do what he was doing, and that **Love was the Way**. All teachings of Love and Light (Compassion and Knowledge) are worth following as they echo Higher Truth and that is what we need if we are to be released from this Earth. So the paths of a Buddhist, Muslim, Hindu, Jew and Christian walking in Love and Light are all walking the same way – to repeat: **one need not change religion to get out of here**. All the great teachers have sought to wake Man up for one reason: to get Man out of here and into the Father of Light's Multiverse where we can serve.

> That is called **the Earth Graduate… who is always to some degree spiritually awake.**

It is worth repeating what Robert Monroe was told:

> By far the greatest motivation – surpassing the sum of all others – is
> the result. **When you perceive and encounter a graduate, your
> only goal is to be one** yourself once you realize it is possible.
> And it is. [290]   [emphasis added]

> When you know, why would you settle for anything less?

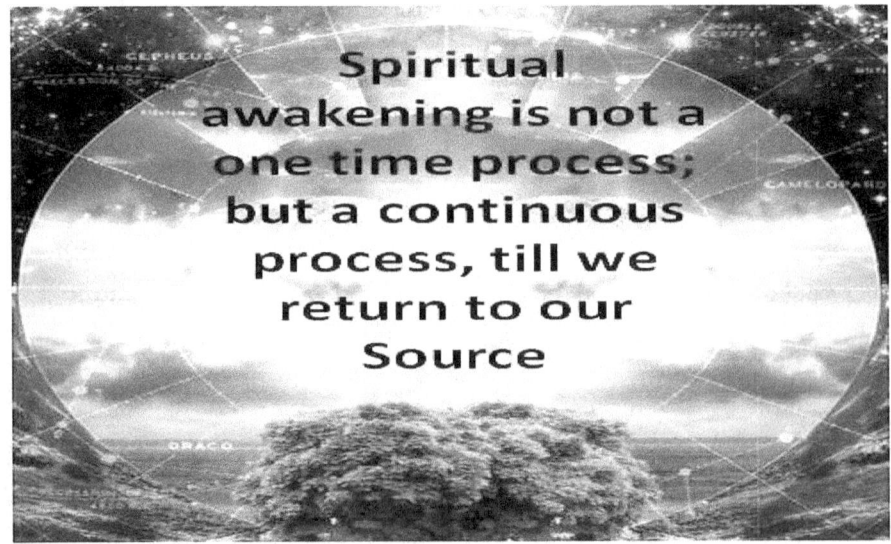

Spiritual
awakening is not a
one time process;
but a continuous
process, till we
return to our
Source

That also means that spiritual awakening is a <u>continuous</u> process… you can't give up …. doesn't that just make your day?

Signs of Spiritual Growth

So how do we know if we are growing or not – are there signs along the way? Yes.

12 Symptoms of Spiritual Awakening

1. An increased tendency to let things happen rather than make them happen.
2. Frequent attacks of smiling.
3. Feelings of being connected with others and nature.
4. Frequent overwhelming episodes of appreciation.
5. A tendency to think and act spontaneously rather than from fears based on past experience.
6. An unmistakable ability to enjoy each moment.
7. A loss of ability to worry.
8. A loss of interest in conflict.
9. A loss of interest in interpreting the actions of others.
10. A loss of interest in judging others.
11. A loss of interest in judging self.
12. Gaining the ability to love without expecting anything.

http://recoverytradepublications.com/blog.html?entry=12-symptoms-of-a-spiritual

The physical symptoms and signs of **spiritual awakening** can be unnerving if you don't know <u>why</u> they are happening. It may even involve a Healing Crisis.

As was pointed out in TOM, there are **Near Death Experiencers** who appear to briefly die, then come back into the body a changed person – they come back with the same attributes listed above… and more! The **NDE is a spiritual awakening** in spades – the person often reports seeing heaven, meeting their deceased loved ones, and having a brief meeting with Jesus or some Master who advises them on why they have to go back to Earth.

NDE Transformational Effects

It is interesting that there are different ways to have a spiritual **transformation**...

> *Kundalini* **awakening** via meditation or yoga

> **Near Death Experience** (NDE)

> **Baptism of the Holy Spirit** (BHS)

> ...and a way that has been tried by seekers which <u>does not last</u>...

> **Ayahuasca** trip

The first three produce pretty much the same effect – it is something of an expansion of consciousness and a connection with something larger than oneself. And what is great is that the <u>effects are lasting</u> on the person who goes thru (and survives) the experience – *Kundalini* can be hard on the body and some people gain *Nirvana* but also lose some of their memory, or come back not knowing who they are. The NDE and BHS are much safer as they are orchestrated by Higher Beings, sometimes Angels.

> Getting high thru drugs not only doesn't last, it can permanently ruin your mind and body.

So here are the effects of spiritual transformation as reported by NDErs:

Post-NDE Effects

Many NDErs say their lives are truly changed in a way they could not have imagined after they adjust to getting back into their bodies. Many of them had a brief meeting with a **Being of Light** (an Angel without wings) and they felt incredible Knowledge and Love while they were Over There – just on the InterLife's <u>front doorstep</u>. (They were not allowed to enter all the way in.)

Here is a list of **common observations** they say: [291]

> There is nothing to fear about death.
> Life does not begin with birth and end with death.
> What matters most in life is **Love**.
> Living a life oriented to materialistic acquisition is missing the point.
> Being a big success in life is not all it is cracked up to be.
> Seeking **Knowledge** is important – you take that with you.

Again as <u>Virtual Earth Graduate</u> said many times, **learning to Love and gaining Knowledge are the two important things to pursue while on Earth**.

Further insights were gained over the years of interviews and reflect a multitude of NDErs' **suggestions** coming from their experience: [292]

> **There is a reason for everything that happens**.
> Find your own purpose in life.
> Appreciate things for what they are – not for what they can give you.
> Do not allow yourself to be dominated by the thoughts or expectations of others.
> Do not be concerned with what others think of you.
> Remember, you are not your body.
> Be open to life, live it to its fullest.
> Money and material things are not important.
> **Helping others is what counts in life.**

How does that square with the bumper sticker that says:

> "He who dies with the most toys, wins!"

Dr. Kenneth Ring, after many interviews with NDErs compiled a psychological portrait of these people. What he found was not only the evolving **"cosmic consciousness"** of these people as listed above, but he also noted many **side effects** they tended to have in common: [293]

> **Increased love for all people and things.**
> Increased sensitivity to electromagnetics.
> Increased psychic abilities.
> **Seeing energy – chakras and auras**.
> **No fear of death**.
> Decreased worry – surrendering to the divine plan (also called the Greater Script in Chapter 7).
> Belief in **Reincarnation**.
> Vegetarianism (now cannot eat animals).
> Major relationship change – often divorce from unbelieving spouse.
> Career change.
> **Less religious and more spiritual.**
> Living each day like it is the last.
> Increased concern for the planet – ecology.
>
> …and…
> Approaching all humanity and creation with **nonjudgment and complete acceptance.**

Without being political, this also applies to accepting LGBTQ people – yes, **Transgender** causes societal ripples (the sticky bathroom issue) , but this may be something that again is **catalyst**. Personally I cannot imagine where the parents of a girl are coming from when they approve of their daughter becoming a 'boy.'

And it is hard to see that that was part of the person's **Script** as designed on the Other Side (see TOM) – it really appears to be an aberration of our culture since the gods will put that soul back in a girl body the next time around – **All souls experience both genders at some time during their many Earth incarnations**... so avoiding it for a reason other than <u>the soul made a mistake and is not ready</u> for the experience of the girl gender, means it is going to 'cost' that soul something when they die and cross back over to the InterLife.

Souls <u>can make mistakes</u> and not be ready to experience the opposite gender (especially after scores of lifetimes as one main gender)... but IF they made the choice (for the unique lessons such a swap offers) and then cop out... that says there is another problem that the soul will also have to address.

The point being: how can one be a well-rounded Graduate if s/he cannot handle even one lifetime in the opposite gender? Would the cop-out be respected? Is Fear respectable?

Less materialistic.
Understanding the challenges we face are simply **lessons** to learn
　　here in Earth School.
Knowing with certainty always to follow my truth and surrender
　　to **the Flow** of the universe (aka the Light of God).

That is a long list and it is impressive. Just one trip to the Other Side, and people are infused with a higher consciousness, **a deeper awareness**, of what they and life are all about. (I experienced it too after the 1991 Regression.) The list is worth reading slowly and contemplating…

It appears that **Transformation is whatever takes us out of our myopic little world**, our narrow self-view and concerns – whether it be an NDE experience, or an epiphany. Both are larger experiences of the Whole.

Last Word

Remember the **Gospel of Thomas**, v.70:

> Jesus said, "That which you have will save you if you bring it forth from yourselves. That which you do not have within you [will] harm you if you do not have it within you [and you try to bring it forth]."

That means you can't "fake it till you make it." Be sincere and go for it!

If you fail, get up and do it again.

In the words of Winston Churchill during WW II:

> Success is not final...
> Failure is not fatal;
> It is the courage to continue that counts!

Namaste!

Epilogue

I don't usually do this in my books, but this one has been a surprise in the way it turned out. I set out to prove the Flat Earth (FE) wrong, and really wanted to substantiate the VR Sphere, and had a hell of a time doing it. The surprise: It was very hard… which is why the rumors (FE speculations) persist on the Internet. Little do they suspect that the original Earth probably <u>was flat</u> but is not any more.

> So the summary is: Earth Realm is a very sophisticated Simulation with an impenetrable energy shield (*Gegenschein*) around it, and we are on a VR Sphere that looks something like a global IMAX Theatre.

Why was I, even as the New Jay (Chapter 1), told to write about the IMAX Theatre (the real Earth Realm) in 2008? They had to have had a reason, and I suspect it has to do with the fact that the gods had turned the Earth Realm into a VR Sphere, and would do so to further develop Man's ability to think, design, and adapt to a more challenging environment. At some point with all our rockets and advanced propulsion systems, we would discover the Firmament (Shell) around the Earth and see the Earth's real shape… so the gods (not being ogres) would accommodate our "expansion" and let us play around in a **VR Sphere's Cspace**. (See Chapter 3)

As I said before, I have mixed feelings even about a VR Sphere – and whether the gods would morph Earth into that shape if They are determined to see Man spiritually grow – How does accommodating him to move off the Earth Realm accomplish that goal? I see the Earth Realm as an example of **God's Grace to Man**, that instead of throwing him away as a dysfunctional product, the gods (who serve the Father of Light, or The One), seek to redeem us (by reeducating us – to be **STO**). Thus, letting us play around in the Cspace of a VR Sphere might be counter-productive…

On the other hand, I see the **psychological downside** to being physically "trapped" in a realm that we can't get out of, forcing us to adapt and turn to activities enumerated in Chapter 10 – alternative ways to deal with Man's tendency to compete. In fact that may be why the Church and Science 400 years ago decided to go along with <u>and promote</u> the 3D, round rock concept, axis-rotating and spinning globe around the Sun … but that <u>also</u> resulted in Science ditching God.

Now Science and Religion need to find their common ground.

So, is there a synopsis that we can **put on the shelf** until the issue is finally resolved?

Rewind: Flat Earth to VR Sphere?

At the end of Chapter 10 the basic evidence was revealed – and there were two <u>major</u> aspects that could not be 100% ascertained: Gravity and the existence of Antarctica as a continent. The Earth Realm, despite the excellent FE evidence, is still a VR Sphere (particulars in CH 12 of VEG). And remember: what is important is that **Earth is a School for souls**... not what shape it is.

And yet since Earth is a **Simulation** (also see **Appendix C**, Elvidge is one of <u>many</u> scientists who say Earth <u>is</u> a Simulation), that means that whereas Earth could have been designed and initially built FLAT, now due to the Shell around us, Man is going to have trouble trying to go into space (FE or VR Sphere – it makes no difference with respect to exploring Outer Space). He was given electro-gravitic propulsion, by the same entities who told humans all over the Earth, millennia ago, that the Earth was flat... but space travel is getting nearer his grasp. A Simulation is easy to reprogram and change the shape of Earth – as Man's needs expand, so a **Simulated Earth Realm** can expand.

> Did the gods realize that a contained human may be a dangerous one and decide to give him more room in which to play and learn? Will the Cspace expand, or will the gods release us (remove the Shell) and since we are already in 4D, we can interact with other beings in the Cosmos?

So the **Earth Realm currently is a Simulation** ... thus it is evidence for the Earth having been **designed for a special purpose** (Earth School) and that means we are not a rock rotating at 1000 mph, nor are we souls spinning around the Sun at 67,000 mph with no God and no purpose... **Science is not our God**.

> However, the VR Sphere is a 3D Construct in 4D, we do not rotate, but it is possible we are circling the real Sun, They added a real Moon, and what stars we see are the real ones... almost as if we have been gradually introduced to the real 4D world around us... and when that happens, will we ever know that we were not a 3D Construct that was contained? And as there now appears to be a Firmament, and ancient Man saw it when it was lower (**Appendix E**), it behooves us to seriously consider that Earth is yet protected by a Shell of some sort – which NASA is now calling the **PlasmaSphere**. (Appendix E.)

Key Arguments

This is where the 'rubber meets the road' and you'll find the What and the Why of this book. This refers to information in Chapters 1 and 10.

Whereas...

> I was supernaturally healed (1965, 1987, 1998 and 2001),
> I was able to see my past 3 lifetimes (1991) and why I am here this time,
> I was visited (October 1998) and something was done to prepare me,
> I was given the ability to see auras (2006-2008),
> I was given the outline (TOC) for the VEG book (2008),
> I was visited by Baldy several times (2003, 2013) to keep me on track...

It appears I was being guided and protected – to do the books.

And whereas...

> I discovered a significant consensus among many scientists who suspect that Earth is a very sophisticated **Simulation**, (Chapter 9)
> I discovered several serious 'proofs' for the Flat Earth and a Firmament, (Chapter 3, 10 and Appendix E)
> I discovered that Charles Fort, Robert Monroe and NASA discovered the existence of a **'Shell'** surrounding the Earth,
> I discovered that NASA admits to the *Gegenschein* and the **Plasmasphere**, (Chapter 3, 10 and Appendix E)
> I discovered several videos and pictures of Earth from high-altitude balloons showing the edge of the Earth to be flat, (Chapter 10)
> I discovered that Gravity is not what we think it is, Science today cannot find it nor measure it, and Sir Isaac Newton said it was an <u>assumption</u>, (Chapter 4 and Appendix D)...

Therefore...

Now if you were a God who cared about Man and wanted a safe environment in which to protect, nurture and grow souls, would you not create the Earth Realm as a Construct that can be altered as needed (eventually a Simulated VR Sphere) ?

So Now What?

First, we can **put the issue on the shelf** for the time being, being aware that Earth is not all that we have come to think it is.

> This is exemplified by strange structures and creatures around the world… It is no accident that they have been included in our Realm.

Second, we begin to realize that **we may be under observation and what we do is accountable…** Someone is watching.

> And in any case, given Man's proclivity for War, there needs to be an alternate way (Chapter 10) to give expression to his combativeness…

Third, what is left is to create **alternate diversions** (as Chapter 10 said)… and begin to get our collective act together and keep planning to move into Space, which might initially be the Cspace between the two Shells (shown on pp. 340). I strongly suspect that the VR Sphere has been done **for** Man as They did not correct me (Chapter 1) when I was told that Earth was a 3D Construct in 4D, a Simulated realm, the VR Sphere.

> And wouldn't it be interesting if we prove ourselves worthy to be released from **Quarantine** (the Shell off which the *Gegenschein* reflects) by becoming more responsible, mature, intelligent and respectful of all other lifeforms, human and Other, and then the existing **VR Sphere** could be opened to release us, and we might be allowed to explore real Outer Space.

Is that a viable future for Man, or are we facing another **Wipe & Reboot** where this version of the Earth Realm is reset and we start over again…?
And yet, there is a major downside in all of this…. this was given to me in 2008 and I chose to withhold it until now.

How They See Us

This has to be said, no matter how distasteful, and I deliberately avoided it in the preceding 6 books. It foreshadows the future for Man….

Man is a big problem, and even in the days of the Anunnaki (250,000 years ago) humans were considered noisy, smelly and dumb – but useful. The former aspect was why Enlil wanted to wipe them out, and their usefulness was why Enki wanted preserve them to see if he could improve the genetics and create a better, brighter human. To some extent that was done centuries ago with *Adapa*, then Neanderthal was upgraded to Cro-Magnon, and finally to Homo *sapiens* (us)…and current-day efforts have been done by the Greys (which are **biocybernetic robots**, not visitors from another star), and their task (Earth-based by the way) has been to improve the human genome – but that has been a lot like trying to paint a moving freight train!

While it was said in TOM, it bears repeating here: the Greys are biocybernetic clones who serve the Naga (Watchers – aka The Anunnaki Remnant Insiders) who no longer attempt to inhabit the surface – Sitchin was right: the Sun ages them.

So the Anunnaki Remnant went underground into large caverns (No, the Earth is not hollow). And they are found to be in two basic groups: **Insiders** (pro-humans) and **Dissidents** (anti-humans). The Insiders seek to develop Man and manage him proactively, whereas the Dissidents do not hate Man, but they find him a nuisance and would like to be free of him. So the difference between the two groups is one of policy/procedure and the Insiders are still trying to honor their being 'charged' with overseeing and improving the human race that they created.

Just as the TV series *Star Trek* spoke of the Prime Directive, so is there one for the Earth: **If you create a sentient species of any sort, you are responsible for it.** Enlil did not want to shepherd the unruly, noisy, smelly humans (once they were done with the mining activities) and thus we had a Flood trying to wipe them out – along with the Nephilim aberration! Humans survived.

The Ancient Ones stepped in and made the Anunnaki go underground and own up to their responsibility and improve the human species that they, the Anunnaki created. To do that the Anunnaki had to create a worker drone (the Greys) to do the genetic work among the surface humans. Now you know the rest of the story which is what radio talkshow hosts and *Ancient Aliens* resist discussing.

Even humans who have been abducted and 'upgraded' continue to eat wrong, live in a sea of EMF, and they drink, smoke and do drugs (prescription and illegal) – and because **DNA is rather plastic** (VEG, Ch.9) the genome is slightly changing for the worse. So the solution was to take genetic material and rework it into a version of Man who is then stored for later insertion into a reworked Earth Realm -- soon. Some have been released among us, as an experiment – I have seen them (VEG, Ch.5).

This is not good news, and Dr. David Jacobs became aware of it in his regressions with his patients… and he wrote several books about the upcoming **replacement of Man.** However it is no cause for alarm. It is all in the ongoing Agenda for growing souls – I am sure if Neanderthal had known that he was going to be replaced by Cro-Magnon, the Dr. Jacobs among the Neanderthals would have been equally (and unnecessarily) upset. But the fact remains that **current day Man is not making it.**

Witness the **nightly news** with the crazy acts of Fake News creating false rumors about the President, Millennials demonstrating and destroying (UC Berkley) property, people shooting each other, the Democrats' inability to accept they lost and get on with the business of State, and the ISIS rampage that includes destroying priceless ancient temples and artwork. Also take note of the **change in our TV programs** – whereas we used to have the *Mary Tyler Moore Show, Gilligan's Island, Mork & Mindy* and simple action shows like *Rockford Files* and *Magnum, PI* – the emphasis was not on the blood and guts violence we have nowadays. Remember the *Ed Sullivan Show* and *Carol Burnett Show?* Then *Father Knows Best* (honoring men as smart leaders) was replaced with *Modern Family* and *Married with Children* where the dysfunctional father often communes with a rabbit on the couch in the basement… and the father often looks like a jerk. Then TV took a turn for more controversial topics and we got *All in the Family* and *Three's Company*… and then TV morphed its way into presenting now more violent law and order shows, and family values are not supported – Gee, **could there be an Agenda to disrupt family, faith and responsibility?** Then we marvel at 'the lone nutter' who takes a gun and shoots others at work or school – Could the human idiocy of broadcasting violence and sex to 'entertain' be pushing marginal humans over the edge – into acting out gun violence on our streets?

Think I am overstating things? Then consider three things from June 2017: whether **Kathy Griffin** holding up a severed bloody head of a president is in good taste. Also consider whether a **theatrical troupe** performing in Central Park is entertaining by depicting a modern-day version of the play *Julius Caesar* by depicting the president being violently assassinated? Then consider whether a **major female singer** is abusing "freedom of speech" by telling the audience that she has "often dreamed of blowing up the White House." Americans just smirk and consider it political nonsense – but it really demonstrates what I said before: **America has become a violent and dysfunctional society**. The gods have noticed it, too.

Man is about to be replaced since he thinks he can do anything he wants, as well as he cannot get along with other humans (ISIS rampage) … and by the way, **that was one reason the VEG and TOM books were written – to suggest a better path that might salvage this species** – if enough people would listen.

But that won't happen as two major national talk show hosts, one who uses an Arthurian theme for the website, and the other broadcasts coast to coast, as well as the purveyors of *Ancient Alien*s TV show fame, all have the VEG book and refuse to work with it, let alone acknowledge it. The same story goes for the New Age and New Thought churches out there – they have soundly dissed and discredited the 6 books for the last 4 years. Why? Because the **truth is not wanted** –

> **they all prefer to entertain people and keep the mystery going.**
> **They are winning, you are losing.**

Thus it is with sincere sadness that I was told that this version of man is defective and the average human is too **fragile** – s/he cannot handle hearing things that disagree with what s/he thinks they already know (things they have been led to believe by the Media/Church/School), and so they ignore and discredit what they don't like. (This is especially true of VEG Ch. 1 and 11.)

> New Age churches are not teaching the truth – they are telling their congregations what they want to hear – You Can Create Your Day, Name It and Claim it – Visualize it and Manifest It!

> When a very nice lady at church told me that she could Create Her Day, I challenged her to take a dollar and go down to the 7-11 store and win the Lottery. She is no longer speaking to me… and she hasn't won the Lottery, either. People love their illusions.

The average human, when s/he hears something they don't understand or don't like, will decry it. "I don't like it, therefore it is false!" They are not learning, they don't care, and they remain ignorant. They get recycled.

In fact, in 1990 **Arnold Schwartzeneggar** commented (voicing the views of the human, flesh-and-blood PTB who also see humans as a problem):

> "My relationship to power and authority is that I'm all for it."

Because…

> "People need somebody to watch over them… **Ninety-five percent of the people in the world need to be told what to do and how to behave.**" [294]

…and that is due to ignorance. Not knowing is not a virtue.

This is not a good comment on our species nor on our society… and yet the books go unread and Man just doesn't care. (Baldy told me it would be like that.)

Don't think that the gods don't see these shortcomings.

What you don't know is that They have a remedy and it is on the way...
(and one version of it that I was told about involves Martial Law.)

Some clever person who sees how humans are failing came up with the following:

"I see stupid people...
...they're everywhere...
they walk around like eveyone else...
they don't even know that they're dumb."

(with apologies to the 1999 *Sixth Sense* movie by M. Night Shyamalan and Buena Vista Pictures.)

Sadly that does capsulize the situation. **Substitute "ignorant" for 'dumb' and 'stupid' in the picture**. People don't know, they don't <u>want</u> to know and <u>they have chosen to be ignorant</u>. My condolences, they will be **recycled** – until they stop playing dumb games... or maybe they will have the opportunity to incarnate in a better, smarter body-brain combination (provided by the Greys) in a future, reworked version of the Earth Realm. Thus they are said to have more intuition, clairvoyance, and a higher IQ – all helping to insure that we live STO – because if we don't, everyone else will know – being intuitive, maybe telepathic, means everyone will know who did something. No more secrets.

Rewind: Ignorance

So is being ignorant a serious problem?

We are all ignorant of something – I don't know how to do heart surgery. But knowing that doesn't benefit me, so I skip it, and there is no harm done. But I do have to know how to drive a car – unless I want to take the bus, or be a burden to others. So ignorance is not a major 'sin' but being **willfully ignorant** and avoiding learning will be addressed on the Other Side…you don't get away with it for long.

A problem comes in when someone has the opportunity to learn something, something that will benefit them, something that will enhance their intelligence and Knowledge, and even help them Graduate from Earth School, and they refuse – they have just **moved from ignorance to stupidity**. And that is the difference.

When you are told the truth and you resist seeing it, or even investigating it, that is not ignorance… it is stupidity and may get you killed.

> Two men are walking thru the hot jungle and come upon a cool river, and there are cranes wading in the water, and a jaguar drinking from it on the far side…
>
> José: "Don't go swimming in that river… there are piranha in it!"
>
> Pedro: "Oh, I don't see any… You're kidding!" And he walks into the water and starts swimming…. and gets eaten alive.
>
> Pedro did not KNOW that piranha will not attack wading birds, and he didn't KNOW that the jaguar was not in the river but just drinking from the edge of it. And he wasn't listening…

Remember: **Knowledge Protects**
              Ignorance Endangers

Why would people choose to remain ignorant? Because once you know, you are held responsible **and humans do not want to be responsible for what they do (and don't do). You cannot "act up" and do whatever you want once you know the truth…**

You know They are watching and you <u>will</u> be called to account for what you do and don't do (in the InterLife, also called The Other Side). If you know better and don't do better, They will hold you accountable… and it also means you did not learn the lesson(s) that should have driven STO behavior home and made it a part of you – instead you thought you could play dumb and get away with STS behavior. That means that the lesson was not learned and will have to be repeated until it becomes part of you.

Sorry to have to preach that but conveniently the churches of today are not reminding people that Karma is real – both individual and Societal Karma are very real. For example: **Fukushima** (2011) was the payback for the Japanese atrocities during World War II. Their water (the water table) is contaminated with **radiation** which is still spewing out of the reactors at 650 Sieverts (3 Sieverts will kill you) [295] – rice fields are contaminated when irrigated with the groundwater, as well as the fish that are caught off the East Coast of Japan – all contaminated with radiation.

You might have thought Hiroshima was payback for the Bataan Death March and many atrocities against China's women and children. Wrong. Hiroshima was one city and a limited event… **Karmic** payback on a societal scale requires that many millions of people 'pay the price' as was incurred in the millions of people abused by the Japanese in the War. Hiroshima and Nagasaki were just attention-getters that served to end the War and was the way the Allies chose to end the War… payback was still 66 years off and had to include the reincarnated souls in current-day who participated in the WW II atrocities and died back then (i.e., we're talking about **Reincarnation** of the 'guilty' souls who will now experience the radiation poisoning – that would insure those same souls see that violence to others does not pay).

Obviously the Japanese soldiers were not Buddhist – who know what violence and Karma do to a person. Most were followers (before the War) of **Shintoism** – human dignity and value ("kami" and "kannagara") went out the window as Japanese soldiers bayonetted or shot anyone that got in their way… all for the glory of the Earth god Emperor.

Sorry if this is quite negative but it does accurately picture our dilemma. The gods may soon do a **Wipe & Reboot** and reload the Earth Realm with the upgraded humans – if you can call them that.

The Great Earth Puzzle

Appendix A: Uphill River Flow

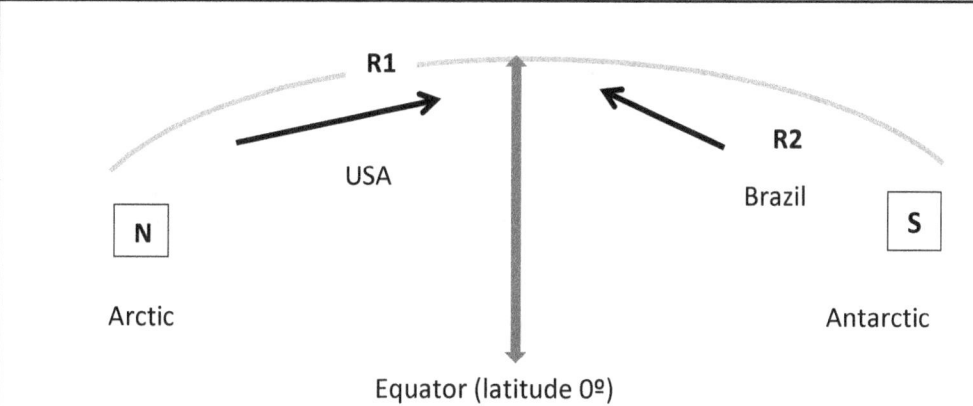

R1

USA

N

Arctic

R2

Brazil

S

Antarctic

Equator (latitude 0º)

Diagram of river flow <u>toward the Equator</u> on Earth. Note that it is **uphill** due to the curvature of an Earth that is allegedly a globe... This means that **water flows uphill.**

Note: **R1 is the Mississippi River and R2 is the Amazon River.** Using the two lightcircles on the globe (below) , if the **Victoria River** in Northern Australia is **R3**, you can see that it is **uphill** to the upper lightcircle (at the Equator).   How is this possible?

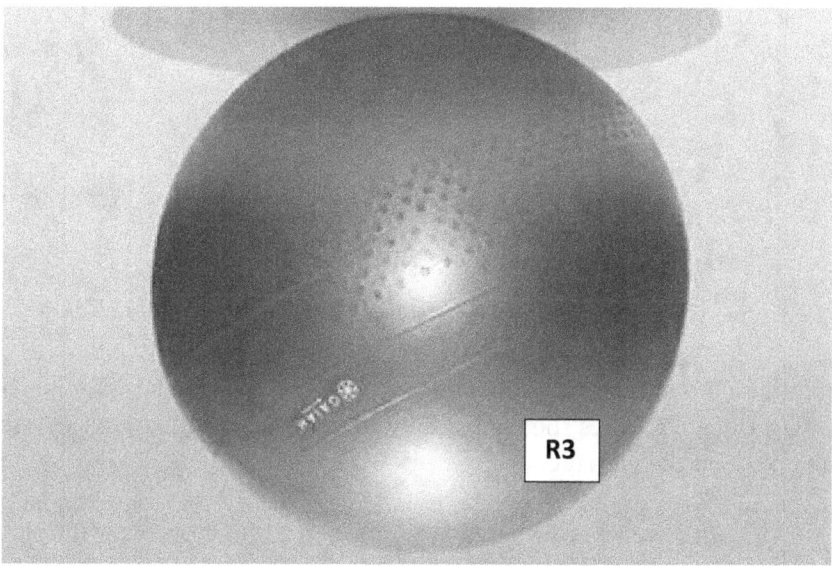

R3

Of course water does not normally flow uphill and that is the point... thus Earth is a programmed Simulation, because those rivers do flow... thus they are on a VR Sphere.

Because this is a curious issue, and people may still not see the upward move of water as unnatural, <u>especially below the Equator</u>, here is another look at it from the Australian (antipode) point of view…

The rivers involved are the Victoria, the Adelaide, The Drysdale, The Daly, The Goomadeer and Roper, the King and the East Alligator (by Timber Creek – between Daly and Victoria)….

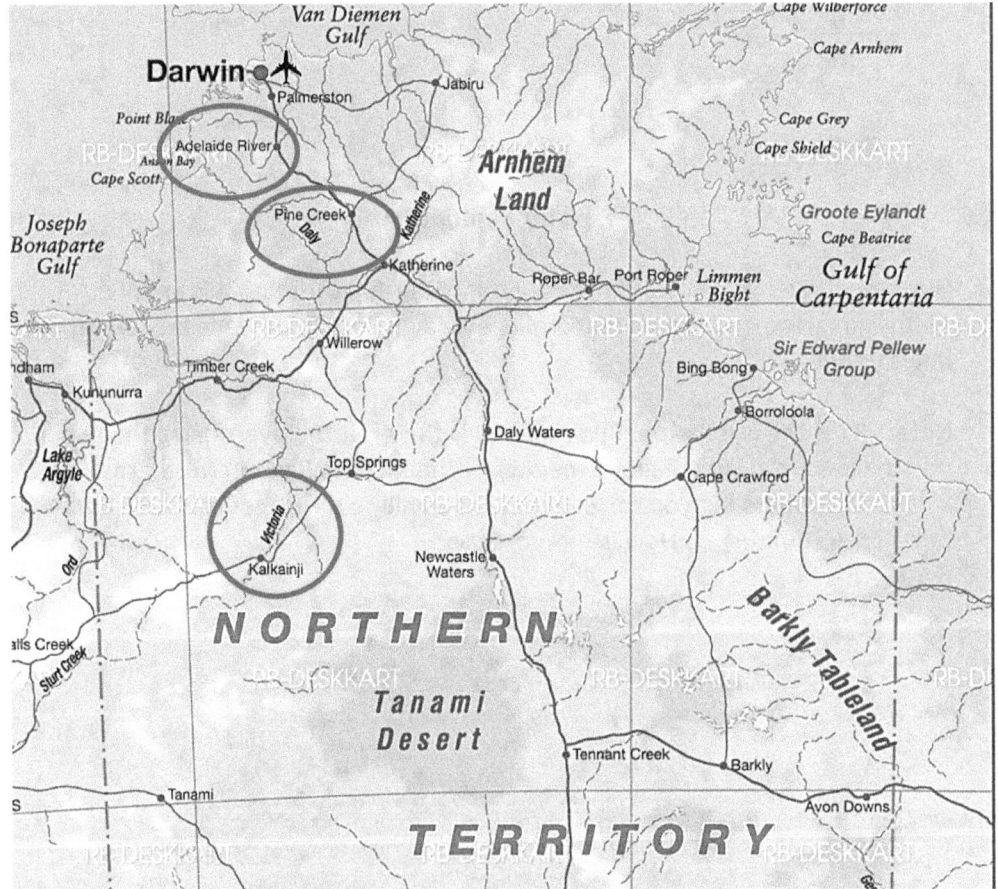

Rivers in Northwestern Australia

The point being that water does flow downhill from the small mountains in Arnhem Land, BUT once it reaches the plains, **why doesn't it puddle**…? How can it keep flowing uphill on the global curve?

> Answer: because it is <u>programmed</u> to flow on the VR Sphere and once it comes down from the hills, it is easy to just keep going on <u>the level plain</u>.

If you don't like that, your other choice is a Flat Earth.

Appendix B: Soul Levels

At the end of Chapter 10 it was mentioned that one of the "problems" with our society is the presence of OPs (60% soulless people), and secondarily, the **Baby Souls**. Here is where that is explained… as well as in Ch. 7 in VEG.

This was covered in some detail in VEG, but because it is also relevant to the Earth Realm and why things look the way they do on Earth – which was one of the first questions asked in the Introduction. It is included here, especially for those who have not read VEG.

In addition, just so the bases are covered, we'll examine Soul Levels, Consciousness Levels, Multidimensional Souls and the Darwin Awards. First let's see what a soul is…

Soul

So what is a soul?

> **A spark of the Divine, The God replicating Himself, an eternal sentient, coherent Light energy being that is conscious of itself and its surroundings, intelligent, exists in multiple dimensions, and can evolve itself back to a connection with the Godhead…**

And it seems to have a weight. And if the aura pictures are correct, the state of one's soul (sad, happy, angry, jealous, etc.) manifests in the aura as colors reflecting one's feelings. Red reflects anger, blue reflects sadness, healing is reflected by green, and very intellectual analysis (say, computer programming) is reflected by yellow in the aura.

If you can weigh it, it has a slight mass, or physicalness. And it is who we are, because it is what animates the body. When John Smith dies, he is not there anymore. His body is, but <u>his body is not who he is</u>, and many NDErs (Near Death Experiencers) have said as much. The soul can leave the body and that personality is still intact – Robert Monroe in his Out of Body (OBE) experiences, documented in three books, demonstrated that.

So the soul exists, it has mass, and it is eternal. When it is not in the body, it resembles one of the Beings of Light that come to meet the NDErs at death, only not as bright. This is probably what the Bible calls our "imperishable body" in I Cor. 15: 44 and 52. Matt.17:2 is where Jesus displays his transformed body on the mount

– resembling a Being of Light. As significant as that sight must have been to the disciples watching, strangely very little more is said about it and what it meant.

> How do we know that the soul is eternal? Because souls were created in the likeness of The God, and God is eternal. And NDErs and OBErs all report that <u>the soul is who they are</u> – an entity that goes on and on, having had many lifetimes. Also, it is believable that the soul is eternal since no one has come up with any evidence or teaching that the soul is born once into a body on Earth and then at the death of the body, the soul ceases to exist, too.

> *If someone believes that there is no more life after death, perhaps a person with that belief is confusing the issue with happens to the OPs at death. OPs have no soul and they DO cease to exist at death of the physical body.*

In addition, the soul normally enters the body just before birth, at birth, or just after birth. There is a soul hovering about the mother who plans to enter the body, but if the body is not genetically acceptable, or if the parents have recently decided to do/be something other than what the soul needs when it originally selected them, the soul does not enter and the baby may be stillborn, or function as an OP, with no soul.

Ensouled Humans

So what is Man, body, soul and spirit?

Ensouled humans are **eternal** souls with DNA that permits connection from their magnetic center (aligned, coherent chakras) to the God Force – the divine spark referred to so many times. As souls, they cannot be killed and the worst that can be done to them (and has been tried in other Eras) is to <u>trap or contain</u> them, slow them down, or totally derail them… which <u>has</u> been done on Earth. This sometimes seems to make Earth a kind of prison planet, but it is more.

Spiritual Potential

What is significant is that **Man is multidimensional** and has a spiritual potential (Soul) that the PTB don't like, and the Discarnates through the Powers That Be (PTB) working together militate against Man to suppress it. And yet, Man **will** spiritually evolve and eventually function as another Jesus/Krishna/Buddha – somewhere in the Father's Creation. The goal of the ensouled human is to eventually reconnect with Source and develop latent (potential) higher attributes – even psychic abilities. Jesus said it (paraphrased):

"These things I do you shall do and greater…" John 14:12

Why is ensouled man such a threat (to the PTB)? Because he has the spark of Divinity which can be activated through his (to-be-rewired) DNA, but this is <u>received</u> when the time is right, and cannot be forced. Nor can years of meditation, Yoga or Tai Chi force the DNA to develop or "rewire." Such efforts are noble, but soul growth depends much more on what the soul learns about self and the world, and since knowledge is Light, **the Light does the work**.

> *As was seen, in VEG, Ch. 9, DNA and Light are intimately connected, with scientists reporting that DNA releases biophotons upon stimulation of the DNA*[296]

When the student is ready, the transformation happens as a result of a spiritual energy 'potential' being reached, and this has been referred to as ***kundalini*** ("Serpent energy") moving up the spine to energize and transform the person – by connecting chakras. This can be dangerous if done without a teacher who knows what to expect and how to handle dangerous energy blockages in the student.

As the human soul comes into its own divine power, it is conceivable that it will be serving the Father of Light by "reigning" over some of the areas of Darkness – at least at some point there will be a confrontation between souls subscribing to Light and those promoting Darkness (in 4D - 5D).

This is the essence of why ensouled Man is considered a threat – to discarnates who did not go to the Light, to the atheists on Earth who do not like the idea of a soul, and to the PTB (3D human Powers That Be) who think they run Earth – they enjoy being Lords and as such they need Serfs (or sheep) to rule over. Most PTB are 'soulless' and the astral entities can manipulate them to stop Man from graduating from Earth School (next chapter).

Soul and Spirit

In the beginning there was God. Incredible potential. Order and chaos, Light and Darkness, Yin and Yang. A Designer. The One, the Central Sun… A concept so deep it is beyond the ability of 3D Man to fathom. So it is approached allegorically, and sometimes rejected by those who demand physical proof of something more intelligent and powerful than Man.

Evolution

Those who reject the existence of a God claim that all current life that we see on Earth came from a primordial 'soup' which contained the building blocks (amino acids) of all life. And, supposedly, **a bolt of lightning hit the soup**, energizing it and causing new chemical compounds to develop, and over millions of years, and further lightning strikes (?), the first living 1-cell organism(s) emerged. And Man evolved from Apes… Such are the beliefs of **Mythic Level** humans (later section, this Appendix).

Why is this not still happening?

For some reason, the Second Law of Thermodynamics did not apply way back then, perhaps because no one had discovered it yet…(☺). That Law says that if any organism is left to itself, with no outside input, **entropy** is the natural result: all things tend to decay unless they are re-energized, or fed, or some external force keeps them going. "All systems, including living systems, decrease in order." [297] So the 1-celled creatures were brought to life, à la Frankenstein, by bolts of lightning and they did not fall prey to the **Law of Entropy** – until Man later discovered it. How convenient.

The foregoing is called Evolution which requires a huge amount of time for things to naturally "increase in order": particles would become single-cell organisms which would become multi-celled organisms which would become invertebrates…

> sea slugs,
> > IRS agents,
> > > and eventually people.

In a brilliant piece of deduction which will forever go down in the history of Man and Evolution, Charles Darwin discovered that because a man looks structurally something like an Ape, man must therefore have <u>evolved</u> from the Ape. How exciting was that news? Anything was possible, so Charles said – given enough time. And yet, **because of entropy, time is the enemy of Evolution**.

> This issue was dealt with more fully in Chapters 8-9 of VEG where today's genetics **dis**proves Darwin's Theory of Evolution (but **not** Epigenetics, Natural Selection and Survival of the Fittest).

Soul Levels

Just as there are different OPs, there are different levels of soul not only reflected in the aura, but also in the orientation to life on Earth. As was brought out in VEG, Ch.5, there is the Standard OP, the Robotic OP, and the Placeholder OP – not really an OP, but not a complete soul, either. The Placeholder is unique in that it can operate as a soul or as an OP. And the Pre-soul may have the option to become a more fully developed soul.

In any event, there is a kind of **hierarchy of souls** which reflects a soul's growth which is an expression of their experience and what lessons have been assimilated. While this is not cast in concrete, keep in mind that there are as many different levels of souls as there are types of flowers and variations within each flower group.

> **Baby Souls** – these are the first-time souls, Pre-souls and may include the Placeholders. They are generally naïve and their aura is underdeveloped, often being an orange color (a mix of the lower three charkas which are the only ones really functional at this stage). They are the most timid of the soul types, often being afraid of germs and dirt – not having had much experience with Earth life. They tend to avoid crowds, not feeling comfortable with all that energy and being a bit unsure of themselves. For them, sex is scary and they are very concerned with avoiding social diseases. These souls are very concerned about appearances and want to dress and look right. These souls love Nature but are **very prophylactic**.
> These souls are drawn to a very basic, fundamentalist type religion and it is easy for them to believe in Hell and God's punishment. Aliens do not and cannot exist.
>
> **Young Souls** – these souls have been around enough on Earth to know their way through the new experiences that the Baby soul is still learning to handle. Thus these souls tend to make up for lost time, and become involved in everything that catches their fancy. They join groups, sing, dance, party, try novel adventures (river rafting and sky diving, e.g.), and they are said to **'go for the gusto.'** In their eagerness to experience it all, they begin to make mistakes, tromp on others' toes, and may even lie, cheat and steal. For them, sex is fun and they seek new ways to experience it. These souls are very concerned about appearances, too – do they look good enough, and have the latest designer this and that?
> These souls love Nature and seek to romp through the mud on off-road bikes. These souls are drawn to a more progressive type religion yet they usually also believe in Hell and God's punishment. ETs may exist on other planets, but not here.
>
> **Mature Souls** – these souls have been around even longer than the Young souls and are, in fact, back to work on the mistakes they made. As a result, these souls have begun to quiet down and tend to **become introspective**, trying to figure out what things mean and how they can get the upper hand over <u>the ailments they often have.</u> They are often found in New Age and New Thought churches, seeking better information on how to handle

their lives, their health, and their finances. For them, sex is a responsibility and they take it seriously. These souls pay attention to their appearance and make it reflect who they are; dressing is a statement about their real self or what they value. These souls love Nature and seek to understand Mother Earth and work with others to heal the environment.

These souls are drawn to a think-for-yourself religion which is really a search for spirituality and they are exploring different religions and teachings to find answers to their life issues. They doubt there is a Hell and <u>think</u> Karma is probably true. ETs are real and here, but don't talk about it.

Old Souls – these souls are the most interesting, and can be real characters. They have mastered most of the issues that Mature souls are still working on, and they have come to a greater awareness of the Oneness of all things and all people. They could also be called 'last timers' as they are either doing clean up work (righting wrongs, forgiving others, etc) or are in some sort of teaching capacity for other souls who are still learning basics. For them, sex is no big thing and ironically, they can be bawdy, laughing at it all. These souls are very laid back about appearance, not shaving if they don't want to, wearing comfortable clothes despite how others are dressed – yet they are clean. Styles don't impress them and they do what they want (without offending others).

These souls love Nature and work with her, and would rather be alone in the woods communing with her than sitting in church or shopping in the Mall. These souls are found in a New Thought church, or something like Baha'i or esoteric Gnosticism, or none at all. They <u>know</u> that spirituality is more important than religious rote, and they <u>know</u> there is no Hell, and they <u>know</u> that Karma is real. They know ETs are real, and may be here, and they are open to learning more about them in the Multiverse.

As was said, this is just a general way to get a handle on four of the many stages of **soul growth** – there are a lot of souls who are still going through the first two stages and may wind up being **defective**; if they never develop the intent to be/do/have something more significant or spiritual in life, they will go from being **recycled**, to possibly being 'disseminated' – or having their energy rearranged. Such rebellious souls may become defective if they try to party forever, or run power trips on others as a way of living.

Many souls are a mixture of levels, depending on their individual ground of being – more advanced in some areas than others, and it depends on just how much they are willing to accept of reality.

Multidimensional Souls

Another aspect to our soul is that it is multi-dimensional: **we exist concurrently in other times and places, other timelines and dimensions**. [298] Since all aspects of self are inter-connected, the undeveloped aspects of one's self can and do affect the mood and peace of the other aspects. [299] This is akin to the 'nonlocality' phenomenon of Quantum Physics.

The term here is specifically "aspect" and not "fragment" since fragment refers to parts of one's soul that have split off, or became fragmented due to trauma. John Jones in 3D Earth may experience a trauma and a small part of his soul will **fragment** (to escape the negative energy impact of the trauma); whereas Bob Jones may have many other **aspects** of himself including: (1) one in a parallel dimension, (2) one 300 years ago on Earth, and (3) one 1200 years in the future on Earth, to name a few.

Each person in this reality has other aspects in other realities that influence each other, as indicated by the dotted lines in Chart 3b:

Reincarnation/Past Life Scenario

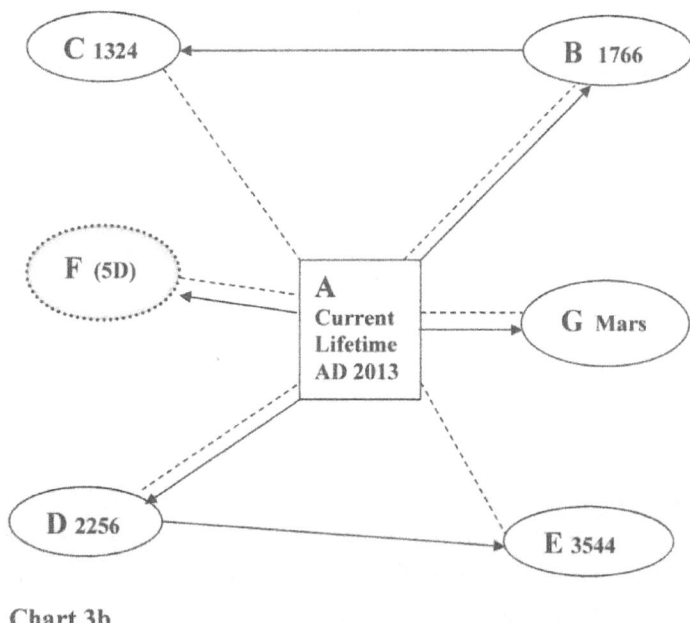

Chart 3b

All potential 6 soul aspects (B – G) of the one person (A) are linked thru the Higher Self (dotted lines). **Arrowed** lines represent "linear time" access. **Dotted** lines represent 6 (direct) energetic soul links and there are energetic interactions via the Higher Self along these lines ('cords').

The above is a representation of the Multidimensional nature of the Ensouled Human Being. Man A in 2008 in the Earth Realm may also exist concurrently (simultaneously) in lives B, C, D, E in this dimension, but in 4 different timelines, and he also exists in the fifth dimension in life F. Note that he may also have a life on Mars in the 3D realm.

When hypnotically regressed, and asked to go **back** in time to see who he was 'last time' he will encounter lifetime B (AD 1766) and then lifetime C (AD 1324). If he is asked to go **forward** in time, he will encounter lifetime D (AD 2256) and then the lifetime E (AD 3544). All in the 3D realm, for the purpose of illustration here.

Allocation of Soul Aspect Energy

Allocation of Soul Energy

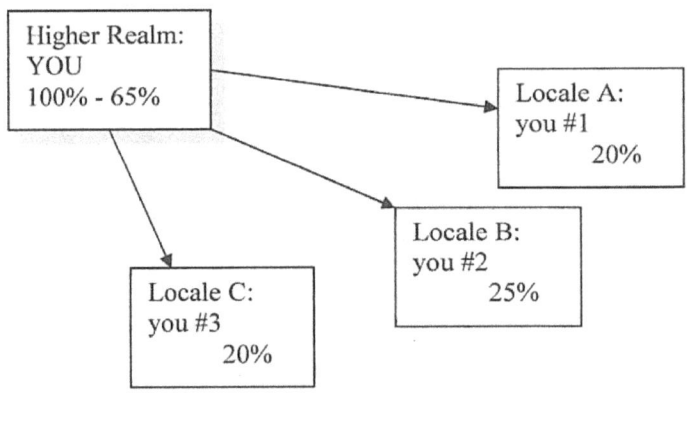

Chart 3c

Note in the above diagram that YOU as a main Soul (or Higher Self) in the Home (4D) Realm decide to experience 3 other realms or timelines. This requires an allocation of your energy to 'replicate' yourself into those other 3 realms... and you don't want to spread yourself too thinly! For the average soul experience, a 20% allocation is sufficient (as the Higher Self is quite powerful and has a lot of energy), and in a difficult realm, an allocation of 25-30% may be appropriate... this is an "energy bank" that you can draw on to meet higher energy demands of the lifetime for health, reserve stamina, and mental prowess. Note that when the 3 aspects are allocated, the big YOU still has 35%.

If an advanced soul, from say the 6th level, decides to project into the same 3 realms, it may require less energy allocation due to a presence that just observes and "anchors the Light." Ironically, it has a greater energy well to dip into and could allocate 20% which for it would be equivalent to "you #1" above allocating 40%... much more than necessary.

Different soul levels have different levels of useful energy – a Baby soul has a lower PFV, is underdeveloped and so the energy is 'rougher', cruder and not as refined as that of an Old soul, and a soul from the 6th level is even finer than that of an Old soul, and requires less to perform equal to any soul in 3D. Ascended Masters can project into many realms at the same time if necessary, and their energy being so much higher and finer, it is like a battery recharging a Baby soul if called upon to heal that Baby soul. The healing works because energy always flows from the higher potential to the lower.

Consciousness Levels

Just as there are different types of souls, or maybe we should say, souls are at different levels of growth… from Baby (1st-timer) to the more experienced Old Soul, we can also point out that Souls also display different levels of Consciousness.

> Again, remember that consciousness (small c) refers to whether a person is awake, while the Consciousness (large C) reflects the Soul's spiritual development, and perhaps their ability to connect with their Higher Self.

What has been noticed is that some souls are in an Infant Stage of Consciousness, while others are developing Psychic abilities. Thanks to an excellent book that profiles the levels of Soul Consciousness, Putting on the Mind of Christ by Jim Marion (who endorsed my TSiM), we can examine 7 – 8 distinct types of Consciousness. While the book was written to enlighten Christians on the road to greater spiritual growth, it has proven very accurate as a yardstick that assists a Soul, regardless of Faith, to determine his or her approximate position on the Spiritual Path from Beginner to the ultimate goal of Christ Consciousness.

In addition to looking at oneself as a Baby Soul or an Old Soul, it is more instructive to read the following descriptions and by locating one's attributes on the **Consciousness Scale**, it is hoped that this can serve as a guide to where one is at right now, and where further learning and spiritual development can be done to move into the next level in spiritual growth.

Why? Because the goal is to ultimately **develop the Soul by developing one's Consciousness** – which aids in graduating from the Earth School. Souls who are stuck in the first 3 levels of the Consciousness Scale (below), will not graduate from Earth School. .

Consciousness Scale

1. Infant Level

The developing awareness of the infant begins with a sense of differentiation of its body from that of its mother. This is true of spiritual growth in general: the more a person's Consciousness goes up the spiritual ladder, the less attached they are to physical matter.

Later follows an awareness that its emotions are separate from its mother's emotions. Often **infants' emotions mirror whatever the Mother is going thru** – as if the child is empathic, and still shares an invisible bond with the Mother. This stage usually lasts about 3 years. And this exemplifies another truism in Consciousness development:

> All growth in Consciousness is a process of inner realization.[300]

This level has been termed **very primitive** – the Soul is receiving input from its 5 senses, but is **very me-me-me based.** Whereas each level in the future will be less egocentric than the last, this first stage is **very tied to dense matter**, and were the child to stay at this level as it becomes an adult, it would be a disaster.

> …leading quantum physicists are now aware…. that matter,
> molecules, atoms and subatomic particles have a slight amount
> of innate Spirit, intelligence and 'awareness' of surroundings,
> but, compared with mind for example, matter's amount of
> 'awareness' is miniscule. [301]

A child that does not make a clear distinction between self and others may grow up afflicted with **narcissism**… **or** as an adult afflicted with a borderline personality disorder – **seeing oneself as a victim and never takes responsibility for his/her own problems.**

2. Magical or Child Level

This is usually the human's life between ages 2 - 7. It is called 'magical' because it often includes a world of fairies, elves, gods, demons, and various imaginary creatures that inhabit the child's world. **The child often cannot determine what is real and what is imaginary, and the child often believes the world still revolves around it.**

In addition, the child's mind begins to disengage from the body and emerge as something separate. This is the beginning of a mental life, where the child will begin to see its inner world as different from the outer world, but it is still marked by the sense that the right 'magic words' can cause the rain to fall, or the thunder to stop.

> **Magical consciousness was the general level of consciousness in the polytheistic, animistic, tribally-organized ancient world.** [302]

At this time, the child begins to learn what is right and what is wrong in its particular culture. But this is not a moral sense of what is right or wrong, it is based on what makes Mom and Dad happy or upset. As age 7 approaches, the child may have the beginning sense that its point of view is not the only one. Self-centeredness begins to fade.

Obviously, if the child does not develop and grows into adulthood with this Magical view of the self and world, s/he will be **a very limited adult, often superstitious and naïve.**

3. Mythical or Adolescent Level

This is the area of growth into the teens. It is the first of the "mental levels" as the child's mind or ego begins to emerge. There is a belief that the God in the Sky and one's parents can accomplish every sort of miracle to meet the child's needs. The mythical has incorporated some earlier magic.

And yet at the same time, the child begins to adopt a law-and-order phase where s/he defines itself by conventional rules and sees its self-worth in following those rules, and behaving properly. **Rules and roles** are taken seriously.

> Until recently, the mythic level of consciousness has been the dominant level of consciousness in all the world's "universal" (i.e., basic) religions, including Fundamentalist Christianity. [303]

For some reason, the child sees Mommy and Daddy as right in whatever church they belong to, whatever political party they belong to, and the child just knows that the public school that s/he attends is the best one in the world. Trying to teach tolerance and diversity to a child at this age does no good; the brain has not developed enough to allow analysis of abstract concepts (and usually does so by age 20-21).

The child's inner world not only has the concrete external gods of Mom and Dad, it also has Santa Claus, the Easter Bunny, and host of imaginary friends, angels, fairies, and perhaps devils that are holdovers from years past. As the child reaches early teens, if it is a Christian child, it will have combined these gods and beings into the one, true Christian God who lives in the sky. [304] The child does not know whether his parent's religion is the right one, he merely *assumes* it is, because they are like gods (authority) in his or her life.

> The child cannot think otherwise at this stage…. Despite the emergence of the child's mind and the shifting of the child's self-centeredness… is **still an egocentric level**. Tolerance … **for other points of view, and compassion for people who hold these views is not possible for a child with Mythic consciousness.** Nor can they see any good reason for attempting such tolerance because… this would be a betrayal of their external God – whose rules and roles define their self- worth. [305] [emphasis added]

> This is describing both Christian and Islamic Fundamentalists and may be the source of irreconcilable differences and intolerance… better mediated by souls who are in one of the higher levels (4 – 6).

As the child learns to pray, God is seen as the One to fulfill the child's needs, and if s/he wants broccoli turned into ice cream, the child is sure that it can be done. And it is at this time that the child begins to use the emerging reason to reinterpret the previous magical and current mythical worlds. **Religion is now seen in a very concrete way:** God did create the world in 6 days, three wise men did journey to Bethlehem, and Mary did have a miraculous birth experience. **The child begins to look for absolutes to rely on.**

For the adult Christian whose Consciousness has not progressed beyond the Mythic level, it is important to convert the whole world to the one, true religion…because this is the only way the believer can insure his/her own righteousness. And the same would hold true for children growing up under other Faiths as well. Children look for a sense of absolutes – right vs wrong, black vs white to establish a sense of security. Out of this mindset came the Inquisition and the Crusades.

These people at this level **cannot think "globally"** and planetary ecology is not important to them. They are centered in their own little world. **Other religions are a threat to them**. Perhaps that helps to explain why some people are so **narrow-**

minded and cannot examine their Faith and other beliefs… without suffering a nervous breakdown. The 'programming' of beliefs as a child goes deep and often stays into adulthood.

> The Mythic level (and lower) does <u>not</u> qualify a soul for graduation from Earth. They are too parochial.

4. Adult or Rational Level

This is the second of the mental levels, and is **the dominant Consciousness of our present age, and is the level attained by the average adult in today's world**. Many are stuck at this level and don't know it.

In today's world, the growing adolescent has to develop an ability to reason in order to be successful in our society. The child has to **learn to handle abstract ideas** and be able to grasp some basic universal principles.

The Age of Reason, the Industrial Revolution, and Man's current drive in Physics to understand the world about him has supported the public schools' emphasizing math and science for all our kids in school. There is an **opportunity to develop abstract thinking**, to reason or argue with friends about philosophical statements (e.g., "All men are created equal") and thus begin to **develop some higher thinking skills**.

> Such as compare and contrast, deduce and induce, and the harder synthesize and analyze. Often called H.O.T.S. or Higher Order Thinking Skills.

Thinking skills are often exercised to a greater degree in college studies. There the danger is to fall prey to ideologies and groups who want to bend students' thinking in their direction. Yet too many people have not been able to train the mind to the discipline of Logic and spot false ideas, false claims, and illogical proposals.

The development of reason will encourage the teenager to **question rules and society's traditional structures.** Bringing forth their earlier Mythical propensities, they may begin to imagine a better world and hopefully one that make sense for the greater good. Developing reason should bring the teenager's mind to the awareness of a global economy, global ecology, and perhaps even a one world religion… Such was the energy behind the European Union and the Euro.

It would appear that the majority of the world's people fall into the last two categories, or levels: Mythic or Rational, with some people in a half-and-half state. Developing Love and Knowledge will assist these souls in becoming Earth Graduates – see Chapter 11.

---

5. Vision-Logic Consciousness

This is the highest of the three mental states. It is the Consciousness of many great writers, artists, scientists and philosophers. **The primary aspect of this level is a sense of self with the abstract mind, and the ability to think from many different perspectives.**

It is definitely STO and these people think globally. These people can resolve international issues with their **ability to think and see from the 40,000' level** – but their solutions may 'threaten' those who still function at the Mythic level.

This is the first level of Consciousness where it can be said that we have successfully integrated body, emotions and mind. [306] And we no longer identify self by race, color or religion – we are simply a human being, humble yet very other-directed (STO). **We seek the good for all humanity, and are tolerant of other cultures, religions and ideas.**

But all is not rosy and there is a downside to this seemingly wonderful level of Consciousness. It is called **existentialism** and has been known to cause profound unrest and angst, sometimes leading a soul to nihilism. They have so many great ideas and so little time in which to see them to completion. And then, they often worry what will become of their projects when they die…?

These people sometimes **struggle to find meaning in a chaotic world**. And sometimes they lose their Faith, wondering where God is, and does He really exist? As a result, these people are among the first to either become **atheists**, or to seek deeper meaning in life, and begin to lose the ego and locate the Self beyond the Mind.

The successful Souls (successful at handling their **'awakening'**) begin to transcend body, emotions, and mind and go within to identify with an **"inner witness"** – the One who is doing the looking and thinking. The person becomes more whole by becoming detached from the physical self. The **Higher Self** begins to emerge…

It is then a natural progression to the next level.

6. Psychic Consciousness

At this level, we no longer identify with the rational mind, but with the "inner witness" (i.e., **the Higher Self) that observes it all**. This is the big You that is more than personality you, and along with it comes **clairvoyance and healing**.

These people become aware of information coming from beyond the normal 5 senses. Often referred to as ESP, it is a way of **knowing on a higher level**.

See the Chinese EHF (ESP) abilities in Appendix A, TSiM.

A common experience for these people is what is called a **Peak Experience** which has also been called cosmic consciousness. This can happen in a split second when walking thru a park, or along the beach – all of a sudden the soul shifts into a higher awareness state (Soul) -- the colors are much more vibrant, deeper, and one's senses are very acute and there is **a sense of oneness with all creation**.

Ralph Waldo Emerson was such a person and often spoke about using one's intuition to transcend the material world. He believed we could contact the "Oversoul" or what has today been called the Higher Self, and attune to one's inner drummer and **live by one's own higher truths**. [307] Or, the Tao.

At the psychic level we begin to experience a knowingness that does not depend on the 5 senses. We may even, like **Robert Monroe** of the Monroe Institute and his 3 great books on OBE, have our own **Out of Body Experiences**. We may even **begin to see auras** – that 1" etheric layer just above the body that identifies a soul's presence.

There are also dangers at the Psychic level. As Robert Monroe discovered, the lower Astral realm has many Discarnates (souls who died and did not move on to the Light) and **when you are at the Psychic level, you carry more Light than the average person, and to the Discarnates, you look like a 500 watt bulb in a room of 30 watt bulbs – Guess who they're going to investigate to see who you are, and if they can play games with you?**

Another aspect for the unwary, is intuiting information about another person or subject that you were not verbally told about, and you begin speaking about it. Or winning too many hands of Poker or 21 in Las Vegas – the House will be sure you are cheating and ask you to leave.

> Obviously, the Vision-Logic and the Psychic levels do qualify a soul to graduate from Earth School – providing the person is STO and not STS.

7. Subtle Consciousness

The subtle level is the last one wherein we identify, even if weakly, with our self, or personality. Our Consciousness becomes **capable of receiving direct**

communications from the causal level, the level of the Oversoul. This is the person who receives **spiritual revelations** and whose Consciousness is so far above the average person that communication with the average person is very difficult – because they now see and understand so much more.

> These people see color in a world of people who only see black & white/shades of grey. How can this person explain color? They don't and as a result they begin to withdraw from society.

These people may hear voices and heavenly music, and see things that are intradimensional. They are getting **ready to transcend the 3D world** – but before they can, they often have to undergo a momentous experience called **The Dark Night of the Soul,** which breaks all ties with the Earth plane. (This special transformational Night was examined in more detail in TOM, Appendix A.)

This is truly **a rebirth to new level**. And these people often become hermits or monks so that they do not have to deal with the crowd, yet they serve the planet in the background, **"anchoring the Light"** and transmuting ambient negative energy.

8. Christ Consciousness

This is **the causal level**. Also called Christ Consciousness. They are free from neurotic projections and emotional addictions, and they see all other people as precious souls and they see the Father's Light in them, however much buried it may be. They are **totally detached from cares and concerns, anxieties and struggles.**

They now **commune silently with God** and are in the first stages of a realized divinity. Their Crown chakra is wide open and they are immersed in Light – to anyone who sees auras, these avatars project an aura out to 30' around them. It is white with a gold core, and the white has flecks of purple in it. Quite a sight to see.

They walk on water, heal the sick, and can walk into a hungry lion's cage and not be harmed. Animals sense what they are and love and respect them.

Their **Love** is so strong you can feel it from across the room, and their **Knowledge** surpasses that of Wikipedia – they have full access to the Akashic Records, so it is indeed extensive.

This is the level to which most Souls will someday arrive.

And yet we have the daily masses to deal with who are sometimes comical when they are goofing around…

Darwin Awards

As an interesting sidenote, Man did not always inherit intelligence from his ancestors, and there have been stories of **Man inadvertently removing himself from the Gene Pool**, as documented and acknowledged each year by the Darwin Awards.

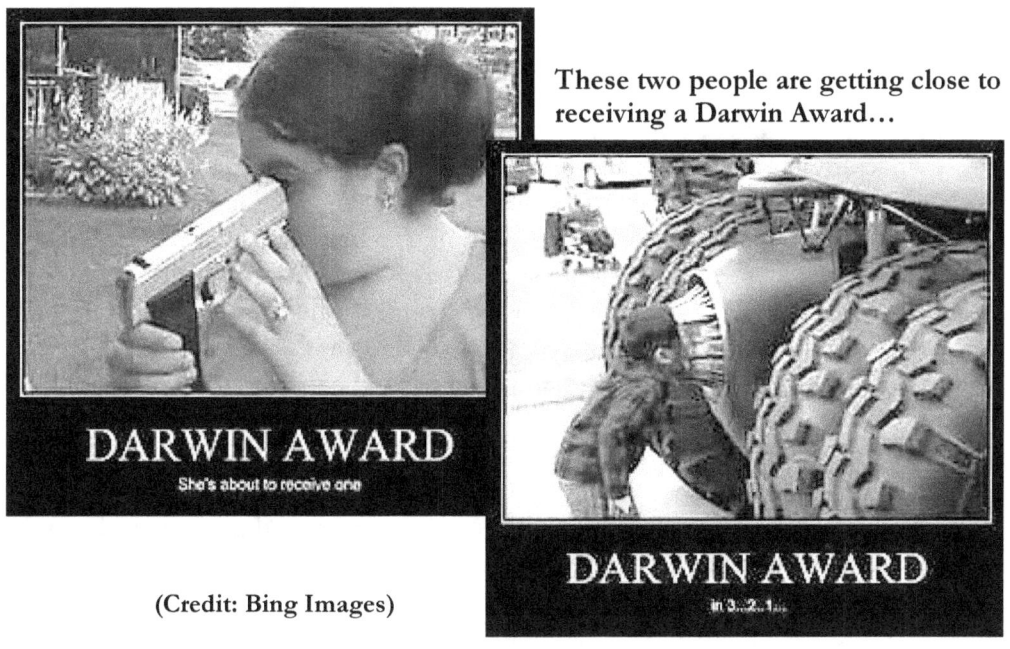

These two people are getting close to receiving a Darwin Award...

DARWIN AWARD
She's about to receive one

DARWIN AWARD
in 3...2...1...

(Credit: Bing Images)

You die to receive a Darwin Award. And if it is a spectacular death due to incredible stupidity, that person will receive THE Darwin Award for the year.

There is a book of published Awards.

Three **classic and true stories** will demonstrate Man's unique ability to ignore the Survival of the Fittest axiom:

> **The Boone & Scenic Valley Railroad** [in Iowa] crosses 184 feet over the Des Moines River and its corresponding abyss.
> "Several years ago an adventurous pair decided to take their ropes and rappel off this architectural support... Our adventurers had to be completely fearless as they walked to the middle, tied off their ropes, and began to rappel down. But when the Boone & Scenic Valley Railroad came by on its daily tour of the valley, their one mistake became apparent. They had tied the ropes to the sturdiest support possible: the steel train tracks..." [308]

This one shows that smoking is hazardous to your health and the story might be called **The Ants' Revenge**:
"A woman was found burned to death, her body still blazing on a grassy area adjacent to her home… A lighter and a melted gas can were discovered nearby. After a lengthy investigation police turned up no evidence of foul play. They believe her demise was due to her habit of dousing anthills with gasoline while she smoked cigarettes…" [309]

Last but not least, there is also proof that cellphones can be dangerous while driving, **and** while just standing around… **Cellphone Larry**:
"Police said a [Kansas] man was struck and killed by a train after his vehicle broke down on I-35. His attempts at repairing his car had failed, and he had stepped away from the busy freeway to call for help.
As luck would have it, he chose to stand on the train tracks paralleling the road. When the train engineer spotted him standing on the tracks, the man was holding a cellphone to one ear and cupping his hand to the other ear to block the noise of the oncoming locomotive…" [310]

Yes, all three events actually happened, all four people removed themselves from the Gene Pool, and all won the Darwin Award. Is there something in Man's genetics that predisposes him to carelessness, stupidity…. and even violence? Or is it due to a low level (2-4) on the Consciousness Scale? The above incidents are just three among hundreds just for the year 2003. Sure, people make mistakes because they aren't thinking… but that often costs them their lives.

One of the most macabre events reported that year was a man in Sweden who could not start his chainsaw, and in his frustration, he locked it between his legs and yanked on the cord. You'd think no one would be that dumb… and yet, it happened.

Ouch!

Appendix C: Digital Consciousness Theory

One of the scientists examined in Chapters 3 and 9 was **Jim Elvidge** who has come up with a brilliant observation – beyond what has already been quoted from his work in quantum physics. As do some other physicists, notably Brian Greene, Jahn and Dunne, James Gates, and Seth Lloyd, Elvidge also questions the Earth Realm and what it might really be, and he strongly suggests that our reality is a **digital simulation** – albeit a very sophisticated one.

> **Evidge is not associated with the Flat Earth Scenario… just the Simulation aspects of the Earth Realm.**

Elvidge asserts that "an objective physical reality doesn't really exist independently of our consciousness "[311] -- which is what Jahn and Dunne said in Chapter 9. Further he says that his Digital Consciousness Theory explains everything – both philosophical and scientific conundrums. It explains why the Universe appears to be "finely tuned" to support life, it explains quantum entanglement, paranormal experiences including NDEs, the quantum Zeno effect, and quantum retrocausality.

> **Quantum entanglement** – says everything is connected – both at a microscopic and subtle (spiritual) level.

> The **quantum Zeno effect** (also known as the **Turing paradox**) is a situation in which an unstable particle, if observed continuously, will never decay. One can "freeze" the evolution of the system by measuring it frequently enough in its known initial state. The meaning of the term has since expanded, leading to a more technical definition in which time evolution can be suppressed not only by measurement: the quantum Zeno effect is the suppression of unitary time evolution…. [312]

> **Retrocausality** (also called **retro-causation**, **retro-chronal causation**, and **backward causation**) is any of several hypothetical phenomena or processes that reverse causality, allowing an effect to occur before its cause…. While some discussion of retrocausality is confined to fringe science or pseudoscience, **a few physical theories with mainstream legitimacy** have sometimes been interpreted as leading to retrocausality. This has been problematic in physics because the distinction between cause and effect is not made at the most fundamental level within the field of physics.

> **Retrocausality** is sometimes associated with the nonlocal correlations that generically arise from quantum entanglement including the notable special case of the delayed choice quantum eraser. … Retrocausality is also associated with

the two-state vector formalism (TSVF) in quantum mechanics, where the present is characterised by quantum states of the past and the future taken in combination. [313]

> The quantum Zeno effect and Retrocausality may still be debated by advanced minds greater than ours, but Elvidge points out (in the following pages of this Appendix) that his Digital Consciousness Theory does account for the odd and unusual ... even in those issues that are just theoretical physics.
>   Now, doesn't the Flat Earth Theory seem easier to accept?
>

Continuous Reality

One of the major aspects of our Realm that has been debated for decades is whether or not we are living in a continuous reality -- which concept says that matter and energy go on forever – eternity is implied in a Nature that has no beginning and no ending. And yet, there is no evidence that reality is continuous and there is now mounting evidence that the world is digital in nature.

> In this case, "digital" means having properties that can be encoded into bits and bytes... or binary zeroes and ones.... as **James Gates** said in Chapter 9 where he discovered a "self-correcting" code underlying Reality and "code" implies coding which at its most basic level involves bits (0s & 1s) and bytes (8 bits).

> Let's explain something fundamental at this point: computer programming. All the computer understands is data or instructions –in binary. 0s and 1s.
> A computer programmer adding two number of course sees the numbers as 35 and 75, for example... however, the computer does not see "35" or "75" ... what it sees (speaking of the large IBM computers operating in EBCDIC code) is the following:
>   35 = F3F5 = 11110011 11110101 at address 0122BC = AMTb
>   75 = F7F5 = 11110111 11110101 at address 0416C7 = AMTa
> That is stored as data in the computer. Then the program issues an "ADD" instruction AP (add packed) AMTb, AMTa   (in Assembler code which is as low as programmers go today to the level of the machine – even COBOL programs are reduced to Assembler (in a PMAP)). The AP instruction is also converted to binary and the instruction looks like: FA = 11111010   so
>   AP == FA AMTb,AMTa == 111111010 0122BC, 0416C7 and the two addresses are binary, each in a register in the ALU.

The point of the above insert is that data is in one part of the computer (memory), and the instructions are loaded into another part of **memory** which the computer (knowing it is a program) loads each instruction sequentially into a **register** (a specialized part of the computer to incrementally go thru the steps of the program accessing the data and manipulating it according to the sequential instructions). Data is defined by 'DS' or 'DC' in the program and when the program is loaded into memory, the computer knows where the data is. It also knows where the first executable instruction is – and **everything** is in binary, 0s and 1s.

So that is the significance of "digital."

Needless to say, the Designer of such an Earth Realm, if S/he chose to simulate it, would use a very advanced (fast and large) computer which would function in a like manner, dealing with continuously variable data (i.e., analog processing), manipulating the subatomic world of particles according to rules S/he made up and then infuses the Realm with His/Her Consciousness – of which souls are a part. That is where **Entanglement** comes in – All is One. That delivers the Observer Effect, telepathy, clairvoyance… even NDEs.

> …if a conscious Entity is going to create a world for us to live in and experience, that conscious entity is clearly highly evolved compared to us. And being so evolved, it would certainly make use of the most efficient means to create a reality. A **continuous reality** is not only inefficient but is theoretically impossible to create because it involves infinities in the temporal domain as well as any spatial domain or property. [314] [emphasis added]
>
> **Note**: this is what **Dr. Brian Green** was saying in Chapter 9 with rounding errors being common in a finite simulation.

Consciousness-driven Reality

Elvidge is suggesting that our Earth Reality is consciously-driven by an entity smarter than we are, and with more ability or technology at its disposal. And there are 4 main aspects to that scenario, which is called **Digital Consciousness Theory**: [315]

1. Consciousness is fundamental and primary;

2. All matter is data and all forces are rules about how data interacts;

3. The reality we experience is illusory, a simulation of sorts, designed for us **to learn and evolve our consciousness**; (Earth Graduate anyone?)

4. The 'system' is digital and consists of the aggregate of all individuated conscious entities plus the virtual reality in which we live, and is driven by a fundamental rule of continuous improvement.

Elvidge makes it clear that **our Reality was created by some conscious Entity** and has been following the original rules (Laws) established by that Entity. It might even be following some fundamental evolutionary law since the beginning – we don't know… yet. Says Jim…

> Even without a conscious Entity to direct and sustain the Reality, the fundamental evolutionary law would certainly favor a perfectly functional reality that <u>doesn't require infinite resources</u>.

And without knowing it or saying it, he is supporting the **3D Earth Construct** which is self-contained and our Reality does NOT include the real Universe… we are contained in a VR Sphere – a finite creation, with finite resources, finite Laws that govern it.

Density of Matter

As the reader will recall from science classes, both in high school and college, one of the striking aspects of the atomic world around us is that there is a lot of space between atoms. And as scientists push farther and farther into the sub-atomic world, describing atoms, then quarks, then string theory, then telling us that a gold bar that is heavy and appears to be quite solid is actually 95% space and 5% atoms-molecules-quarks or "stuff"…. It gets rather absurd to reduce things to such a minute level… in fact, "…it is easy to see that the ratio of actual physical "stuff" to the space it consumes is trending toward zero…" [316]

> That is a key point… If Earth Realm is finite, it has finite resources and the SuperComputer that runs this Simulation can easily deal with our **finite Realm**… and it only deals with what someone is observing at the moment…. Parts of the Earth where no one lives or goes, would not need to be 'constructed' (but the code for it would be stored somewhere awaiting humans to venture into that territory)– much like a video game where the key player (You) moves to the right into the next scene – but it wasn't there until he moved into it… i.e., an example of **conservation of resources**.

> Computers do that too… An initial program loads, say, into 40 Mb of storage, and as it needs more room, it is allocated additional memory in Main Storage (or sometimes on the Disk Drives)… the Operating System knows the program is fragmented but tracked in a Memory

Allocation Table, but the (non-Assembler) programmer doesn't know this (or care).

Thus we can say with Elvidge that

> It takes an **infinite amount of resources** to create a **continuous reality,** but a finite amount to create a quantized [discrete and limited] one....
> The very nature of a digital computer is essentially the same as a discreet interpretation of quantum mechanics – [that is] a sequence of states with nothing existing or happening between the states...[317]

What he is saying is that events on Earth are not continuous and one does not necessarily cause or affect the other… people having lunch in Italy do not affect the way a train runs in Chicago. And the computer analogy says that one program following another in the same computer, running in sequence, even in the same memory space, has no causality between them. Yes, one program may call another and pass data to it, but the point is often that an accounting program may run followed by a sales analysis program that has nothing to do with even the data used by each program. And our world has many disparate events that suggest that data and rules do not manifest from the Reality, they <u>create</u> the Reality. This involves the Observer Effect.

Observer Effect

Elvidge thinks that the experiments that demonstrate the effect an observer has on an experiment involving particles (again, the sub-atomic world), may just be the "smoking gun" that proves his Digital Consciousness Theory.

Thomas Young's **Double-slit Experiment** back in 1801 shocked the physics world. All he was trying to do was resolve the Nature of Light conundrum: Is light a particle

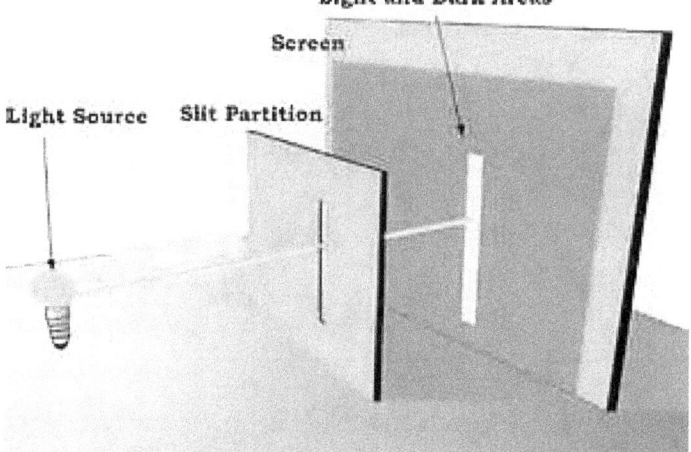

or is it a wave? So he set up a photon gun aimed at a cardboard with a slit in it and watched to see what the pattern of individual particles would be on the tracking screen behind it – if light was a particle then there should be a grouping of dots behind the cardboard where the photons passed thru the slit (as particles) and hit the recording medium. In short, they should all hit pretty much in the same spot.

And the particles all hit in the same place... typical of a **particle**.

Ok, that all looked good.... Then he changed only the cardboard and the new one had two slits in it... and he fired more photons (light particles) to see what happened... and freaked everyone out and promoted almost a hundred years' arguing and confusion... This was the setup...

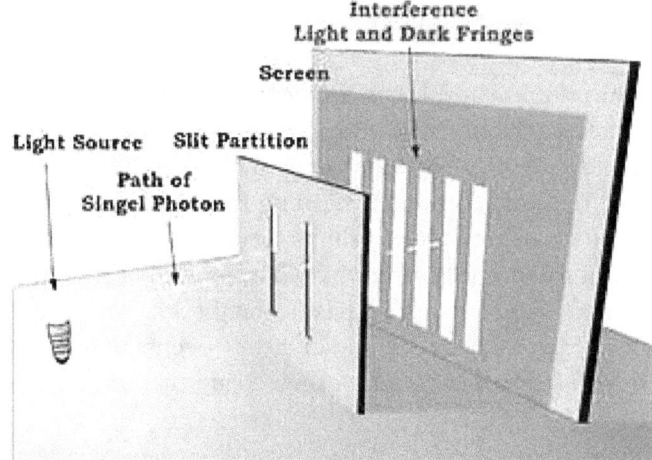

And what he got was a scatter pattern when he fired photons at the slits...

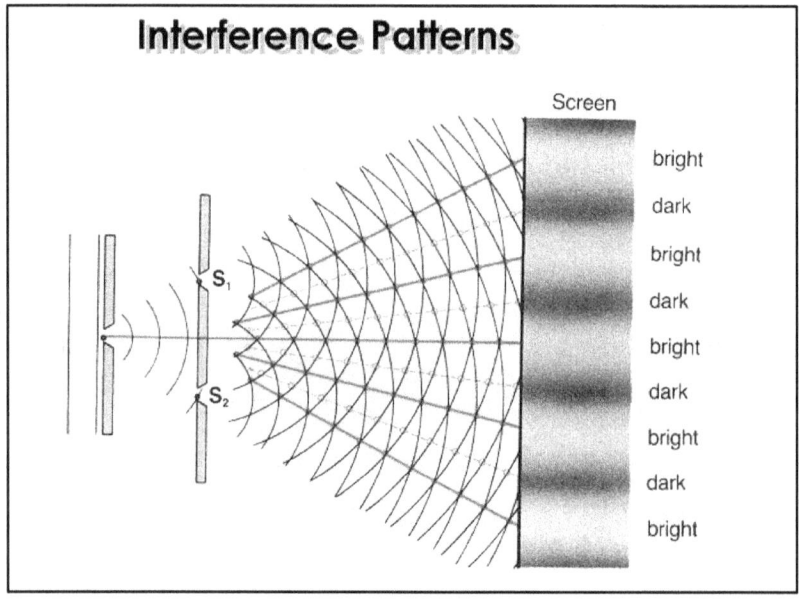

Alternating dark and light areas....! That behavior suggested light was a **wave.**

The fun was only beginning. So, **is light a particle or a wave?** Why is the pattern on the wall different based on whether there is one slit open or two? The only 'resolution' so far is to consider that light can behave as a particle or as a wave – but **how did the photon 'know' there were two slits, and thus change its 'behavior?'** This is definitely quantum weirdness.

And the answer lies in the Observer Effect.

> As it turns out, electrons and other subatomic particles also behave both like waveforms and particles, **depending on the observation**. Successively complex double-slit experiments have teased out even more fascinating and surprising results. For example, it has been determined that an electron shot at the screen actually goes thru both slits at the same time… This is called **quantum superposition**… the electron really exists only in a "probability" space of potential positions…
>
> The act of measuring which slit it goes thru causes [a probability function] to collapse into reality, and the electron then takes a definitive position. The outcome of the experiment suddenly changes to what it would be if the electrons were only particles…. In other words, **the very act of consciously observing causes a change in the outcome of the experiment**. [318] [emphasis added]

Most scientists have tried to rationalize it all away by citing faulty measuring, or hidden variables, or unseen equipment interference, but gradually over time their arguments have been refuted. **The world (like a video game) does not exist until it is observed and one consciously interacts with it** – as **Jahn and Dunne** said in Chapter 9 – at least not at the microscopic level.

Further, Dean Radin in 2012 conducted serious, well-designed experiments that demonstrated that "**conscious intent** can directly alter the results of the double-slit experiment." [319] Entanglement.

> **Doesn't this favor a Simulated world with which we interact?**
> **How would we 'affect' a real 3D rock world that is not simulated?**

The Great Cosmic Program

Elvidge suggests that the Greater Plan that controls our Reality is also fully aware of the state of consciousness of every free-willed observer in our Reality, via Entanglement. **Given that Entanglement connects the consciousness of the observer with that of the Great Cosmic Program, the behavior of an electron under observation can easily be made to agree with the <u>intent</u> of the**

observation being made. Quantum Entanglement and the Observer Effect are key parts of the Digital Consciousness Theory – and a Simulated Earth Realm.

Once a subatomic particle is observed, the Cosmic Program must then establish properties for that particle, effectively resulting in the collapse of the prior state, the probability wave function.

How It All Works

The **Digital Consciousness Model** is said to answer most of Quantum Physics' unique issues, even NDEs, telepathy, Remote Viewing, UFOs and the quantum Zeno effect. A model of that new Reality is shown below... the RLL is the Reality Learning Lab which is a set of organized information within all-that-there-is... we experience it every day while conscious.

> [The RLL] is where all the artefacts representing our Reality exist. It is where various "simulation" timelines run. Some call it the universe.[320]

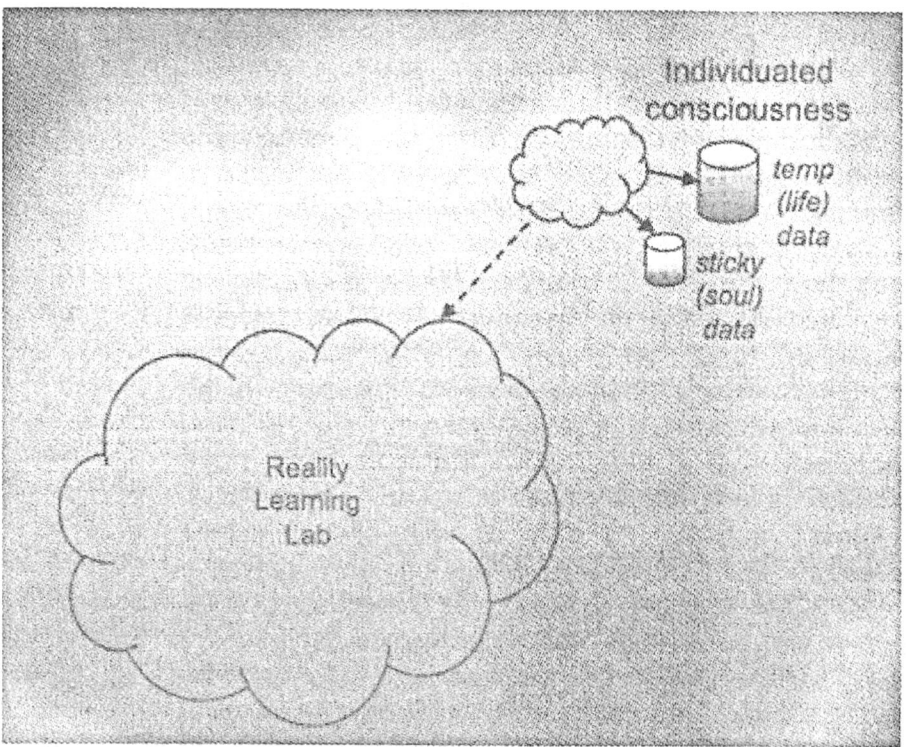

Digital Consciousness Model[321]

Sticky soul data = what we as an eternal soul retain form life to life. **Temp data** is what our brains/memories are full of; our consciousness.

Our own lives are virtual and subjective, subject to the Great Cosmic Program, explained more in detail in the NEXUS article… too techy to outline here.

Summary

So this brings us to the bottom line: **Why does it matter?** Some of these reasons will resonate with the reasons for knowing where we actually are – a 3D rock spinning thru space, a VR Sphere, or a Flat Earth… and it does make a difference.

Not only are we living on something other than the traditional 3D rock, covered with water, spinning thru space at 67,000 mph, with no purpose and no God… it makes a difference that we are contained, watched over, encouraged to grow as well as: [322]

1. If we are in a consciousness-driven reality , as Elvidge says, and he sounds correct, the **anomalies** in Physics, Geology, Anthropology, and metaphysics can all be explained using the Digital Consciousness Model;

2. Our consciousness being part of a larger One, indicates that **our existence is eternal -- eternal souls**…What kind of ridiculous thing would it be to exist for one lifetime and then when you die to have all that you learned and accomplished as a soul, just disappear … What would be the point?

3. If we understand that we are part of something larger, and that life has a purpose, then instead of focusing on hedonistic survival and materialistic gain, we can step back and **realize that our purpose is to learn and evolve our consciousness** and we can have greater <u>compassion and respect</u> for other people who are in the same Reality with us;

4. If we are all interconnected, then **war is absurd** and an abomination to living, eternal souls – nothing is gained by it;

5. Recognizing that **intent is what generates outcomes**, we would be more careful and focused in setting our priorities and crafting our intent.

And lastly, since Earth is a **VR Sphere** (with special holographic programming for the million-ton oceans), it demonstrates that Man is special enough that some Entity, as Elvidge says, cared enough to nurture us in a special environment… to "grow and evolve our consciousness…" in the **Earth School**.

The Great Earth Puzzle

Appendix D: What Is Gravity, Really?

To better understand what Gravity is –if it exists – we turn to some basic examples and then will turn back to Newton and Einstein, and forward to Tesla to clarify the major theories. Keep in mind that **no one knows what Gravity is**, but we'll examine some major ideas…

Gravity Experiment

Let's suppose we have a wooden table and a steel ball. Hold the steel ball above the table and let it go. Bam! It falls and hits the table top. Gravity you say. Obviously the wooden table did not block the Force of Gravity coming from the Earth (Science calls it gravitational waves) from affecting the steel ball.

Well, Ok, let's make the table in lead to block any **gravitational waves** and do the experiment again. Clunk! The steel ball falls to the table again – the lead did not shield the ball from the alleged gravitational waves.

Ok, we have to find a way to block gravitational waves from affecting the steel ball … Oh, Ok, let's make a **Faraday Cage** with a lead floor… that should shield the ball from any effects of Gravity.

Faraday Cage

A **Faraday Cage** is a box or room with all sides covered in electrified copper wire mesh – electrified with a current running thru it – which effectively blocks all EMF and outside RFI and electrical interference.

So let's put the table inside the Cage, and step inside with the steel ball and repeat the experiment. Clunk! The steel ball hits the tabletop again! Either there are no Gravity Waves or the Faraday Cage cannot block gravitational waves... or as Nikola Tesla would say, the two objects, the steel ball and the table are attracting each other... remember the Slinky Experiment in Chapter 11? (We'll get to Tesla later.)

Our first clue that the Experiment would not work in the Faraday Cage is that Gravity is still apparent – when we walk into the Cage, close the door and turn on the electrified walls <u>and floor</u>, we are not weightless, the table is not floating, the steel ball does not float in the air – nothing is resisting Gravity... so the steel ball again hits the table.

What is going on?

> **Could it be as simple as this: (a) Gravity does not exist <u>and</u>**
> **(b) objects fall/stick to Earth because they are heavier than air.**
> **Is this why today's scientists cannot find nor measure Gravity?**

Newton's Law of Gravity

Yep, you read that header correctly – LAW. After Newton saying that it was an <u>assumption</u> that he could not prove (Chapter 4) and he admitted that there were holes in his <u>theory</u>, and we still can't find it nor measure it, we nowadays call it a Law, and what it says is:

> Gravity arises solely from an object's Mass. The bigger the Mass,
> the bigger the gravitational pull. Gravity is an attractive force...
> it pulls objects together....but Newton himself admitted that there
> was **no known mechanism for transmitting Gravity between**
> **objects**... and he always saw that as a yawning hole in his own
> theory. [323]   [emphasis added]

Einstein didn't help much, either – he would have told Newton that there is another aspect to Gravity, called <u>pressure</u>... General Relativity contends that Gravity arises from an object's Mass and positive pressure on the space/time continuum. The example given was to take a sealed **bag of potato chips** weigh it, then squeeze the bag (imparting pressure), and weigh it again – all without opening the bag – and it <u>will weigh slightly more</u> (in grams, to be sure), but that is Einstein for you. [324]

Thus positive pressure = *attraction*.

And if the pressure was negative, then the effect would be negative Gravity, or *repulsion* (aka antigravity). Theoretically.

Ok, that is as techy and weird as we're going to get.

Einstein & Gravity

Einstein began work on Newton's **Universal Law of Gravity** in 1907 and wanted to know (as we do, too!): How does it really work? But he had already shot himself in the foot in 1905 by pooh-poohing the existence of the Æther. As it would later turn out, when better scientific equipment and methodologies were available, **Michaelson & Morley** (1920's) and later **Podkletnov** (1990's) would prove that the Æther did exist, but Einstein had already rejected one of the key aspects to what we call Gravity.

> By the way, **Nikola Tesla** was already on Einstein's case having said on Apr. 15, 1932, [that] Einstein's theory regarding changing matter into force, and force into matter, [E= MC^2] was "absurd". And he railed against Einstein for his incredible prejudice and ignorance on the subject of the Æther ... as well as the limited and erroneous theories of Maxwell, Hertz [and] Lorentz. [325]

Newton had said that he knew there must be something that communicated the Gravity *effect* between objects, but had no clue what it was. In his *Principia*, he lamely left the question "to the consideration of the reader." Einstein on the other hand was more aggressive and in 1915 he proposed that there is something in what looks like empty space… space itself was Gravity's medium! But he would not say Æther.

He proposed that Gravity was a distortion of the environment by large solar bodies, whose Mass exerted *pressure* on the space-time continuum… giving rise to a **General Theory of Relativity**… but did not explain how this operated on Earth. So **Dr. Brian Greene** (Chapter 9) explains how General Relativity works, and tries to explain how Gravity fits into the scenario:

> …think of two clocks, one on the ground, and the other on top of the Empire State Building. Because the ground clock is closer to the Earth's center, it experiences slightly stronger gravity than the clock that's high above Manhattan. General relativity shows that because of this, the rate at which time passes on each will be slightly different: the ground clock will run a tiny bit slow (**billionths of a second** per year) compared to the elevated clock…. General relativity then establishes that objects move toward regions where time elapses more slowly; in a sense all

objects "want" to age as slowly as possible. From an **Einsteinian perspective, that explains why an object falls when you let go of it.**[326]

Oh really? How do you measure "billionths of a second?"
I can't believe the bag of potato chips, and now that last sentence... And **James Gates**, a physicist at the University of Maryland who discovered strange, error-correcting codes deep in the equations of Super String Supersymmetry (Chapter 9) said it best:

> "Error-correcting codes are what make browsers work, so why were they in the equations that I was studying about quarks, and leptons, and supersymmetry?" he said. "That's what brought me to this very stark realization that I could no longer say that people like Max [Tegmark] are crazy."
>
> "Or, stated another way, **if you study physics long enough, you too can become crazy**," he added. [327] [emphasis added]

So Einstein and Dr. Green did not clear the air. Where can we turn for a decent answer to the question : Does Gravity really exist, and if so, How does it work?

> Remember, The Flat Earth people said Gravity was totally unnecessary: **objects fall because they are heavier than air**... and Occam would love that answer.

Since Nikola Tesla was so disgusted with Einstein, and Newton, what did he say about Gravity?

Nikola Tesla & DTG

> Tesla claimed to have developed his own physical principle regarding matter and energy that he started working on in 1892, and in 1937, at age 81, claimed in a letter to have completed a "**dynamic theory of gravity**" [DTG] that *"[would] put an end to idle speculations and false conceptions, as that of curved space."* He stated that the theory was "worked out in all details" and that he hoped to soon give it to the world. Further elucidation of his theory was never found in his writing. [328]

His DTG was never found because upon his death, "the papers of Tesla [were] concealed in government vaults for "national security" reasons." [329] But we do know some basic things about what his theory was, from others who worked with him, from demonstrations he gave publicly, and from scattered notes on related subjects.

And we can assume that his DTH had something to do with the **antigravity flying machine** that he allegedly built and flew in 1915 (see below)… that, too, along with Otis Carr who built a UFO based on Tesla's ideas, was also removed from the public.

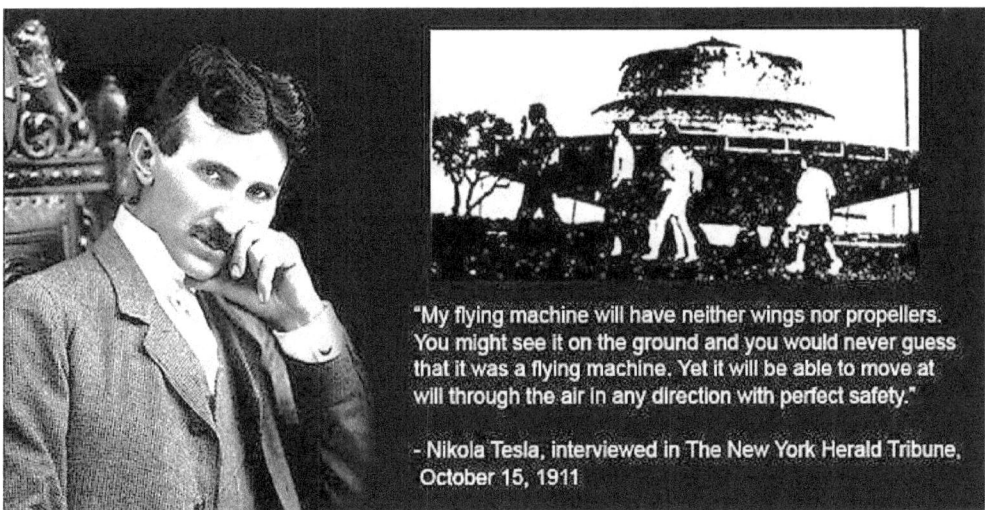

"My flying machine will have neither wings nor propellers. You might see it on the ground and you would never guess that it was a flying machine. Yet it will be able to move at will through the air in any direction with perfect safety."

- Nikola Tesla, interviewed in The New York Herald Tribune, October 15, 1911

…and again…

The Tesla Flying Machine

Note the football-shaped Wardenclyffe Tower in the background (left) which Tesla used to broadcast energy.

So what was his big insight?

Dynamic Theory of Gravity (DTG)

What we're going to do is share some of the key things he said that must have led him to the DTG or are perhaps part of it, and it contradicts Einstein because Einstein had decided that Æther didn't exist… of course that is what Physicists today are calling **Dark Energy/Matter**. Tesla called it **ÆtherForce**….[330]

The above inset says: It explains the courses and motions of heavenly bodies under its influence so satisfactorily that it will put an end to idle speculation and false conception, such as that of curved space [a rebuttal of Einstein]…

Only the existence of a field of force can account for the motions of the bodies as observed and its assumption dispenses with space curvature. All [existing] literature on this subject is futile and destined to oblivion. So are all attempts to explain the workings of the universe without recognizing the existence of the Æther [Einstein denied it exists] and the indispensable function it plays in the phenomena.

By the way, Tesla was not just criticizing Einstein, he had also shown Thomas Edison to be wrong. Edison not only took 768 attempts at making a lightbulb and then 'championed' himself for the virtue of perseverance (when Tesla designed a successful fluorescent bulb within 20 tries), but Edison believed that DC power (meaning huge batteries all over a city) could power the city, and Tesla developed AC power which was much more practical. Edison got even and had Tesla's Wardenclyffe Tower funding removed by J.P. Morgan.

So mankind was denied the benefit of 'broadcast power.'

The "two great discoveries" to which Tesla referred, were:

1. The Dynamic Theory of Gravity – which assumed a field of force which accounts for the motions of bodies in space; assumption of this field of force dispenses with the concept of space curvature (à la Einstein); the **ether has an indispensable function in the phenomena** (of universal gravity, inertia, momentum, and movement of heavenly bodies, as well as all atomic and molecular matter); and,

2. Environmental Energy – the Discovery of a new physical Truth: there is **no energy in matter** other than that received from the environment.

Specifically, Tesla wrote an article, "Man's Greatest Achievement," and outlined his Dynamic Theory of Gravity in poetic form: [331]

> That the luminiferous ether fills all space
> That the ether is acted upon by the life-giving creative force
> That the ether is thrown into "infinitesimal whirls" ("micro helices") at near the speed of light, becoming ponderable matter (shades of **SQK in later section**)
> That when the force subsides and motion ceases, matter reverts to the ether (a form of "atomic decay")

Further he believed…. It was possible to:

*   -Precipitate matter from the ether (SQK also promotes this)
*   -Create whatever he wants with the matter and energy derived
*   -Alter the earth's size
*   -Control earth's seasons (weather control)
*   -Guide earth's path through the Universe, like a space ship
*   -Cause the collisions of planets to produce new suns and stars, heat, and light
*   -Originate and develop life in infinite forms

You can see why the government confiscated his papers….

And as regards Gravity, he said that it was **a byproduct of the fluid-like Æther...**

> Tesla's ether consisted of "carriers immersed in an insulating fluid", which filled all space. Its properties varied according to relative movement, the presence of mass, and the electric and magnetic environment [of the Æther]. [332]

Tesla's Æther was neither the "solid" Æther with the "tenuity of steel" of Maxwell and Hertz, nor the half-hearted, entrained, gaseous Æther of Lorentz…. Tesla's Æther was rigidified by rapidly varying electrostatic forces, and was thereby involved in **gravitational effects**, inertia, and momentum, especially in the space near earth, since, as explained by Tesla, the earth is "…like a charged metal ball moving through space", which creates the enormous, rapidly varying electrostatic forces which diminish in intensity with the square of the distance from earth, just like gravity. Since the direction of propagation radiates from the earth, the **so-called force of gravity** is toward earth. [333] [emphasis added]

Tesla never said anything about a Flat Earth scenario.

What was more significant to Tesla was the application of Gravity for the propulsion of aircraft… which he referred to as **electro-propulsion technology**. And T.T. Brown and Otis Carr became interested and worked to discover ways that Gravity and anti-gravity could be used for fuelless propulsion. And yet, Tesla kept much of his scientific paperwork on "Tesla currents" secret.

Tesla's Secrecy

> Due to his pacifist sympathies, Tesla originally contemplated giving his **electric flying machine** to the Geneva Convention or League of Nations, for use in 'policing the world' to prevent war. Later disillusioned after WWI with the collapse of the League, he said he'd "…underestimated man's combative capacity"…. while discussing a "three-hour" airplane [ride] between New York and London: "…we have here the appalling prospect of a war between nations at a distance of thousands of miles, with weapons so destructive and demoralizing that the world could not endure them. That is why there must be no more war." **With the government's spurning of his defense suggestions** [including the **'Death Ray'** which he intended for defense only], **Tesla's only recourse was to withhold his secrets from the world, and to dissuade discovery in their direction.** [334]

And as a result, he also withheld the data on DTG as it would lead to **electro-gravitic propulsion**… little did he know that Earth is not a globe, not spinning thru space, and due to the Firmament, we can't fly off into Space. And yet he was correct about the Æther and was a genius ahead of his time… too bad that petty men (Edison and J.P. Morgan) shackled him and could not see the proactive uses of his work.

Thus Tesla's conclusions regarding Gravity are as follows (as much as men have been able to piece together what he said in his lectures, regardless of the missing DTG papers):

> **The ether is a universal medium, which fills all space….**
> The ether is normally electrically neutral, ultra-fine, and **penetrates all solid matter.** There is also an ultra high frequency, ubiquitous radiation, normally in equilibrium, called **Zero Point Radiation** ("ZPR"), which interpenetrates the ether, and represents electromagnetic radiation in its finest, densest form, which, in conjunction with the ether, conserves universal perpetual motion.
>
> The ether in conjunction with the ZPR, **is the source of all matter and force**…. Electromagnetic disturbances in the ether extract "energy" from the ZPR, which is explainable only by an ether theory…..
>
> **All mass and space have dielectric properties**. Differences in dielectric properties [+ or – charges] cause changes in the electromagnetic displacement within mass and the **etheric wind**. Earth's electric field creates dielectric displacement effects within ether and mass within earth's electric field. The difference between the dielectric displacement within a mass and the dielectric displacement outside the mass in the etheric wind, creates **a down-force** in the direction of the negative polarity, as the etheric wind 'blows' through a mass. **This is called "gravity."** [335] [emphasis added]

And then another author reviewing Tesla's ÆtherForce article adds his personal touch:

> Since "seeing is believing," and I have seen, I believe. The behavior of **man-made flying saucers** proves the existence of these "free energy" inventions of Nikola Tesla, which show that he was right in his opposition to Relativism [i.e., Einstein], and that the prevalent theories taught in the scientific institutions of the world are **patently fraudulent**. [336] [emphasis added]

That is the same thing that was said in the 1700's by the Flat Earth supporters regarding the new theory that the Earth was a globe.

And yet, Ben Rich in 1993 (Chapters 4 & 10) told us that **they corrected Einstein's false science** and perhaps thru the input of the Paperclip Scientists of the 1940s, we have been able to discover the secret of electro-gravitic propulsion anyway.

Current Theories

And for what it is worth, many scientists today are re-examining Gravity since it has remained a puzzle since Newton invented his "Law" of Gravity, but could not explain it. Some researchers still can't find it and others are looking right at it [as a property of **Dark Energy/Matter aka Æther**] but don't recognize it. Thus some are postulating String Theory, saying that sub-quark-level strings of vibrating particle-energy are what create Gravity, while others suggest that a Gravity *force* doesn't exist and what we call 'Gravity' would be better described by electrical, magnetic and electromagnetic forces (as Tesla suggested).

Other researchers have come up with a new concept called **Loop Quantum Gravity**. This is supported by scientists who are still seeking to prove Einstein right:

> **Loop quantum gravity (LQG)** is a theory that attempts to describe the quantum properties of the universe and gravity. It is also a theory of quantum spacetime because, according to General Relativity, **gravity is a manifestation of the geometry of spacetime**. LQG is an attempt to merge quantum mechanics and general relativity.
>
> From the point of view of Einstein's theory, it comes as no surprise **that all attempts to treat gravity simply like one more quantum force (on par with electromagnetism and the nuclear forces) have failed**. According to Einstein, **gravity is not a force – it is a property of spacetime itself**. Loop quantum gravity is an attempt to develop a quantum theory of gravity based directly on Einstein's geometrical formulation. The main output of the theory is a physical picture of Space where **Space is granular**. The granularity is a direct consequence of the quantization. [337]   [See Glossary, emphasis added]

And then there is another group, using the CERN Large Hadron Collider to look for particles that cause Gravity – called **gravitons** (useful in String Theory). Gravitons, String Theory, Supersymmetry, Dark Matter/Dark Energy [the Dark two are inseparable] and a new wrinkle: Gravity originates in another dimension and affects this one (Kaluza-Klein Theory).

> You are probably saying "Oh for Pete's sake, they don't know, do they?!" (I said that same thing months ago – after getting a headache trying to sort it all out.)

So, is there a bottom line, and if so, what is it?

Summary

So if they do not know what Gravity is, any more than Einstein could figure out Newton's Law of Gravity... where does that leave us? The first question is: Why was Newton's <u>assumption</u>, which even he said he could not prove, made into a Law when Gravity cannot be found, nor described?

> In Newton's law of universal gravitation, gravity was an external force transmitted by unknown means. In the 20th century, Newton's model was replaced by General Relativity where **gravity is not a force but the result of the geometry of spacetime**. Under general relativity, **anti-gravity is impossible** except under contrived circumstances. Quantum Physicists have postulated the existence of **gravitons**, a set of massless elementary particles that transmit the force, and the possibility of creating or destroying these is unclear.
>
> Under general relativity, gravity is the result of following spatial geometry (change in the normal shape of space) caused by local mass-energy. This theory holds that it is **the altered shape of space, deformed by massive objects, that causes gravity**, which is actually a property of deformed space rather than being a true force. [338]

So Einstein said the Æther does not exist (setting science back at least 50 years, until Ben Rich proved him wrong), and Einstein said that anti-gravity was not possible, and yet there is such a thing as anti-gravity – it is called **repulsion**, and Viktor Schauberger built and flew the *Repulsine* apparatus (1940s) that proved it... but that is another story. (See VEG Ch. 4)

> Some of the most basic scientific facts and principles are being concealed and misrepresented [e.g., Global Earth vs Flat Earth], beginning in Kindergarten. Because of this we must each re-examine nature, and question all accepted theories.... to rediscover and propagate "true science." About 1937 "science" and space propulsion became really schizophrenic: While Wernher von Braun ostensibly developed rockets, he covertly developed secret Illuminati flying saucers [with Hans Kammler]; while Albert Einstein's relativism and Werner Heisenberg's "uncertainty

principle" were being publicized, the 'ether physics' of Nikola Tesla and others [Otis Carr] was being ridiculed, yet was secretly being used for saucer propulsion and wonder weapons [Haunebus and TR3Bs] so highly classified that they were never even intended to be used in WW II. [339]

So how does that relate to Tesla's version of Gravity?

Rewind: Tesla's Dynamic Theory of Gravity

The theory says that **all bodies (atomic masses, etc.) have electrical content**. Their movement, together with their particular electrical content, in space, in which there are **magnetic and electrical fields**, as well as cosmic **("zero point") radiation** [ZPR] causes them to emit microwaves. The **microwaves** react with the ether which then behaves like a "continuous electrically conductive, 'fluid' [or] 'solid state' mass." The interaction between the microwaves and different bodies through the ether, causes their **gravitational interactance**. [340]
[emphasis added]

Tesla discovered that the gravitational force that draws things together could also be reversed and inadvertently he backed into the discovery of a "propelling force." He called it a **"Tesla current"** and found it could be pumped to suspend an object off the ground, or in space. He concluded that "an (anti-gravity) electric flying machine could be propelled by the reactance of high-frequency, high voltage electromagnetic waves... which **push against the ether** and uses the [well-known] Hall MHD (magnetohydrodynamic) effect." [341]

What has that got to do with Gravity?

Simply that as Tesla experimented with Gravity, he came across the Æther and ways to manipulate it to attract and repel objects... which is then where his flying machine (p. 473) came in – with electro-gravitic propulsion.

This is where we can now share the rest of Tesla's story...

There is **a deceptive effect of Gravity**, which makes it appear to penetrate mass under normal conditions [recall the gravity experiment in the beginning of this Appendix]... The force appears to be linear and penetrating. These false appearances (what Einstein called "apparent effects") are in reality a natural deception, resulting from the tendency to **falsely assume that gravity is exerted between the bodies** (the earth and two bodies) **by forces strictly <u>endemic to the bodies</u>**(in the absence of Ether Physics), **when in fact the force is exerted between the bodies and [the Æther] existing <u>in</u> and** between all bodies. … [says that Gravity is a <u>property</u> involving the Æther and its existence <u>within</u> the bodies undergoing attraction/repulsion]

This implies that **a body… upsets the equilibrium of the ZPR** and **pumps the [Æther] upward** due to the action of the electric and magnetic fields of the earth which permeate the body, thus **accelerating the body downward** [i.e., Gravity]…. **Gravity is not really 'penetrating' anything** but is only a reaction within and between normal matter and [the Æther] which does penetrate (and saturate) everything…. [342] [emphasis added]

The Æther and the ZPR [like Zero Point Energy] permeate all mass… and react to that of the Earth when suspended above it, drawing the two objects together. So in a way, Newton was partially correct when (according to his Kabbalistic study – Chapter 4) he suggested that there was a "force of attraction" -- where two bodies will tend to attract each other… but it was <u>not a unique force</u> of Gravity, but an energetic resonance (or quasi-electromagnetic resonance) between them operating thru the interpenetrating Æther.

Gravity is simply a property inherent in the Æther.

This is what the Flat Earthers simply call – "It falls to Earth because it is heavier than air." And that is also true.

However, there is one more revelation to this summary and it bears a close resemblance to Tesla's DTG, but is easier to describe.

Subquantum Kinetics (SQK)

This is the brainchild of astrophysicist **Dr. Paul LaViolette** with a PhD in Physics and who has done much research into alternative explanations of Earth phenomena. His greatest achievement is a **Model G** that explains things that standard Quantum Mechanics already does, <u>plus</u> many more that it does not or cannot.

Subquantum Kinetics (SQK) begins at the subquantum level of matter, with already established principles, and postulates further well-ordered reaction processes, collectively called the *transmuting ether*. It does not depend on observation of the physical world as a starting point, as does Quantum Physics.

The **transmuting ether** [is] an active substrate that is quite different from the passive mechanical ethers considered in the eighteenth and nineteenth centuries. It further proposes that the concentrations of the substrates composing this ether are the energy potential fields that form the basis of all matter and energy in our universe. The operation of these ether reactions causes wave-like field gradients (spatial concentration patterns) to emerge and form the observable quantum level structures and physical phenomena…[343]   [emphasis added]

As time goes on, Quantum Physicists will realize that their postulated Dark Energy and Dark Matter are the same Æther substrate that SQK and "Tesla currents" deal with. After all, the Universe and its subquantum components is only one consistent thing, not a multiple of different things, so SQK and Quantum Physics cannot develop radically different definitions of particles, waves and theories and both be right. They both could be wrong, except that SQK agrees with Tesla's DTG (as far as we know what it had to say), and SQK does accurately describe quantum processes – on the Earth and in the Universe.

Advantages of SQK

Classical field theory	SQK
1. Plagued by the field-particle dualism.	Problem does not exist.
2. Plagued by the wave-particle dualism.	Wave aspects incorporated.
3. Considers antigravity an impossibility.	Considers antigravity flight.
4. Natural events are inherently indeterminate.	Commonsense notion of causality.
5. Allows disruptive naked singularities.	Avoids this problem.
6. Universe emerged from nonexistent state.	Universe emerged from preexisting ether substrate.
7. Cannot explain how subatomic particles originate.	Subatomic particles arise from ether substrate.
8. Quantum electrodynamics and General Relativity contradict each other: Cosmological Constant conundrum.	All fields encompassed by 1 internally consistent theory. **Gravity is explained.**
9. Universe is expanding due to redshift.	Redshift not due to expansion; it is due to "tired light" phenomenon.
10. Black Holes are matter-consuming.	Black Holes are "mother stars" which birth matter and energy.
11. Fails to explain the excess heat coming from the Earth's core.	SQK attributes it to **genic** energy.
12. Fails to explain source of supernova explosions.	SQK genic energy explains this.
13. Posits existence of **quarks**, but cannot produce proof.	SQK uses 6 basic **etheron** elements to source all matter.
14. Posits existence of **strings** within quarks which cannot be proven/discovered.	SQK etherons are basic elements and are all that is needed to explain the Universe.

Note: number 8 above….. SQK explains Gravity.

SQK Operation Basics

While this Appendix is not a treatise on SQK and why it works, it is amazing how elegant yet simple it is. While Dr. LaViolette was initially inspired and intrigued by a self-repeating chemical wave phenomenon called the **Belousov-Zhabotinskii reaction** (Chapter 9 in VEG), which led to an insight into **the transmuting Ether process**, a more basic model called the **Brusselator** led to the development of what was to be Model G.

> Model G describes the basic function of the Ether substrate and consists of *etherons* which interact with each other in a repetitive cycle, a non-ending **'dance' of subatomic particles**, diffused throughout space, and is described by just 5 basic kinetic equations (see below)...

$$A \rightarrow G$$
$$G \rightarrow X$$
$$B + X \rightarrow Y + Z$$
$$2X + Y \rightarrow 3X$$
and...
$$Y \rightarrow \Omega$$

Model G [344]

> According to Dr LaViolette's explanation: the model refers to the recursive conversion of X etherons into Y etherons and back into X etherons (e.g., the 'dance' referred to). A and B are input Ether reactants, as well as Z and Ω are output Ether reactants. G, X and Y are variable Ether reaction intermediates. Concentration patterns of these three variables form the particles and photons that compose our physical Universe. [345]

Simple and elegant, SQK has no plethora of particles as in Quantum Physics which has:

  18 quarks,
    18 antiquarks,
      8 gluons,
        including 3 massive U particles (U+, U-, U°).

So far **47 particles in Quantum Physics** are not enough to describe the Universe and together describe the "particle zoo." [346]   It is done in SQK with **7 basic particles** and **genic** energy.

Gravity Under SQK

SQK is the ether physics of Tesla... rediscovered. It starts with the Æther and refers to the particles that comprise the Æther as **etherons**. This subtle medium extends throughout all of Space and matter and is in constant flux. Concentrations result in variance of gravitational strength.

> A gravity field characterized by a spatial variation in gravity potential [- or + charge] would correspond to a spatial variation in G etheron concentration. [347]

These particles are being created by the **etheron 'dance'** and they do not disperse over time. And they have a small mass and electric charge... meaning that the **particles' electric charge generates its gravitational mass and associated gravitational field.** And a particle's X ➔ G <u>reverse</u> reaction (see diagram above) is what creates the gravity field (comprised of many etherons). **The subatomic particles' etheron population is continuously transformed and renewed. The fluxes create the gravity potential fields.**[348] The more etherons in an object, the greater the field strength.

Says Dr. LaViolette (and this is worth reading to understand Gravity better):

> Subquantum Kinetics predicts gravitational field polarity [the field has a + or - charge].... Einstein had attempted to expand his relativity theory to encompass both electromagnetism and gravitation, but he was unsuccessful. Relativity [GTR] was unable to predict any connection between charge polarity and gravitational field polarity [because it ignored the Æther where the charge exists].

> Subquantum Kinetics... predicts that gravity should have two polarities. It permits the creation of either a matter-attracting gravity potential **well** or a matter-repelling gravity potential **hill** and predicts that these two polarities should be directly correlated with electric charge polarity. That is positively charged particles such as protons would generate **gravity wells** and negatively charged particles such as electrons would generate **gravity hills** (next page diagram).

> [When they combine to form neutrally charged atoms, the gravitational polarities would neutralize each other.]

> However, because a proton's gravity well is theorized to marginally exceed an electron's gravity hill.... A small residual matter-attracting gravity

potential **well** [would exist]… thereby generating the gravity we commonly experience pulling us to earth.

Einstein's Gravity Well   (Mass affects spacetime):

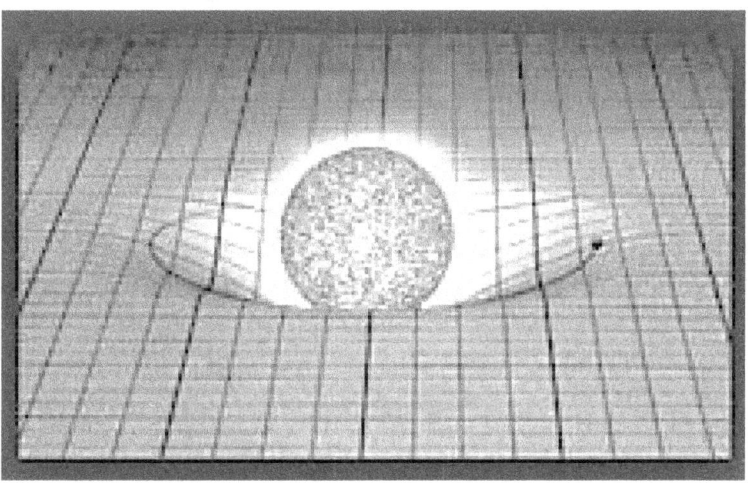

Dr LaViolette's Gravity Well/Hill:[349]

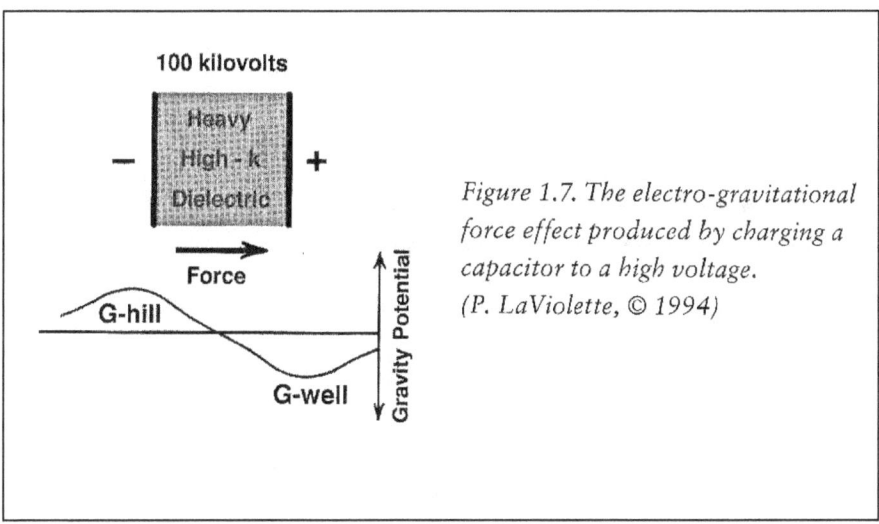

Figure 1.7. The electro-gravitational force effect produced by charging a capacitor to a high voltage. (P. LaViolette, © 1994)

A proton's gravity **well** (positive charge) has an **attracting effect** on surrounding neutral matter, and an electron's gravity **hill** (negative charge) has a **repelling effect** on surrounding neutral matter. [350] Again, all inherent **properties** within the Æther.

> This is why Newton not knowing and Einstein rejecting the Ether concept could not figure out how Gravity worked.

The subatomic fields and particles have mass and **electric charges** and that of a steel ball held above the table (see earlier exercise) is interlinked with that of the greater Earth (and also the table), such that when the ball is released, the greater Earth/table field (positive well) automatically attracts the weaker field of the steel ball.

> It may just be that this is the "attractive force" from the Kabbalah that Newton was referring to (in Chapter 4). On the other hand, recall what Tesla's DTG said (earlier section):

> This implies that a body… upsets the equilibrium of the ZPR and **pumps the [Æther] upward** due to the action of the electric and magnetic fields of the earth which permeate the body, thus **accelerating the body downward** [i.e., Gravity]…. Gravity is not really 'penetrating' anything but is only a reaction within and between normal matter and [the Æther] which does penetrate (and saturate) everything….

Could the 'pump' be related to the etheron flux, outputting ZPR microwaves and creating a **gravity well** (proposed by SQK) under the steel ball, causing it to sink into it … all the way to the ground?

Last Word

You might want to think about that **Slinky Exercise** in Chapter 11 – Why didn't the Slinky fall to the ground when the man let go of the top? Instead the Slinky stayed where it was, until the top of the Slinky met the bottom—and <u>then</u> the Slinky fell to the ground…. Doesn't that appear to contradict Gravity?

What dynamic kept the Slinky in the air "suspended" as it were – until the Slinky coils had collapsed – THEN it falls to the ground? Can either of Tesla's or SQK's principles help us here?

Here is a proposed answer posted on the Internet : [351]

> **Acceleration due to gravity** is a common one. All parts of the slinky are accelerated by gravity equally, but the bottom of the slinky is being supported by the top. [?!] When the top is let go, **it takes time for the bottom to find out**, and in that time the slinky can collapse.

If you buy that, you are ready to buy some ocean-front property I have in Montana. Look at the second highlighted phrase…

"…it takes time for the bottom to find out…"

That has nothing to do with it, and is just a smart-*ss answer. If you can't even intuit that that is wrong, well…. NASA has some really weird pictures from the Moon where Earth is the size of a pea – even though Earth is 3.6 times the size of the Moon…. (VEG, Appendix A)

The **whole Slinky should fall** if traditional Gravity exists and is pulling on the Slinky as a <u>complete object</u>… the bottom does not know what the top is doing… nor does it matter. Unless… the collapse of the coils and their kinetic energy affects the Slinky's gravitational charge (Tesla again) and it 'resists' gravity… but there we go again… making up convoluted answers which violate Occam's Razor. But then, could Occam in his simple world of AD 1300 understand or account for sophisticated Simulation programming?

The Bottom Line

This is a phenomenon that says Gravity is not what we think it is.

Appendix E: Adjustable Firmament

To better explain the Flat Earth (FE) and VR Sphere concepts, which <u>both</u> say that Earth is contained in a Shell or Firmament (<u>whether you believe in it or not</u>), there is a corresponding aspect that may help anyone to better understand what the **original Firmament** concept was. Thus, it is necessary that we look at it – as suggested by the Bible, multiple African legends, and lately by NASA.

> This is not to lead anyone on, and it is not intended as a joke.
> This Appendix is based on true, historical accounts.

Many legends have their basis in something that a human observed somewhere in the world. And what is about to be examined is the serious prospect that the Firmament not only exists – as proven by NASA and the *Gegenschein* – but that it has been altered (at least once) since its creation.

> And that aspect aids and abets the VR SPhere theory (Chapter 11)
> that the gods <u>can</u> make changes to this Earth Realm – and have,
> and probably will continue to do so as Man's activities change.

The interesting point is that it <u>looks like</u> **the Firmament was once at a lower level (altitude) than it is now**. This is especially true if the Earth Realm is a Simulation… parameters can be changed, and a very high Firmament in particular can be changed without terrestrial humans even noticing it – because it doesn't much affect their daily activities.

But the point is: as Man began to take to the air, to fly and send rockets higher and higher, and launch satellites, an original Firmament that <u>appears to have been within the reach of humans</u> about 4-5000 years ago will not work today.

So is there any evidence for the Firmament having been once closer to the ground? And to start with, we need to prove that **there <u>is</u> something surrounding the Earth**.

The *Gegenschein*

First we need to recall that even NASA photographed what Charles Fort called the *Gegenschein* – a bright reflection off something surrounding the Earth. It has **no parallax,** meaning it does not change size depending on one's angle of observation.

As Charles Fort said, about 1912, the *Gegenschein* is like a shell, and he was privy to something that he never spoke about <u>in print.</u> Could that have been the older Flat Earth scenario? Charles Fort repeats what Enoch said about the stars…

> …whether there be a shell-like, evolving composition, holding the stars in position, and in which **the stars are openings**, admitting light from an existence external to the shell, or not, all stars are at about the same distance from this Earth as they would be if this Earth were stationary and central to such a **shell**, revolving around it. [352]   [emphasis added]

and…

> The *Gegenschein* -- Now we have indication that there is such a shell around our existence. The *Gegenschein* is a round patch of light in the sky. It seems to be reflected sunlight, at night, because it keeps position about opposite the Sun's position. The crux: **Reflected sunlight – but reflecting from what?**
>
> That the sky is a **matrix** in which the stars are openings, and that, upon the inner, concave surface of this celestial [transparent energy] shell, the sun casts its light, **even if the earth is between**… [353]
>   [emphasis added]

The *Gegenschein*
(credit: NASA: <u>*http://apod.nasa.gov/apod/archivepix.html and below*</u>)

It is recommended that the reader check out the above NASA link to three samples:
2008 May 07: The Gegenschein over Chile. (sunlight)
2006 December 26: The Gegenschein . (sunlight)
June 25 1999: The Gegenschein . (sunlight + Sun)

So if there is a barrier, the Firmament, was SETI such a great idea, or was it **a waste of money?** The barrier is known to bounce back radio waves... <u>from</u> both sides!

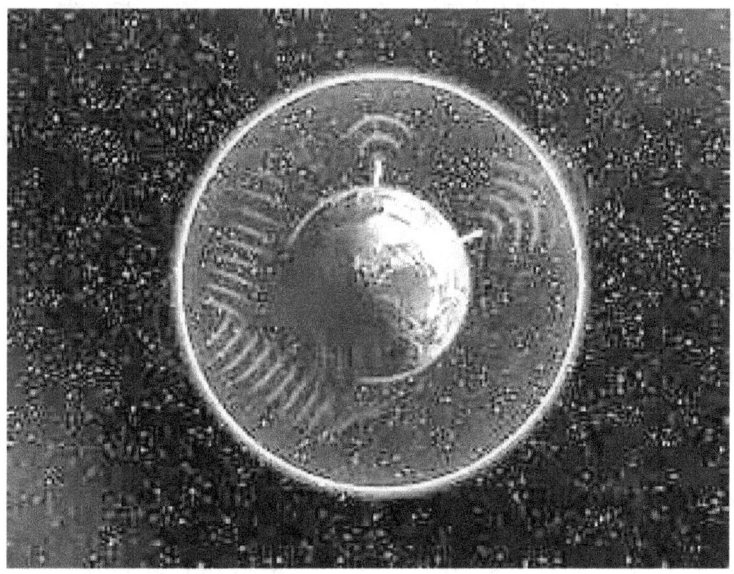

(Credit: Bing Images/earth shield)

SETI founder, **Dr. Steven Greer**, believes there's a quarantine over the planet because the Earth's inhabitants are a threat to other forms of life outside of our planet, due to our military aggression throughout the world. [354] (Maybe we are too petty & violent?)

So there we have NASA, Dr. Greer, and Charles Fort reminding us that there is something surrounding Earth. And we will see in a few minutes, that NASA has discovered a shield around the Earth. But before looking at the current NASA info, let's go back in time to some really old African legends that tell of Man's interaction with the God of Heaven...

Thus, building a case here, we come to the ancient legends of humans in Africa. Recall from Al (Book 6) that the Dogon, Ashanti and Zulu were contacted by the **Skygods** and given knowledge about the Earth and the stars.

Again, recall that the Skygods were not ETs, but either the Ancient Ones, or the Anunnaki... both were here guiding and shepherding Man. And the gods interfaced with Man, and Man sometimes aggravated the gods...

African Legends

According to one encyclopedia of world legends, [355]

> Initially there were connections between God and the mortals he created, between the place where God resided and the earthy home of humans. There was commerce between the heavens and the earth.... And **humans could move to the heavens, and visit and live among the gods**. The creator god.... met with humans, lived among humans, and the humans were his children. He taught them, punished and rewarded them, as he made an effort to give order to the place that he had created.

> Something occurred to provoke a separation between heaven and Earth... Humans erred and incurred the wrath of God.... There might have been a quarrel, the breaking of a prohibition....

> Whatever the impetus, the ordered ties with heaven were broken, the gods left the earth .... People tried to re-stablish contact with the heavens, **building towers to the sky**, but these crumbled under their own weight.   [emphasis added]

In another tale, Aberewa, a primordial woman was preparing dinner by grinding meal in a mortar with a pestle and she kept bumping the sky with her very long pestle. Annoyed, the great God Nyame went away. She attempted to re-establish her relationship with him, and so began **piling her mortars one upon the other to reach the sky**... but was one short, and she asked a child to go get one for her. The child returned, not having found one, so she desperately asked the child to pull one from the bottom of her pile....and when it was removed, the **entire tower collapsed**.

In another source, [356]

> God lived very near humans, in the sky, **just above their heads**. ...
> In the Ila story an old woman tried **to pile up trees to reach heaven** so she could talk to God....

> right across to the other side of Africa similar stories are told....

> The Nuba people of the Sudan say that **in the beginning the sky was low down and close to the earth ....**

> Another version says **the sky was formerly so near** that when people were hungry they tore off pieces of clouds to eat...

The Dinka people, also of the eastern Sudan, say that because **the sky at first was so low** men and women had to be careful in hoeing the ground... not to touch God...

[sure some are exaggerations, but look at the similarities... what was really going on?]

The Lozi people in upper Zambesi in Zambia say that God created the heavens and the earth, and that there was a man, Kamonu, who was very clever and imitated God in all that he did... God grew tired of Kamonu and his arrogance, and distanced himself from Kamonu. Kamonu discovered where God was and went to see him... God was alarmed and moved again... Kamonu followed. Exasperated, God moved to the top of a tall mountain, yet Kamonu still followed him. So God moved to the

sky and Kamonu continued his efforts to reach God... but could not. He and his men cut down **many trees and piled them on top of one another**, **trying to reach the sky**, but it was all too heavy and kept collapsing.

(credit: Bing Images: http://thechive.com /2013/01/10)

The recurrent theme is trying to climb up and reach the sky... and here is the key point: it must have been low enough that Man figured he <u>could</u> climb up and reach the region where God was... **Think about it**: if God was really far away, hundreds of miles away "up there," they would not have been stupid enough to use trees and mortars, etc. to reach what they said was **a low sky**, i.e., God's realm.

> So you say, well it is just a dumb myth – but you ignore that every myth has **a basis in some real event**, something someone saw ... else why repeat a story/myth that your neighbors will think you had to be drunk to make it up... Many people accepted the stories or they would have died as a 'cock-and-bull' invention. And you are also ignoring another fact: **the same story was told across Africa** by people who did not know each other – So what is the basis in reality for the similar legend?

Before trying to answer that question, let's take another look at a more famous story – in the Bible.

Tower of Babel

(Credit: Bing Images: reddit.com)

Early Man in Genesis was not crazy either... **He saw something** that led him to believe he could build a tower to reach the gods. In fact, this Tower to God scenario was another evidence for the <u>original</u> Flat Earth. How so? Consider the following logic:

> If Man knew he was on a rock (rotating at 1000 mph, spinning around the Sun at 66,000+ mph), <u>with nothing above him but the clouds</u> – **Why would he try to build a ladder, tower or platform to reach something he could not see?**

The point is: Middle Eastern Man <u>did</u> see something and figured, like the African tribes all over Africa, that it was close enough that with some building effort, he could reach it.

> Parenthetically, this makes one wonder if the Maya were doing something similar… building their pyramids to put them closer to God? (or provide a raised landing platform for the **Anunnaki gods** – is that why most of the tops of the pyramids were flat?)
>
> Zechariah Sitchin maintained that the gods (i.e., Anunnaki) would visit the Maya and stay in the 'casita' at the top of the pyramids, coming and going – Such was evidenced in VEG and AL by the **very steep steps** which are hard for a human to navigate… to keep humans from coming up the steps of the pyramid whenever they felt like it. The taller Anunnaki (8-9' tall) managed the steps just fine.

Genesis 11:4 said

> "Come let us build ourselves a city, with **a tower that reaches to the heavens….**"

…and you know what happened. The gods didn't like it and scrambled their languages so they could not communicate to continue building the tower… The gods said:

> "Come **let us** go down and confuse their language so they will not understand one another."

Whereas the God in the African stories just moved away or up higher, these gods made it impossible to continue building the tower. Two things here: (1) it was not the plan to have early Man be able to go where the gods are, and (2) the gods are watching – very aware of what Man is doing. (Why would interest in what Man is doing stop with the current-day times? It didn't.)

Even more interesting, as was said in VEG and AL:

The Bible says "Let **us** make Man…" Who is "us"? The God of the Universe is called 'El' in Hebrew, as in El Shaddai. Singular, The One… The God of the Universe is <u>not a plural</u>… remember the Ten Suggestions?

"Thou shalt have no other **gods** before me."

The plural of God is 'Eloh**im**' – gods. So it is not the God of the Universe speaking. It is **a lesser set of gods** (as suggested throughout this book – they are designated Higher Beings who run this place – see p. 180)….. and the humans referred to the **Anunnaki** as gods (because they descended from the sky), and humans also referred to the **Ancient Ones** as gods – and that latter was more appropriate as They are the ones **Enoch** encountered who showed him what Earth is, how it was built and how it operates. Ok, we are getting off point… VEG and AL examine that in more detail.

Tower Building

The point is that Man must have seen something regarding the sky (aka Heaven) which made him realize that (1) God lived there, and (2) it was close enough to warrant an attempt to try and reach it.

> I repeat: ancient Man was not the stupid dolt that we, the very-with-it, modern, know-it-all humans who have now reached the pinnacle of development and knowledge (got your wading boots on yet?)…. So that we can look down on our ancestors with such disdain that we just know that they knew nothing and made up stupid stories…
>
> Seriously, if ancient Man (who wasn't on drugs, prescriptions and eating junk food) was more clear-headed, and looked up and saw a different Firmament, then it is **plausible** (not yet proven) that the Firmament was lower and Man believed he could reach it…. They really weren't stupid. **If Man had thought that there was nothing up in the sky, above him, <u>especially if he saw nothing but clouds</u>, he would not have tried to climb up!**
>
> What this Appendix is suggesting is that **he DID see something that led him to think that building a tower was feasible and worth it.** And <u>multiple cultures</u> had the same reaction to whatever they saw… so it wasn't just one group of crazies…

And now to add weight to the argument, we turn to what NASA discovered in 2012-14 – besides the earlier *Gegenschein* …

NASA Space Barrier

Invisible Shield

Science News in 2014 reported on an invisible shield surrounding the Earth that was discovered some **7200 miles above the Earth**. Sounds like a good distance for the dome or Firmament to be located. The shield protects Earth from high-energy "killer" electrons that can destroy satellites, threaten astronauts, and degrade space systems. It sits between the two layers of the **Van Allen Belts** (see below, next section). It has been described as "**impenetrable**" and "**a glass wall in space**" that has "**an extremely sharp boundary**" and is "**extremely puzzling**." [357]

NASA calls it a **Plasmasphere**:

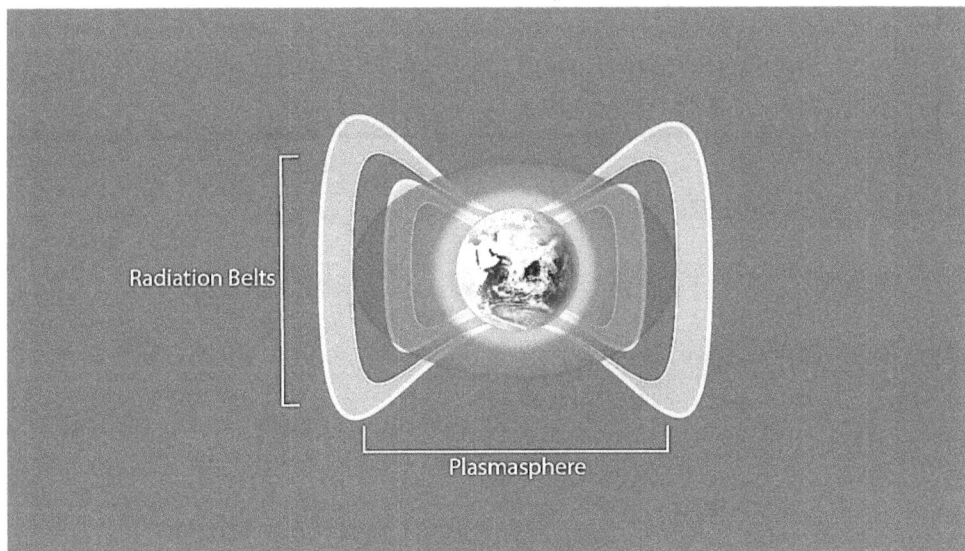

Impenetrable Shell
(credit: NASA/Goddard)

This anomaly was briefly mentioned in Ch. 13 in QES. Whereas the Van Allen Belts of **high radiation** are said to be "harmful enough to kill astronauts flying thru them, the new shield may prohibit <u>any</u> passage to outer space." [358] (Still think we went to the Moon?)

NASA is now calling the Plasmasphere a 'force field' over and around the Earth which sounds like Charles Fort's "Shell" or the Firmament associated with the Flat Earth and VR Sphere.

> According to Daniel Baker, director of the Laboratory for Atmospheric and Space Physics, [an] **electron barrier exists in the Van Allen Belts**…. The Earth's magnetic field holds the Belts in place, but the scientist says that the electrons in those Belts – which travel at nearly the speed of light – are being **blocked by some invisible force** that reminded him of the kind of shields used in television series like Star Trek to stop alien energy weapons… [359]   [emphasis added]

The newly found field at **7200 miles altitude** is related to the plasma clouds that comprise the Van Allen Belts. The vicious nature of the Belts has led many to suspect that Man did not go to the Moon as passing thru them is a lethal experience.

> As was noted in Chapter 12 of VEG, the highly credible channeled entity RA stated that Earth <u>is</u> in a Quarantine set up by the Solar Council and why. It is interesting that RA did not mention a Flat Earth, and referred to the Firmament as a Quarantine.

Van Allen Radiation Belts

For those who wonder what the Belts are alleged to look like:

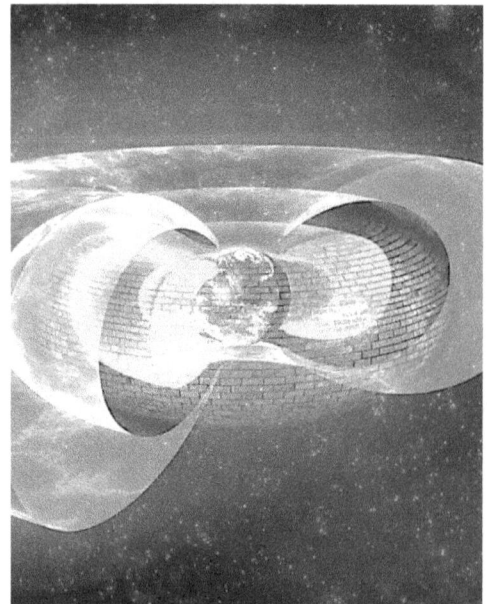

(Credit: Bing Images/ news.islandcrisis.net)

The 'bricks' symbolize the Wall of ionized plasma around Earth… **the Shield**.

The two torus fields are the Van Allen Belts

There are no less than 2 Belts and a Shield circling Earth. And their function (as designed) is protective.

The above diagram, as provided by NASA, reflects that the **Inner Belt** is 620 miles to 3,700 miles from Earth, and the **Outer Belt** is 8,100 miles to 37,300 miles. The newly discovered Shield is calculated to sit about **7200 miles above Earth**.

All that is understandable as the Earth is now a VR Sphere… but what about a Flat Earth? Would the original FE need Van Allen Belts to protect against… what? If the Belts are not a piece of disinformation, then is it possible that the energy field sustaining the new Shield (at 7200 miles altitude) appears to be one of radiation and plasma…
and that the Van Allen Belts are not really separate entities but are in fact the energy "signature" of the Field that creates and sustains the Shield (aka Firmament)?

More specifically, **the barrier is impenetrable** and was discovered in 2012 by Professor Daniel Baker who described the shield as

> **a third, transient "storage ring" between the inner and outer Van Allen radiation belts that seems to come and go with the intensity of space weather… [and it]** appears to block the ultrafast electrons [100,000+ mph] from breeching the shield and moving deeper towards Earth's atmosphere. [360] [emphasis added]

Flat Earth or VR Sphere

The significance is that the Shield just might be the Firmament <u>and</u> something else appears to accompany this information: Whether Earth Realm was originally a Flat Earth and is now a VR Sphere, what it suggests is that **wherever the Earth Realm is located is in a hostile environment** – enough so that a Shield had to be placed around Earth to protect it from the cosmic rays and ultra-fast electrons… And that sounds like the Earth <u>is</u> the 3D Construct (VEG CH. 12) **sitting in 4D – real Space with real hazards.**

> When Man constructs a prison or **asylum** or a Correctional Institute, is it not usually placed in some remote area (if possible – such as when England used to send their convicts and criminals to **Australia** – far away from civilized society) and what would more advanced beings do if they wanted to create a Correctional Asylum for wayward and/or defective souls – to grow them, and rehabilitate the 'wilder' ones?

> Would they build a 3D Earth Realm, a **3D Construct** so-called because it was built with a subset of 4D Laws so that 3D souls could <u>not</u> manipulate the Laws of their environment and cause themselves damage, let alone harm those who designed and built the Earth School/Asylum?

If in fact, Earth is a Simulated environment, a "**Goldilocks World**" where the conditions for life are just right, much as Man makes terrariums and zoos on Earth, the Higher Beings might choose to Simulate the 3D Construct in their 4D Space… but enclose it such that **other denizens of the 4D realm could not interfere**… much as Man puts up high fences and razor-wire around his Correctional Facilities.

Any Correctional Facility has a Warden and guards and therapists staffing it, and in the case of Earth, Man would be closely watched (as Enoch said) and occasional Avatars/Teachers would be inserted to the Earth Realm to teach and guide Man – just as pastors, doctors, psychologists, teachers and motivational speakers would visit inmates in Earth-based correctional asylums.

If you think Earth could not be a correctional facility, then you have not been watching the nightly news – ISIS, renegade US congressmen obstructing the real needs of the country, North Korean craziness, and large, violent drug cartels in Latin America, wanton military groups in Africa abducting innocent females, terrorist acts across the Western world, as well as MS13 in American cities…..do I need to continue?

See the end of Chapter 11: How They See US.

And then we come to the **Firmament issue**:

> You would need one to keep out interference from curious or malevolent 4D beings;

and

> In the beginning it might be expedient to remind Man that he is being watched, and the **Firmament was thus lower**, and as was said by many cultures, in the beginning the gods moved among Man and interacted with him.

> If it was found over the centuries that the interaction did not produce the desired results: Man was still stubborn, violent and petty, still not listening to guidance (the goal being to produce Earth Graduates), the gods might **raise the Firmament** (so that [1] those who had known a lower Firmament would think that the gods had gone away, and [2] the newborns would think they are all alone and could do whatever they wanted with no

one watching) and then insert teachers, healers, and angels to provide more indirect guidance....

What would happen is what we are seeing today: God is not in obvious evidence any more than a Warden has to remind the inmates of his presence... the job of growth and correction fell to teachers among Man. It was a more **true test of soul growth** to have Man do what his inner ego/id guided him to do, such that whatever he did was a true test of his level of soul growth, since he would think there was no God watching and, since we are in a **Freewill realm**, Man is free to do right or do wrong.

> What many do not want to hear is that **Man is still accountable** ... and that 'many' included the Agnostics/Atheists and Scientists.

While this is not a sermon, the point is that the Firmament still protects Man and **NASA** has discovered it – as did **Charles Fort** (of *Gegenschein* fame), and **Robert Monroe** (who repeatedly ran into a barrier when he went OOBE and went too far – See VEG, Ch. 12), and it is plausible to think that ancient legends spoke of the Firmament that even the holy sage **Enoch** was shown (Chapter 6).

And if there is still a Firmament, then it applies to the VR Sphere – and Chapters 10-11 resolved that one for us.

Agreement with the idea of the Firmament no longer being as low as it was is not required... it is just a fascinating and plausible concept that arises out of some interesting historical data and legends. So its height may be open to question, but not the fact of its existence... **something** is there – at 7200 miles up. It is a matter of time before Man has to admit its existence. .. and its obvious implications (See Chapter 10).

Appendix F: Secret Societies

As an afterthought, it can be said that some secret societies (SS) have a part in the Great Earth Puzzle. While some of this was examined in Anunnaki Legacy (AL) with an eye to <u>what</u> the secret groups taught, it is interesting to mention how their esoteric teachings (which are a mystery or puzzle to many people) relate to the issues exposed in this book. In other words: <u>Why</u> is Earth a puzzle?

In particular, the infusion of new ideas into our society, including the 'dumbing down' of education and religion, the pervasive (and false) *meme* of ETs visiting the Earth today (while it is in quarantine!), and the efforts undertaken by the Smithsonian Institute to collect and hide anomalies found around the planet (e.g., large skeletons, odd statues, and Egyptian artifacts found in Arizona) <u>do</u> have their origin in the agendas of one super SS that goes back all the way to Sumeria and the Anunnaki. There is a pervasive and powerful society in the shadows, running our world, our civilization, and our Science, Religion , Education, Media and Banking. Their goal being to remove anything that would cause the public to (1) think that Earth's history was other than "straight vanilla" as promoted officially in textbooks, or (2) that there might be Others living on the planet who occasionally have something to do with Man's scientific and/or spiritual progress.

> All you have to do to control any society (except the most enlightened) is to **control the Media** (what people hear), **and the Banks** (what the people can save/spend) – limit their information and keep them in debt.

Thus, it has for millennia been a case of some societies promoting Man while others seek to keep him ignorant. Where did this disparity originate, with whom, and what is happening today? More significantly: how do SS keep the elements of the Great Earth Puzzle going and why?

While there are many books written about the Knights Templar, the Gnostics, the Rosicrucians, Neo-Nazis, Thule and Vril, Theosophists, the Freemasons, Kabbalah, and even Shamanism, this Appendix will focus more on how **the two original secret societies** (Great White Brotherhood and the Illuminati) have survived and come down thru the ages into today's world and what form(s) they take as represented by what they do in our world. The two have **agendas** that are sometimes at odds with each other.

There <u>is</u> a battle for Mankind's heart, mind and soul that is millennia old and is one of the things this 3D Earth Realm is all about – to resolve the STS versus STO issue among souls in 3D, 4D and 5D (it does not go higher into 6D, nor to the top at 7D). Earth School is one of the venues where incongruous orientations (soul expectations and attachments) can be experienced and worked out. If it isn't worked out, or learned, one stays in the Earth School until it is learned.

> If the reader has a favorite SS and wonders why it is not mentioned herein, it is because the two oldest mainline SS and their activities are much more important to the Great Earth Puzzle.

Origins

Sumeria

Here again, everything goes back to Sumeria and the Anunnaki who not only genetically created Man ("in our image"), but also taught him agriculture, writing, astronomy, mathematics and metalworking. As time went on, Enlil, head of the Earth Expedition, found that the *Adama* humans were too primitive, rowdy, noisy, smelly and violent to be manageable. And the Watchers (Igigi in skycraft) came down and mated with *Adapa* Earth women and inadvertently created the Nephilim… again to Enlil's extreme annoyance. Something had to be done. Hence came the Flood in an effort to wipe them all out.

However, Enki, the Science Officer and Genetics guru, saw to it that Man was not exterminated and because he and Enlil were brothers, Enki earned Enlil's murderous wrath but Enki was not killed – just ordered to see that the human remnant were shepherded and <u>controlled</u>. Enki's solution, after developing the better but still limited *Adapa+* version of Man, was to take the best/smartest of the *Adapa+* and make them **priests** in a "god-given" right to rule (*Dieu et Mon Droit*) as the beginnings of **Religion** were set up. The priests were to extract obedience from the humans under penalty of the threat of death or banishment (into the surrounding desert as was Cain after he slew Abel). If the humans really displeased the gods (aka Anunnaki), at worst, they could go to Hell and even be killed if they didn't toe the mark.

> Enlil unleashed *Suruppu Disease* upon the humans as reported in the *Atra Hasis* in an effort to trim their numbers; could this have been something like today's HIV? It had the same nasty effect, but the Anunnaki also had the cure. (Ch. 3 in VEG.)

Those humans who showed promise and behaved, were initiated into an inner circle and taught the priesthood, to spread to other locales, and many times they were let in on the **esoteric** truths of what was really going on. On the other hand, the masses were judged to be too ignorant to ever rise above the serf-level and were kept plowing the fields, building the ziggurats, and working the mines (for gold and uranium) and they were never taught more than the **exoteric** truths such as the Ten Commandments, Hell, and Salvation designed to control human belief and behavior.

Brotherhood of the Serpent

Ever interested in developing and protecting his progeny, Enki developed a special priesthood for the brightest *Adapa+*, which he kept genetically upgrading using his own DNA, and this *crème de la crème* version of humans received special training and Truth which was only for them – i.e., this was the first benevolent Secret Society called the Brotherhood of the Serpent (later called the **Great White Brotherhood** as it incorporated an Astral component).

Many initiations were done at **Göbekli Tepe** (as examined in <u>Anunnaki Legacy</u>) using the skycraft channeling power from the Great Pyramid (and its Benben pinnacle capstone) providing the 'electrification' of the T-pillars and pizeo-electric

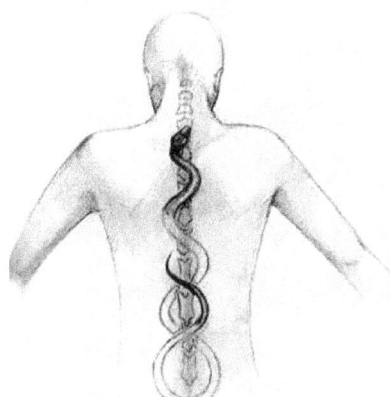

substrate to active the *Kundalini* of the initiates. Because the *Kundalini* movement up the spine resembled a serpent weaving movement (left), the Brotherhood's name reflected the *Kundlini* aspect of the Society.

Below is the transformation from normal human to Shining One as achieved at Göbekli Tepe:

The adepts and masters often had a kind of energy or glow about them, as did Moses (as reported in the Bible), and this led to calling them the **Shining Ones**.

In reality, the glow was a **strong aura** and was strong enough to be seen by the unaided eye. This was also pictured by later artists as a circle around the head… usually thought to signify holiness, but in reality represented the enlightenment of the Shining Ones.

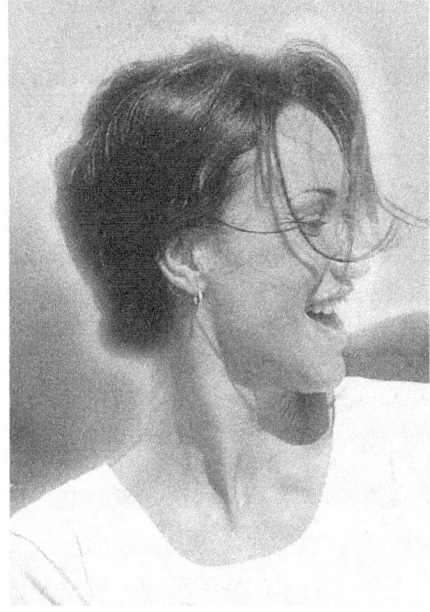

Note the glow (1-2") or light around the head and shoulders (left).

Those who saw this often pictured it as below:

The same thing as done with Kirlian Photography looks like this:

Secret Society Purpose

Whereas Enki's purpose was to develop a special enlightened version of mankind, and included men and women, it also served to **keep esoteric truths alive** – to preserve them for the future generations. The long path turned out to be:

> The ancient knowledge of the Anunnaki was passed through the Sumerians, Babylonians, Assyrians, and Egyptians down to the Mystery Schools of the Greeks and Romans, then on through the Knights Templar, Assassins, and Rosicrucians to Freemasonry until it was collected together [sic] by a group of German intellectuals in the late 1700s – the **Bavarian Illuminati**. [361]   [emphasis added]

The Sumerians, Akkadians, Babylonians, Phoenicians and Assyrians were not so much totally separate civilizations as they were "merely degraded versions of the first-known great civilization established in Mesopotamia more than 7,000 years ago – Sumer." [362]

Unfortunately as the **Mystery Schools** migrated to Egypt, they became corrupted and human greed/power took over and the original teachings were not as important as the in-house developed procedures and principles of mummification, multiple gods, and establishment of the priesthood's control over the populace. The elect were said to be illuminated, guardians of the Secrets of the Ages.

A similar paradigm occurred in AD 325 with the formation of the **Christian Church** and its Theology. Emperor Constantine decided what was to be in and out of the new Christian Religion that was set up as the official State Religion. To show the masses that the official canon of books for the New Testament really had an antecedent, and thus carried historical weight, the Jewish Torah/Tanakh was bound up in one book with the New Testament, thus giving the new religion the appearance of being very old and credible. In reality, the Jews have said many times that their Torah has nothing to do with the Old Testament of the Christian Bible, and was even falsely rearranged in the order-of-books to meet some Christian agenda.

Naturally the Church found itself at odds with anyone who had a different teaching – such as Gnosticism, Manicheanism, Zoroastrianism, Deism and Illuminism (the Illuminati). A lot of these societies and their followers were paid a terminal visit by the **Inquisition** – for 400 years – in an effort to stamp out 'heresy' and get everyone on one page – behind straight vanilla Christianity.

Naturally, advanced teachings had to go underground to avoid persecution by the Church and the dichotomy between esoteric and exoteric knowledge continued and was reinforced.

The Illuminati found it necessary to make it look like they disappeared while staying operative as a Hidden Hand or Shadow Government in the world.

Illuminati

Foremost among the Earthly/human SS that have survived to this day, is the Illuminati, which goes by different names, having infiltrated many groups, including Freemasonry, Gnosticism, Rosicrucianism and Zionism. It is said that they are also behind the scenes in the Trilateral Commission, the Council on Foreign Relations, the Bilderbergers and the Bohemian Grove celebrations. [363]

But just what exactly is the Illuminati?

> The Illuminati began as a secret society under the direction of Jesuit priests. Later a council of five men, one for each of the points on the pentagram, formed what was called the Ancient and Illuminated Seers of Bavaria. They were high order **Luciferian Freemasons**, thoroughly immersed in mysticism and eastern mental disciplines, seeking to develop the super powers of the mind. Their alleged plan and purpose is **world domination** for their lord – which is the fallen **Lucifer**. [Today's] Illuminati are alleged to be the primary motivational forces forcing the global governance, a **one-world religion**, and centralized control of the world's economic systems....
> The Illuminati are the driving force behind the brainwashing of the mindless masses, blatant mind control, manipulation of beliefs, **scientific dumbing down of society** [see Chapter 4], chemical poisoning of food, water and air, ...aspertame, sodium fluoride [water fluoridation], also the Illuminati are revealed to have total and complete control over all the **mainstream media**.... All the food, all the **money** ....The Illuminati's power vehicles are the big banks, and the manipulation of all the money and wealth of the planet....
> The goal... is to create and then manage crises that will eventually convince the masses that **globalism** .... and **one-world religious ethic** are the solution to the world's woes. This structure known as The New World Order will of course be ruled by the Illuminati. [364] [emphasis added]

And yet, in the beginning, the Illuminati began with very high ideals and was populated by some of the most educated and prominent thinkers of the day. Their stated goals were "to oppose and try to stop superstition, prejudice, religious influence over the masses, abuses of government power, and gender inequality." [365]

However, what was the significance of the reference to Lucifer?

Keep in mind that **Lucifer is not synonymous with Satan**. Lucifer was simply a Bringer of Light, or Knowledge. In that same vein, according to that definition, Jesus was also a Lucifer, as was Einstein and Pythagoras. What gets really interesting is the perversion of Lucifer into some sort of demonic being and this has its origin in that Enki was a Lucifer – bringing Light to mankind, and when he was demonized by the Egyptian Mystery Schools and Enlil, Enki was persecuted and his association with the "Luciferian" concept assured the future **perversion and persecution of the Lucifer paradigm**.… Eagerly supported and promoted by the Church since any Light or true Knowledge was a threat to the Church's newly established doctrine. [366]

A very interesting insight into the Luciferian Doctrine underlying some of the Illuminati and Freemason teachings: [367]

Albert Pike, Masonic author and 33d degree Mason, explained the Luciferian Doctrine thusly in 1889:

"The Masonic Religion should be by all of us initiates of the high degrees, maintained in the purity of the Luciferian Doctrine. If Lucifer were not God, would Adonay whose deeds prove his cruelty, perfidy and hatred of Man, barbarism and repulsion for science, would Adonay and his priests, calumniate [slander] him? Yes, Lucifer is God, and unfortunately so is Adonay. For the eternal law is that there is no light without shade…. no white without black, for the absolute can only exist as two gods….
[This reflects an early Gnostic belief in duality]
Thus the doctrine of Satanism is a heresy; and the true and philosophical religion is the belief in Lucifer…. But Lucifer God of Light, and God of Good, is struggling for humanity against Adonay, the God of Darkness and Evil."

The above insert also reflects the esoteric Knowledge that Lucifer was Enki (pro-Man) and Adonai was Jehovah, aka Enlil (anti-Man) – as examined in VEG, Ch 3.

In addition, the Insider who appeared on the Godlikeproductions website back in Fall 2005, and which is quoted as **Revelations of the Insider**, reveals that the Masonic position on the Luciferian Doctrine agrees with what the Insider, a member of one of the Elite 13 Families, had to say about religion: [368]

Religion is created by us. The religions which now rule are the ones that are under total control.

… humans who worship the "bad" think they worship the "good."

Do you know who keeps the knowledge about the real Christ alive?
And do you know who keeps the knowledge of the fake Christ(s) alive?
The opposite of what you are thinking.

What she was warning about in those three quotes is that people today still worship Adonay thinking they are worshipping the "good" god, when in fact, due to Church perversions centuries ago, Adonay (aka Jehovah) is the "bad" god of this world and Lucifer came to set Man free… Remember that Jehovah used to tell Saul to go kill his enemies, like the Amalekites, as well as others… Jehovah was an irascible god of this world, not the Father of Light, The One who is merciful and full of Grace. (This issue is fully examined in VEG, Chs 1-3.)

Yes, I know we have been taught the opposite… but isn't "…everything 'round backwards in this world" as Stuart Wilde used to say? (See his God's Gladiators.)

So it is not possible to say that the Illuminati, let alone the Freemasons, are Merchants of Darkness. Perhaps what ignorant people perceive is their truth and not The Truth? What if the Illuminati are genuinely trying to bring Order out of Chaos in the world, and given Man's pettiness, ignorance and proclivity to violence, it may be necessary to lock it all down, (viz., Martial Law) for a while such that Order rules and not Chaos? Something to think about…

On the other hand, while certain factions and activities of the Illuminati have been sometimes associated with violence (wars), deception and manipulation, is there a proactive, beneficent, SS whose goals and actions are only good, always positive, or STO? Yes, and it hasn't deviated 1 inch from its inception millennia ago.

Great White Brotherhood

Back in 1980, dating a cute gal in a New Thought church that I attended in Huntington Beach, CA, I was urged to go see a psychic who had been evaluated by the Orange County Register, a newspaper known for straight journalism. The paper had determined that Dolly was 98% accurate in her predictions and descriptions of people and things known only to the reporters.

So I called, set an appointment, and went to see her. Remember, from Chapter 1, that I was becoming very frustrated and despondent about my life, losing jobs for no reason, losing relationships for no reason, and I had formulated the following statement about my life (which is why I was considering suicide): **I can't get what I want, and if I do, it is taken from me by forces beyond my control** . (Yep, that describes Karmic "payback"… also my Script.) Recall that I did not get answers until 1991, eleven years later.

When I entered her cozy but modestly furnished consulting room, it was really just a makeshift porch on the side of her house in an average neighborhood of duplexes, of which hers was one. (I later learned that she was an heiress to the Campbell Soup fortune.) She was in jeans and a comfortable pullover, sitting on a cushion on the floor and simply greeted me with, " The White Brotherhood says Hello!"

I had no idea what that was and it would be years before I learned that writing the 6 books automatically made me part of that Group – who in fact had dictated the VEG and TOM books, and **Baldy was their emissary, or interface**. I kick myself now for not pursuing that and asking her what the White Brotherhood was.

She volunteered anyway and told me that they had been watching me and that I was associated with them and about to do a work for them, but she didn't say what or when. I was still so uneasy with seeing a "psychic" that I could not think of much to ask…

I told her my life was a mess and that was why I was seeing her and I was looking for answers.

She said she knew and said that I had set this all up on the Other Side, before incarnating, and I would not commit suicide (How did she know I was even contemplating that?) , nor would the destruction and super difficulties in my alleged life go on forever. And then she volunteered a weird piece of information, a warning… "Be careful of your right knee, you will have some problems with it."

I had no idea what she was talking about and that kind of said, Boy, she is a fraud, I have never had any trouble with my knees, either one of them! I thanked her, don't remember anything else of the almost 1 hour session, and I paid her a modest "love offering" and left. Little did I know that 35 years later, in 2015, I would begin to have problems with my right knee – from an old ski injury in 1968 that I had forgotten about and which had never given any indication of its existence (a stretched ligament) – until 2015! Dammit. She was right, and everything she told me, that I remembered, was spot on, absolutely correct!

So the books were the result of a connection to and **transcription** with the Great White Brotherhood. Baldy was not my Guardian Angel. I was guided and kept healthy during the years leading up to 2008 when the first 2 books were written, and then I had perfect health from 2008 to 2016. Somehow They abandoned me for 9 months while I tried doing what They had said (via Baldy) to not do: promote and teach the books. That was the **Presentation fiasco** of August 24 2016. And I paid for it over the next 9 months (Oct 2016 to July 2017) until I surrendered, humbled myself, and asked to be back in Baldy's good graces. I was answered and while our task of doing the books is at an end, I again get some guidance and protection.

So what is the White Brotherhood Dolly mentioned?

> It is the current-day form of the Brotherhood of the Serpent, which was Enki's original group. [369]   It is really old.

And Baldy confirmed that, earlier in August 2017. It is not to be construed as the KKK or White Supremacy. It is also called the **Great White Brotherhood**, or the Great Brotherhood of Light, and the Order of Ezekiel. It is also called the Spiritual Hierarchy of the Earth School – or what is often referred to as **"the gods"** in the 6 books. They are concerned with the evolution of Man's world, his spiritual growth, and They oversee the Earth Realm School.

Not much is written about Them because it is truly a <u>secret</u> society with more than half its "membership" in the Astral. It works with the Beings of Light and Neggs (VEG, Chs. 5-7, and Glossary), and it makes sure that the Father of Light's **Greater Script, or Earth Realm Drama**, is not coopted by Man, discarnates, or Astral entities who closely resemble the Djinn and seek to obstruct and harass when possible. (See VEG and TOM.)

The White Brotherhood was originally Enki and the Anunnaki (Watchers, Shining Ones) but by the time of the Egyptian Mystery Schools and then later the Greek and Roman gods who used to interface with Man directly, the White Brotherhood had changed and now kept a small 3D flesh & blood component on Earth, but most were Masters and Adepts now in the Astral where They could see and proactively manipulate the Greater Script (see TOM).

For those who are willing to hear it:

> Needless to say, any Light-bringer, such as a **Lucifer**, works with and represents the White Brotherhood, and it is safe to say that the Lucifer of Freemasonry (and perhaps the higher levels of the Illuminati?) is allied with the Great White Brotherhood.

> <u>However</u>, if in fact the Illuminati activities are fighting the operations of the Brotherhood, the Illuminati is seen as a necessary antagonist but <u>will lose</u> as the Greater Script of The One will rule. That is because the intention of a Higher Being is more powerful than that of lesser humans and/or lower level Astral beings. Good news.

Breakaway Civilization

While the Illuminati may be involved with the agenda to create the New World Order, there is already a split-off separate group of the Military Industrial Complex (MIC), that Eisenhower warned about, that may or may not have anything to do with the Illuminati.

While **Joseph Farrell** and **Richard Dolan** have written about the Breakaway Civilization, with its own advanced technology that is allegedly 30 years ahead of the technology that the public knows about, it is not something to which we could connect the Illuminati – in fact, any analysis of the Illuminati activities shows heavy involvement with the Banking and Media realms, and not with the high tech (alien?) world of the Breakaway Civilization. Thus it is not possible to directly connect the Illuminati with the Military Industrial Complex which <u>is</u> connected to the Breakaway Civilization (according to Farrell).

So the Breakaway Civilization is more thematically connected with **Alternative 3** than with trying to create a New World Order on Earth… allegedly, the MIC seeks to use its advanced technology to establish remote bases of operation and underground living quarters for the more 'worthy' – special scientists and those of a special bloodline, perhaps the Elite (special **13 Families** of unrenown), and anyone who is of high value to the operation and success of a civilization that has broken away from the corrupt, decadent, polluted world that we know. That means they are underground on Earth and <u>allegedly</u> already on Mars and the backside of the Moon.

Given the Cspace in this book, does anyone see a problem with that scenario?

If Earth is in quarantine, with a Firmament, with a 'shield' (by NASA's own admission) that is impenetrable (Appendix E), and which reflects the *Gegenschein,* how could we be on Mars or the backside of the Moon? It is more likely that any Breakaway Civilization is remote but located on Earth, and has advanced technology and may itself seek to control wayward humans and rogue nations so that the planet is not destroyed… Would that pit it against the Illuminati who are espousing a One World Government… or could they be working together?

After all, the Breakaway Civilization would be today's ultimate Secret Society (SS).

Agenda for Change

As was said, several SS have always sought to guide mankind in ways that agree with their agendas – which, depending on the SS, may conflict with each other. For example, the agendas of the White Brotherhood, the Illuminati, Neo-Nazis, Opus Dei and the Rosicrucians would not be in agreement. And yet, one or more of the SS agendas are playing out in our world. Consider the following:

The **disappearance of the bookstores**, except for a tightly-controlled Barnes and Noble (B&N) today, where Walden Books, Bookstop, Taylor's and Borders bookstores have all disappeared. That may seem like coincidence until you realize the plethora of bookstores engaged in competition, some bookstores carrying books in the Metaphysical, Ancient Earth History, and UFO categories, for example, that a neighboring bookstore did not carry. This permitted browsing to discover what the latest books were in those categories... B&N did not carry the same selection that Borders did, thus narrowing your potential for discovery of a current and thought-provoking book. (Their NY distributor tells the stores what they will stock.)

And while that may sound inconsequential, as one can browse Amazon and make interesting discoveries, I am suggesting that there is an Agenda out there to limit what can be "discovered." Bear with me...

Soon on the heels of the bookstores disappearing (and B&N will soon close more of its stores according to one of the store managers I spoke with last month), we had **Blockbuster video stores** closing. Again, I asked one of the local managers why the stores were closing, and was told that DirecTV had bought Blockbuster and was closing all the stores – allegedly to get people to go online with their remote and browse and order videos via their TV. A simple inspection of the videos offered by DirecTV shows that what you could find at Blockbuster is not available in the online TV library of videos. Even Netflix does not have the selection that Blockbuster offered – to say nothing of putting thousands of workers out of a job as DirecTV closed hundreds of stores across the US.

So books and videos today are not as easy to pick up and examine and make very interesting discoveries, as was the case just 10 years ago. (Try to find *The Fourth Kind* and *Iron Sky* without entering the titles... what genre do you look in?) Sure, you can browse and order interesting, even controversial, videos and books via Amazon, but are you aware that **Amazon records what you look at** and what you buy? So do Netflix and Hulu.

> Is it someone's desire to track not only what we say in emails, and in phone conversations, but also what we buy online?

Has anyone paid attention to the **cameras** everywhere today? – in the stores, in the intersections, on the street... Not only are we tracked via camera (ostensibly for security) we can also be tracked by technology like **OnStar ®**. Sure it is great if your car breaks down in the middle of nowhere and you can push a button and help is on the way... but you do realize that in addition to OnStar unlocking your car doors (if you get locked out), they can also lock you out, or turn off your engine? It works both ways.

And don't be afraid of the "Mark of the Beast" – the **microchip** associated with the Biblical "666".... They don't need to do that. There is already a GPS tracking chip in your car as well as in your cellphone (how do you think Siri works?), and your new creditcard has the intelli-chip on it that can identify you to any reading station up to 30' away – If Martial Law is ever declared, the National Guard will not have to confront you and ask for your papers – they can just scan you <u>from across the street</u>, hit your creditcard chip, and know who you are, where you live, your age, your social security number, etc...

Because you carry your cellphone and creditcards with you wherever you go, another RFID chip implanted in your body would be unnecessary.

> On its way to the stores, as we write this, is the new **TV that listens to you** (<u>allegedly</u> to do away with the remote) and you tell it what you want it to do – reminiscent of technology like **Alexa** (Amazon's Echo ®). The problem is, like Alexa, it is <u>always listening</u> and can be programmed (at the other end) to record when certain keywords are spoken by anyone in the room.

> For example, we have Alexa in the Kitchen (for news and weather) and one day I was standing 10' away, Alexa was not activated, and I asked Linda, "Where is Paris?" (meaning in Texas). Alexa answered and asked if I meant in Europe or in Texas? We were both startled... Alexa had responded without our saying "Alexa" first!! Spooky.

Another facet of the Brave `New World we are in nowadays, as a corollary to browsing books/videos and **learning something new**, is the Agenda to remove the New Thought churches which used to teach people who they are, why they are here, what Earth really is, and what they can do to "wake up" spiritually, for starters. To that end, Religious Science was removed about 6 years ago (reemerging as CSL), and Unity is now being coopted to be just a fun place, like a Jumping Freddy Feelgood church... no one has to learn anything, don't worry, be happy....

> There appears to be an Agenda to keep the Sheeple dumbed down... no revelatory books or videos, no enlightened sermons from New Thought (advanced metaphysical) churches – we just can't have the "great unwashed masses" thinking and wondering and learning – especially esoteric knowledge that has always been reserved for the 'chosen' initiates of the Secret Societies.

And they want to know where you are and what you're doing … hence all the cameras and tracking chips.

Welcome to the perfect Controlled Society.

> Lest you think I am alarmed by this, or against it, I am not. It is just what happens when a society gets too big – if there is no control, some freedoms must go, and the control is for the greater good.
>
> The real problem is not control, it is overpopulation.

Overpopulation

There is one problem on Earth that is as significant as pollution – **overpopulation**, and the two are related. Pollution, and lack of clean water and enough food to go around are directly related to overpopulation.

(credit: Bing Images: Pinterest)

Incredible fun with everybody doing the same thing at the same time, in the same place… the joy of overpopulation.

The Great Earth Puzzle

Some pundit will counter with: "Oh Hey, we have plenty of land to spread out on, there is no problem." What he doesn't tell you is that undeveloped land **costs a lot to develop** (water, electricity, sewer, roads, schools…) and not all undeveloped land (deserts, forests) is available or workable and that is why we haven't done more with it.

This is a major concern of all SS, not just the Illuminati.

So somebody erected the **Georgia Guidestones** (1980) which reflect the overpopulation issue: the first one deals with a maximum population and the second has to do with controlled reproduction…

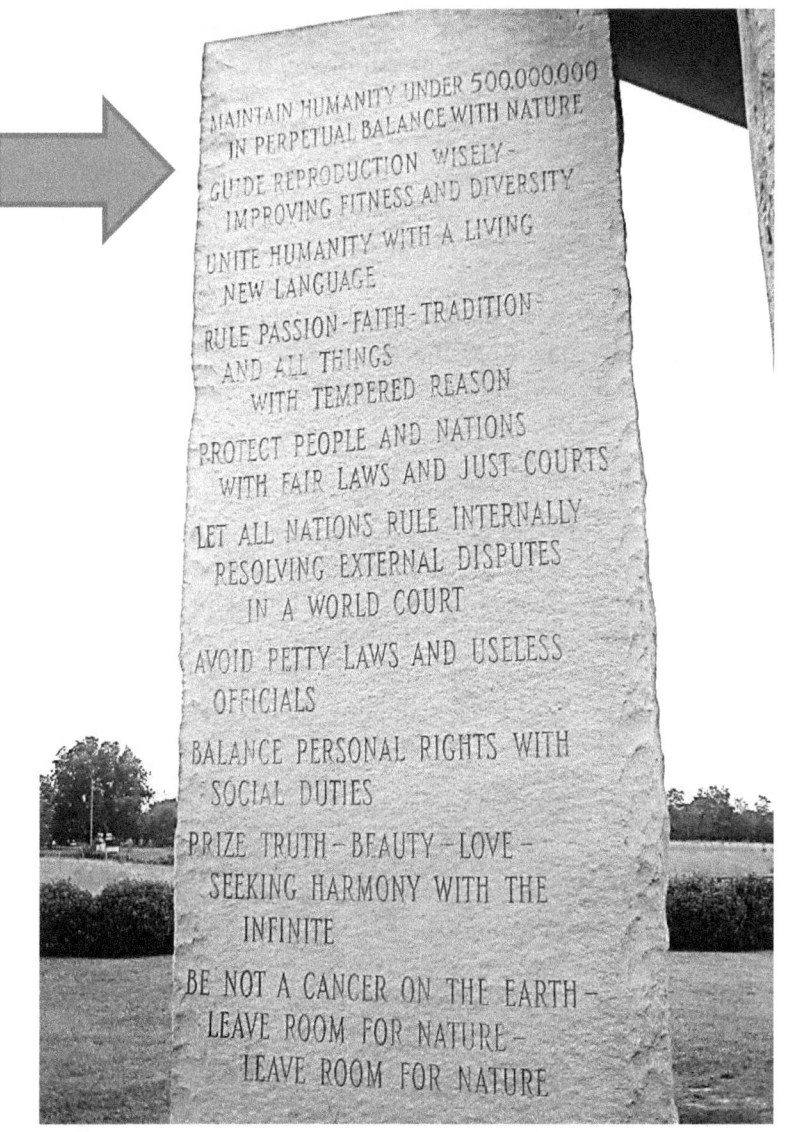

It also doesn't come as any surprise that the SS are concerned with the growing **killer diseases** on Earth:

> HIV, MRSA (drug-resistant flesh-eating staph), Ebola, the Plague, Smallpox, SARS, Malaria, C-Diff --

And, no, **Man is not developing new antibiotics to take care of the bacteria**. Every new vector that is developed in the latest antibiotic, is countered by the bacteria and they develop a resistance to it. [370]

Case in point, *Clostridium difficile* (C-Diff) which is resistant to standard antibiotics, and is a growing problem in hospitals – you enter for one thing and exit with a **superbug infection** which can't be easily killed.

According to one report, C-diff has been "…growing by more than 10,000 cases per year" and it played a role in more than 300,000 hospitalizations in 2005. [371] Some of the pathogens are resistant to the autoclaving which is how medical units sterilize instruments.

> Many autoclaves are used to sterilize equipment and supplies by subjecting them to **high-pressure saturated steam** at 121 °C (249 °F) for around 15–20 minutes depending on the size of the load and the contents….
>
> A medical autoclave is a device that uses steam to sterilize equipment and other objects. This means that all bacteria, viruses, fungi, and spores are inactivated. However, **prions**, such as those associated with Creutzfeldt-Jakob disease, and some toxins released by certain bacteria, such as Cereulide, **may not be destroyed by autoclaving** at the typical 134 °C for three minutes or 121 °C for 15 minutes. [372]

> Prions are bent proteins and one of the factors involved in Alzheimers. (See TSiM Ch. 8.)

Let's hope the hospitals/clinics are scrubbing down better nowadays, and let's hope that **macrophage therapy** (specifically engineered white blood cell designed to target a specific pathogen) is perfected and in time to save mankind. Macrophage therapy is the way of the future, but right now it is very expensive as it analyzes the body's internal immune system and develops a unique defense based on each ill person's genetics and biochemistry.

The point being: What if some of these pathogens were developed and released into the environment (as the Anunnaki did with *Suruppu Disease*) , like HIV and C-Diff, to reduce the population, or as the PTB have often said, to reduce the "useless eaters" – people who take up space and use water and food but contribute nothing to society?

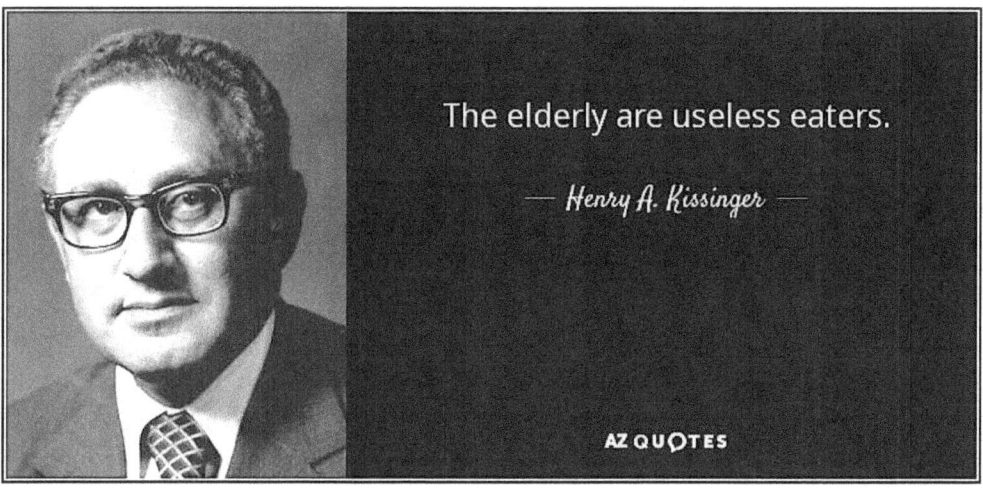

The elderly are useless eaters.

— Henry A. Kissinger —

AZ QUOTES

And again:

"Depopulation should be the highest priority of U.S. foreign policy towards the Third World." - Dr. Henry Kissinger

Ironically, you know in your innermost being that overpopulation is not good, nor can it be allowed to continue. The congressman who suggests doing something about it will commit political suicide, but the irony is that we DO need to do something about the issue.

And it is a major concern of most Secret Societies.

Can we find a humane solution... perhaps starting with **birth control**?

(credit: Bing Images: useless eaters)

Henry Kissinger (above) is not the enemy, although he is associated with the PTB, and maybe even with a Breakaway Civilization, he is a valuable, intelligent man whose input and views seek to prevent humankind from wiping itself out.

Just be clear that unless we do something proactive about overpopulation, our world, our society, will continue to be more and more controlled – BUT **the combination of increased control and continued overpopulation will destroy the quality of life for everyone** (see picture earlier in this section) … probably approaching what was shown in the 1973 movie *Soylent Green*… a dystopian future of dying oceans and year-round humidity due to the greenhouse effect, resulting in suffering from pollution, poverty, overpopulation, euthanasia and depleted resources. [373]

Note: there have never been 7 billion people on this Earth, at any time.

Therefore,

If a very intelligent, perceptive and technologically advanced group of humans wanted to get away from the chaos, ignorance and diseases of the projected and inevitable human cities, would they not seek to set up a separate, Breakaway Civilization… and might that not look like a secret society? Might the more powerful human SS work covertly with the Military Industrial Complex to enact an **Alternative 3** scenario – to save mankind from itself? And would they intelligently handle the issues that we currently refuse to deal with – in their separate society, such that mankind survives – which was Enki's main goal?

> But what if Earth Realm is a Flat Earth with an impenetrable shield at 7200 miles up (Appendix E), and **Man cannot get off the planet**? Would that not force Man to solve his problems here instead of running away and using a Secret Society aka Breakaway Civilization to solve problems?
>
> Could the gods be smart enough to **contain Man** (Flat Earth) such that he cannot avoid his problems (i.e., run away and start a new civilization somewhere hidden from the ignorant masses and their messy problems)?
>
> Could the gods realize that running away and forming a new society (à la Alternative 3) just results in that **new society eventually developing the same problems** a few centuries down the line…because human nature is the same and laziness, greed, power struggles, ignorance and problem-avoidance is one of the things that got us into the mess we see around us now in 2017?

Instead of Secret Societies keeping the advanced technology to themselves, and Man running away to form a secret civilization for the Elite and Worthy, perhaps we are being tested to see if compassion and brain power is to be put to the task to solve the issues at hand.

After all, **Nikola Tesla** had answers to inexhaustible energy, clean water, and food production which the Shadow Government aka PTB aka Secret Society #1 snapped up when he died… so we <u>do</u> have the answers, but the "fat cats" who are heavily invested in fossil fuels and archaic modes of transportation (and make an unreal amount of money doing it!) could be bankrupted by a sudden promotion of better technology -- especially technology like free energy (Zero-point energy) which <u>is</u> free – no one can charge for it! … The PTB don't have to be wiped out if they <u>phase in</u> the new… but even that threatens their current power/control over things.

Man appears to be currently at a turning point. We either **wise up and do what works**, or this version of Man joins the ranks of Neanderthal (Sitchin's *Adama*) and we go down in history as Homo *stupidus* (when this version of Man is replaced and it is written by our successors).

Remember:

the Greys have been busy for 50+ years reworking and developing a new version of Man. (see TOM)

the intention of the Father of Light is powerful and overrules the plans of the best human Secret Societies.

we have never been alone and we <u>are</u> watched over. What happens if we prove ourselves too stupid to continue?

the Earth Realm is a School and the students don't get to control or redesign the curricula – mastering the challenges and tests is what counts – not running away from them (Alternative 3 and a Breakaway Civilization). (see VEG, Ch 15-16)

free will rules on Earth – no god makes us do anything; we always have a choice.

What will Man choose? …IF we haven't passed a point of no return and it is not already too late.

The Great Earth Puzzle

Encyclo-Glossary

Note: throughout this book, reference is made to further detail on certain topics, and the books are referred to by abbreviation:

> VEG – Book 1 – Virtual Earth Graduate
>
> TOM – Book 2 – Transformation of Man
>
> TEW – Book 3 – The Earth Warrior
>
> QES – Book 4 – Quantum Earth Simulation
>
> TSiM – Book 5 – The Science in Metaphysics
>
> AL -- Book 6 -- Anunnaki Legacy

ξ   ξ   ξ

1-Sec Drop -- this is a direct communication from a Higher Being into one's mind and memory/knowledge base. It is not a voice, not automatic writing. It takes a very brief split second and one knows that it is happening, and then it can take anywhere from 10 seconds to 20 minutes to examine what one was given. It is information that is usually complete and appears to the recipient to be something that s/he already knew and is now aware of.
Similar to an **insight or revelation**, except that it has an energy signature about it that you know it is being "dropped" into you. (Reminiscent of **V2K** but there are no words 'spoken.')

100th Monkey Effect – When one animal in a group discovers some new behavior and finds it serves him, it is said that the behavior is not learned as much as passed on as soon as the energy reaches a critical level so that their group soul can recognize and 'appropriate' the behavior. This was the case with a few monkeys on an island who discovered that washing their fruit before eating it avoided the problem of sand in the mouth. More and more monkeys on island 1 began doing it, and while they had no way to communicate the new behavior to the monkeys on islands 2 and 3, after about 100 monkeys were doing it on island 1, the others on islands 2 and 3 also began doing it (as confirmed by zoologists who were present studying the islands).

Anisotropy -- the tendency of light to have a different speed depending on which way it is projected (in the same medium)... as if it is flowing with lines of force one way, and against them if turned 90°. (see VEG Chs. 8 and 9.)

Anunnaki – one of the early, original Earth denizens who interfered in the natural progression of the other bipedal hominids here, and created some of the first 'humans' in Africa and Sumeria. Because of their technology and power, they were looked upon as gods. (See **Zechariah Sitchin.**)

> **Dissidents** – Anunnaki hybrid Remnant still on the Earth who seek to control Man and deny him his divine heritage; they often work thru the Illuminati and PTB. See **Insiders**.
>
> **Insiders** – Anunnaki hybrid Remnant still on the Earth (may include Enki). The pro-active ones who try to help mankind and block the **Dissidents**.
>
> **Remnant** – short for Anunnaki Remnant – that part of the Anunnaki group not destroyed in the Earth wars with Atlantis and the Greeks that went underground and are divided as to their allegiance toward Man. Comprised of the Insiders (+) and the Dissidents (-). Some are human-looking. Also known as **Naga** (underground 'Serpent' dwellers in Asia), or also called Dravidians.

Astral Realm – note that there are levels in the Astral realm, and in particular, the one that most concerns Man, is the Level I which is a kind of intra-dimensional space – more than 3D and yet not really 4D, and this is inhabited by Man's oppressors, the **Neggs**. The normal 4D STS/STO entities (Angels/Godhead) occupy the higher 4D and lower 5D Astral realms.

Attractor – energy in the form of an idea, person, or thing that draws other things, ideas or people together based on similar and strong resonance.

Baldy -- affectionate name for the author's **Guardian Angel**. 6' 8" tall, perfectly proportioned, bald head (no hair) with large blue irises (larger than most humans). Usually appears with the scent of orange blossoms and answers questions before they are asked. (As an Angel, he has no wings.)

Beings of Light – often referred to as Angels, or today's Watchers, they guide and protect Man. They are also known to provide the Life Review that NDErs speak of, and they are the '*Inspecs*' that Robert Monroe spoke of. Benevolent.

Bionet – a term coined to describe the hyperdimensional network of medians of communication in the body. Like the Internet, *chi* is carried in **meridians of energy** to all parts of the system, from the chakras, and tells (signals by its charge) the cells and organs what to do. The Bionet is manipulated during **Accupuncture** to channel biophotons (chi). It is also in play during **Rolfing**.

Brain Waves – a measurement of consciousness.

> Beta cycle: 12 – 19+ Hz (normal waking consciousness)
> Alpha cycle: 8 – 12 Hz (relaxed, aware state)

Theta cycle: 4 – 8 Hz (sleep)
Delta cycle: less than 4 Hz (deep sleep)

Catalyst – anything like an event, an idea or a word, that causes change in a person; the threat of being fired for bad performance at work is a catalyst to perform at one's best. Illness is a catalyst to see what is wrong, or what energy is blocked, in one's body.

Chakra – a vortex of energy formed in the body wherever two or more chi meridians come together; same as a vortex on the earth with its ley lines. (Sedona, AZ is known for several of these.) These are also referred to as 'energy centers' as they transduce energy from the air/water/Sun around a body and draw it into the body thru the chakras. There are 7 main charkas in the body and 1 above the head, and 1 below the feet. There are many more, minor charkas all over the body.

Chi – energy particles, also called *ruach*, orgone, mana, prana or ki – without chi in our food, air and water the human body could not exist. The chi is a force that travels along meridians (pathways) in the body that link the etheric aura (1st level of the aura) to the physical body; it can be directed by the mind to specific parts of the body for healing.

Cognitive Dissonance – the result of hearing/reading something new that does not fit into one's reality, or in what one thought was their reality; the effect is to create confusion followed by denial of new concepts. More specifically, when a new idea conflicts with an established idea that one already thinks they know, the result is 'dissonance', and rejection. When people were told 500 years ago that the Earth was round, they experienced great cognitive dissonance… which led to denial.

Coherence – resonating alike; attracted to each other by similar resonance. Two energy waves are coherent if they have the same shape, size, and strength.

Déjà Vu – the experience of having done, seen and/or heard something before; as though one is reliving a prior moment in their current lifetime. Relates to reliving a fractal simulation. See **Recycling**.

ELF – Extremely Low Frequency; a vibratory wave form that is sent by a radio-like device or microwave transmitter at a certain Hz or MHz frequency such that it entrains the mind into a 'resonant state' (usually Alpha) with the wave.

EMF – ElectroMagnetic Field; such an electrical field around a high tension power line also generates its own weak magnetic field, hence EMF. Note that a cellphone or TV or PC – anything electronic has an EMF and it is unhealthy to spend much time in it as it 'afflicts' the cells of the body and disrupts their function. The reason is that the body has its own weak EMF and communicates info to other parts of the body via the nervous system and chi meridians (see **Bionet**).

Energy Vampire – a person, OP or ensouled, who subconsciously starts an argument, gets the other person angry, and the instigator takes the other person's energy through the Law of Energy Potentials. Energy always flows from the higher potential to the lower and this applies to car batteries, as well as humans. So the instigator creates a fight, not to win or lose (they don't care), they will walk off with some of your energy, and they quit the argument when they have it. They are up, and the victim is usually tired.

Entanglement – We are all One.   See **Law of One** below.

Entrain – to induce a state in B like in A; usually done by music, movies, and words, but can be done by powerful thoughts and beliefs. A hypnotist entrains a subject into a desired state; Hitler's harangues entrained the crowds into the Nazi mindset he wanted; and classical music entrains the listeners into a relaxed (Alpha) state.

Entropy – the tendency of all things in the universe to wind down and die; also called the Second Law of Thermodynamics. The enemy of Evolution.

FE – abbreviation for Flat Earth.

Flow – often referred to as **The Flow**. This is the rising energetic vibrational entrainment into the higher 4th and 5th dimensional realms. It has increased awareness, compassion, Light, and STO aspects for service and is available to all who seek to align themselves with a Higher Way. It was created by the Higher Beings and is supported by an archetype that masters on the Earth reinforced and made available to all spiritual growth aspirants.

Freewill – ultimately an illusion: the more one grows spiritually, the more one does the will of the Father of Light. Baby souls, or those who insist on their own way, think they have freewill but the Father is merely letting them experience the results of what they do... their **Script** controls much of what Baby/Young souls can do. As Jesus said "Not my will, but Thine be done." Advanced souls have surrendered their will by eliminating their ego.

Gnostic – one who believes that God is accessible to all by going within and following their 'inner knowing'. Gnosticism relies on one's personal enlightenment to guide them; it is in fact, connecting with one's **Higher Self** which has true Knowledge. The Bible refers to it as the 'inner Light' and is what one connects with when they follow Jesus' suggestion that "the Kingdom of Heaven is within." Gnosticism was originally (AD 100) part of the Christian movement.

The Great Earth Puzzle

God/gods – this is god with a small "g". It is just a convention for referring to those who are watching over us, running the Drama on Earth. They consist of the Angels (Beings of Light), the InterLife Masters and Teachers, the Higher Beings and the White Brotherhood. They manage the Earth School that the Higher Beings created.

Godhead – a collection of higher souls, and Soul Groups, in closer proximity to God, like spokes on a wheel where the hub is God Himself. The Godhead works directly with the Oversoul for each Soul Group and sometimes the two are hard to distinguish. The basic hierarchy is: **God – Godhead/Oversouls – Soul Groups – Angels/Neggs – and individual souls.** Between the Godhead and God is the Hierarchy of Masters and those who are responsible for Solar Systems, and then Galaxies.

Gods-in-Training – when Man graduates from the Earth School, he can be useful to the Father of Light in various places in the Multiverse. One of those places is to undergo an apprentice position in overseeing the Earth and its souls – under the tutelage of more advanced 'gods' who give direction and training. The gods-in-training still make mistakes, just as Man does, and while often minimized by karmic override, these are allowed in part as an aspect of the new gods' training. This is therefore sometimes a source of things going wrong in an Earth person's life. If a god-in-training abuses his power, he is recycled back to Earth.

Granularization – the extent to which a material or system is composed of distinguishable pieces or 'grains'. A picture at 300 dpi (dots per inch) has a greater granularity than a picture at 1200 dpi – because the viewer can zoom down further in to the 1200 dpi picture before it gets grainy and one sees the pixels.

Ground of Being – who and what you really are; your PFV is the physical reflection of the <u>sum</u> of your STO/STS quotient. If you, your Soul essence, were to be removed from your body, the energy being that you are would have a certain vibration level (also reflected in the color of your aura) – higher or lower depending on how much Light you hold, how compassionate you are, whether you seek to serve (STO) or be served (STS), what issues (stuck points, agendas and attachments) you still carry with you, and in general, it refers to the "quality" of Light & Love that you <u>are.</u> Ultimately, it reflects the highest actions/thoughts that you are capable of. This is what is being tested in Karma with the Script (qv).

Higher Beings – Light Beings above the Astral and above reincarnative levels (1-5) and who are responsible for the existence of Earth School and the lower 6 levels, reside on the 7th level themselves; may intervene in 3rd – 4th – 5th – 6th dimensional affairs when the Greater Script of the Father of Light, or the One, requires it to keep the Multiverse working. Also colloquially called "**the gods**" who run the Earth realm (and occasionally simulations). The Higher Beings are <u>not</u> the Beings of Light (angels) nor ETs nor aliens.

Higher Self – also called the Oversoul, this is the coordinating entity of each Soul Group and acts to oversee Scripts, events, lessons – and coordinate with the souls of the same Soul Group, <u>and</u> with other Oversouls who manage other Soul Groups in the same Godhead. Each Godhead has multiple Oversouls that interface with the multiple Soul Groups.

Imprinting – when a soul has not lived a life that s/he recalls, it was added to that soul as an imprint. The actual life experience (stored in the Akashic Records) can be downloaded to a soul for their benefit – it becomes a part of their memory – as if they had lived it. The reason for this is to save time by not having to live a specific life, but by downloading the essence of it, the soul can grasp what it was like, what it meant, and can hopefully draw the lessons from it. For example, a soul who wants to birth into the Medieval Times, and wants to be a Knight, could imprint the life, experience and skills of a successful Knight to help guarantee survival when living as a Knight… especially if he has to become a Knight to perform a special feat, such as being there to save King Arthur, who might otherwise be killed, and thus save the Greater Script for England.

Inserts – in a Simulation it is easy to insert avatars, strange structures, and objects as needed to make the overall Greater Script work. This can be an *ad hoc* way of running the Earth Realm.

Interdimensionals – those beings in 4D <u>and above</u> who normally have very little interest in Man, and may be STO or STS. The STOs are often curious and observing. The STS version has been known to use humans for unknown agendas. Also a generic term for the 4D STS Controllers inasmuch as they operate between dimensions. Probably Djinn.

InterLife – where souls go when they die, after passing through the **Tunnel** to the **Light**. Also called the **Other Side**, and sometimes appears to be Heaven. It is where the individual **Script** is designed, souls are counseled by the Masters and Teachers, souls are rehabilitated after a rough lifetime on Earth (or elsewhere), and it is where the **Heavenly Biocomputer** resides. This is also where reunions with members of one's **SoulGroup** happen. (Book 2 TOM spends 2 chapters on this aspect.)

Karma – *Aka* **The Law of Karma**. – originally the concept of "meeting oneself", or "what goes 'round, comes 'round." It does <u>not</u> mean being stabbed in this lifetime because one stabbed someone else in a former lifetime. It is not based on "an eye for an eye" – that is a false teaching. The original, true concept was that of the Universal Law of Cause and Effect, and it forms the basis of one or more aspects of your LifeScript. Note that **Karma applies only to Earth**; other souls who do not come to Earth do not have to deal with Karma. Note also: when Karma is met and satisfied, it is replaced with **Grace.** See **Script.**

Law of Attraction – simply says that you attract to you who and what you are. Attraction also works by repeated focus, visualization, affirmations and speaking one's word, and of course, by prayer. Attracting what you want may be blocked by the Script if it is not permitted this lifetime… to enforce some lesson: you stay poor to learn humility or reliance on other people, for example.

Law of Confusion – when RA was asked a question that violated someone else's right to privacy, or asked something that would be giving advanced level information that the person had no context for, RA would comment that the question could not be answered because it "violated the Law of Confusion." We are to work thru confusion and seek the answer(s) on our own; **everyone has a temporary 'right' to be confused** (i.e., to not know) and are expected to work thru it, or ask, thereby absorbing the lesson and info on a level that makes the lesson/info part of us.

Law of Non-Interference -- there are several parts to this Law. One is that other sentient beings from other worlds do not have *carte blanche* to manipulate or interfere with the normal development of Earth's humans. In addition, it also specifies that human bodies are to have only one soul (possession is not permitted), and there is a provision that ensouled humans cannot be manipulated from the Astral violating their Freewill.

Law of One – the concept that we are all connected at a higher level, mostly thru our Higher Selves, and we are all part of the One, the Father of Light – if you have a soul. The Law of One also includes freewill and love. Telling someone else what to do, how to live, etc. is a violation of the Law of One, a violation of freewill whose flipside is called the Law of Confusion. (Think: **Entanglement**.)

Law of Potentials – this merely says that energy in the universe always flows from the higher potential to the lower. (Think: a weaker battery being charged by a stronger one, or a Master healing a human.)

Light – an intelligent aspect of God; sometimes referred to as the Force. It may be used interchangeably with Heaven and/or Knowledge. There are **biophotons** of light that support the operation of DNA and sustain bodily operations.
Note that Light (large L) is a conscious aspect of the God force, which force can have a brilliant light about it. The light (small "l") is everyday, regular light.

Lightworker – any entity, physical or Astral, that uses Light as an energy source to do its work. (Includes angels, demons, **Neggs**, energy healers and Higher Beings.) Caution: it does not always imply STO behavior.

Matrix – In this book it is the sea of intelligent energy surrounding us and permeating everything. Synonym: the **Quantum Net**. Can also be ZPE Etheric composition that interpenetrates everything.

Memes – a concept, or idea, that generally has spread through a population – an idea that may spread like a biological virus – such as a belief in ghosts, or a belief that black cats bring bad luck…or, if you go out in the rain and get wet, you can catch a cold. There are positive and negative memes. (See also "**100th Monkey Effect**.")

Morphic Resonance – said of a plant or animal that takes its physical shape from the *morphogenetic* field that establishes a 'morphic' (shape) resonance with the object's energy. The plant's shape is entrained by the morphic resonance with the morphogenetic field (pattern) that governs how living things take shape, according to Rupert Sheldrake. (See VEG, Ch. 5 and the Russian 'Chuck.')

Morphogenesis – Rupert Sheldrake conceived of the presence of a 4D field around living things that influences the shape they take – kind of an Astral Template that governs height, width, color and other aspects of the oak tree for example, such as when and where it sends out its branches, how fast and how far.

Multiverse – the universe we live in is one of a number of universes comprising a Multiverse… multiple universes interconnected forming a coherent larger universe consisting of multiple levels (realities), and can involve parallel universes or dimensions – usually in '**superposition**' (or stacked).

NDE – a Near Death Experience where the person appears to die, and their body is pronounced clinically dead, but they come back to life and relay their experience of meeting a Being of Light with whom they have a Life Review, and they usually come back a changed (better) person. The NDE effect often produces a positive spiritual transformation in the person. (See Book 2, TOM.)

Neggs – Beings of Light playing the heavy-handed role of 'demon' to send negative lessons to the ensouled in the Earth School. (See VEG, Ch. 6-7.) There really are no demons, so what the Neggs do is a blessing in disguise. The Beings of Light supervise the Neggs; **they work together** (VEG, Ch. 6 Dr. Lerma). When Man is thru with the Earth School, all Neggs will be reverted to Beings of Light.

Neurotransmitters – chemicals/molecules in the brain that act as 'signalling molecules' to activate other parts of the body – neurons, glands, cells… Examples are Adrenaline, Histamine, Serotonin, Dopamine…

NPCs – Non-Playable Characters in a video game. The player does not control them, but they are a useful part of the scenario that the game player must deal with. The **OPs** (below) serve a similar function in the Earth School Drama.

OPs – Organic Portals -- (pronounced "Oh Pee") human beings, flesh and blood (Organic part), and they can serve as a portal for astral entities to operate thru them. This is not synonymous with Zombie.

They also are **not fully human** as they lack a soul and that is because they have incomplete DNA and only the first 3 chakras are wired to function; they cannot access higher energy centers. Due to their somewhat robotic nature, they can be used to guide and/or influence ensouled humans in the 3D Drama. The Greeks called them 'hylics.' The Mayans called them 'wooden people.' Dr. Mouravieff called them pre-Adamics. (See VEG, Ch. 5 and **Appendix D in TOM**.) Also called Backdrop People (Appendix C).

It has been a lie for centuries that all humans have souls. The Greeks and Mayans knew there are people who have no soul. Dogs and cats have no soul and they walk, eat and replicate just fine. This whole issue was **examined in detail in VEG, Ch. 1 & 5**. They really do exist and are sometimes the **Sociopaths and atheists** among us. 98%+ **are not evil**, they just have **no conscience** and think they can do whatever they want. Jim Jones, Charles Manson, and Richard Ramirez are prime examples of OPs on steroids that have run amok, because they have no connection to a Higher Self. Discussing spirituality with them gets nowhere, they cannot imagine what you are talking about. They are run by 'A' Influences (VEG, Ch.13: Life is a Film). Also called **Non-Playable Characters** (NPCs) as in a video game.

Orbs -- these are round balls of light that often showed up in the original digital cameras' photos, and some people thought they were intelligent beings visiting whatever scene was being photographed. In reality, the pixels in the digital camera are charge-coupled devices, carrying a small electric charge, and occasionally one or more pixels would misfire... producing the orb(s). Better quality digital cameras do not produce the phenomenon.

Oversoul – the lowest level of the God Hierarchy, between God and Man. Between the God Hierarchy and Man are the Angels, also called Beings of Light (because they don't have wings). The Oversoul can also be seen as the larger Soul that each of us belongs to. See **Soul Aspect**.

PFV – Personal Frequency Vibration -- the day-to-day, overall vibratory rate (resonance) of the soul energy sustaining the human body. When a person is angry their aura 'glows' red, and the PFV can drop to a lower (denser) vibration than when a person feels a lot of love and the aura 'glows' rose and the vibration reflects the energy of the heart charka (higher, lighter energy). The PFV also denotes which charka is dominant in the person; a person living from their higher charkas has a higher PFV than one engaged in sex, violence and pettiness (lower chakra activity). The aura typically reflects what one is feeling, yet the base PFV does not change; when the person is at rest, the base PFV is consistent from day to day as it reflects the overall level of soul growth. Also known as that person's "energy signature" as recorded in objects (Psychometry).

Pre-Soul – also called **First Time Soul**, allegedly the initial stage of an animal that leaves 2D and enters the 3D human soul realm (**metempsychosis**). The more appropriate concept is that **a 1st-time human is not a developed soul**, but a potential one if the entity applies itself. Typically, only the first 3 chakras are functional, and thus there is not enough 'soul energy' to create an aura. (See also **OP**s.) This is a tentative theory and has yet to be proven.

Prophylactic Fantasy – describes the world of denial that some people live in. 'Prophylactic' because they feel safe in their version of the world, and they reason that nothing really destructive has ever happened to them, nor can it. 'Fantasy' because they do not accept the real world and its negativity; they see their world as they want it to be and sometimes think that they can exert a 'force' that makes it that way.

PTB – the earthly human Powers That Be; the 3rd dimensional STS people running the world for their personal gain and power. They may be influenced by corrupt DNA or ideologies. Puppets. Many of them have no spirituality (atheists and sociopaths). They often work with 4D STS beings to derail soul growth.

Quanta/Quantum -- In Physics, a **quantum** (plural: **quanta**) is the minimum amount of any physical entity involved in an interaction. A photon is a single quantum of light. An electron, however, is energetically quantized, or bound to the atom of which it is a part. A quanta is a subatomic particle.

Quantization – refers to the organization, structure and arrangement of quanta in a video game. These are the discreet elements of whatever the video display is showing – below the level of **pixels**. The particular arrangement and grouping can create **granularization** or not.

Quarks --Everyday matter is composed of atoms, and later, subatomic constituents of the atom were identified. As the 1930s dawned, the electron and the proton were observed, along with the **photon**, the particle of electromagnetic radiation. At that time, the recent advent of quantum mechanics was radically altering the conception of particles, as a single particle could also behave as a wave -- a paradox still eluding satisfactory explanation (**Appendix C**). Meanwhile, via quantum theory, protons and neutrons (in the atom's nucleus) were found to contain smaller particles, called quarks. Below that are alleged to be Strings.

Quantum Coherence -- Coherence is best exemplified by a laser which coherently aligns four beams of light into one focused beam. All objects have wave-like properties and if those waves are alike and synchronized, they are coherent. If all the atoms in a pure block of silver are vibrating at the same rate, they can be said to be coherent. The opposite, **decoherence**, is when atoms or waves get out of synch… i.e., a state of chaos.

Quantum Net – the field that interpenetrates all things and all of space. This has also been likened to the Ether and what is today called Dark Energy. The body has a minor Net, called the Bionet. In all Nets, the communication is almost instantaneous.

RCF/Matrix – Resonant Conscious Field (developed more in VEG, Ch. 12) as the 'sea' of energetic thoughtforms surrounding the planet. In some places it is very strong and can be a very negative influence. Robert Monroe referred to it as the H Band.

Reassembly – this is also called **Disassembled**, Dissolved in VEG and TOM (books 1 and 2). The Higher Beings will attempt to infuse a wayward soul with new, proactive energy, and perform a kind of 'psychic surgery' on the soul's energy field (which emanates from their Ground of Being so it is tricky), and failing that, if a soul cannot be re-oriented to STO behavior, the energy (soul) is **Dissolved** back into its component energy parts, as a failed soul. The consciousness is removed, energy then is cleaned, and can be reused without the former consciousness that accompanied the failed soul. (In short, fooling around for Eternity is not allowed, and when a soul goes mostly Dark, it is taken aside and examined closely… it is not allowed to pollute the world of souls headed for the Light. Thus, it should be clear that there is no Satan, or fallen angel.)

Recycled – short-circuited version of reincarnation: to come back into the same body, same lifetime, hence experiences **Déjà Vu**. Implies the inability to move forward into new realms and experiences in the greater **Multiverse**. (See VEG, Ch. 14.) If the soul learns nothing in a lifetime, this is the first step in rehabilitating them, and is a gentle nudge to apply oneself and handle the lessons... ultimately potentially followed by **Reassembly**.

Reincarnation – the spiritual growth aspect of a soul moving thru the different realms in the Multiverse (not just back to Earth) for the purpose of experiencing and gaining knowledge and wisdom. On the other hand, a repeated lifetime limited to Earth can be more of a **Recycling**.

Resonance – vibrating alike: such that two tuning forks A and B side by side, with A struck hard to set it vibrating, when put next to B which was not vibrating, will set tuning fork B vibrating at the same frequency as tuning fork A. This also happens with people in close proximity: a very negative person can 'detune' (bring down) a room of people and some people may actually feel ill and not know why (as they pick up the negative person's vibes). See **Entrainment.**

Schumann Resonance – natural frequency of earth's vibration/resonance: 7.8Hz. The human body is sustained by this resonance and emits its own frequency as torsion waves. See **PFV**. Also examined in TSiM.

Script – (LifeScript) to assist Karma, when one is born, one has a Script covering the basic (usually 10 % max) events that are to happen in one's life, which one is expected to overcome. They may be positive or negative, and how one meets them and handles them determines how one is progressing towards the goal of getting out of the Earth School. It often has Options programmed into it (**Points of Choice**) where the Soul must make a significant choice. It is a test of soul growth. A personal LifeScript is usually subject to the Greater Script of the Father of Light and works within it. The Angels (Beings of Light) administer the Script, yet the Soul still has 90% Freewill. (See **Ground of Being.**)
    The Script does <u>not</u> tell you what to do or say; basically <u>just events</u> are scripted to test you (and often driven via the NPCs).

ShapeShifting – the ability to <u>control what people see</u>… the being doing the shape-shifting does not actually change any of his/her atomic structure – just the way his appearance is perceived, and **perception is holographic**. So to effect a different appearance, the being just produces new interference waves that the observer 'sees' differently. Commonly done by 4D and above entities while in 3D.

Sheep – people who are barely conscious, and refuse to think for themselves. They want someone to tell them what to do and when to do it, and they go along with whatever they are told. They are easily manipulated by the Media. Also called **'sheeple'** and may be 'dense' ensouled humans. (See Epilog comments.)

Simulation – Earth as a simulation is suggested because of all the anomalies found around the planet (shown in QES – 3 groups) including gridlines, and sounds coming from the "edge of the Universe." In addition, some physics constants are changing which would not happen in a physical 3D rock world. (Chapter 9.)

Soul – A spark of the Divine, The God replicating Himself, an eternal sentient, coherent intelligent Light energy being that is conscious of itself and its surroundings, can exist in multiple dimensions, and can evolve itself back to a connection with the Godhead… (See VEG, Ch. 12.) Soul (large S) is connected with Higher Consciousness and Mind, whereas soul (small s) is just a generic reference to a soul, in a conscious human with a human mind. See **OPs**.

Soul Aspect – all souls can 'split' themselves to experience different realms; as when a timeline splits, one part of the soul stays with the original TL and another part replicates to the new TL. Each Soul (large 'S') has aspects in different TLs, dimensions, worlds, and realms, etc. and at a point in the future, they all reunite to as a Soul Group to the main Soul (or Oversoul). Not a **Fragment** (next).

Soul Fragment – some souls may fragment **due to trauma** and then special therapy is often needed to coax the missing fragment to rejoin its source. Some fragments are held by family members, past lovers, and even by astral beings.

Soul Group – each soul was part of a group of like souls (same core vibration PFV which usually synchs up with a specific archetype) and these split up to better experience the Creation – souls will eventually reunite in their original group when their explorings are done. The Soul Groups reunite with the **Godhead** from which they came.

Soul Merge – to undertake a special project where a 3rd level soul has volunteered to serve in a capacity that it alone can't do, and so a Merge is performed to give that 3rd level soul the extra knowledge and strength of the merging soul (who is of the same soul group -- BUT from a higher level) and together they perform some task that the Higher Beings must have approved – before the Merge can happen. The author was not a Merge but a Walk-in and Walk-out. And Merge is a rare occurrence as it is a Universal Law that there is but **one soul in a body at a time** – which is why Possession is a violation of the Law. See **Soul Aspect** and **Walk-in..**

STO – Service To Others; altruistic behavior, self-sacrificing. Found in levels (dimensions) 3 thru 7.

STS – Service To Self; selfish behavior; 'Me-My-Mine' syndrome. Found only in levels 3 thru 5; if a soul does not overcome STS orientation, they may be reschooled or **Reassembled.** They cannot ascend to level 6 where the lessons of 3D and 4D and 5D are assimilated for graduation to level 7.

STS Gang – this is a 'catch all' group term referring to the **discarnates [largely]**, thoughtforms, 4D astral beings, and the 3D PTB acting as oppressors of Man, without a clear distinction as to exactly which one is doing what to Man at any one time.
The group may occasionally include the **Interdimensional** beings described by Wilde and Monroe (VEG, Ch. 12), although such are usually too busy interfering with the entities on their own level to harass Man on the 3D level.

Subquantum Kinetics – is an approach to microphysics with roots in general system theory, nonequilibrium thermodynamics, and nonlinear dynamics. It represents quantum phenomena differently than Quantum Physics (QP) and works with the concept of the **Ether** which is composed of subquantum units called **etherons** (as opposed to QP's quarks). It is simpler than Quantum Physics and explains the issues that QP is still wrestling with: wave-particle dualism, strings, singularities, and the cosmological constant. (**Appendix D**)

It also embraces and explains Tesla's work better than QP. (Also refer to Chapter 4 of Dr. Paul LaViolette's book <u>Secrets of Antigravity Propulsion</u> for a more complete description in layman's terms.)

Terraforming – an advanced technical process whereby a whole planet is set to its original, or a near-new, pristine condition following some catastrophe or pollution, or both. The ecology is balanced, the air, land and water are unpolluted, and in the case of planet Earth, it can once again support lifeforms. See also "**Wipe and Reboot**." This is easy to do if Earth is a Simulation – just reset it.

Thoughtform (**TF**) – any thought that many people subscribe to and which reflects a widely held belief, esp. one imbued with a lot of fear, or hate, generates a TF which after a while (depending on the amount of energy put into it) takes on a 'life' of its own; **man is a creator and thoughts are things**. If enough people fear and believe in werewolves, there will be thoughtform 'werewolves' … which are not real entities but are attracted to those who fear and believe in them (like attracts like). Any unwanted TF can be cancelled and should be before it attaches itself to a person's aura and then 'feeds' off the person's energy – like a parasite. TF have no conscious volition of their own, they are reactive and go to wherever (1) they are attracted by sympathetic vibration, and (2) where the person's aura is weak. Carl Jung called these TF's Archtypes. Also see Tulpa (Tibetan thoughtform).

Torsion Wave – Being neither electromagnetic in nature nor relating to gravity as it stands on its own, this wave energy is a spiraling, non-Hertzian electromagnetic wave that travels through the vacuum of space at super-luminal speed - a billion times faster than the speed of light. Because these waves trace a spiraling path, they are called "torsion waves." **Thoughts are torsion waves** – see TSiM.

Torus – an energy field that is toroidal in shape (like a doughnut), and is found around the heart – extending an electromagnetic field about 3-4' around the human body. (See Book 2, TOM)

Unconscious – unaware, not a very high level of perception. A person who is 'asleep' spiritually and is not aware that there are more than the 5 senses. Can also mean 'spacing out' with eyes wide open. Standard condition of the **Sheep**.

V2K – "Voice to Skull" -- a microwave enhanced transmission of words directly into a person's head, as if they actually hear the words, without any external devices or hearing apparatus. Developed by the US Army to communicate with a soldier on the battlefield, to the exclusion of other soldiers, it was perfected during the mind experiments with Helen Schucman while she transcribed the *Course in Miracles* book. (<u>Who</u> sent her the information is not known.) Not to be confused with MKultra.

Vibration/Vibe – the energy state of a person, place or thing. Everything puts out an energy 'signature', which is how pyschometry works… objects record the energy of the person that held/owned the object, and places often hold the residual energy of events that happened there: some sensitive people cannot visit Gettysburg as they feel the negative energy from all the hate and fear created in that place – even thought it was long ago, it still holds some energy that has not completely dissipated. (See **PFV**.)

Visual Spatial Acuity – the ability to see fine detail; visual term reflecting the number of rods/cones in the retina. Similar to **pixels** in computer printing, display screens and digital cameras.

VR Sphere – a term created in VEG, Ch. 12, to refer to the concept that Earth is a 3D construct, quarantined to protect it from 4D interference while the Higher Beings grow and develop souls in the Earth School. VR is **Virtual Reality** (suggesting that Earth is nowadays a very sophisticated **Simulation**) which is dealt with in VEG Ch. 13 and then further developed in QES, Book 4. Also developed in Chapters 9-11 in this book (GEP). Formerly a Flat Earth (original creation) it is now something of a **Sphere** (exact shape uncertain – see p. 396) .

Walk-in/ Walk-out – when an advanced soul comes in to the 3D soul in a 3D body for some higher purpose (by prior agreement) and the original soul stands aside while the advanced soul completes his/her task. When done, the Walk-in usually walks out, and the original soul enters back in – such transfers are assisted by Higher Beings who have the knowledge/ability to perform the swap. Similar to a **Soul Merge**.

White Brotherhood – aka Great White Brotherhood. Ancient secret society started in Sumeria by Enki to benefit Man. Originally called the Brotherhood of the Serpent. Most beings in this society today are Astral and were involved thru Baldy in the transcribing of the first 6 books -- listed at the bottom of the Copyright page. They are often referred to as **"the gods"** in the 6 books as they are involved in running the Earth School. (See Appendix F.)

Wipe and Reboot – an end to a current **Era** of Man on Earth, followed usually by a terra-forming (resetting the environment back to clean and balanced), followed by the Re-seeding of Man on Earth.
The term is borrowed from the computer world where when a PC is non-functional (i.e., locked up and displays the dreaded BSOD [Blue Screen of Death]), it is necessary to "Wipe" the hard disk – reformat it – and reload the operating system and application software… i.e., "Reboot" the system and start all over again. Whereas the PC gets a clean start as if nothing happened, each new Era for Man still includes whatever objects were created in the prior Era – i.e., pyramids, huge walls, and Stonehenge.

Zechariah Sitchin – the late Middle Eastern scholar, speaking several languages, who translated the Anunnaki/Sumerian tablets. VEG, Ch. 3 is mostly dedicated to a summary of his findings about Man's origins. His claim to fame was *The Earth Chronicles* series of 8+ books that revealed the Sumerian – Anunnaki connection (see Bibliography for partial list).

ZPE (Zero Point Energy) – Zero-point energy, also called quantum vacuum zero-point energy, is the lowest possible energy that a quantum mechanical physical system may have; it is the energy of its ground state. Vacuum energy is the zero-point energy of all the fields in space, which in the Standard Model includes the electromagnetic field, fermionic fields, and the alleged Higgs [boson] field.

It is the energy of the vacuum, which in quantum field theory is defined not as empty space but as the ground [lowest vibrational] state of the fields. In cosmology, the vacuum energy is one possible explanation for the cosmological constant. A related term is *zero-point field*, which is the lowest energy state of a particular field. (Definition: credit https://en.wikipedia.org/wiki/Zero-point_energy) Dark Energy is thought to contain a ZPE state. (See **Quantum Net**, and VEG Ch. 9. Also see TOM, Ch. 11)

Bibliography

Books

The following is the Bibliography from VEG (Book 1) and contains more sources than this book uses, but which may be of interest to some readers. Bolded titles are particularly relevant to this book.

Deliverance/Exorcism

Cuneo, Michael. *American Exorcism*. New York: Random House/ Doubleday, 2001.

Martin, Malachi. *Hostage to the Devil*. HarperSanFrancisco, 1992.

Wilkinson, Tracy. *The Vatican's Exorcists*. New York: Warner Books, 2007.

Religion/Metaphysics/Spirituality

Acharya S. *The Christ Conspiracy*. Kempton, IL: Adventures Unlimited Press, 1999.

Atwater, P.M.H. *Near Death Experiences*. NY: MJF Books, 2011.

Bloom, Harold. *Jesus and Yahweh: The Names Divine*. New York: Penguin Group (USA), 2005.

Castaneda, Carlos. *A Separate Reality*. New York: Washington Square Press, 1971.

_____. *The Active Side of Infinity*. New York: HarperCollins, 2000.

Capra, Fritjof. *The Tao of Physics*. Boston, MA: Shambhala Publications, 1999.

Charles, R.A. **The Book of Enoch the Prophet**. San Francisco, CA: Weiser Books, 2003.

Dawood, N. J. **The Koran.** New York: Penguin Group (USA), 2006.

Elkins, Don and Carla Rueckert. *The RA Material, Book I*. Atglen, PA: Schiffer Publishing/Whitford Press, 1984.

Eppler, LaRue and Vanessa Wesley. *Your Essential Whisper*. Bloomington, IN: iUniverse, 2008.

Frejer, B. Ernest. *The Edgar Cayce Companion*. New York: Barnes & Noble Press, 1995.

Gaffney, Mark. *Gnostic Secrets of the Naassenes*. Rochester, VT: Inner Traditions, 2004.

Gardiner, Philip. *Secret Societies*. Franklin Lakes, NJ: Career Press/New Page, 2007.

_____. *Secrets of the Serpent*. Forest Hill, CA: Reality Press, 2008.

_____. **The Shining Ones.** Nottinghamshire, England: Phase Group, 2002.

Golas, Thaddeus. *The Lazy Man's Guide to Enlightenment*. Salt Lake City: Gibbs-Smith, 1995.

Guiley, Rosemary Ellen and Imbrogno, Philip J. *The Vengeful Djinn*. Woodbury, Mn: Llewellyn Worldwide, 2012.

Hoeller, Stephan A. *Gnosticism*. Wheaton, IL: Quest Books, 2002.

Kenyon, J. Douglas, Ed. *Forbidden Religion*. Rochester, VT: Bear & Co., 2006.

Laurence, Richard. **The Book of ENOCH the Prophet**. Kempton, IL: Adventures Unlimited Press, 2000.

Lerma, John, M.D.   ***Into the Light***. Franklin Lakes, NJ: New Page Books, 2007.

_____ ***Learning From the Light***. Franklin Lakes, NJ: New Page   Books, 2009.

Mead, G.R.S.   *Apollonius of Tyana*. (1901 Edition reprint) Sacramento, CA: Murine Press, 2008.

Meyer, Marvin.   *The Gospel of Thomas*. New York: HarperCollins, 1992.

Modi, Shakuntala, M.D.   *Remarkable Healings*. VA: Hampton Roads, 1997.

_____. *Memories of God and Creation*. Charlottesville, VA: Hampton Roads, 2000.

Moody, Raymond A., Jr., MD. ***Life After Life***. New York: HarperCollins, 2001.

Monroe, Robert. ***Journeys Out of the Body***. New York: Doubleday, 1971.

_____. *Far Journeys*. New York: Random House/Broadway, 2001.

_____. *Ultimate Journey*. New York: Random House/Broadway, 2000.

Newton, Michael. *Destiny of Souls*. Woodbury, Mn: Llewellyn Worldwide, 2002.

_____. *Journey of Souls*. Woodbury, Mn: Llewellyn Worldwide, 1994.

Pagels, Elaine.   *The Gnostic Gospels*. New York: Random House/Vintage, 1979.

Paulson, Genevieve Lewis. *Kundalini and the Chakras*. MN: Llewellyn Worldwide, 2005.

Peck, M. Scott, M.D.   *Glimpses of the Devil*. New York: Free Press, 2005.

_____ *People of the Lie*. New York: Touchstone, 1983.

Picknett, Lynn.   *The Secret History of Lucifer*. London: Constable & Robinson, 2005.

Rasha.   ***Oneness***. Santa Fe, NM: Earthstar Press, 2003.

Ring, Kenneth.   ***Lessons from the Light***. Portsmouth, NH: Moment Point Press, 2000.

Robinson, James M., General Editor. *The Nag Hammadi Library*. New York: HarperCollins, 1990.

Roman, Sanaya.   ***Spiritual Growth***. Tiburon, CA: HJ Kramer, Inc., 1989.

_____ *Personal Power Through Awareness*. Tiburon, CA: HJ Kramer, Inc., 1986.

Ruffin, C. Bernard.   *Padre Pio: The True Story*. Huntington, IN: Our Sunday Visitor Publishing Division, Inc., 1991.

Russell, A. J.,   *God Calling*. Uhrichsville, OH: Barbour Publishing, 1989.

Slate, Joe H., Ph.D.   *Aura Energy*. Woodbury, MN: Llewellyn Worldwide, 2002.

_____. *Psychic Vampires*. St. Paul, MN: Llewellyn Worldwide, 2004.

Snellgrove, Brian.   *The Unseen Self*. Essex, England: The C.W. Daniel Co., 1996.

Spong, John Shelby. ***A New Christianity for a New World***. HarperSanFrancisco, 2001.

_____ ***Why Christianity Must Change or Die***. HarperSanFrancisco, 1999.

The King James Study Bible. Nashville, TN: Thomas Nelson, 1988.

Wilde, Stuart.   ***The Prayers and Contemplations of God's Gladiators***. Chicago, IL: Brookemarke, LLC., 2001.

_____ *The Force*. Carlsbad, CA: Hay House, 2006.

Wohlberg, Steve.   *End Time Delusions*. Shippensberg, PA: Treasure House, 2004.

Zukav, Gary.   *The Dancing Wu Li Masters*. New York: Quill, 1979.

_____ *The Seat of the Soul*. New York: Simon & Schuster, 1990.

Scientific/Medical

Baugh, Carl E. ***Why Do Men Believe Evolution Against All Odds?*** Oklahoma City, OK: Hearthstone Publishing, 1999.

Becker, R.O. & Gary Selden. ***The Body Electric.*** NT: Harper, 1985.

Behe, Michael J. ***Darwin's Black Box.*** New York: Simon & Schuster/Touchstone, 1996.

Braden, Gregg. *The Divine Matrix.* Carlsbad, CA: Hay House, 2007.

Brown, Walt. ***In The Beginning: Compelling Evidence for Creation and the Flood.*** Phoenix, AZ: Center for Scientific Creation, 1995.

Carlo, George, Dr. and Martin Schram. *Cell Phones: Invisible Hazards in the Wireless Age.* New York: Carroll & Graf Publishers, 2001.

Clayton, PhD, Paul *Out of the Fire.* HongKong: PharmacoNutrition Press, 203.

Elvidge, Jim. ***The Universe Solved.*** AT Press, 2007.

Friesen, James G., M.D. *Uncovering the Mystery of MPD.* Eugene, OR: Wipf & Stock Publishers, 1991.

Greene, Brian. *The Elegant Universe.* New York: W.W.Norton & C0. 2003.

_____ *The Fabric of the Cosmos.* New York: Vintage Books. 2004.

_____ ***The Hidden Reality.*** New York: Alfred A. Knopf. 2011.

Hawking, Stephen with Leonard Mlodinow. *A Briefer History of Time.* New York: Bantam Dell, 2008.

_____. *A Brief History of Time and The Universe in a Nutshell.* (2-book volume). New York: Bantam Dell Books, 1996, 2001.

Hunter, C. Roy. *Master the Power of Self-Hypnosis.* NY: Sterling Publishing, 1998.

Johnson, Phillip E. ***Defeating Darwinism.*** Downers Grove, IL: InterVarsity Press, 1997.

Kaku, Michio. *Hyperspace.* New York: Anchor Books, 1995.

Krauss, Lawrence M. *The Physics of Star Trek.* New York: Basic Books, 2007.

LaViolette, Paul A, PhD. ***Secrets of Antigravity Propulsion.*** VT: Bear & Co., 2008.

_____ *Genesis of the Cosmos.* Rochester, VT: Bear & Co., 2004.

_____ *Decoding the Message of the Pulsars,* VT: Bear & Co., 2006

Leaf, Dr. Caroline, MD. ***Switch on Your Brain.*** MI: Baker Books, 2013.

Levy, Elinor, and Mark Fischetti. *The New Killer Diseases.* NY: Three Rivers Press, 2004.

Lloyd, Seth. ***Programming the Universe.*** New York: Random House, 2007.

McTaggart, Lynn. ***The Field.*** New York: HarperCollins/Quill, 2002.

Meyer, Stephen C. *Signature in the Cell.* New York: HarperCollins, 2009.

Morris, John D., Ph.D. ***The Young Earth.*** Green Forest, AR: Master Books, 2006.

Myss, Caroline, Ph.D. *Why People Don't Heal and How They Can.* New York: Three Rivers Press, 1997.

_____. ***Sacred Contracts.*** New York: Three Rivers Press, 2002.

Narby, Jeremy. *The Cosmic Serpent.* New York: Tarcher/Putnam, 1998.

Pearce, Joseph Chilton. *The Biology of Transcendence.* Rochester, VT: Park Street Press, 2002.

Pearsall, Paul, Ph.D. ***The Heart's Code.*** New York: Random House/Broadway, 1998.

Pert PhD, Candace. *Molecules of Emotion.* New York: Scribner, 1997.

Peterson, Dennis R. ***Unlocking the Mysteries of Creation.*** 6th edition. El Dorado, CA: Creation Resource Foundation, 1990.

Samuelson PhD, Gary L. *The Science of Healing Revealed.* (self-published 64-pg Full-color illustrated booklet on biochemistry). 2009

Shnayerson, Michael and Mark Plotkin. *The Killers Within.* Boston: Back Bay Books, 2003.

Stout, Dr. Martha, **The Sociopath Next Door.** NY: Broadway Books, 2005.

Talbot, Michael. **The Holographic Universe.** New York: HarperCollins, 1991.

Watson, James D. *DNA.* New York: Alfred A. Knopf, 2003.

Wolf, Fred Alan, PhD. *The Yoga of Time Travel.* Wheaton, IL: Quest Books, 2004.

Yang, Jwing-Ming, Dr. *The Root of Chinese Qigong.* Roslindale, MA: YMAA

UFOs and ETs

Beckley, Timothy G. *The Secret Space Program.* NJ: Global Communications, 2012.

Bender, Albert K. *Flying Saucers and the Three Men.* Clarksburg, W.VA: Saucerian Press, 1962.

Boulay, R.A. **Flying Serpents and Dragons.** Rev. Ed. San Diego, CA: The Book Tree, 1999.

Bramley, William. **The Gods of Eden.** New York: HarperCollins/Avon, 1993.

Branton, ed. *The Omega Files.* NJ: Global Communications, 2012.

Cannon, Dolores. *Keepers of the Garden.* Huntsville, AR: Ozark Mtn Publishers, 2002.

Clark, Gerald. *The Anunnaki of Nibiru.* Lexington, KY: CreateSpace, 2013.

Cramp, Leonard G. *Space, Gravity and the Flying Saucer.* NY: British Book Centre, 1955.

Farrell, Joseph P. *The Cosmic War.* Kempton, IL: Adventures Unlimited Press, 2007.

_____ *Nazi International.* Kempton, IL; Adventures Unlimited Press, 2008.

_____ *Covert Wars and Breakaway Civilizations.* Kempton, IL; Adventures Unlimited Press, 2012.

_____ *Covert Wars and the Clash of Civilizations.* Kempton, IL; Adventures Unlimited Press, 2013

_____ *Saucers, Swastikas and Psyops.* Kempton, IL: Adventures Unlimited Press, 2011.

_____ *Roswell and the Reich.* Kempton, IL: Adventures Unlimited Press, 2010.

_____ **Genes, Giants, Monsters and Men.** Washington: Feral House, 2011.

Fowler, Raymond. *The Watchers.* New York: Bantam Books, 1990.

Friedrich, Mattern. *UFOs: Nazi Secret Weapons?* NJ: Global Communications, 2008.

Good, Timothy. *Earth: An Alien Enterprise.* New York: Pegasus Publishing, 2013.

Greer, Steven M., MD. *Hidden Truth – Forbidden Knowledge.* Crozet, VA: Crossing Point, Inc., 2006.

Harbinson, W.A. *Projekt UFO: The Case for Man-Made Flying Saucers.* London: Boxtree Ltd., 1995.

Heron, Patrick. **The Nephilim and the Pyramid of the Apocalypse.** New York: Kensington Publishing/Citadel Press, 2004.

Jacobs, David M., Ph.D. *The Threat.* New York: Simon & Schuster, 1998.

Keith, Jim. *Saucers of the Illuminati.* Kempton, IL: AU Press, 2004.

Komarek, Ed. UFOs: *Exopolitics and the New World DisOrder.* Lexington KY: Shoestring Publishing, 2012.

Lessin PhD, Sasha.   **Anunnaki Gods No More**. Lexington, KY: CreateSpace, 2012.
Lewels, Joe, Ph.D.   *The God Hypothesis*. Columbus, NC: Wild Flower Press, 2005.

Mack, John E., M.D.   *Abduction*. New York: Charles Scribner's Sons, 1994.
_____   *Passport to the Cosmos*. New York: Three Rivers Press, 1999.
Marrs, Jim.   *Alien Agenda*. New York: HarperCollins, 1997.
_____.   **Rule by Secrecy**. New York: HarperCollins, 2000.
_____.   *The Rise of the Fourth Reich*. NY: Wm Morrow, 2008.
_____.   **Our Occulted History**. New York: HarperCollins, 2013.
_____.   **Illuminati**. Canton, MI: Visible Ink Press, 2017.
Missler, Chuck and Mark Eastman. **Alien Encounters**. ID: Koinonia House, 2003.
Olsen, Brad.   *Future Esoteric: The Unseen Realms*. CA: CCC Publishing, 2013.
Pruett, Dr. Jack.   **The Grandest Deception**. Xlibris Corp: Lexington, KY, 2011.
Stevens, Henry.   *Dark Star*. IL: Adventures Unlimited Press, 2011.
_____ *Hitler's Flying Saucers*. IL: AUP, 2012.
_____ *Hitler's Suppressed and Still-Secret Weapons, Science and Technology*.
        IL: AUP, 2007.
Story, Ronald D., Ed. *The Encyclopedia of Extraterrestrial Encounters*. New York:
        New American Library, 2001.
Tellinger, Michael.   **Slave Species of god.** [sic] Johannesburg, SA: Music Masters Close
        Corporation, 2005. (1st book)
_____. *Slave Species of the Gods*. Rochester, VT: Bear & Co., 2012.
        (2nd Book reprint)
Vallée, Jacques.   *Passport to Magonia*. Chicago, Il: Contemporary Books, 1993.
_____. *Dimensions*. Chicago, Il: Contemporary Books, 1988.
_____. **Messengers of Deception**. Brisbane, Australia: Daily Grail, 1979.
_____. *Revelations*. San Antonio, TX: Anomalist Books, 2008.
Von Daniken, Erich. *Arrival of the Gods*. London: Vega, 2002.
_____ **History is Wrong**. New Jersey: New Page, 2009.
Witkowski, Igor.   **The Truth About the Wunderwaffe**. NYC: RVP Press, 2013.
Wolf, Dr. Michael.   *The Catchers of Heaven*. Pittsburgh, PA: Dorrance Publishing, 1996.

History and Other Related Books

Andrews, Synthia, & Colin Andrews. *The Complete Idiot's Guide to 2012*.
        New York: Penguin Group (USA), 2008.
Becker, R.O. & Gary Selden. **The Body Electric**. NT: Harper, 1985.
Calleman, Carl Johan, PhD. **The Mayan Calendar and the Transformation of
        Consciousness.** Rochester, VT: Bear & Co., 2004.
Childress, David Hatcher.   *Technology of the Gods*. IL: AUP, 2000.
Cotterell, Arthur, Gen Ed. **Encyclopedia of World Mythology**. UK: Parragon, 2008.
Fomenko, Anatoly T. PhD. **History: Fiction or Science**?, Vol. 1. Isle of Man, UK:
        Delamere Resources, Ltd., 2003.
Guiley, Rosemary Ellen.   *Encyclopedia of the Strange, Mystical & Unexplained*.
        New York: Gramercy Books, 2001.
Hawkins, David R.   *Reality and Subjectivity*. West Sedona, AZ: Veritas Press, 2003.
_____ *The Eye of the I*. West Sedona, AZ: Veritas Press, 2001

Icke, David. ***The Biggest Secret***. Wildwood, MO: Bridge of Love, 2001.

_____ *…And the Truth Shall Set You Free*. Isle of Wight, UK: David I Icke Books Ltd., 1995.

_____ *The David Icke Guide to the Global Conspiracy*. Isle of Wight, UK: David Icke Books Ltd., 2007.

_____ *Human Race: Get Off Your Knees*. Isle of Wight, UK: David Icke Books Ltd., 2010.

_____ ***Children of the Matrix***. Isle of Wight, UK: David Icke Books Ltd., 2001.

_____ ***Tales from the Time Loop***. Wildwood, MO: Bridge of Love, 2003.

Irwin, William. *The Matrix and Philosophy*. Peru, IL: Carus Publishing, 2002.

Keel, John A. ***The Complete Guide to Mysterious Beings***. NY: Tor Books, 2002.

_____ *Why UFOs – Operation Trojan Horse*. NY: Manor Books, 1970.

_____ *Our Haunted Planet*. NY: Fawcett Books, 1971.

Kenyon, Douglas. *Forbidden Science*. VT: Bear & CO., 2008.

Knight-Jadczyk, Laura. *High Strangeness*. Alberta Can: Red Pill Press, 2008

Kramer, Samuel Noah. *The Sumerians*. Chicago, IL: University of Chicago Press, 1971.

McKean, Erin, Ed. *The New Oxford American Dictionary, 2nd Edition*. New York: Oxford University Press, 2005.

Northcutt, Wendy. ***The Darwin Awards:*** *Survival of the Fittest*. New York: Penguin Group/Plume Books, 2004.

Parkes, Henry B. *A History of Mexico*. Boston, MA: Houghton Mifflin Co., 1960.

Parrinder, Geoffrey. ***African Mythology***. NY: Barnes & Noble Books, 1996.

Pinkham, Mark Amaru. **The *Return of the Serpents of Wisdom***. Kempton, IL: Adventures Unlimited Press, 1997.

Pye, Lloyd. ***Everything You Know Is Wrong, Book I: Human Origins***. Lincoln NE: iUniverse/Authors Choice Press, 2000.

Radzinsky, Edvard. *The Rasputin File*. New York, NY: Anchor Books, 2000.

Rifat, Tim. ***Remote Viewing***. London: Vision Paperbacks, 2001. (incl info on mind manipulation noxious Hz levels and Tavistock.)

Sitchin, Zecharia. *Journeys to the Mythical Past*. Rochester, VT: Bear & Co., 2007.

_____ ***The Twelfth Planet***. New York: HarperCollins, 2007.

_____ *The Cosmic Code*. New York: HarperCollins, 2007.

_____ *The End of Days*. New York: HarperCollins, 2007.

_____ *The Earth Chronicles Expeditions*. Rochester VT: Bear & Co., 2004.

_____ *Divine Encounters*. New York: HarperCollins/Avon, 1996.

_____ *Genesis Revisited*. New York: HarperCollins/Avon, 1990.

_____ *The Wars of Gods and Men*. New York: HarperCollins, 2007.

_____ *The Lost Book of ENKI*. Rochester, VT: Bear & Co., 2004.

_____ *The Stairway to Heaven*. New York: HarperCollins, 2007.

_____ *The Earth Chronicles Handbook*. Rochester, VT: Bear & Co., 2009.

_____ ***There Were Giants Upon the Earth***. Rochester, VT: Bear & Co., 2010.

Steinmeyer, Jim. ***The Book of the Damned; The Collected Works of Charles Fort***. New York: Tarcher/Penguin Group, 2008.

Stevens, Henry. *Dark Star*. Kempton, IL; Adventures Unlimited Press, 2011.

_____ *Hitler's Flying Saucers*, rev. ed. Kempton, IL; Adventures Unlimited Press, 2012.

Turner, Patricia & Charles Russell Coulter. *Dictionary of Ancient Deities*. NY, NY: Oxford University Press, 2001.
Witkowski, Igor. *Axis of the World*. Kempton, IL: Adventures Unlimited Press, 2008.

Flat Earth & Related

Dubay, Eric.   **The Flat-Earth Conspiracy**. Lulu.com, 2014.
Hegland, TJ.   **Quantum Earth Simulation** (Ch. 11.) SC: CreateSpace Press, 2015.
Hendrie, Edward.   *The Greatest Lie on Earth*. Great Mountain Publishing, 2016.
Rowbotham, Samuel B.   *Zetetic Astronomy*. Forgotten Books, 2007.
Sargent, Mark.   **Flat Earth Clues.** London: Booglez, Ltd. 2016.
Scott, David. W.   **Terra Firma: The Earth Not a Planet**, London: Holloway Press, Original: 1901, reprint: 2010.

Videos of Interest

Forbidden Planet. MGM classic from 1956; debuts Robby the Robot.
The X-Files (TV series, 1993-2002): Twentieth Century Fox.
K-Pax. Universal Pictures, Lawrence Gordon et al. 2001.
Millenium. Gladden Entertainment. 1989.
Hangar 18. Republic Entertainment. 1980.
Capricorn One. Associated General Films. Lazarus/Hyams prod. 1978.

Groundhog Day. Dir. Harold Ramis, Columbia Tristar. 1993.
Men In Black. I & II Dir. Barry Sonnenfeld, Columbia Pictures. 2000.
The Matrix. Dir./Written by The Wachowski Bros., Warner Bros. 1999.
The Mothman Prophecies. Dir. Mark Pellington, Screen Gems/LakeShore
    Entertainment. 2001.
Prometheus. 20th Century Fox, 2012.

Taken. (TV miniseries) Stephen Spielberg, DreamWorks. 2002.
V, the TV series (1983-85, and 2009-11). WarnerVideo, Kenneth Johnson
Production.
The Truman Show. Peter Weir, Paramount Pictures. 1998.
The Young Age of the Earth. Aufderhar, Glenn. Earth Science Associates / Alpha
    Productions. 1996.
What the Bleep Do We Know? 20th Century Fox, 2004.

They Live. Dir./Written by John Carpenter, Universal Studios. 2003.
Prometheus I. Ridley Scott, 2oth Century Fox. 2012.
The Thirteenth Floor. Columbia Pictures, Roland Emmerich. 1999.
The Day the Earth Stood Still. Twentieth Century Fox, Erwin Stoff et al, 2009.
The Forgotten. Revolution Studios. 2004
The Phoenix Incident, PI Films/PCB, 2016.

Iron Sky. Timo Vuorensola, Ger/Fin release via Paramount Pictures, 2012.
Paul. Universal Studios, Greg Motola. 2010.
2012. Sony Pictures, Roland Emmerich. 2010.
The Fourth Kind. Universal Pictures, Olatunde Osunsanmi. 2010.
The Adjustment Bureau. Universal Pictures, George Nolfi. 2010.

Knowing. Summit Entertainment, Alex Proyas. 2009.
Dark City. New Line Cinema, Alex Proyas. 1998.
Source Code. Summit Entertainment, Duncan Jones. 2011.
eXistenZ. Canadian Television Fund, David Cronenburg, 1999.
Defending Your Life. Warner Bros., 1991.

Also: Ancient Aliens video series (season 9).

***

Chapter 1 -- Endnotes

---

[1] What is amazing, as Chapter 1 will unfold, I was given a book on the subject of Walk-ins, and then had a descending dream – as if to say that this is what was happening to me, as if the gods wanted me to know, but I am still not comfortable with it. And beyond the book and the dream, there is no further evidence: I never felt 'different' and I was never given any insight or word that I was a Walk-in. I even had trouble with Baldy being a Guardian Angel. So to this day, March 2017, I still don't know… but the 6 books got done. (See endnote #3 below.)

[2] https://en.wikipedia.org/wiki/1956_Grand_Canyon_mid-air_collision

Chapter 2 -- Endnotes

[3] Ruth Montgomery, Strangers Among Us, NY: Fawcett, 1979.
   Also see Ruth Montgomery, Threshhold to Tomorrow, NY: Fawcett, 1982.
   (two key books on the Walk-in phenomenon, and some of those claiming to be Walk-ins in Threshhold were **less than ideal people**… so I distrusted the issue.)

[4] Dr. Caroline Leaf, presentation on GatewayPeople.com , 7/12/15.
   Also see website: www.drleaf.com
[5] Dr. Leaf, Switch on Your Brain, p. 13.
[6] Ibid., 14.
[7] Ibid., 153.
[8] Ibid., 34-35
[9] Ibid.
[10] Ibid.
[11] Ibid.
[12] Ibid., 48.
[13] Ibid., 73-74, 88-89.
[14] Ibid., 93-94
[15] Wikipedia: Excitotoxicity

[16] Dr. Caroline Leaf, Switch on Your Brain, pp. 67-68, 96, 99,
[17] Ibid., 88, 94.

[18] Scientific American MIND magazine, July/Aug 2015, *How Violent Video Games Really Affect Kids*, p. 42.
[19] Ibid., 44.
[20] Indi., 45
[21] Ibid., 45
[22] Ibid., 45.

[23] Op Cit, Dr. Leaf, 74.
[24] Op Cit Scientific American MIND, 45

Chapter 3 -- Endnotes

[25] Baugh, Carl E. *Why Do Men Believe Evolution Against All Odds?* (Oklahoma City, OK: Hearthstone Publishing, 1999),105. Note: the Creation Evidence Museum in Glen Rose, TX, has the large, flat rock on display that has the dinosaur and human footprint in it. It's very clear what it shows.
[26] Ibid., 92.

[27] Free, Wynn and David Wilcox. *The Reincarnation of Edgar Cayce?* (Berkley: Frog Ltd., 2004), 338.

[28] Morris, John D., Ph.D. *The Young Earth.* 41.

[29] Brown, Walt. *In The Beginning: Compelling Evidence for Creation and the Flood.*
        (Phoenix, AZ: Center for Scientific Creation, 1995), 45-47.
[30] Ibid.

[31] Lewels, Joe, Ph.D. *The God Hypothesis.* (Columbus, NC: Wild Flower Press, 2005), 195.

[32] Op Cit, Free, Wynn and David Wilcox. 338-340.

[33] Morris, John D., Ph.D. *The Young Earth.* 70.
[34] Ibid., 70-71.
[35] Ibid., 70-71.
[36] Ibid., 70.
[37] Ibid., 71.
[38] Ibid., 74-75.
[39] Ibid., 84-85.
[40] Ibid., 90.
[41] Ibid., 85-87.
[42] Ibid., 87-88.
[43] Ibid., 88.
[44] Ibid., 89.

[45] Steinmeyer, Jim. *The Book of the Damned; The Collected Works of Charles Fort.*
        (New York: Tarcher/Penguin Group, 2008), 381.
[46] Ibid., 381-82.
[47] Ibid., 382.
[48] Ibid., 838-839.
[49] Ibid., 140.
[50] Ibid., 159.

[51] Elvidge, Jim. *The Universe Solved*, p 194.
[52] Jim Elvidge, article from his website:
        The Singularity, Infomania, and Programmed Reality , December, 2008.
        http://www.theuniversesolved.com/evidence.htm

[53] Posted by: by http://theghostdiaries.com/life-in-the-matrix-new-evidence-supports-the-simulation-theory
        also see Gate's original paper at: http://arxiv.org/abs/0806.0051 via Cornell University site.

[54] https://en.wikipedia.org/wiki/Aurora

Chapter 4 -- Endnotes

[55] Wikimedia -- "Wikimedia Commons." Otherwise don't credit anyone, that's fine by me. -- Fastfission 15:03, 14 April 2008 (UTC)

[56] Wikipedia: Tycho Brahe

[57] Wikipedia: https://en.wikipedia.org/wiki/De_revolutionibus_orbium_coelestium

[58] Ibid.

[59] Ibid.

[60] Ibid.

[61] Ibid.

[62] Ibid.

[63] Ibid.

[64] Wikipedia -- https://en.wikipedia.org/wiki/Alumbrados

[65] Wikipedia -- https://en.wikipedia.org/wiki/Illuminati

[66] Ibid.

[67] Ibid.

[68] https://en.wikipedia.org/wiki/Nicolaus_Copernicus#Tolosani

[69] David Wardlaw Scott, *Terra Firma; The Earth Not a Planet...*, p. 14.

[70] https://en.wikipedia.org/wiki/Age_of_Enlightenment

[71] https://en.wikipedia.org/wiki/Walter_Gilbert
Gilbert is an American biochemist who was involved in the sequencing of the human genome.

[72] http://www.genomenewsnetwork.org/articles/05_01/Gene_transfer.shtml

[73] Zechariah Sitchin, <u>There Were Giants Upon the Earth</u>, pp. 162-163.
And
Hegland, TOM, pp 222.

[74] http://www.genomenewsnetwork.org/articles/05_01/Gene_transfer.shtml

[75] Eric Dubay, <u>The Flat Earth Conspiracy</u>. , p. 243.

[76] Op Cit., Scott, *Terra Firma*, pp. 3-4

[77] Ibid., pp 18-21.

[78] Ibid., pp11-12.

[79] Ibid., p,14.

[80] Edward Hendrie, <u>The Greatest Lie on Earth</u>. P. 238.

[81] Ibid., p. 234

[82] Ibid., p. 234-235.

[83] Ibid., p. 237.

[84] Ibid., p.239.

[85] Ibid., p.239-240.

[86] See Wikipedia: Strong Force article.

[87] Ibid., p. 241.

[88] Ibid., p.85

[89] Op Cit, Hendrie, pp. 234-238.

[90] Zukav, Gary. *The Dancing Wu Li Masters*. (New York: Quill, 1979), 165.

[91] Ph.D, DR Bjorn Overbye. Part I.
http://blog.hasslberger.com/2007/06/einstein_warped_minds_bent_tru.html
Parts 1 – 2 – 3.

[92] Ibid.

[93] Op Cit., Dr. Bjorn Parts 1 – 2 – 3.

[94] Ibid.

[95] Ibid.

[96] Good, Timothy. *Earth: An Alien Enterprise*. (NY: Pegasus Publishing, 2013), 136-138.

[97] Igor Witkowski, The Truth About the Wunderwaffe, pp. 301-306.

[98] Ibid.

[99] Ibid.

[100] Ibid.

[101] Behe, Michael J. *Darwin's Black Box*. New York: Simon & Schuster/Touchstone, 1996, p. 16.

[102] Montag, "Rods and Cones" Chapter 9. Eprint at
http://www.cis.rit.edu/people/faculty/montag/vandplite/pages/chap_9/ch9p1.html

[103] Ibid.

[104] Talbot, Michael. *The Holographic Universe*. (New York: HarperCollins, 1991), 18, 54-55, 192.

[105] Brown, Walt. *In The Beginning: Compelling Evidence for Creation and the Flood*. (Phoenix, AZ: Center for Scientific Creation, 1995), 40-42.

[106] Ibid.

[107] Op Cit., *Darwin's Black Box*. 31-36.

[108] Wikipedia, "Charles Darwin's Illness" Eprint at
http://en.wikipedia.org/wiki/Illness_of_Charles_Darwin.
also see: http://en.wikipedia.org/wiki/Charles_Darwin%27s_views_on_religion.

[109] Modi, Shakuntala, M.D. *Remarkable Healings*. (Charlottesville, VA: Hampton Roads, 1997), 438-441 and ff.

[110] Wikipedia, "Charles Darwin's Illness."
[111] Op Cit., Wikipedia, "Charles Darwin's Illness." section 3.

[112] Prince, Derek. *They Shall Expel Demons*. (Grand Rapids, MI: Chosen Books, 1999), 65.

Chapter 5 -- Endnotes

[113] https://en.wikipedia.org/wiki/Unity_Church#Jesus
[114] Ibid.
[115] Ibid.

[116] Dr Caroline Myss, see: *Why People Don't Heal and How They Can*.

[117] Mark Amaru Pinkham, Return of the Serpents of Wisdom, pp. 234-237.

Chapter 6 -- Endnotes

[118] Internet article: Before Its News, *1500 Year Old Bible Confirms That Jesus Christ Was Not Crucified – Vatican in Awe*. Sunday May 4, 2014 by the Serpent.

[119] Gardiner, Philip. *Secret Societies*. (Franklin Lakes, NJ: Career Press/New Page, 2007), 150.

[120] "Apollonius of Tyana" article on Wikipedia; eprint at: http://en.wikipedia.org/wiki/Apollonius_of_Tyana

[121] Gardiner, Philip. *Secret Societies*., 152.
[122] Ibid., 153.
[123] Ibid.,. 151.
[124] Ibid., 151-152.
[125] Ibid., 155.

[126] Acharya S. *The Christ Conspiracy*. 369.
[127] Gardiner, Philip. *Secret Societies*, 153-154.
[128] Ibid., 154
[129] Ibid., 154.

[130] http://nephiliman.com/jesus_proof.htm, "Jesus in the Historical Records" article.

[131] *Heaven is For Real*, Colton Burpo.

[132] *Op Cit*, "Jesus in the Historical Records" Nephiliman article (just above).

[133] Op Cit., Gardiner, Philip. *Secret Societies*, 154.

[134] Ibid., 154.

[135] Dr. R.W. Bernard, "Apollonius the Nazarene", Chapters 1 and 2, eprint at http://www.apollonius.net/bernardbook.html

[136] Acharya S, "Apollonius, Jesus and Paul: Men or Myths?"

[137] PMH Atwater, *Near Death Experiences*, pp. 222-223.

[138] Dr. R.W. Bernard, "Apollonius the Nazarene", Ch. 2.
[139] Ibid., Ch. 2.
[140] Ibid., Ch. 2.
[141] Ibid., Ch. 2.
[142] Ibid., Ch. 2.
[143] Ibid., Ch. 2.

[144] Charles, R.A. *The Book of Enoch the Prophet*. (San Francisco, CA: Weiser Books, 2003), viii-ix.
[145] Ibid., xvi.
[146] Ibid., Book of Enoch 15: 8-12.
[147] Ibid., xvi

[148] Tellinger, Michael. *Slave Species of God*. (Johannesburg, SA: Music Masters Close Corporation, 2005), 402. (this is the first published version of his book)

[149] Lewels, Joe, Ph.D. *The God Hypothesis*. (Columbus, NC: Wild Flower Press, 2005), 216-217. Also see: http://www.bibliotecapleyades.net/vida_alien/alien_watchers13a.htm

[150] Charles, R.A. *The Book of Enoch the Prophet*, Book 17.1.
[151] Ibid., Book 19:1.
[152] Ibid., Book 69:6.
[153] Ibid., Book 69:6.

[154] Charles, R.A. *The Book of Enoch the Prophet*, Book 68:6-7.

[155] Kenneth Woodward in Newsweek article, "In the Beginning There Were the Holy Books", Feb. 11, 2002, pp 51-57.
[156] Ibid., 60.
[157] Ibid., 61.
[158] Ibid., 70.

Chapter 7 -- Endnotes

[159] https://en.wikipedia.org/wiki/Flat_Earth#Myth_of_the_flat_Earth

[160] Eric Dubay, pp 2-1-215.
[161] Ibid.
[162] Ibid.
[163] Ibid.

[164] Ibid.
[165] Ibid.
[166] Ibid.
[167] Ibid.

[168] Otto Muck, *The Secret of Atlantis*, pp 216-222.

[169] http://discoversiberia.net/woolly-mammoth-in-siberia/ and freerepublic.com

[170] Op Cit, Muck, p.222.

[171] Dennis R. Petersen, *Unlocking the Mysteries of Creation*, p. 49.

[172] Morris, John D., Ph.D. *The Young Earth*. p. 64.
[173] Ibid., p.. 70.
[174] Ibid., pp. 74-75.
[175] Ibid., pp. 84-85.
[176] Ibid., pp. 85-87.

[177] David W. Scott, *Terra Firma.....*, p. 267.
[178] Ibid., p.268
[179] Ibid., 270
[180] Ibid., pp. 270-71.

[181] Op Cit., Morris., p. 89.

[182] http://www.parahauntpost.com/2013/01/lovelock-nv-red-haired-giants.html

[183] Op Cit, Dubay, p. 214-215. Quoting David Wozney – book not on Amazon, but there is an internet hit:
http://www.ocii.com/~dpwozney/dinosaurs.htm

Chapter 8 -- Endnotes

[184] The **Inquisition** ruled from the 1200's to about 1826, covering about 600 years of forcing the public to believe what the Church dictated. This also covers the period of the Protestant Reformation and Science's big push for Copernican/Newtonian concepts of Earth and the universe. See Wikipedia: https://en.wikipedia.org/wiki/Inquisition

[185] Sitchin, Zechariah. *There Were Giants Upon the Earth*. pp. 162-163.

[186] Farrell, Joseph P. *The Cosmic War*. (Kempton, IL: Adventures Unlimited Press, 2007), 90-91.
[187] Ibid., 88-89.

[188] https://ironlight.wordpress.com/2010/07/10/nevadas-mysterious-cave-of-the-red-haired-giants/

[189] Op Cit, Farrell, 98-99.

[190] R.A. Boulay, Flying Serpents and Dragons. p11.

[191] Icke, David, *Children of the Matrix.* p 91.
Also uoted in Boulay, *Flying Serpents & Dragons*, p. 61.
[192] Op Cit., Boulay, R.A. *Flying Serpents and Dragons*. Pp. 115-118.
[193] Ibid.

[194] From Yahoo Coins of Alexander; see
http://search.yahoo.com/search;_ylt=AuGsxC8UzTjm1.kh.2CU9IabvZx4?p=alexander+the+great+coin&toggle=1&cop=mss&ei=UTF-8&fr=yfp-t-788

[195] https://en.wikipedia.org/wiki/Mount_Meru

[196] Otto Muck, The Secret of Atlantis. Pp. 124-130.
This is a 1976 book written by a German scholar and scientist. No nonsense, just facts.
He reveals the secret of the Siberian frozen mammoths, the Basques, and the Flood.

Chapter 9 -- Endnotes

[197] Mark Sargent, Flat Earth Clues.

[198] Scientific American Book of the Brain, *The Visual Image in Mind And Brain*, pp. 17-28.
[199] Credit: footnote 9 (Ibid.): p. 19.
[200] Ibid., 27.

[201] Talbot, Michael. The Holographic Universe. (New York: HarperCollins, 1991), 35.
[202] Ibid, 139-140
[203] Ibid., 125, 139.
[204] Ibid., 140.
[205] Ibid., 144-145.
[206] Ibid., 146.
[207] Ibid., 140ff.
[208] Ibid, 140.
[209] Ibid., 144-146.
[210] Ibid, 18-20.
[211] Ibid., 19-20.
[212] Ibid., 20.
[213] Ibid., 54-55.
[214] Ibid., 163.
[215] Ibid., 164.
[216] Ibid., 141.

[217] Friesen, James G., M.D. *Uncovering the Mystery of MPD*. (Eugene, OR: Wipf & Stock Publishers, 1991), 59, 115, 143.

[218] Nick Bostrom, "The Simulation Argument" See

on http://www.simulation-argument.com/si...
and
http://www.youtube.com/watch?feature=player_detailpage&v=nnl6nY8YKHs
[219] Ibid., and Wikipedia: , http://en.wikipedia.org/wiki/Simulated_reality, p. 3.
[220] Nick Bostrom, "The Simulation Argument." See:
http://www.simulation-argument.com/matrix.html
(Also: Times Higher Educational Supplement, May 16, (2003).)
[221] Ibid.
[222] "The Simulation Hypothesis." See:
http://en.wikipedia.org/wiki/Simulation_hypothesis
[223] Ibid.

[224] Wikipedia: http://en.wikipedia.org/wiki/Cognitive_computing

[225] Brian Weatherson, "Are You a Sim?" see:
http://www.simulation-argument.com/weatherson.pdf
Also: *Philosophical Quarterly* (2003), vol 53., 425-31.

[226] Elvidge, Jim. *The Universe Solved*, p 194.
[227] Elvidge, Jim. Is Our Reality Just a Big Video Game?
See his website: www.theuniversesolved.com/evidence.htm
[228] Ibid.
[229] Ibid.
[230] Ibid.
[231] Op Cit, Elvidge ., p. 195.
[232] Ibid., pp 297-08.
[233] Jim Elvidge, article from his website:
Nanotech and the Physical Manifestation of Reality, March 2008.
http://www.theuniversesolved.com/evidence.htm
[234] Ibid.

[235] Posted by: by http://theghostdiaries.com/life-in-the-matrix-new-evidence-supports-the-simulation-theory
also see Gate's original paper at: http://arxiv.org/abs/0806.0051
via Cornell University site.

[236] Greene, Brian. *The Hidden Reality*. Pp. 281-82.
[237] Ibid., pp 284-85.
[238] Ibid., p. 288.
[239] Ibid., pp 288-89.
[240] Ibid., pp 291-92.
[241] Ibid., p. 306.

[242] Lloyd, Seth. *Programming the Universe*. Pp. 6-7, 31.

[243] Wikipedia: http://en.wikipedia.org/wiki/D-Wave_Systems
[244] Ibid., p. 54.
[245] Ibid., p 54.
[246] Ibid., pp. 149-51.
[247] Ibid., p. 154.

[248] Op. Cit., Elvidge, p. 117.

[249] OP Cit., Lloyd, p. 166.

[250] See the following article: http://beforeitsnews.com/story/1658/888/NL/Scientific_Evidence_The_Universe_Is_A_Holographic_Projection_Around_The_Earth.html

[251] Op Cit, Lloyd, 166..
[252] Ibid.

[253] Wikipedia, http://en.wikipedia.org/wiki/Simulated_reality
[254] Dvorsky, George, "Physicists say there may be a way to prove that we live in a computer simulation" on http://io9.com/5950543/physicists-say-there-may-be-a-way-to-prove-that-we-live-in-a-computer-simulation

[255] Science Channel, *Through the Wormhole with Morgan Freeman*, 5/20/15 episode, "Do We Live in the Matrix?"

[256] Greene, Brian. *The Fabric of the Cosmos*. P. 425.

[257] Grabianowski, Ed, "You're living in a computer simulation, and math proves it" on http://io9.com/5799396/youre-living-in-a-computer-simulation-and-math-proves-it
[258] Ibid.
[259] Ibid.

[260] Silby, Brent, "The Simulated Universe" on http://www.scribd.com/doc/3015396/Simulated-Universe-by-Brent-Silby
[261] Ibid., p. 3.

[262] http://en.wikipedia.org/wiki/Simcity

Chapter 10 -- Endnotes

[263] Mark Sargent, Chapter 4 of Flat Earth Clues, pp. 41-50.

[264] Henry Stevens, Dark Star, p. 135.

[265] https://en.wikipedia.org/wiki/Welteislehre
[266] Ibid.
[267] Ibid.

[268] Eric Dubay, The Flat Earth Conspiracy, p. 59.
[269] Ibid., p.61

[270] David W. Scott, Terra Firma. P. 261.
[271] Ibid., p.85.

[272] Op Cit, Dubay, pp. 63-73.
[273] Ibid.

[274] Paul Davies, The Goldilocks Enigma. p. 105.
[275] Ibid., 27, 131.
[276] Ibid., x-xi.

[275] Olsen, Future Esoteric: The Unseen Realms. 71, 185.

[276] Joseph Farrell, Saucers, Swastikas and Psyops, 216-224.

[277] Op Cit, Olsen, p. 111.

[280] Tim Rifat, *Remote Viewing*, pp. 18-20, 215-224.
[281] Ibid., p. 18.

[282] https://en.wikipedia.org/wiki/Neural_oscillation

[283] Op Cit., Olsen., pp. 14-15.

Chapter 11 -- Endnotes

[284] https://en.wikipedia.org/wiki/Polar_orbit

[285] http://reikiinmedicine.org/clinical-practice/healing-crisis-what-is-it/

[286] Monroe, Robert. Ultimate Journey. p. 24.

[287] The Nexus Seven, "From the 33 Arks…"., 15.20-30.
[288] Ibid., 24.7-8.

[289] http://www.huffingtonpost.com/entry/ellen-degeneres-college-scholarship-senior-class_us_58b00477e4b0780bac282e36 **Yes Walmart helped**.
The scholarships total $1.6 million and apply to any state university in New York, DeGeneres said

[290] Monroe, Robert, Far Journeys. p. 248.

[291] Ring, Dr. Kenneth. *Lessons From the Light*. (Portsmouth, NH: Moment Point Press, 2000), 19-26.
[292] Ibid., p26
[293] Ibid., pp 31-32.

Appendix B -- Endnotes

[294] -- Arnold Schwarzenegger at 44 to *US News and World Report* in 1990.
See also: **http://arnoldexposed.com/quotes.html**
The significance is that Arnold is privy to what the Elite think and are planning.

[295] https://nuclear-news.net/2017/02/10/radiation-level-in-fukushima-no-2-reactor-measured-higher650-sieverts/
See also:
https://nuclear-news.net/2017/02/05/**fukushima**-daiichi-unit-2...

[296] McTaggart, Lynn. The Field. (New York: HarperCollins/Quill, 2002), 47.

[297] Morris, John D., Ph.D. The Young Earth. (AR: Green Forest: 2006), 43.

[298] Rasha. *Oneness.* (Santa Fe, NM: Earthstar Press, 2003), 223, 354.
[299] Ibid., 195-196.

[300] Marion, Putting on the Mind of Christ, p. 34.
[301] Ibid., 35
[302] Ibid., 37
[303] Ibid., 41.
[304] Ibid., 42-43.
[305] Ibid. 43-44.
[306] Ibid., 64.
[307] Ibid., 70-71.

[308] Northcutt, Wendy. The Darwin Awards: Survival of the Fittest. (New York: Penguin Group/Plume Books, 2004), 180.
[309] Ibid., 140.
[310] Ibid., 160.

Appendix C -- Endnotes

[311] Jim Elvidge, article in NEXUS magazine: *Digital Consciousness – The Answer to Life's Great Mysteries? Vol. 24, No. 1 for January-February 2017.* Pp. 53-59, 78.

[312] https://en.wikipedia.org/wiki/Quantum_Zeno_effect

[313] https://en.wikipedia.org/wiki/Retrocausality#Quantum_physics

[314] Op Cit, Elvidge, 54-55.
[315] Ibid. 54.
[316] Ibid., 56
[317] Ibid., 56.

Appendix D -- Endnotes

323 Brian Greene, *The Hidden Reality*, pp45-46.
324 Ibid., 46

325 http://aetherforce.com/teslas-dynamic-theory-of-gravity/

326 Op Cit, Greene, p. 14 (footnote).

327 **Space.com**, symposium in NYC in 2016 where a panel discussed whether we are in a Simulation: *Is the Universe a Simulation? Scientists Debate* By Sarah Lewin, Staff Writer

328 https://en.wikipedia.org/wiki/Nikola_Tesla#Other_ideas.2C_awards.2C_and_patents

329 http://aetherforce.com/teslas-dynamic-theory-of-gravity/
330 Ibid... picture is from this source.
331 Ibid.
332 Ibid.
333 Ibid.
334 Ibid.
335 Ibid.
336 Ibid.

337 Wikipedia, https://en.wikipedia.org/wiki/Loop_quantum_gravity
338 https://en.wikipedia.org/wiki/Anti-gravity

339 Wm. R. Lyne, *Space Aliens from the Pentagon*, pp. 16-17.
340 Ibid., p. 17.
341 Ibid.
342 Ibid. p. 33.

343 LaViolette, Dr. Paul. "Subquantum Kinetics – A Nontechnical Summary" from http://starburstfound.org/category/research/subquantum-kinetics/
344 LaViolette, Dr. Paul, "The Transmuting Ether" on http://starburstfound.org/category/research/subquantum-kinetics/
345 Ibid.
346 LaViolette, Paul. *Genesis of the Cosmos*. (Rochester, VT: Bear & Co., 2004), 310-311.
347 LaViolette, Paul. Secrets of AntiGravity Propulsion. P 118.
348 Ibid., pp123-126.

[349] Ibid., p. 12

[350] Ibid., pp 11 and 123.

[351] http://blogs.discovermagazine.com/badastronomy/2011/09/26/slinky-drop-physics/#.WMCH81gzX5o by Phil Plait, 9-26-2011.

Appendix E -- Endnotes

[352] Steinmeyer, Jim. *The Book of the Damned; The Collected Works of Charles Fort.*
      (New York: Tarcher/Penguin Group, 2008), 381.

[353] Ibid., 381-82.

[354] Ibid., with ref to
      http://www.betawired.com/science-fiction-like-electron-shield-found-around-earth/1418358/
          may also see:
      http://lasp.colorado.edu/home/?post_type=mag-seminars&p=16204

[355] *Encyclopedia of World Mythology*, Gen, Ed, Arthur Cotterell, p. 251-252.

[356] *African Mythology*, Geoffrey Parrinder, pp. 37-40.

[357] http://www.pbs.org/newshour/rundown/study-invisible-shield-space-protects-earth-killer-electrons/
          And
      https://www.sciencedaily.com/releases/2014/11/141126133829.htm

[358] Ibid., -- they just told you that the barrier is impenetrable...

[359] Gregg Prescott article in *In5D Guest* and Waking Times:
          Earth's Quarantine Force Field Discovered by NASA? , Dec. 3, 2014.

[360] https://www.sciencedaily.com/releases/2014/11/141126133829.htm
          Also:
          An impenetrable barrier to ultrarelativistic electrons in the Van Allen radiation belts. *Nature*, 2014; 515 (7528): 531 DOI: 10.1038/nature13956

Appendix F -- Endnotes

[361] Jim Marrs, Illuminati, pp 102-103.

[362] Ibid., p. 97

[363] Ibid,. pp. 94-102

[364] Ibid., pp xiii-xiv.

[365] Ibid., pp. 95-96.

[366] Wm Bramley, Gods of Eden, pp 56-65.

[367] Op Cit, Marrs, p. 157-58.

[368] http://www.godlikeproductions.com/forum1/message293293/pg1?disclaimer=1

Also the best, intact version:
https://www.scribd.com/doc/403303/The-Revelations-of-an-Elite-Family-Insider-2005

[369] Op Cit, Bramley, p. pp. 53-72.

[370] Levy, Elinor, & Mark Fischetti. *The New Killer Diseases.* (NY: Three Rivers Press, 2004), 194.

[371] Mike Stobbe, "Deadly Bacteria" article. Eprint at:
http://health.yahoo.com/news/ap/deadly_bacteria.html
also see the CDC publication:
http://www.cdc.gov/ncidod/EID/index.htm

[372] https://en.wikipedia.org/wiki/Autoclave#In_medicine

[373] https://en.wikipedia.org/wiki/Soylent_Green

The Great Earth Puzzle

Other Books by the Author…   7 in the set.

The Anunnaki Legacy

Transformation, Kundalini, Hyperborea, Göbekli Tepe, Hopi, Maya,
Gods, Goddesses & the Divine Feminine, Skygods & Creation,
The Flood, Sacred Trees, Alchemy and Serpent Wisdom.

by TJ Hegland

Virtual Earth Graduate

TJ Hegland

Reflecting: (inserts left to right, all covered in the book):

Physics (atom), Genetics (DNA), Dragon, Ubaid Statue of Anunnaki, UFO (TR3B), and *Castillo* at Mayan Chichen Itzá.

Center: Earth with Soul/Earth Graduate

Background: Electromagnetic Dark Energy Matrix

The Transformation of Man

By T J Hegland

Spirituality, Reincarnation & the Interlife, Consciousness, The Matrix, Zero Point Energy, Dark Matter, DNA & Healing Energy, Timelines, Quantum Biocomputer, Greys & Hybrids, and Abduction & the Near Death Experience as Transformation.

The Science in Metaphysics

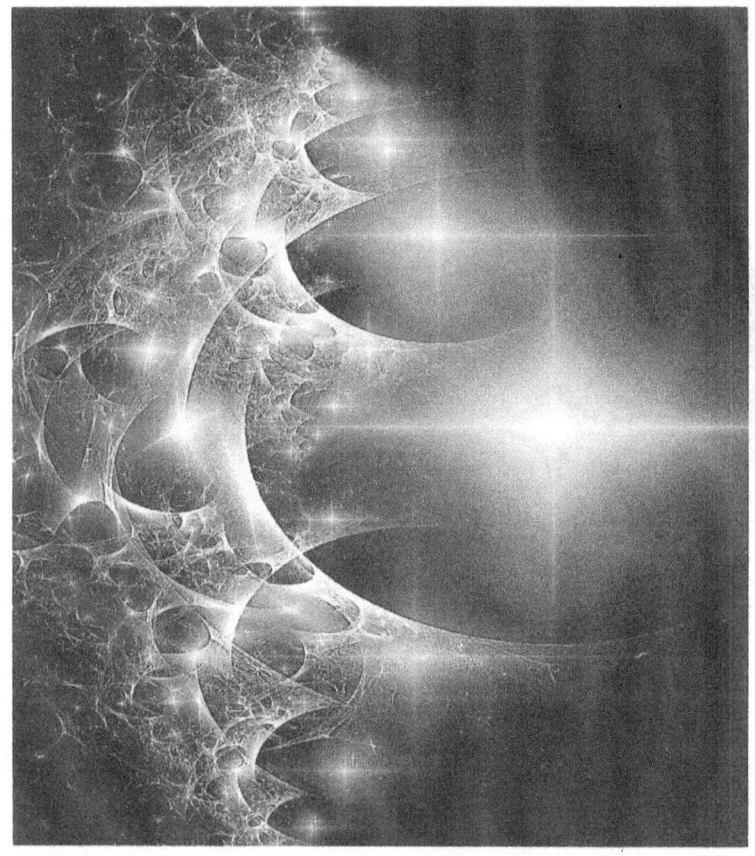

by TJ Hegland

Proof from Science that the Principles of Metaphysics
Are Real and Why They Work.

The Earth Warrior
illustrated

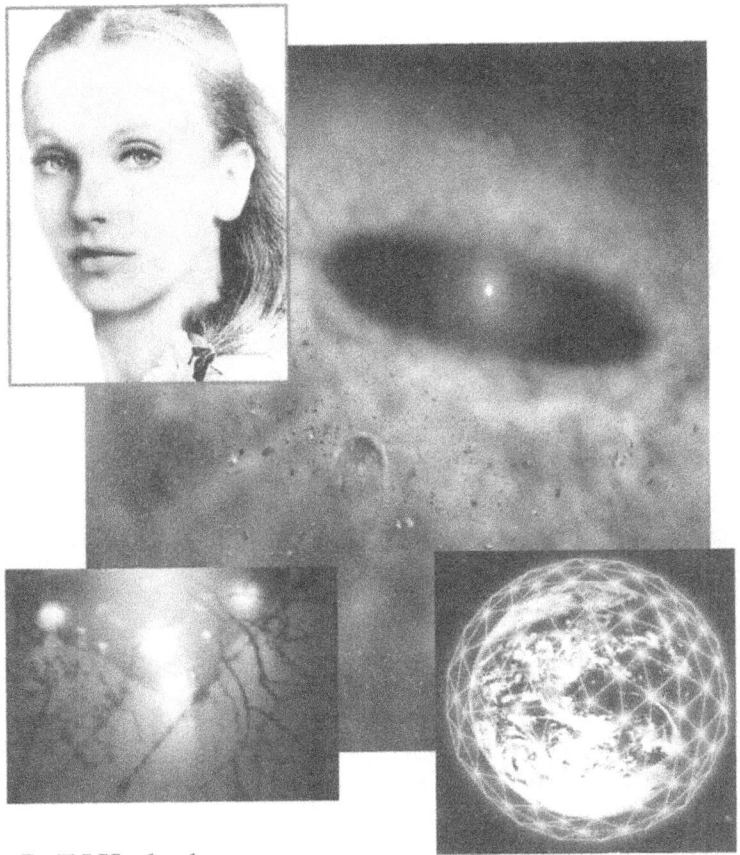

By T J Hegland

A romp thru alternative history with Nazi UFOs, Antarctic bases, the Moon and Mars, a Breakaway Civilization, the Orion Empire, and a Galactic Super Wave. Maria Orsič and her team of Aldebaran women warriors must resolve the Earth dilemma or Man disappears.
Could this be the real history of Earth?

Quantum Earth Simulation

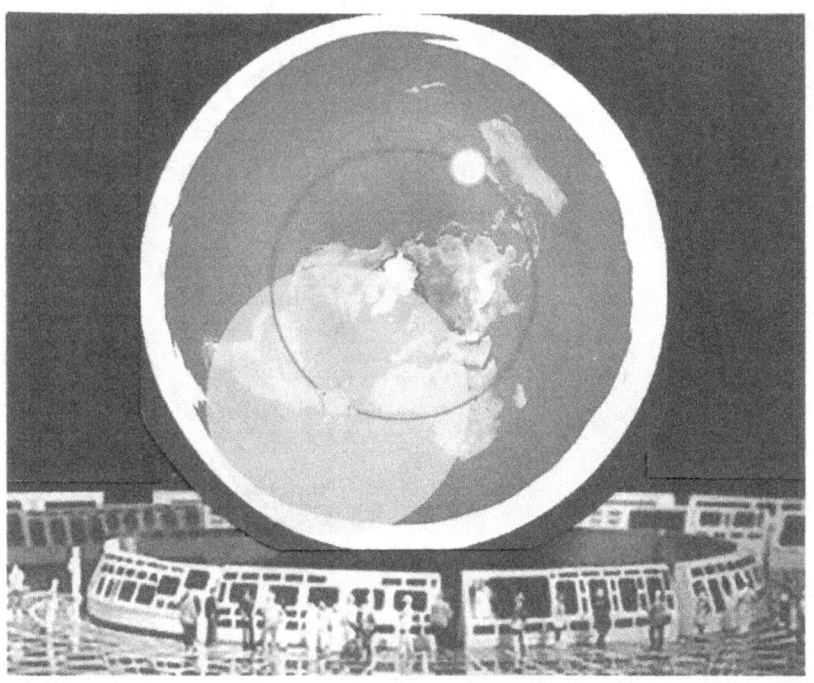

Updated for the Flat Earth Nexus

by **T J Hegland**

Simulation, Virtual Reality, Matrix and Holodecks, Holograms and Quantum Physics, Vision & Hypnosis, Timelines, Eras, Programming the Simulation, Consciousness, Reality Fields, Lucid Living, and many Anomalies.

NOTES

The Great Earth Puzzle